Current Topics in Microbiology
185 and Immunology

W0042450

Editors

A. Capron, Lille · R.W. Compans, Atlanta/Georgia
M. Cooper, Birmingham/Alabama · H. Koprowski,
Philadelphia · I. McConnell, Edinburgh · F. Melchers, Basel
M. Oldstone, La Jolla/California · S. Olsnes, Oslo
M. Potter, Bethesda/Maryland · H. Saedler, Cologne
P.K. Vogt, Los Angeles · H. Wagner, Munich
I. Wilson, La Jolla/California

Rotaviruses

Edited by R. F. Ramig

With 37 Figures and 26 Tables

 **Springer-Verlag**
Berlin Heidelberg New York
London Paris Tokyo
Hong Kong Barcelona
Budapest

ROBERT F. RAMIG

Division of Molecular Virology
Baylor College of Medicine
One Baylor Plaza
Houston, Texas 77030
USA

*Cover illustration: A computer-generated, composite recon-
struction of rotavirus SA11 showing surface views of the
three nested capsid layers at 40 Å resolution. The outer capsid
containing VP4 spikes and a smooth surface comprised of
VP7 is shown in yellow. The inner capsid, composed of VP6,
is shown in blue. The capsid of the core, composed of VP2,
is shown in gray. Red indicates the density of material
(genome dsRNA?) contained within the core as seen
through the channels that perforate the three capsid layers.
Courtesy of Dr. B.V.V. Prasad.*

Cover design: Harald Lopka, Ilvesheim

ISBN-13: 978-3-642-78258-9   e-ISBN-13: 978-3-642-78256-5
DOI: 10.1007/ 978-3-642-78256-5

© Springer-Verlag Berlin Heidelberg 1994
Softcover reprint of the hardcover 1st edition 1994
Library of Congress Catalog Card Number 15-12910

Typesetting: Thomson Press (India) Ltd, New Delhi
23/3020-5 4 3 2 1 0 – Printed on acid-free paper.

Preface

The proliferation of information on rotaviruses in the twenty years since their description as human pathogens has made it difficult to keep up with the latest developments in the field. Numerous aspects of rotavirus biology have been addressed in reviews, but these have been infrequent and quite selective in the material covered. The time appeared right to attempt to gather, into a single source, an overview of what we have learned. This volume is the result of that attempt. I hope that the reader will find in this single source, the core of information covering our current state of knowledge.

I acknowledge my debt to the authors of the various chapters, for it is they who performed the tedious process of reviewing the literature and synthesizing and organizing it into concise works.

R.F. RAMIG

Contents

Contributors

(Their addresses can be found at the beginning of their respective chapters.)

Introduction and Overview

R.F. RAMIG

In the nearly two decades since the first description of rotaviruses in association with human disease (BISHOP et al. 1973; FLEWETT et al. 1973) and their identification as major etiologic agents of diarrhea in infants and young children, significant progress has been made in our understanding of the virus, its molecular biology, and its disease. The virus shares many similarities in its capsid structure, genome structure, and replication strategy with the mammalian reoviruses, and rotaviruses have been classified as a genus within the virus family Reoviridae (MATHEWS 1979). It is the goal of this volume to present our current understanding of the many facets of this virus.

Rotaviruses were first described as viruses possessing a double capsid morphology and a genome of 11 segments of double-stranded RNA. Early studies showed that all rotaviruses, isolated from diverse animal species, shared a common antigen. More recently, with the characterization of increasing numbers of virus strains, viruses sharing the basic morphologic and biochemical characteristics but lacking the common antigen have been identified (BRIDGER 1980; SAIF et al. 1980). These atypical rotaviruses have been isolated from a wide array of animal species, including humans, and have been shown to fall into a number of antigenically distinct groups between which antigenic cross-reaction is not observed. As a result, rotaviruses are now divided into seven serogroups, A-G (SAIF 1990). With the exception of the chapter by Saif and Jiang on nongroup A rotaviruses, this volume will concentrate on the group A, or conventional, rotaviruses, about which the greatest amount of information is available.

The following is an overview of the contents of this volume. In this general overview, references are not provided; they may be found in the individual chapters indicated.

Rotavirus Structure

Rotaviruses are large, nonenveloped, icosahedral viruses with morphologic aspects similar to those of the other Reoviridae. Studies with negative

Division of Molecular Virology, Baylor College of Medicine, One Baylor Plaza, Houston, TX 77030, USA staining electron microscopy yielded conflicting results regarding the capsid symmetry. Recently, electron cryomicroscopy and computer image processing techniques have been applied to rotaviruses and have resulted in a recon-struction of the rotavirion, with resolution to ~ 40 Å, that has resolved these conflicts. The current status of rotavirus structure is reviewed by Prasad and Chiu.

Numerous questions relative to rotavirus structure remain. Clearly many of these questions will be answered by continued application of cryomicro-scopy and image reconstruction techniques, but the ultimate structural characterization of the virus awaits crystallization and solution of the structure at atomic resolution.

Rotavirus Genome Structure

As mentioned above, the genome of rotaviruses consists of 11 segments of double-stranded RNA. In general each genome segment is monocistronic. However, the genome segment encoding the outer capsid protein VP7 (segment 7, 8, or 9, depending on virus strain) has two, conserved, in-frame initiation codons and produces two polypeptides that differ by four or five amino acids at the NH_2-terminal (CHAN et al. 1986). The nucleotide sequence has been determined for each of the 11 genome segments, although all 11 segments have not been sequenced from a single virus strain. All segments contain conserved 3'- and 5'-terminal sequences, relatively short 5'-terminal untranslated regions, and somewhat longer 3'-terminal untranslated regions. Recently, considerable effort has been expended on sequencing segments encoding VP4 (segment 4) and VP7 (segment 7, 8, or 9), the neutralization antigens of the virus. Comparative sequence analysis has revealed conserved and variable regions of these genes that are most certainly related to functional domains within the gene products. The current knowledge about the rotavirus genome is reviewed by Desselberger and McCrae.

Many questions remain about the structure of the rotavirus genome. Perhaps most important among these is the nature of the *cis*-acting elements that function as assortment signals during genome packaging and as replication signals. A system for the reintroduction ('rescue') of genetically engineered genome segments into viable particles would be a keystone event that would allow study of these *cis*-acting signals, in addition to a large number of other studies on a wide variety of aspects of rotavirus biology.

Rotavirus Proteins and Protein Structure

Rotavirus contains six structural proteins, one of which (VP4) is cleaved in highly infectious preparations of virions and another (VP7) that is glycosylated. In addition, there are five nonstructural proteins. The amino acid sequences of these protein species have been deduced from nucleotide sequences, and the sequence data have allowed structure and function of the individual protein species to be inferred. However, in many cases the inferred function has not been biochemically confirmed. The existence of rotavirus cDNAs, and expression systems that yield bona fide gene products from the cDNA, has allowed the beginnings of site-directed mutagenesis for analysis of functional domains within proteins. This approach, however, awaits the development of rescue systems so that genetically engineered segments can be rescued into viable virus particle before the mutations can be analyzed in the context of the infected cell rather than in vitro or in expression systems. Our current knowledge of the proteins of rotavirus is reviewed by Both, Bellamy, and Mitchell.

Numerous questions remain to be answered regarding rotavirus proteins, ranging from function, to structure, to interactions with other viral and/or cellular proteins. Much work remains to be done on rotavirus proteins.

Rotavirus Replication

The entirely cytoplasmic replication cycle of rotavirus is superficially like that elucidated for the mammalian reoviruses. However, as more detail is revealed it appears that RNA synthesis mechanisms are quite similar to those used by reovirus, whereas rotavirus morphogenesis and assembly, particularly for the outer capsid, is quite distinct. Infection is initiated by interaction of virions with specific, saturable, but unidentified, cellular receptors. After binding, virus penetration appears to be catalyzed by the outer capsid protein VP4, but subsequent uncoating remains uncharacterized. Viral mRNA is synthesized by the endogenous RNA-dependent RNA polymerase (transcriptase) that is activated by removal of VP4 and VP7 during uncoating. Following synthesis of viral mRNA and translation of viral proteins, large aggregations of viral protein and RNA accumulate in the cytoplasm. Morphogenesis begins in these conclusions and results in the formation of a series of subviral particles, in which replication of the double-stranded RNA occurs, and culminates in the formation of single shell-like particles at the periphery of the inclusion. Morphogenesis continues as the single shell-like particles bud through the membrane of the endoplasmic reticulum (ER). In this budding process, the resulting transiently enveloped particles gain VP4 and the ER-resident protein VP7. Finally, the VP4/VP7 outer capsid is assembled and the membrane is lost. Virus release is apparently accomplished through cell death and lysis. The details of this replication cycle are reviewed by Patton.

Many of the details of the rotavirus replication cycle remain unknown. Among these are the nature of the cellular receptor, the precise mechanism of penetration and uncoating, the details of double-stranded RNA synthesis, The mechanisms involved in the assortment of genome segments, and the details of the morphogenetic pathway. The precise roles of the structural and nonstructural proteins, and the possible roles of cellular components, in the various steps of the replication cycle also remain unknown.

Rotavirus Genetics

The segmented nature of the genome of rotaviruses suggested that it would have genetic properties similar to those of other viruses with segmented genomes, particularly the mammalian reoviruses. Temperature-sensitive (ts) mutants were isolated from several rotaviruses, and studies with these mutants confirmed the similarity in genetic interactions. Specifically, recombination, the primary genetic interaction between viruses, was shown to occur by a mechanism of reassortment. This allowed studies of rotavirus using reassortants, derived from phenotypically distinct viruses, to identify genes encoding phenotypes of interest. This powerful technique has been applied in many studies of rotavirus biology, biochemistry, and pathogenesis. In addition, the ts mutants are now being exploited to identify specific functions encoded by individual genes and gene products. The current status of genetic studies with rotaviruses is reviewed by Gombold and Ramig.

As in other areas, many questions remain to be answered relative to rotavirus genetics. Perhaps the most interesting question is the mechanism of assortment of genome segments in singly infected cells and reassortment in mixed infected cells. An understanding of this mechanism, or learning how to subvert it will be critical to devising methods to rescue genetically engineered genes into viable virus particles. The possibility of intramolecular recombination has arisen with the demonstration of RNA recombination in a number of other virus families with RNA genomes.

Rotavirus Antigens and Serology

Although the products of all the rotavirus genes are recognized as antigens and elicit an immune response, three major antigenic specificities have been recognized. The common antigen, detected as the cross-reactive antigen among all group A rotaviruses, has been shown to segregate with genome segment 6 and VP6 in reassortment experiments. An additional antigenic specificity found on VP6 is the subgroup antigen. All rotaviruses characterized display subgroup I, subgroup II, or neither subgroup I or subgroup II specificity. The distinctness of the subgroup antigen is demonstrated by the ability to generate anti-VP6 monoclonal antibodies with subgroup I or subgroup II specificity. Two distinct neutralizing antigens exist in rotaviruses and have been shown, in reassortment studies, to segregate with the two outer capsid proteins VP4 and VP7. VP7 appears to be the major neutralizing antigen, as hyperimmune sera neutralize reassortants to high titer in a manner that correlates with the origin of VP7. Hyperimmune sera contain much lower titers of VP4-specific neutralizing antibody, and VP4 antibodies are thought to account for the levels of cross-neutralization frequently seen between viruses with distinct VP7 specificity. Two date, 12 distinct forms (serotypes) of VP7, distinguishable by neutralization, have been identified, and the existence of multiple forms of VP4 antigenic specificity has been demonstrated, although the number of VP4 types is not currently large. The VP4 and VP7 neutralization antigens have been shown to segregate independently in mixed infections, so the potential number of combinations of VP4 and VP7 neutralization specificities is quite large. The details of rotavirus antigenic structure are reviewed by Hoshino and Kapikian.

The study of rotavirus serology is important for rational vaccine design and therefore a pressing question is the number of serologically distinct forms (serotypes) of VP4 and VP7. The possible identification of cross-reactive neutralizing epitopes on VP4 and VP7 would certainly facilitate the formulation of broadly protective vaccines. The development of monoclonal antibody-based serotyping assays for VP4, like those in existence for VP7, will be an important goal.

Immunologic Determinants of Protection from Rotavirus Disease

Prior infection with rotavirus has been demonstrated to protect against disease after reinfection with serologically homologous virus. However, the immunologic mechanisms of this protection remain unclear. Numerous studies have shown that suckling animals can be protected by the passive transfer of neutralizing antibody to either VP4 or VP7. However, in some studies, the development of active immunity following infection has not correlated with the development of type-specific neutralizing antibodies. Thus, the role of serotype in protection is not clear. In addition to humoral immunity, a role for cellular immunity has been suggested. Passive transfer of rotavirus-specific cytotoxic T lymphocytes has been shown to protect against subsequent challenge. Offit reviews the current status of our understanding of immune protection from rotavirus disease.

The study of rotavirus immune protection is clearly in its infancy, and the development of immune correlates of protection are fundamental to the development of effective vaccines. Much work remains to be done in this area, work made difficult by the paucity of homologous virus: animal model systems in which the virulent virus can be manipulated in tissue culture.

Rotavirus Pathogenesis

Rotaviruses induce diarrheal disease following lytic infection of the mature enterocytes lining the villi of the small intestine. The strict enterocyte tropism of rotaviruses and association with only intestinal disease has recently been called into question by the observation that some virus strains can spread to the liver of immunocompromised mice and cause hepatitis and death. This observation may extend to humans, as abnormal liver function in human rotavirus infection is infrequently observed. Rotavirus disease is generally seen only in the young of a species, suggesting that host factors may limit the age of susceptibility to infection. There is also evidence for asymptomatic infection of susceptibly aged animals, an observation that suggests there may be viral determinants of virulence in addition to host factors. Furthermore, the severity of rotavirus disease is greater in underdeveloped regions of the world, suggesting that factors present in these regions may promote the disease. Indeed, there is evidence for increased severity of disease in malnourished animals. The current status of our understanding of the pathogenesis of rotavirus infection is reviewed by Greenberg, Clark, and Offit.

Many more questions remain to be answered relative to rotavirus pathogenesis than have been answered. While the histopathology of the disease is described, most areas dealing with pathogenesis remain open to investigation. Important remaining questions are: What accounts for the age restriction of susceptibility to rotavirus disease? Is this restriction related to expression of a host cell virus receptor? Are other host factors related to age restriction and disease severity? What viral genes play a role in determining the virulence of infection? What viral genes are related to the ability of viruses to spread from host to host? Are specific viral mutations associated with spread to extra-intestinal sites?

Rotavirus Vaccines

Rotavirus disease is associated with significant morbidity, mortality, and health care burden. There is an urgent need for a safe and effective vaccine. Because of this need, various live-attenuated vaccines have advanced directly to human trials. Unfortunately, these vaccines have been generally disappointing in their ability to protect from disease. The design of rotavirus vaccines is made particularly difficult by the complexity of the antigenic and serotypic structure of the virus and the need to vaccinate extremely young children. Given the disappointing results obtained in the early vaccine trials, alternative approaches are under development. Many of these new approaches involve the application of modern molecular techniques. The problems associated with rotavirus vaccination, the potential for development of effective vaccines, and alternative strategies of vaccine development are discussed by Conner, Matson and Estes. Given the very limited success in development of effective rotavirus vaccines to date, it is clear that there is much to be done in this area.

Nongroup A Rotaviruses

Rotaviruses can be divided into seven serologically distinct groups that share a common morphology and segmented genome structure. Relatively little is known about the viruses in serogroups other than group A, primarily because, with few exceptions, none of these viruses can be grown in tissue culture cells. However, serological reagents for some of these virus groups are becoming available and a picture of the epidemiology of these viruses appears likely to emerge rapidly. In addition, amounts of virus sufficient for molecular cloning can be isolated from clinical samples, so that sequencing of genome segments and comparison of cognate segments among the various sero-groups promise to provide significant useful information on these viruses. The current status of knowledge of the nongroup A rotaviruses is reviewed by Saif and Jiang.

The nongroup A rotaviruses will be a fruitful area for future research, particularly if epidemiologic studies reveal that they account for a significant fraction of diarrheal disease of unknown origin.

Conclusion

Progress in our understanding of the molecular biology and replication of group A rotaviruses has been quite rapid. However, significant questions remain to be answered even in these areas of progress. In other areas, little other than descriptive work has been done. Clearly much remains to be done before rotaviruses and the disease they cause are understood in molecular detail. However, given the powerful molecular tools and techniques at the disposal of the rotavirologist, progress promises to continue to be rapid.

Such progress in rotavirus research is necessary to meet the long-term goal of most rotavirus research programs, namely, the control of rotaviral disease. The disease burden, attributable to rotavirus, present in the human and domestic animal populations is quite significant, and the goal of conquering the disease has both humanitarian and economic benefits.

The remainder of this volume presents our current understanding of the various aspects of rotavirus biology and molecular biology, in the hope of providing a comprehensive source of information for researchers, students, and others interested in the rotaviral disease problem.

References

Bishop RF, Davidson GP, Holmes IH, Ruck BJ (1973) Virus particles in the epithelial cells of duodenal mucosa from children with acute non-bacterial gastroenteritis. Lancet 2: 1281–1283
Bridger JC (1980) Detection by electron microscopy of caliciviruses, astroviruses, and rotavirus-like particles in the faeces of piglets with diarrhoea Vet Rec 107: 532
Chan WK, Penaranda ME, Crawford SE, Estes MK (1986) Two glycoproteins are produced from the rotavirus neutralization gene. Virology 151: 243–252
Flewett TH, Bryden AS, Davies H (1973) Virus particles in gastroenteritis. Lancet 2: 1497
Mathews REF (1979) The classification and nomenclature of viruses. Summary of results of meetings of the international committee on taxonomy of viruses. The Hague, September 1978. Intervirology 11: 133–135
Saif IJ (1990) Nongroup A rotaviruses In: Saif LJ, Theil KW (eds) Viral diarrhea of man and animals, CRC Press, Boca Raton, pp 73–95
Saif LJ, Bohl EH, Theil KW, Cross RF, House JA (1980) Rotavirus-like, calicivirus like, and 23nm virus-like particles associated with diarrhea in young pigs. J Clin Microbiol 12: 105–11

Structure of Rotavirus

B.V. Venkataram Prasad and W. Chiu

1 Introduction

The ultimate goal of structural analysis of any virus is not only to delineate its molecular structure but also to understand the structure in terms of its functional aspects. In recent years, three-dimensional structures of several animal viruses have been determined using X-ray crystallography (HOGLE et al. 1985; ROSSMANN et al. 1985; ACHARYA et al. 1989; HARRISON 1990) and electron microscopy (CROWTHER and KLUG 1975; VOGEL et al. 1986; FULLER 1987; PRASAD et al. 1988; BAKER et al. 1988; SCHRAG et al. 1989). These structures have provided greater insight not only into the architectural principles of viruses but also into the molecular mechanisms such as entry of viruses into cells, virus disassembly, assembly, and neutralization (HARRISON 1990; ROSSMANN and RUECKERT 1988; ROSSMANN 1989; FULLER 1987; PRASAD et al. 1990). In some cases, structural analyses of the viral

Verna and Marrs McLean Department of Biochemistry, Baylor College of Medicine, One Baylor Plaza, Houston, TX 77030, USA

Current Topics in Microbiology and Immunology, Vol. 185
© Springer-Verlag Berlin · Heidelberg 1994

components have also been very useful. Structural studies of glycoproteins of influenza virus have provided a good deal of information relating to receptor binding, membrane fusion, and antigen-antibody interactions (WILEY and SKEHEL 1990; COLMAN et al. 1987). It is to be expected that such structural studies on viruses and viral components will prove useful in the rational design of antiviral drugs and vaccine development.

So far, X-ray crystallography has been used to obtain structures, at near atomic resolution, of plant and animal viruses with diameters less than 400 Å. It is difficult, though not impossible, to crystallize larger spherical viruses suitable for crystallography. However, it may not be always possible to get homogeneous preparations of viruses in high enough concentrations for crystallization. In the last few years, electron cryomicroscopy of frozen hydrated specimens, together with computer image processing, has emerged as an alternative technique for determining three-dimensional structures of spherical viruses. This technique does not require the specimen to be in a crystalline state. An additional advantage is that this technique allows structural analysis of the specimen in different chemical and functional states (UNWIN and ENNIS 1984). Though the resolution of the structural analyses of spherical viruses using this technique reported so far has been around 35 Å, prospects of higher resolution seems very encouraging. Recently, it has been shown in the case of tobacco mosaic virus that it is possible to approach a resolution of 10 Å, in order to visualize structural features such as α-helices and RNA (JENG et al. 1989).

Rotavirus is a large spherical virus which has been identified as a major cause of infantile gastroenteritis (DE ZOYSA and FEACHEM 1985; KAPIKIAN and CHANNOCK 1990). Ever since its discovery in 1973 (BISHOP et al. 1973), this virus has been a subject of several structural studies using conventional electron microscopy (MARTIN et al. 1975; ESPARZA and GIL 1978; KOGASAKA et al. 1979; ROSETO et al. 1979). Recently, the three-dimensional structure of rotavirus has been determined using electron cryomicroscopy and computer image processing techniques (PRASAD et al. 1988, 1990). In the following sections we briefly discuss these techniques and follow with a detailed description of the three-dimensional structure and its implication on the functional aspects of the virus.

2 Methods

2.1 Electron Cryomicroscopy

In conventional transmission electron microscopy one would normally use metal shadowing or negative staining techniques in order to enhance image contrast. These preparative techniques may potentially alter the structure of the specimen and sometimes may destroy fragile structural features by chemical modification and dehydration. In 1975, TAYLOR and GLAESER introduced a method of embedding the specimen in a thin layer of ice and imaging at low temperatures using a low electron dose. Since then several laboratories have been involved in the improvement of this technique. These developments have been well documented in the review by DUBOCHET et al. 1988. This cryopreparative technique not only provided good contrast but also preserved the structural integrity of the specimen.

Cryoelectron microscopy mainly involves three steps: (1) quick freezing of the specimen, (2) transfer of the specimen to the microscope, and (3) examination of the specimen in the microscope at low temperature. The first step is carried out using a mechanical guillotine-type plunging device (JENG et al. 1988). The specimen grid is held by a tweezer in the quick freezing device. After the excess liquid on the grid is blotted using filter paper, the specimen grid is rapidly plunged into liquid ethane (cryogen) at its melting point. This rapid freezing produces a thin layer of amorphous ice in which the specimen is embedded. In the second step, the frozen specimen is initially transferred to a cryospecimen holder maintained at liquid nitrogen temperature in a work station. Then the cryospecimen holder is quickly transferred to the electron microscope. The specimen is examined in the microscope at a temperature below-160 °C using low electron doses. Several technical problems associated with this technique have been addressed during the past few years, making this a successful technique (reviewed in CHIU 1986; STEWART and VIGERS 1986; DUBOCHET et al. 1988)

2.2 Computer Image Processing

Because of the large depth of focus of conventional electron microscopes, transmission electron microgrphs, in effect, represent two-dimensional projections of the specimen. Inference of the detailed three-dimensional structure by direct examination of electron micrographs is often a difficult task. Over last two decades computerized procedures have been developed to reconstruct the three-dimensional structure of a specimen from such projections (reviewed in KLUG 1979; CROWTHER 1982). These procedures offer an objective way of extract three-dimensional structural information from electron micrographs. Similar procedures of reconstructions from projections are used in other contexts, for example, diagnostic tomography.

In order to determine the three-dimensional structure of any object it is necessary to combine the information content from different views of that object. These different views are sometimes provided by the specimen lying in different orientations or else they may be obtained by tilting the specimen in the microscope. Three-dimensional reconstruction from electron micrographs is based on what is known as projection theorem. This theorem states that the two-dimensional Fourier transform of a projection of a three-dimensional object is a central section, normal to the direction of the view, of the three-dimensional Fourier transform of the object. If all the different orientations of the specimen can be identified with respect to a common frame of reference, the three-dimensional Fourier transform of the specimen can be built from the two-dimensional Fourier transforms of different views. Fourier inversion of the three-dimensional transform thus obtained gives the three-dimensional structure of the specimen.

For icosahedral viruses, CROWTHER et al. (1970) have developed image processing procedures to determine their three-dimensional structure from micrographs. The first step in computer processing of the micrographs is converting the image into digitized data by microdensitometry. Regions in the micrographs with a sufficient number of particles are digitized using a computer controlled microdensitometer. The digitized regions are put into the computer, where each particle is windowed and centered inside a box. The orientation of each particle is determined in a computer from its Fourier transform using the so-called common lines procedure (CROWTHER 1971). Recently, statistical procedures have been incorporated into this method which have proved useful in processing low contrast data (FULLER 1987; SCHRAG et al. 1989). The orientations of enough particles are determined such that they evenly represent the asymmetric unit of an icosahedron. This information is then combined in Fourier space using cylindrical expansion methods to obtain a three-dimensional Fourier transform (CROWTHER 1971). An inverse Fourier transform would then give the three-dimensional structure.

3 Three-Dimensional Structure

Like other members in the family *Reoviridae* (MATTHEWS 1979) rotaviruses are isometric particles. These large, nonenveloped spherical viruses have two shells of proteins, referred to as the outer and inner shells. These two shells surround a central protein core, the inner core, which contains the genome

Table 1. Rotavirus proteins

Gene Segment	Polypeptide	Molecular weight[a] (kDa)	Location in the virus
1	VP1	125 (Cohen et al. 1989)	Inner core
2	VP2	102 (Kumar et al. 1989)	Inner core
3	VP3	83 (Liu and Estes, 1989)	Inner core
4	VP4	87 (Nishikawa et al. 1988)	Outer shell
5	NS53	58 (Bremont et al. 1987)	Nonstructural
6	VP6	45 (Estes et al. 1984)	Inner shell
7	NS34	36 (Both et al. 1984)	Nonstructural
8	NS35	36 (Both et al. 1982)	Nonstructural
9	VP7	34 (Arias et al. 1984; Both et al. 1983a)	Outer shell
10	NS20	20 (Both et al. 1983b)	Nonstructural
11	NS26	22 (Mitchell and Both 1983)	Nonstructural

[a] From cloned SA11 genes where sequences are available otherwise the data is from the bovine RF strain.

consisting of 11 double-stranded RNA segments (RODGER et al. 1975; NEWMAN et al. 1975). The gene coding assignments and the properties and locations of the proteins encoded by each of the 11 genome segments are well established and are detailed in several recent reviews (see, for example, ESTES and COHEN 1989; ESTES 1990; KAPIKIAN and CHANOCK 1990). Six of these proteins are structural and the other five are nonstructural. Information relevant to this review is summarized in Table 1.

3.1 Outer Shell

Electron micrograph of double-shelled rotavirions, embedded in vitreous ice suspended across a hole of a holey carbon grid, is shown in Fig. 1. It is to

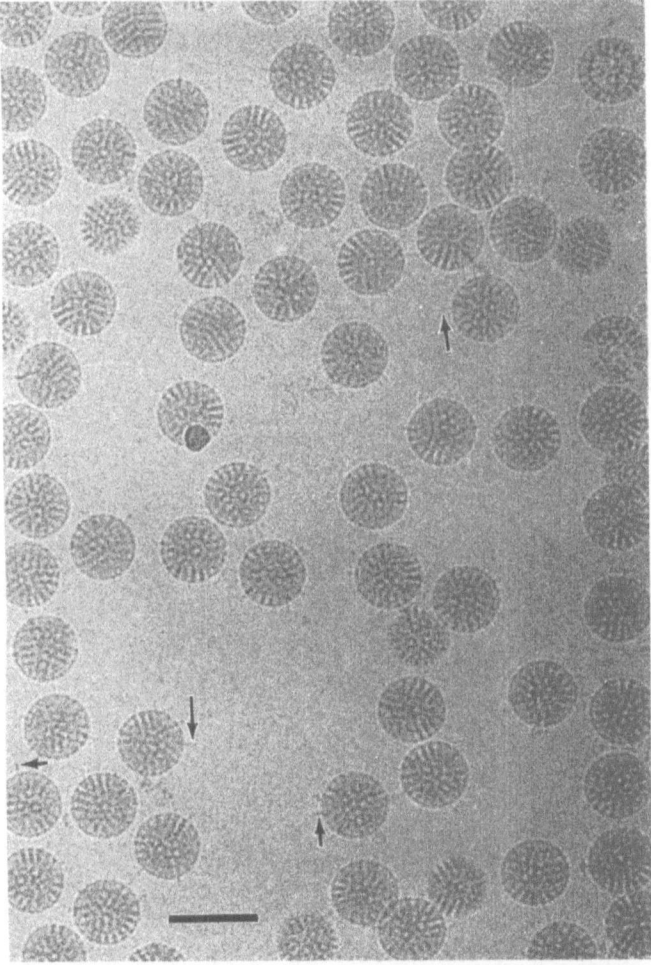

Fig. 1. Electron micrograph of double-shelled rotavirus particles embedded in a thin layer of vitreous ice. Scale bar, 1000 Å

be noted that with respect to images of negatively stained specimens, cryoimages result in an opposite contrast; that is, in a typical electron micrograph of a negatively stained specimen, the stain-excluded region, which represents protein, appears lighter than the stained regions. In the cryoimages the protein appears dark because it has a higher density than the surrounding ice. As in the case of images of negatively stained specimens, mature double-shelled virions have a smooth, circular, outer margin. In some of the images, the particles clearly have a characteristic wheel-like appearance, i.e., a narrow rim set on short multiple spokes. This appearance is the reason for the name rotavirus (in Latin, rota is wheel). A similar smooth outer surface has not been seen in any of the other members of the *Reoviridae* family. One of the features which was not observed by the conventional techniques of electron microscopy is the spikes emanating from the circular outer margin of the rotavirions. The diameter of these particles, excluding the length of the spikes, is 765 Å. The three-dimensional structure of double-shelled rotavirus has been determined from such electron micrographs (PRASAD et al. 1988, 1990).

3.1.1 Icosahedral Symmetry and T Number

Figure 2 shows the surface representations of the three-dimensional reconstruction of double-shelled rotavirus along the icosahedral five and three fold axes. The structure conforms to an icosahedral symmetry. In the reconstruction, instead of a full icosahedral symmetry, i.e, 532 symmetry, a lower 522 symmetry is used. The icosahedral symmetry in the virus is inferred by confirming the presence of a strict threefold symmetry on the final reconstruction.

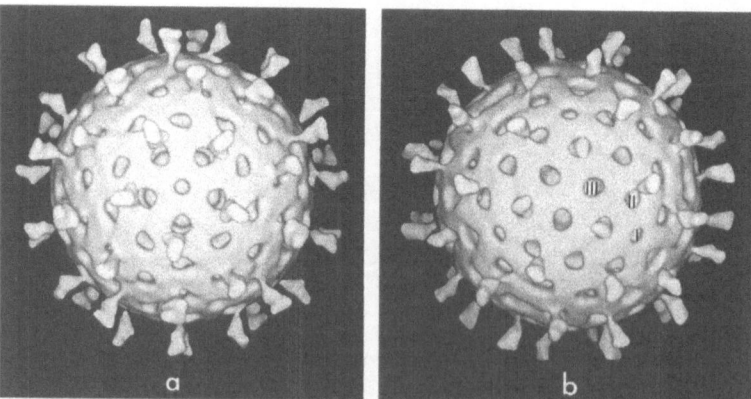

Fig. 2a, b. Surface representation of the three-dimensional structure of double-shelled rotavirus along the icosahedral five fold and three fold axes

An Icosahedral surface lattice is characterized by a triangulation number (T) which defines the relationship between the neighboring icosahedral five-fold axes in terms of six-coordinated positions. The concept of triangulation was introduced by CASPER and KLUG in 1962 to account for the arrangement of more than 60 subunits on the icosahedral surface. Several spherical viruses have been shown to have triangulated icosahedral surfaces. The icosahedral surface lattice of rotavirus shows $T = 13$. These studies unambiguously settled the controversy raised by a number of conflicting reports in which the T number was estimated variously as ranging from 3 to 16 (MARTIN et al. 1975; ESPARZA and GIL 1978; KOGASAKA et al. 1979; ROSETO et al. 1979). However, ROSETO et al. (1979), by using freeze drying and metal shadowing techniques on the single-shelled particles, inferred a $T = 13$ icosahedral lattice.

A $T = 13$ icosahedral surface lattice is a skewed lattice and has a handedness. In such a lattice a five-fold axis is reached from its neighboring five-fold axis by stepping over three six-coordinated positions and taking a left or a right turn. This is demonstrated in Fig. 3, wherein a $T = 13$ lattice with a left-handed and a right-handed configurations are shown. METCALF (1982) has shown that the simian rotavirus exhibits a left-handed surface lattice by using computer modeling of optically filtered images of virus particles prepared by the freeze etching technique. Electron images of metal shadowed single-shelled porcine rotavirus have also suggested a left-handed surface lattice (LUDERT et al. 1986). A $T = 13l$ lattice has been inferred for reovirus, another member of the *Reoviridae* family (METCALF 1982). It would be interesting to see if all the members of rotavirus genus in particular and Reoviridae family in general show the same handedness. The biological significance of the handedness in these viruses remains unknown.

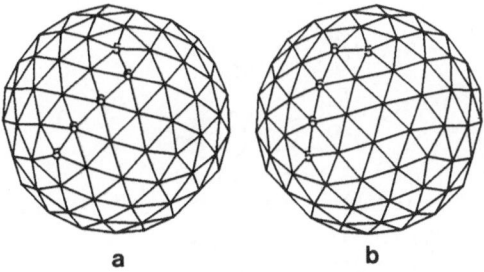

a   b

Fig. 3a, b. Triangualtion number (T) of an icosahedral surface lattice in **a** left-handed configuration ($T = 13l$) and **b** right-handed configuration ($T = 13d$). The icosahedral symmetry axes are marked. The drawing shows one side of the polyhedral surface, which consists of 120 six-coordinated and 12 five-coordinated points. The relationship between the neighboring five fold axes (which designate the T number), in terms of six-coordinated positions, is shown for both handednesses. This relationship holds true for any two neighboring five-coordinated positions in the $T = 13$ icosahedral lattice

3.1.2 Aqueous Channels

A distinctive feature of the rotavirus structure is the presence of aqueous channels at all five- and six-coordinated centers. Based on their location with respect to the icosahedral symmetry axes, these channels have been classified into three types. Type I channels are those running down the icosahedral five fold axes, type II are those on the six-coordinated positions surrounding the five fold axes, and type III are those on the six-coordinated positions around the icosahedral three fold axes. In each virion there are 132 channels: 12 type I, 60 type II, and 60 type III. The vertices of the triangles in Fig. 3 correspond to the locations of these channels in the structure.

One of the advantages of structural analysis using electron microscopy images of an ice embedded specimen is the ability to retrieve the internal structural details. This is in contrast to structure determination from images of negatively stained specimens, in which it is possible to obtain only the surface features. This is demonstrated in Fig. 4a, which shows a central section, from the three-dimensional map of the double-shelled virus, normal to one of the five fold axes. This section cuts across type III channels providing their side view. In order to ensure reliability in visualizing internal details, for comparison a similar section from an independent reconstruction of single-shelled particles is presented in Fig. 4b. Details of the structural features of the inner shell are discussed in Sect. 3.3. Type III channels are about 55 Å wide at the outer surface of the virus. They constrict before widening and have their maximum width at the position close to the surface of the inner shell proteins. The entire depth of these channels is about 140 Å. Similar features and dimensions are seen in the other two types of channels, except that type I channels have a narrower ($\sim$ 40 Å) opening at the outer surface of the virus.

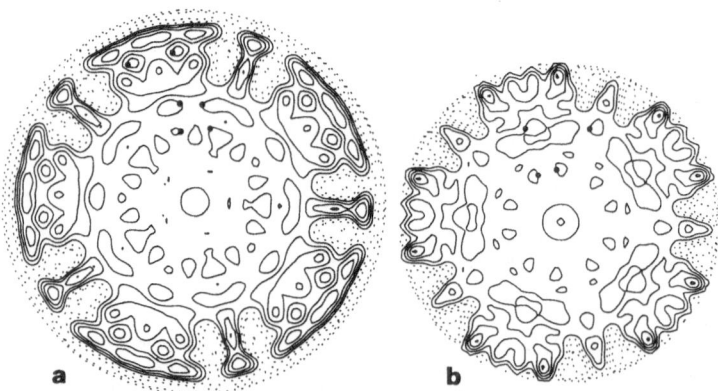

Fig. 4a, b. Section perpendicular to the five fold axis and passing through the center of the virus from both **a** double-shelled and **b** single-shelled particles. *Continuous lines* correspond to the regions of higher scattering density (i.e., proteins), *broken lines* represent the solvent

The biological role of these aqueous channels is not yet clear. However, biochemical studies have shown that the rotaviruses exhibit RNA transcriptase activity when the outer shell proteins are removed. These single-shelled virions remain intact during the process of transcription (COHEN et al. 1979; MASON et al. 1980; BICAN et al. 1982; SANDINO et al. 1986). It has been speculated that these channels may be involved in importing the metabolites required for viral RNA transcription and exporting the nascent RNA transcripts for subsequent viral replication processes (PRASAD et al. 1988). It is interesting to note that such channels are seen even in the case of bluetongue virus, which also belongs to family *Reoviridae* (PRASAD et al. 1992).

3.1.3 Surface and Spikes

The protein mass density on the outer shell is distributed uniformly on the local and strict two fold and three fold axes, giving rise to what appears as a smooth surface. In the immediate vicinity around the channels the mass density is slightly elevated giving, a crater-like appearance to the channels. This makes the regions between the channels (corresponding to the centres of the triangles in Fig. 3) depressed. The depressions between type II and type III channels are the most prominent (see Fig. 2). This canyon-like feature is about 50 Å wide, 120 Å long, and 30 Å deep. It is yet to be determined if there is any functional significance to this structural feature.

From the surface, 60 spikes of density extend to a length of 120 Å (Fig. 2). These spikes are located at the outer edge of the type II Channels surrounding the icosahedral five-fold axes. From the surface of the capsid up to a length of 45 Å, the spikes are thin with a maximum width of about 35 Å Beyond this point they have a well defined globular domain of 55 Å in diameter. At the distal end of these spikes, lying across the globular domain, is the bilobed structure, 40 Å wide and 70 Å across.

The outer shell of the rotavirus is made up of two proteins, VP7 and VP4 (KAPIKIAN and CHANOCK 1990). VP7, a glycoprotein, is the major component, and antibodies against this protein neutralize virus. VP4, though a minor component, is implicated in several of the important functions of the virus such as hemagglutination, virulence, neutralization, and virus entry into cells. From their earlier studies, PRASAD et al. (1988) predicted that the surface spikes are made up of VP4 with VP7 molecules making up the rest of the outer surface. One of the ways of confirming such a predictions is to carry out immunolabeling studies using monoclonal antibodies against VP4 and VP7. Immunoelectron microscopy on rotaviruses using conventional techniques has been reported (KAPIKIAN and CHANOCK 1990). Theses studies were carried out using monoclonal antibodies to the major outer capsid protein, VP7, and the inner capsid protein, VP6. PRASAD et al. (1990) recently used electron cryomicroscopy and image processing techniques on immunolabeled viruses to identify the location of VP4 in the three-dimensional structure of the virus.

3.2 Structure of Rotavirus Anti (VP4) Fab Complex

Figure 5a shows as electron micrograph of rotavirus complexed with anti-VP4 antibodies (IgGs) embedded in vitreous ice. The increased density around the periphery of the virions strongly indicates that the antigenic part of VP4 is accessible from the outside. The bivalent nature of the IgGs makes these virions clump together. This is avoided when Fabs are used instead of IgGs, as shown in Fig. 5b. These micrographs are suitable for carrying out three-dimensional reconstruction and demonstrate the power of electron cryomicroscopy in visualizing antibody-virus complexes in their native state, without having to use gold markers or any other chemical labels.

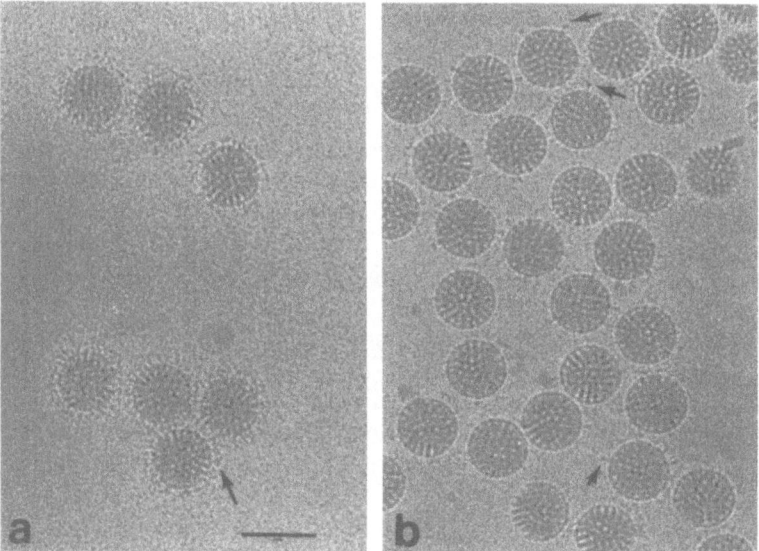

Fig. 5a, b. Electron micrographs of **a** anti(VP4)IgG complexed and **b** anti(VP4)Fab complexed double-shelled rotavirus particles embedded in a thin layer of vitreous ice. *Arrows* in **a** indicate IgGs, in **b** they indicate Fab bound regions. *Scale bar*, 1000 Å

3.2.1 Spikes are Dimers of VP4

The three-dimensional structure of rotavirus complexed with anti-VP4 Fabs is shown in Fig. 6. Two molecules of Fab bind to the distal end of each spike. The mass density of the spike and the fact that the antibody used in these studies is monoclonal clearly indicate that each spike is a dimer of VP4. This means that each virion has 120 molecules of VP4. VP4 in many virus strains is a hemagglutinin (ESTES and COHEN 1989). Recently, it has been shown that the baculovirus – expressed VP4 exhibits the ability to hemagglutinate.

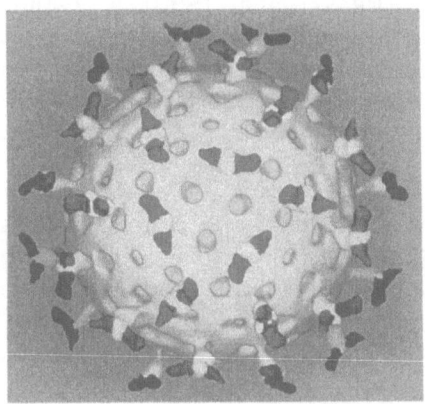

Fig. 6. Surface representation of the three-dimensional structure of anti(VP4)Fab bound double-shelled rotavirus along its icosahedral three fold axes. The Fab molecules are shown in a *darker shade* and the virus surface is represented in a *lighter shade*

Although this implies an oligomeric nature of VP4, direct biochemical data showing that VP4 is a dimer are not yet available. It is interesting to note that hemagglutinins of reovirus and influenza virus also show an extended configuration and oligomeric nature (FURLONG et al. 1988; WILSON et al. 1981). Another feature that is common to VP4 and the hemagglutinin of influenza virus is that both of them require proteolytic cleavage for infectivity.

In the presence of trypsin, VP4 is cleaved to produce VP5 (60 kDa) and VP8 (28 kDa). It has been well established that proteolytic cleavage of VP4 greatly enhances the infectivity of the virus in vitro (CLARK et al. 1981; ESPEJO et al. 1981; ESTER et al. 1981). The monoclonal antibody (2G4, see SHAW et al. 1986; BURNS et al. 1988) used by PRASAD et al. (1990) has been shown to bind to the VP5 region of VP4 (MACKOW et al. 1988). This localizes the VP5 segment of VP4 to the distal end of the spikes. The homology of the amino acid sequence in this neutralization region with the proposed fusion regions in Sindbis and Semiliki Forest viruses may implicate this region in viral penetration (MACKOW et al. 1988).

The location of VP4 in the three-dimensional structure of the virus makes it a good candidate to initiate cell attachment and be involved in viral entry processes. Such an idea, however, contrasts with other studies, which have implicated VP7 in cell attachment (FUKUHARA et al. 1988; MATSUNO and INOUYE 1983; SABARA et al. 1985). So far, however, there is no conclusive evidence as to which of the two, VP4 or VP7, is the cell attachment protein. Recent studies of RAMIG and GALLE (1990) on rotavirus replication in cultured liver cells suggest a cell surface function of VP4, either viral attachment to or penetration into the cell. Since VP4 exhibits hemagglutinating activity both in the free state (MACKOW et al. 1989) and when it is bound to the virus (ESTES and COHEN 1989), it is more likely that VP4 is a cell attachment protein. It has been proposed that the exposed region of the distal end of the spike is involved in initial attachment to the cell and is the region binding to Fab in cell penetration (PRASAD et al. 1990). A possible mechanism then by which 2G4 antibodies neutralize viral infectivity is perhaps by blocking this region critical for viral entry into the cell.

3.3 Inner Shell

Treatment of intact double-shelled virion with chelating agents such as EDTA removes the outer shell, exposing the inner shell proteins. The resulting single-shelled particles are indistinguishable from those produced in the

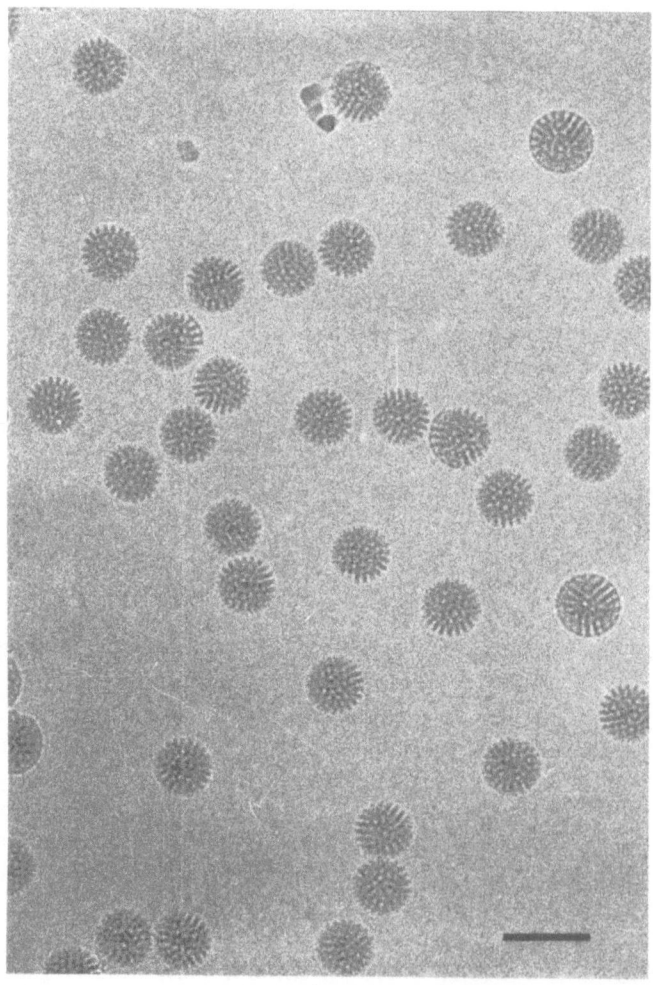

Fig. 7. Electron micrograph of single-shelled particles embedded in vitreous ice. *Scale bar,* 1000 Å

infected cells (COHEN et al. 1979). Figure 7 is an electron micrograph of single-shelled particles embedded in vitreous ice. These particles are 705 Å in diameter. In contrast to the double-shelled particles, they have a bristly surface.

The three-dimensional structure of the single-shelled virion has been determined from such micrographs (PRASAD et al. 1988). Surface representations of the three-dimensional structure along the icosahedral axes are displayed in Fig. 8. The distribution of protein mass on the surface of the single-shelled virion is not so uniform as observed on the surface of the double-shelled virion (Fig. 2). The protein mass is mainly concentrated into 260 morphological units positioned at all the local and strict three-fold axes of a $T = 13$ icosahedral lattice. The positions of these three-fold axes correspond to the centers of the triangles shown in Fig. 3. this strongly suggests a trimeric clustering of the capsid protein. These 260 morphological units are arranged in such a way that there are holes, or channels at all five- and six-coordinated centers. These units, shaped like knobs, are connected at the lower radii, forming prominent saddles at all the local and strict two-fold axes (corresponding to the sides of the triangles in Fig. 3). These knobs impart the bristly appearance seen in the electron micrographs (Fig. 1b).

Biochemical studies (BICAN et al. 1982) have indicated that the protein on the outer surface of the single-shelled particles is VP6 (45 kDa, ESTES et al. 1984). Besides being a subgroup antigen (ESTES and GRAHAM 1985), VP6 has been implicated in virion-associated RNA transcriptase activity (BICAN et al. 1982; SANDINO et al. 1986). It has been interpreted that the 260 morphological units seen on all the three-fold axes correspond to 260 trimers

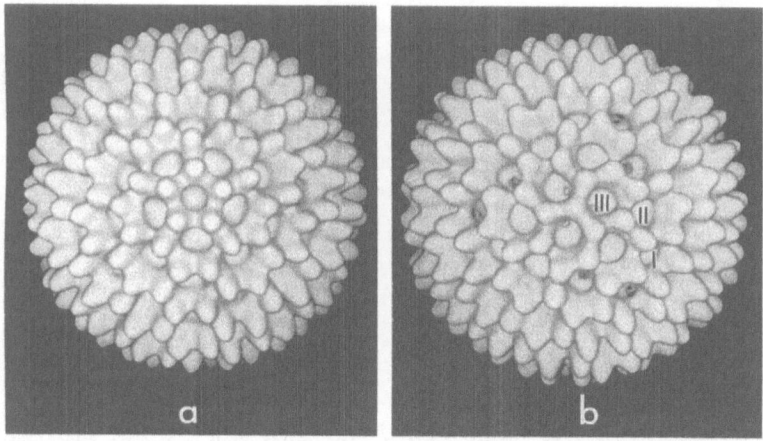

Fig. 8a, b. Surface representation of the three-dimensional structure of single-shelled rotavirus along the icosahedral five- and three-fold axes. One set of neighboring five-fold axes and the six-coordinated positions linking them, and types I, II and III channels are shown

of VP6. This interpretation is further supported by biochemical studies proposing a trimeric aggregation of VP6 in the virion (GORZIGLIA et al. 1985). Thus, there are 780 molecules of VP6 per virion. This agrees well with the volume of the protein mass in the inner shell, as calculated from the three-dimensional structure.

3.3.1 Interaction Between Outer and Inner Shell Proteins

Independent reconstructions of double- and single-shelled particles have made possible a description of the interactions between the outer and inner shell proteins (Fig. 9). The outer shell is partially stripped such that the interaction between the two shells can be readily seen in the figure. It is apparent in Fig. 4 that the interface between outer and inner shell proteins occurs at the level at which the channels begin to widen. The proteins between the outer and inner shells interact in such a way that the channels in these shells coincide. The knobby portions of the VP6 trimers protrude into the outer shell such that close contract between the two shells occurs mainly around the local and strict three-fold axes. The tips of these knobs, particularly of those around the type I channels, are in fact visible through the outer shell (small holes immediately around the icosahedral five fold axes in Fig. 9). The outer shell proteins spread across the saddles seen between the knob by morphological units. The interaction between the shells is such that there exist small gaps right above the saddles. Portions of the saddles are visible through the outer shell channels (Fig. 9). It is possible that small molecules or small enzymes may be able to interact with portions of the inner shell even in the presence of outer shell.

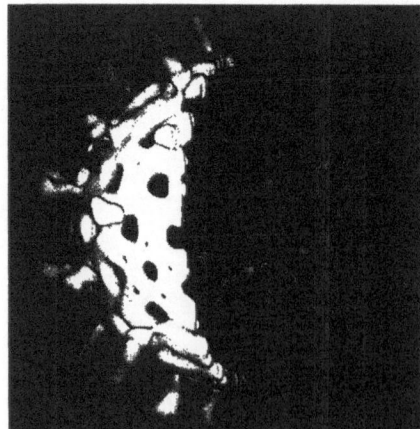

Fig. 9. The outer shell in the double-shelled virion is partially peeled off so as to expose the inner shell proteins. Notice the interaction between knob by portions of the VP6 trimers with the outer shell

3.3.2 Inside the Inner Shell

As mentioned earlier, one of the advantages of three-dimensional electron cryomicroscopy is the ability to obtain not only surface information but also information at the lower radii. Figure 10 shows a radial density plot obtained from the three-dimensional density map of the entire virion. The mass density breaks up into three distinct shells between the radii of 210 Å and 500 Å. These three shells are between the radii 500 Å and 340 Å, 340 Å and 270 Å, and 270 Å and 210 Å. At a radius of 210 Å, the mass density decreases but rises again immediately.

Though he composition of first two shells is most certainly known, the composition of the third shell is a conjecture. The total mass density in the first shell, i.e., the outer shell, accounts for 780 and 120 molecules of VP7 and VP4, respectively. The mass density in the second shell, i.e., the inner shell, accounts for 780 molecules of VP6. The third shell is a continuous density bed with indentations at all the five- and six-coordinated positions. These are the bottom parts of the channels seen in the inner and outer shells. The knobby morphological units of the inner shell sit on this density bed. Figure 11 shows a three-dimensional surface representation of this third shell. While type I and type II channels terminate here, it seems that type III

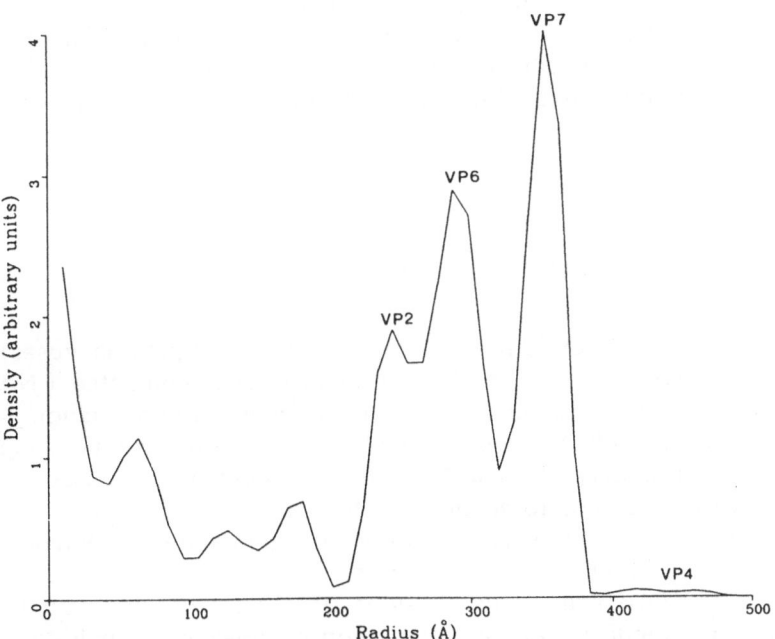

Fig. 10. The radial density profile of the rotavirus structure. The locations of the various structural proteins (*VP2, VP6, VP7* and *VP4*) are indicated

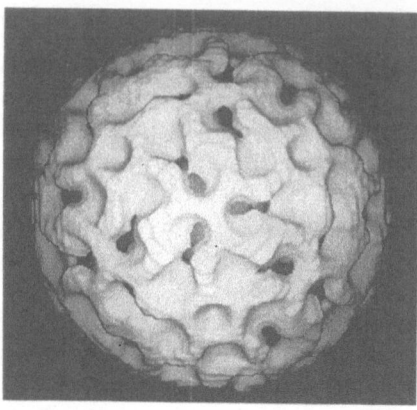

Fig. 11. Surface representation of the "core", which includes the mass density up to the radius of 260 Å. Some portions of the "subcore," represented in a *lighter shade*, are seen through the small holes. It is proposed that the surface of this subcore is predominantly made up of VP2

channels continue beyond this shell. It is tempting to speculate at this point that the third shell, between the radii 210 Å and 270 Å, is composed of VP2 molecules. The mass density of this shell can accommodate approximately 180 molecules of VP2. Further inside this shell, within the radius of 210 Å, we think that VP1 VP3, and the genome constitute another structural entity. The composition of the third shell and further inside must be considered tentative, pending further biochemical and structural studies. We propose that the structure within the radius of 210 Å be called the "subcore" and the structure including the "VP2" shell, up to the radius of 270 Å, be called the "core". These studies advocate that matured rotavirus is a triple-shelled structure instead of the conventional description of this virus as a double-shelled virus.

4 Concluding Remarks

Rotaviruses are large (about 12 times larger in volume than picornaviruses) and uniquely complex ($T = 13$, multiple protein layers, segmented RNA genome) structures. Understanding the basic principles of construction in such large viruses is a challenging task and promises to be an interesting topic for further structural studies. In 1962, CASPER and KLUG proposed the theory of quasi-equivalence to account for the surface morphology of all isometric viruses. In recent years, however, the structures of polyoma (RAYMENT et at. 1982) and SV40 viruses in the papova virus family and the proposed architecture of adenovirus (BURNETT 1984) have cast doubt on the applicability of this theory to virus structures more complex than simple plant viruses. In the case of rotavirus, because of the relatively smooth distribution of the protein mass in the outer shell, it is difficult to ascertain whether the

VP7 molecules are arranged according to the principles of quasi-equivalence. It is possible that VP7 molecules are arranged in groups of five and six surrounding the six- and five-coordinated channels, respectively, giving rise to the observed elevated mass density around each channel. However, either from the symmetry arguments or from the theory of quasi-equivalence, the location of the spikes, which have been argued to be dimers, of VP4, could not have been predicted unequivocally. In the case of the capsid structure of single-shelled rotavirus, the position of the morphological units, which are in all probability trimers of VP6, follow the quasi-equivalence theory. However, the present degree of resolution is insufficient to describe the various quasi-interactions between these units.

In summary, electron cryomicroscopy and image processing techniques have provided a better insight into structure-function relationships in rotaviruses. The three-dimensional structures of both double- and single-shelled particles, which represent the final and intermediate stages of the assembly processes, exhibit $T = 13$ icosahedral symmetry in a left-handed configuration. Interesting features include the spikes on the surface of the outer shell and the large aqueous channels linking the outer surface with the inner core. The bilobed structure at the distal end of the spike may be involved in both attachment to the cell and in viral penetration. The aqueous channels may play an important role during transcriptional activity of the virus. Structural studies have enabled unambiguous localization and enumeration of the number of copies per virion of structural polypeptides in the outer and inner shells. It has been possible to identify the regions of interaction between the outer and inner shell proteins of the virion. It is proposed that VP2 molecules form a shell between the inner shell, composed of VP6, and the subcore; consisting of VP1, VP3, and the genome.

Acknowledgement. This work was supported by NIH grant GM41064 and RR02250. We thank Drs. M.K. Estes and R.F. Ramig for many helpful discussions.

References

Acharya R, Fry E, Stuart D, Fox G, Rowlands D, Brown F (1989) The three-dimensional structure of foot and mouth disease virus at 2.9 Å resolution. Nature 337: 709–716.

Adrian M, Dubochet J, Lepault J, McDowell AW (1984) Cryo-electron microscopy of viruses. Nature 308: 32–36

Arias CF, Lopez S, Bell JR, Strauss JH, (1984) Primary structure of the neutralization antigen of simian rotavirus SA11 as deduced from cDNA sequence. J Viriol 50: 657–661

Beker TS, Drak J, Bina M (1988) Three-dimensional structure of SV40. Proc Natl Acad Sci USA 85: 422–426.

Bican P, Cohen J, Charpilienne A, Scherrer R (1982) Purification and characterization of bovine rotavirus cores. J Virol 43: 1113–1117

Bishop RF, Davidson GP, Holmes IH, Ruck BJ (1973) Virus particles in epithelial cells of duodenal mucosa from children with viral gastroenteritis. Lancet 2: 1281–1283

Both GW, Bellamy AR, Street JE, Siegman LJ (1982) A general strategy for cloning double-stranded RNA : nucleotide sequence of the simian-11 rotavirus gene 8. Nucleic Acid Res 10: 7075–7088

Both GW, Mattick JS, Bellamy AR (1983a) Serotype specific glycoprotein of simian 11 rotavirus: coding assignment and gene sequence. Proc Natl Acad Sci USA 80: 3091–3095

Both GW, Siegmen LJ, Bellamy AR, Atkinson PH (1983b) Coding assignment and nucleotide sequence of simian rotavirus SA11 gene segment 10: location of glycosylation sites suggests that the signal peptide is not cleaved. J Virol 48: 335–339

Both GW, Bellamy AR, Siegman LJ (1984) Nucleotide sequence of the dsRNA genomic segment 7 of simian 7 of simian 11 rotavirus. Nucleic Acid Res 12: 1621–1626

Bremont M, Charpilienne A, Chabanne D, Coben J (1987) Nucleotide sequence and expression in Escherchia coli of the gene encoding the nonstructural protein NCVP2 of bovine rotavirus. Virology 161: 138–144

Burns JW, Greenberg HB, Shaw RD, Estes MK (1988) Functional and topogrraphical analyses of epitopes on the hemagglutinin (VP4) of the simian rotavirus SA11. J Virol 62: 2164–2172

Burnett RM (1984) Structural investigations of hexon, the major coat protein of adenovirus. In: Jurnak FA, McPherson A (eds) Biological macromolecules and assemblies, vol I: virus structures. Wiley New York, pp 337–385

Casper DLD, Klug A (1962) Physical principles in the construction of regular viruses. Cold Spring Harbor Symp Quant Biol 27: 1–32

Chiu W (1986) Electron Microscopy of frozen hydrated bilogical specimens. Annu Rev Biophys Biophys Chem 15: 237–257

Clark B, Roth JR, Clark ML, Barnett BB, Spendlove RS (1981) Trypsin enhancement of rotavirus infectivity mechanisms of enhancement. J Virol 39: 816–822

Cohen J, Laporte J, Charpilienne A, Scherrer R (1979) Activation of rotavirus RNA polymerase by calcium chelation. Arch Virol 60: 177–186

Cohen J, Charpillienne A, Chilmonczyk S, Estes MK (1989) Nucleotide sequence of bovine rotavirus gene 1 and expression of gene product in baculovirus. Virology 171: 131–140

Colman PM, Laver WG, Varghese JN et al. (1987) Three-dimensional structure of a complex of antibody with influenza virus neuraminidase. Nature 326: 358–363

Crowther RA (1971) Procedures for three-dimensional reconstruction of spherical viruses by Fourier synthesis from electron micrographs. Phil Trans Soc Lond [B] 261: 221–230

Crowther RA (1982) Image reconstruction from electron micrographs of macromolecular structures. In: Davies DB, Saenger W, Danyluk SS (eds) Structural molecular biology. Plenum, New York

Crowther RA, Klug A (1975) Structural analysis of macromolecular a assemblies by image reconstruction from electron micrographs. Annu Rev Biochem 44: 161–182

Crowther RA, DeRosier DJ, Klug A (1970) The reconstruction of a three-dimensional structure from projections and its application to electron microscopy. Proc Soc Lond [A] 317: 319–340

de Zoysa I, Feachem RG (1985) Interventions for the control of diarrheal diseases among young children: rotaviruses and cholera immunization. Bull WHO 63: 569–583

Dubochet J et al. (1988) Cryo-electron microscopy of vitrified specimens. Q Rev Biophys 21: 129–228

Esparza J, Gil F (1978) A study of the ultrastructure of human rotavirus. Virology 91: 141–150

Espejo RT, Lopez S, Arias C (1981) Structural polypeptides of simian rotavirus SA11 and the effect of trypsin. J Virol 37: 156–160

Estes MK (1990) Rotaviruses and their replication. In: Fields BN, Knipe DM, Chanock RM, Melnick JL, Roizman B, Hope RE (eds) Virology. Raven, New York, pp 1353–1404

Estes MK, Cohen J (1989) Rotavirus genestructure and function. Microbiol Rev 53: 410–449

Estes, MK, & Graham, DY (1985). In Atassi MZ, Bachrach HL (eds) Immunobiology of proteins and peptides-III. Plenum, New York, pp 201–214

Estes MK, Graham DY, Mason BB (1981) Proteolytic enhancement of rotavirus infectivity: molecular mechanisms. J Virol 39: 879–888

Estes MK, Mason BB, Crawford S, Cohen J (1984) Cloning and nucleotide sequence of the simian rotavirus gene 6 that codes for the major inner capsid protein. Nucleic Acid Res 12: 1875–1887

Fukuhara N, Yoshie O, Kitaoka S, Konno T (1988) Role of VP3 in human rotavirus after target cell attachment via VP7. J Virol 62: 2209–2218

Fuller SD (1987) The $T = 4$ envelope of Sindbis virus is organized by interactions with a complimentary $T = 3$ capsid. Cell 48: 923–934

Furlong DB, Nibert ML, Fields BN (1988) Sigma 1 protein of mammalian reoviruses extends from the surfaces of viral particles. J Virol 62: 246–256

Gorziglia M, Lanea C, Liprandi F, Esparza J (1985) Biochemical Evidence for the oligomeric (possibly trimericz) structure of the major inner capsid polypeptide (45K) of rotavirus. J Gen Virol 66: 1889–1900

Greenberg HB, Valdesuso J, VanWyke K et al. (1983) Production and preliminary characterization of monoclonal antibodies directed at two surface proteins of rhesus rotavirus. J Virol 47: 267–275

Harrison SC (1990). Principles of virus structure. In: Fields BN, Knipe DM, Chanock RM, Melnick JL, Roizman B, Hope RE (eds) Virology, vol I. Raven, New York, pp 37–71

Hogle JM, Chow M, Filman DJ (1985) Three dimensional structure of poliovirus at 2.9Å resolution. Science 229. 1358–1365

Jeng T-W, Talmon Y, Chiu W (1988) Containment system for the preparation of vitrified-hydrated virus specimen. J Electron Microsc Tech 8: 343–348

Jeng T-W, Crowther RA, Stubbs G, Chiu W (1989) Visualization of alpha-helices in tobacco mosaic virus by cryo-electron microscopy. J Mol Biol 205: 251–257

Kalica AR, Greenberg HB, Wyatt RG, Flores J, Sereno MM, Kapikian AZ, Chanock RM (1981) Genes of human (strain Wa) and bovine (strain UK) rotaviruses that code for neutralization and subgroup antigens. Virology 112: 385–390

Kalica AR, Flores J, Greenberg HB (1983) Identification of rotaviral gene that codes for hemagglutination and protease-enhanced plaque formation. Virology 125: 194–205

Kaljot KT, Shaw DRD, Rubin DH, Greenberg HB (1988) Infectious rotavirus enters cells by direct membrane penetration, not endocytosis. J Virol 62: 1136–1144

Kapikian AZ, Chanock RM (1990) Rotaviruses. In: Fields BN, Knipe DM, Chanock RM, Melnick JL, Roizman B, Hope RE (eds) Virology. Raven, New York, pp 1353–1404

Kitaoka S, Fukahara N, Tazawa F, Suzuki H, Sato T, Konno T, Ebina T, Ishida N (1986) Characterization of monoclonal antibodies aganist human rotavirus hemagglutinin. J Med Virol 19: 313–323

Klug A (1979) Image analysis and reconstruction in the electron microscopy of biological macromolecules. Chem Scripta 14: 245

Kogasaka R, Akihara M, Horino K, Chiba S, Nakao T (1979) A morphological study of human rotavirus. Arch Virol 61: 41–48

Kumar AA, Charpllienne A, Cohen J (1989) Nucleotide sequence of the gene encoding for the RNA binding protein (VP2) of RF bovine rotavirus. Nucleic Acid Res 17: 2126

Liu M, Estes MK (1989) Nucleotide sequence of the simian rotavirus SA11 genome segment 3. Nucleic Acid Res 17: 7991

Ludert JE, Gil F, Liprandi F, Esparza J (1986) The structure of rotavirus inner capsid studied by electron microscopy of chemically disrupted particles. J Gen Virol 67: 1721–1725

Mackow ER, Shaw RD, Matsui SM, Vo PT, Dang MN, Greenberg HB (1988) Characterization of the rhesus rotavirus VP3 gene: location of aminoacids involved in homologous and heterologous rotavirus neutralization and identification of a putative fusion region. Proc Natl Acad Sci USA 85: 645–649

Mackow ER, Barnett JW, Chan H, Greenberg HB (1989) The rhesus rotavirus outer capsid protein VP4 functions as a hemagglutinin and is antigenically conserved when expressed by a baculovirus recombinant. J Virol 63: 1661–1665

Martin ML, Palmer EL, Middleton PJ (1975) Ultrastructure of infantile gastroenteritis virus. Virology 68: 146–153

Mason BB, Graham DY, Estes MK (1980) In vitro transcription and translation of simian rotavirus SA11 gene products. J Virol 33: 1111–1121

Mason BB, Graham DY, Estes MK (1983) Biochemical mapping of the simian rotavirus SA11 genome. J Virol 46: 413–423

Matsuno S, Inouye S (1983) Purification of an outer capsid glycoprotein in neonatal calf diarrhea virus and preparation of its antisera. Infect Immun 39: 155–158

Matthews REF (1979) The classification and nomenclature of viruses. Summary of results of meetings of the international committee on taxonomy of viruses in Hague. Intervirology 11: 133–135

Metcalf P (1982) The symmetry of the reovirus outer shell. J Ultrastruct Res 78: 292–301

Mitchell DB, Both GW (1988) Simian rotavirus SA11 segment 11 contains overlapping reading frames. Nucleic Acids Res 16: 6244

Newman JEE, Brown F, Bridger JC, Woode GN (1975) Characterization of a rotavirus. Nature 258: 631–633

Nishikawa K, Taniguchi K, Torres A, Hoshino Y, Green K, Kapikian AZ, Chanock RM, Gorziglia M (1988) Comparitive analysis of the VP3 gene divergent strains of rotavirus simian SA11 and bovine Nebraska calf diarrhea virus. J Virol 62: 4022–4026

Prasad BVV, Wang GJ, Clerx JPM, Chiu W (1988) Three-dimensional structure of rotavirus. J Mol Biol 199: 269–275

Prasad BVV, Burns JW, Marietta E, Estes MK, Chiu W (1990) Localization of VP4 neutralization sites in rotavirus by three-dimensional cryo-electron microscopy Nature 343: 476–478

Prasad BVV, Yamaguchi S, Roy P (1992) Three-dimensional structure of single-shelled blue-tongue virus. J Virol 66: 2135–2142

Ramig FR, Galle KL (1990) Rotavirus genome segment 4 determines viral replication phenotype in cultured liver cells (Hepg2). J Virol (in press)

Rayment I, Baker TS, Caspar DLD, Murakami WT (1982) Polyoma virus capsid structure at 22. 5Å resolution. Nature 295: 110–115

Rodger SM, Schnagl RD, Holmes I (1975) Biochemical and biophysical characterization of diarrhea viruses of human and calf origin. J Virol 16: 1229–1235

Roseto A, Escaig J, Delain E, Cohen J, Scherrer R (1979) Structure of rotaviruses as studied by the freeze drying technique. Virology 98: 471–475

Rossmann MG (1989) Neutralization of small RNA viruses by antibodies and antiviral agents. FASEB J 3: 2335:2343

Rossmann MG, Reuckert RR (1987) What does the molecular structure of viruses tell us about viral functions? Microbiol Sci 4: 206–214

Rossmann MG, Arnold E, Erickson JW, Frakenberger EA, Griffith JP, Hecht HJ, Johnson JR, Kramer G, Luo M, Mosser AG, Rueckert RR, Sherry B, Vriend G (1985) Structure of a human common virus and functional relationship to other picornaviruses. Nature (London) 317: 145–153

Sabara M, Gilchrist JE, Hudson GR, Babiuk LA (1985) Preliminary characterization of an epitope involved in neutralization and cell attachment that is located the major bovine rotavirus glycoprotein. J Virol 53: 58–66

Sandino AM, Jashes M, Faundez G, Spencer E (1986) Role of the inner protein capsid on in vitro human rotavirus transcription. J Virol 60: 797–802

Schrag JD, Prasad BVV, Rixon FJ, Chiu W (1989) Three-dimensional structure of HSV-1 nucleocapsid. Cell 56: 651–660

Shaw RD, Vo PT, Offit PA, Coulson BS, Greenberg HB (1986) Antigenic mapping of the surface proteins of the rhesus rotavirus. Virology 155: 434–451

Stewart M, Vigers G (1986) Electron microscopy of frozen-hydrated biological material. Nature 319: 631–636

Taylor KA, Glaeser RM (1975) Electron diffraction of frozen, hydrated protein crystals. Science 186: 1036–1037

Unwin PNT, Ennis PD (1984) Two configurations of a channel-forming membrane protein membrane protein. Nature 307: 609–613

Vogel RH, Provencher SW, Bonsdorff vof C-H, Adrian M, Dubochet J (1986) Envelope structure of Semiliki Forest virus reconstructed from cryo-electron micrographs. Nature 320: 533–535

Wilson IA, Skehel JJ, Wiley DC (1981) Structure of hemagglutinin membrane glycoprotein of influenza virus at 3Å resolution. Nature 289: 366–373

Wiley DC, Skehel JJ (1990) Viral membranes. In: Fields BN, Knipe DM, Chanock RM, Melnick JL, Roizman B, Hope RE (eds), Virology vol 1. Raven, New York, pp 63–85

Note Added in Proof

Recently more structural studies have been carried out on rotavirus using electron cryomicroscopy and computer image processing techniques. These

include studies by YEAGER et al. 1990 on rhesus rotavirus at 35 Å resolution
(Yeager M, Dryden KA, Olson, NH, Greenberg HB, Baker TS (1990).
Three-dimensional structure of rhesus rotavirus by cryo-electron microscopy
and image reconstruction. J Cell Biol 110: 2133–2144) and by SHAW et al.
1993 on a variant of SA11 strain and a reassortant of rotavirus at 28 Å
resolution (Shaw AL, Rothnagel R, Chen D, Ramig RF, Chiu W, Prasad BVV
(1993). Three-dimensional visualization of the rotavirus hemagglutinin
structure. Cell 74: 693–701). The three-dimensional structure of the rhesus
strain of rotavirus exhibits the same structural features described in this
review. The studies by Shaw et al. have confirmed that the spike is a dimer
of VP4. These studies showed that the exchange of native VP4 with a bovine
strain resulted in a poorly infectious reassortant. No spikes could be detected
in the three-dimensional structure of this reassortant. The difference map
between the native and the reassortant revealed a large globular domain of
VP4 buried within the virion that interacts with the inner shell protein VP6.
These authors have suggested that the assembly of VP4 may precede that of
VP7 and that VP4 may play an important role in the budding of the single-
shelled particle through the rough endoplasmic reticulum during virus
maturation.

The Rotavirus Genome

U. Desselberger[1] and M.A. McCrae[2]

1 Introduction

Since rotaviruses were discovered as pathogens, most notably as the causative agent of acute gastroenteritis in infants and in the young of a variety of animal species, a wealth of data on genome structure, gene-protein assignments

---

[1] Clinical Microbiology and Public Health Laboratory, Level 6, Addenbrooke's Hospital, Hills Road, Cambridge CB2 2QW, United Kingdom
[2] Department of Biological Sciences, University of Warwick, Coventry CV4 7AL, United Kingdom

Current Topics in Microbiology and Immunology, Vol. 185
© Springer-Verlag Berlin · Heidelberg 1994

and structure-function relationship have been accumulated. Those data are of great interest to the molecular biologist and are also important for the understanding of the epidemiology of and host responses to rotavirus infection and for the development of vaccines.

In the following the present knowledge of the rotavirus genome structure and of its coding potential will be reviewed. The structure of the genome products in virions and infected cells and their functions are discussed in detail by the contributions of W. Chiu, V. Prasad, G. Both, and J. Patton in this monograph.

The genome of rotaviruses consists of 11 discrete segments of double-stranded (ds) RNA. The genomic RNA can be easily extracted from semi-purified rotavirus particles and separated by polyacrylamide gel electrophoresis (PAGE) (Fig. 1). The RNA segments are numbered 1–11 according to their order of migration, RNA 1 being the slowest migrating RNA segment and RNA 11 the fastest. Rotaviruses have been isolated from various animals and from humans, and at least five different groups (A–E) have been differentiated according to the lack of cross-reactivity with polyclonal antisera (PEDLEY et al. 1983, 1986). As far as tested there is no cross-

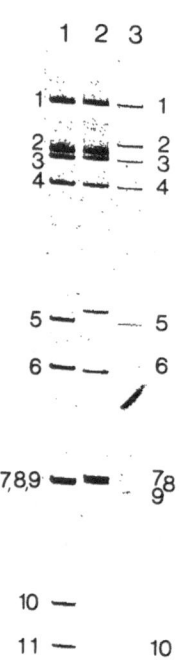

Fig. 1. RNA profiles of human group A rotaviruses of subgroup I serotype 2 (*lane 1*) and subgroup II serotype 1 (*lanes 2 and 3*). RNA segments (1–11) are indicated. 3% polyacrylamide gel; silver staining

hybridization between the genomes of the different groups (EIDEN et al. 1986; McCRAE 1987). Since published sequence information is predominantly available for the group A rotaviruses, this review will deal exclusively with them. (For some groups B and C rotavirus sequences see *Note added in proof.*)

2 Physicochemistry of Rotavirus RNA Segments

All RNA segments of group A rotaviruses have now been sequenced at least once and the sequenced sizes are between 3302 (segment 1) and 663 (segment 11) nucleotide pairs. A list of citations for sequences of rotavirus RNA segments and in part for sequences of their translation products is given in Table 1. A complete listing of rotavirus sequence entries which have been deposited in the two major sequence data bases is given in Table 2. (Entries after late 1990 are listed in *Table 6.*)

From the sequences a total average size of the genome of approximately 18 550 base pairs (bp) can be calculated. The segments of isolated rotavirus RNA have been visualized and measured by electron microscopy (RIXON et al. 1984): the sizes thus obtained were close to the sizes determined by sequencing and more accurate than size measurements from comparison of the relative migration of RNAs or reverse transcribed complementary DNAs (cDNAs) on gels (for review see RIXON et al. 1984).

Size measurements of sequenced RNA segments (segments 7–9) by electron microscopy also allowed determination of the translation or axial distance between bp in the rotavirus RNA duplexes: it amounts to 2.80 ± 0.11 Å per bp (as compared to 3.29 ± 0.05 Å for dsDNA of $\phi \times 174$) and is consistent with the 11-fold helix of the A form of double-stranded nucleic acid (TAYLOR et al. 1985). Subsequently, the A form has also been found for RNAs of other dsRNA viruses (e.g. infectious bursal disease virus; MUELLER and NITSCHKE 1987). Sedimentation studies of isolated RNA segments of bovine rotavirus were consistent with the model of "worm-like" or flexible cylinders with a persistence length of 1125 Å and a hydrated diameter of 30 Å; the persistence length of dsRNA was found to be nearly twice as great as that of dsDNA (KAPAHNKE et al. 1986). At the indicated persistence length of rotavirus RNA in aqueous solution free RNA cannot be bent into a capsid of an inner diameter of 50 nm (PRASAD et al. 1988) without assuming intimate protein-RNA interactions in the virion (ESTES and COHEN 1989). The structural core proteins VP1, VP2 and VP3 are obvious candidates for this function, but non-structural proteins may also play a role (GALLEGOS and PATTON 1989). Despite the presence of RNA of positive polarity in the dsRNA, isolated genomic RNA is noninfectious.

Table 1. Nucleotide sequences of group A rotavirus RNA segments (Autumn 1990)

RNA segment number	Size range (bp)	Strain	Reference
1	3302	A/bovine/RF A/bovine/UK	COHEN et al. (1989) MITCHELL and BOTH, personal communication TARLOW and McCRAE (1990)
2	2687–2717	A/bovine/RF A/human/Wa	KUMAR et al. (1989) ERNST and DUHL (1989) MITCHELL and BOTH, personal communication TIAN et al. (1990)
3	2591	A/simian/SA11	LIU and ESTES (1989) MITCHELL and BOTH, personal communication TIAN et al. (1990)
4	2359–2364	A/simian/SA11	LOPEZ et al. (1985) LOPEZ and ARIAS (1987) NISHIKAWA et al. (1988) MITCHELL and BOTH (1989)
		A/bovine/C486 A/human/RV-5 A/simian/RRV A/human/Ku A/human/Wa A/human/DS-1 A/human/P A/human/VA70 A/human/M37 A/human/1076 A/human/McN13 A/human/ST3 A/bovine/NCDV A/bovine/UK A/porcine/OSU	POTTER et al. (1987) KANTHARIDIS et al. (1987) MACKOW et al. (1988b) TANIGUCHI et al. (1988b) GORZIGLIA et al. (1988a) GORZIGLIA et al. (1988a) GORZIGLIA et al. (1988a) GORZIGLIA et al. (1988a) GORZIGLIA et al. (1988a) GORZIGLIA et al. (1988a) GORZIGLIA et al. (1988a) GORZIGLIA et al. (1988a) NISHIKAWA et al. (1988) KANTHARIDIS et al. (1988) NISHIKAWA and GORZIGLIA (1988)
		A/porcine/Gottfried	GORZIGLIA et al. (1990)
5	1578–1611	A/bovine/RF A/simian/SA11	BREMONT et al. (1987) MITCHELL and BOTH (1990) TIAN et al. (1990)
6	1356	A/simian/SA11 (SgI)	BOTH et al. (1984b) ESTES et al. (1984b) SMITH et al. (1989)
		A/human/Wa (SgII) A/bovine/RF (SgI) A/human/S2 (SgI) A/human/1076 (SgI) A/equine/Fl-14 (SgI + SgII) A/porcine/Gottfried (SgII)	BOTH et al. (1984b) COHEN et al. (1984) HOFER et al. (1987) GORZIGLIA et al. (1988b) GORZIGLIA et al. (1988b) GORZIGLIA et al. (1988b)
7[a]	1075–1104	A/bovine/UK A/simian/SA11 A/porcine/OSU	DYALL-SMITH et al. (1983) BOTH et al. (1982) RUSHLOW et al. (1988)

Table 1. (*Continued*)

RNA segment number	Size range (bp)	Strain	Reference
8[a]	1062	A/simian/SA11 (St3)	BOTH et al. (1983a) ARIAS et al. (1984)
		A/bovine/UK (St6) A/human/Hu5 (St2)	ELLEMAN et al. (1983) DYALL-SMITH and HOLMES (1984)
		A/human/Wa (St1)	RICHARDSON et al. (1984) MASON et al. (1985)
		A/human/S2 (St2) A/bovine/NCDV (St6) A/bovine/RF (St6) A/porcine/OSU (St5)	GUNN et al. (1985) GLASS et al. (1985) CHARPILIENNE et al. (1986) GORZIGLIA et al. (1986a) RUSHLOW et al. (1988)
		A/simian/RRV (St3) A/human/D (St1) A/human/Mo (St1) A/human/M37 (St1) A/human/DS1 (St2) A/human/HN 126 (St2) A/human/P (St3) A/human/VA70 (St4) A/human/ST3 (St4) A/porcine/Gottfried (St4) A/porcine/YM (St11) A/human/KU (St2) A/human/69M (St8) A/human–WI61 (St9) A/human/B37 (St8) A/bovine/B223 (St10)	GREEN et al. (1987) GREEN et al. (1987) GREEN et al. (1987) GREEN et al. (1987) GREEN et al. (1987) GREEN et al. (1987) GREEN et al. (1987) GREEN et al. (1987) GREEN et al. (1987) GORZIGLIA et al. (1988c) RUIZ et al. (1988) TANIGUCHI et al. (1988a) GREEN et al. (1989) GREEN et al. (1989) HUM et al. (1989) XU et al. (1990)
9[a]	1059	A/simian/SA11 A/bovine/UK A/porcine/OSU	BOTH et al. (1984a) WARD et al. (1984) RUSHLOW et al. (1988)
10	750–751	A/simian/SA11 A/human/Wa A/bovine/UK A/bovine/NCDV	BOTH et al. (1983b) OKADA et al. (1984) BAYBUTT and McCRAE (1984) WARD et al. (1985) POWELL et al. (1988)
11	663–667	A/human/Wa A/bovine/UK A/simian/SA11 A/porcine/OSU A/lapina/Alabama	IMAI et al. (1983b) WARD et al. (1985) MITCHELL and BOTH (1988) WELCH et al. (1989) GONZALEZ and BURRONE (1989) GORZIGLIA et al. (1989)

The average genome size of group A human rotavirus is 18 557 base pairs (bp).

Some of the references only contain amino acid sequences predicted on the basis of unreported nucleotide sequences. Accession and identification code in databases of most published nucleotide sequences are listed in Table 2.

The subgroups (Sg) and serotypes (St) of the VP6 and VP7 gene products, respectively, are indicated. Since 1990 a differentiation of VP7-specific serotypes (G types) and VP4-specific serotypes (P types) has been adopted.

[a] This gene protein assignment is for A/bovine/UK rotavirus; for gene protein assignment of other strains see Table 4.

Table 2. Accession and identification numbers of rotavirus gene entries in the EMBL and Genbank databases (Autumn 1990)

RNA segment number	Accession number	ID line	Accession number	Locus ID	Virus strain	Reference
1	J04346		J04346	ROBVP1	A/bovine/RF	Cohen et al. (1989)
	X55444		M32805	GPRVP1	A/porcine/RF/Gottfried	Fukuhara et al. (1989)
					A/bovine/UK	Tarlow and McCrae (1990)
2	X14057 and X14507	ROBTRFVP2	X14507	ROBTRFVP2	A/bovine/RF	Kumar et al. (1989)
	X14942	HRVVP2	X14941	ROBVVP2	A/human/Wa	Ernst and Duhl (1989)
	X52589	REBRVVP2			A/bovine/UK	Tian et al. (1990)
3	X16062	X16062	RSVS3		A/simian/SA11	Liu and Estes (1989)
	X16949	RV1DIRB3	X16949	X16949	B/rodent/IDIR	Sato et al. (1990)
4	Y00127	ROC486G4	Y00127	ROBC486G4	A/bovine/C486	Potter et al. (1987)
	M22306	ROBSEG4A	M22306	ROBSEG4A	A/bovine/UK	Kantharidis et al. (1988)
	Y00336[a]	ROSA11G4	Y00336	ROTSA11G4	A/simian/SA11	Lopez et al. (1985)
	M11158					Lopez and Arias (1987)
	X14204	RASA11VP4	X14204	X14204	A/simian/SA11	Mitchell and Both (1989)
	M18736 and J03567	RORVP3A	M18736	RORVP3A	A/simian/RRV	Mackow et al. (1988b)
			J03567			
	X13190	REROVP3	X13190	PRRVOVP3	A/porcine/OSU	Nishikawa and Gorziglia (1988)
	M21014	ROHVP3	M21014	ROHVP3	A/human/KU	Taniguchi et al. (1988a)
	M33516	PRVVP4	M33516	PRVVP4	A/porcine/RF/Gottfried	Gorziglia et al. (1990)
	M32559		M32559	ROHSEG4A	A/human/RV5	Kantharidis et al. (1987)
5	M22308	ROBNCVP5	M22308	ROBNCVP5	A/bovine/RF	Bremont et al. (1987)
	X14914	RSA115NS	X14914	TORN553	A/simian/SA11	Mitchell and Both (1990)
6	K02254	REBR6	K02254	ROBMCP	A/bovine/RF	Cohen et al. (1984)
	X00421	ROSRV6	X00421	ROTG6	A/simian/SA11	Estes et al. (1984b)
	K02086	RESEG6	K02086	RO2SEG6	A/human/Wa	Both et al. (1984b)
	Y00437	ROTS2VP6	Y00437	RO1S2VP6	A/human/S2	Hofer et al. (1987)
	D00323	EROVVP6F	D00323	RO1VVP6F1	A/equine/F1-14	Gorziglia et al. (1988b)
	D00324	EROVVP6H	D00324	RO1VVP6H2	A/equine/H-2	Gorziglia et al. (1988b)

Segment	Accession	Name	Accession	Name	Virus	Reference
7	D00325	HROVVP6	D00325	RO1HVP6	A/human/1076	Gorziglia et al. (1988b)
	D00326	PROVVP6	D00326	RO1PVP6	A/porcine/Gottfried	Gorziglia et al. (1988b)
	M27824	ROTVPS6A	M27824	ROTVPS6A	A/simian/SA11	Smith et al. (1989)
	M29287	PRVVP6	M29287	PRVVP6	C/porcine/Cowden	Bremont et al. (1990)
8[a]	J02420	RE7	J02420	ROB7	A/bovine/UK	Dyall-Smith et al. (1983)
	J02353	REG8	J02353	ROTG8	A/simian/SA11	Both et al. (1982)
	X00896	REROTAUK	K00037	ROBG	A/bovine/UK	Elleman et al. (1983)
	M12394	ROVUP7NC	M12394	ROBVP7NCD	A/bovine/NCDV	Glass et al. (1985)
	V01190	REROT1	J02354	ROTG9	A/simian/SA11	Both et al. (1983a)
	V01546	ROSI11			A/simian/SA11	Both et al. (1983a)
	K02028	REVP7	K02028	ROTVP7	A/simian/SA11	Arias et al. (1984)
	X04613	ROOSUVP7	X04613	PRVOSUVP7	A/porcine/OSU	Gorziglia et al. (1988c)
	X06722	ROPRVVP7	X06722	PRVPRVVP7	A/porcine/OSU	Rushlow et al. (1988)
	X06386	REOPR9	X06386	PRVOPR9	A/porcine/Gottfried	No reference
	X06759	ROPRVP7G	X06759		A/porcine/Gottfried	Gorziglia et al. (1988c)
	M23194	RE9	M23194	M23194	A/porcine/YM	Ruiz et al. (1988)
	K02033	ROHVP7A	K02033	RO29	A/human/Wa	Richardson et al. (1984)
	M21843	REROHU5G	M21843	ROHVP7A	A/human/Wa	Mason et al. (1985)
	X00572	ROT3S9	X00572	RO2A577	A/human/HV5	Dyall-Smith and Holmes (1984)
	X13603	ROVVP7S2	X13603	STTT359	A/human/ST3	Reddy et al. (1989)
	M11164	REBRVP7	M11164	RO2VP7S2	A/human/SZ	Both, 1985, unpublished
	X52650				A/bovine/B223	Xu et al. (1990)
	X06759	ROPRVP7G	X06759	PRVPRVP7G	A/porcine/Gottfried	Gorziglia et al. (1988c)
			J04334	ROHVP7B37	A/human/B37	Hum et al. (1989)
9	K02170	REP9	K02170	ROBP9	A/bovine/UK	Ward et al. (1984)
	X00355	ROSA11	X00355	ROTG7	A/simian/SA11	Both et al. (1984a)
	X06722				A/porcine/OSU	Rushlow et al. (1988)
	M21650	RORVP7	M21650	RORVP7	A/simian/Rhesus	Mackow et al. (1988a)
10	M21885	ROB10A	M21885	ROB10A	A/bovine/UK	Baybutt and McCrae (1984)
	K03384	UKBRV10	K03384	ROB10	A/bovine/UK	Ward et al. (1985)
	X06806	ROBNCVP5	X06806	ROBNCVP5	A/bovine/NCDV	Powell et al. (1988)
	K01138	REG10	K01138	ROTG10	A/simian/SA11	Both et al. (1983b)
	K02032	RE10	K0232	RO210	A/human/Wa	Okada et al. (1984)

Table 2. (*Continued*)

RNA segment number	Accession number	ID line	Accession number	Locus ID	Virus strain	Reference
11	K03385	UKBRV11	K03385	ROB11	A/bovine/UK	WARD et al. (1985)
	X07831	ROSA11G	X07831		A/simian/SA11	MITCHELL and BOTH (1988)
	M28347	ROTN526	M28347	ROTNS26	A/simian/SA11	WELCH et al. (1989)
	X15519	ROTPVP11	X15519	X15519	A/porcine/OSU	GONZALEZ and BURRONE (1989)
	V01191	REROTA	J02440	RO211	A/human/Wa	IMAI et al. (1983b)
	J04361	RBVVP11G	J04361	RBVVP11G	A/lapine/Alabama	GORZIGLIA et al. (1989)
	M34380	ARRSEG11	M34380	ARRSEG11	B/human	CHEN et al. (1990)
	D00474	PRVS11	D00474	PRVS11	A/porcine	GONZALEZ et al. (1989)
	M33606	ROB11AA	M33606	ROB11AA	A/bovine	MATSUI et al. (1990)
	M33607	ROH11AA	M33607	ROH11AA	A/human	MATSUI et al. (1990)
	M33608	ROH11AB	M33608	ROH11AB	A/human	MATSUI et al. (1990)

[a]There is evidence to suggest that this sequence is very similar to that of bovine rotavirus C486 (POTTER et al. 1987) and quite distinct from the SA11 sequence of MITCHELL and BOTH (1989).
[b]This gene-protein assignment is for A/bovine/UK rotavines. If differs between isolates. Details are shown in Table 4.
For further entries see *Table 6*.

3 Common Features of Rotavirus RNA Segments

Like many other segmented RNA viruses, rotavirus genome segments have conserved sequences at their 5' and 3' ends (IMAI et al. 1983a). The 5' terminal consensus sequence is 5'(GGC$^{AA}_{UU}$U$^{A}_{U}$A$^{AA}_{UU}$....)3' and the 3' terminal consensus sequence is 5' (.....$_U$U$^{GUGA}_{UGUG}$CC)3'. Within these consensus sequences there is more heterogeneity between the different RNA segments of one strain than, for instance, within the consensus sequences of the different RNA segments of individual influenza virus strains (LAMB 1983).

In most rotavirus genes the 5' and 3' terminal non-coding sequences are short; they are variable in length between genes (5' end: 9–135 nucleotides; 3' end: 17–185 nucleotides), but mostly conserved for one gene of different strains. With the exception of the second AUG of the VP7 gene, all the 5' untranslated sequences are less than 50 bp long, whereas some of the 3' untranslated regions are considerably longer, e.g., 185 bp for gene 10 (25% of total length, BAYBUTT and MCCRAE 1984; WARD et al. 1985). A summary of the findings is given in part of Table 3. (RNAs with rearrangements have longer untranslated sequences at the 3' end, see Section 7 below.)

Most of the RNA segments have only one open reading frame and use the first initiation codon. However, there are exemptions (see below). All rotavirus genes are A + U rich (58%–67%). No polyadenylation signal is found. The plus-sense strands of genomic dsRNAs carry a cap at the conserved 5' end: 5'(m^{7}GpppG(m)GC.....)3' (IMAI et al. 1983a; MCCRAE and MCCORQUODALE 1983); this is similar to the genomic cap structure of other viruses of the *Reoviridae* family (reoviruses: FURUICHI et al. 1975; orbiviruses: MERTENS and SANGAR 1985; cytoplasmic polyhedrosis virus; KUCHINO et al. 1982).

4 RNA Profiles and Electropherotypes

The RNA segments of rotaviruses are easily extracted from semipurified particles, separated by PAGE, and visualized by ethidium bromide or silver staining (HERRING et al. 1982; WHITTON et al. 1983). The resulting "RNA profiles" (also called "genome patterns" or "electropherotypes"; see Fig. 1) have been widely used to characterize isolates from outbreaks (molecular epidemiology: RODGER et al. 1981; SCHNAGL et al. 1981; BRANDT et al. 1983; FOLLETT et al. 1984; reviews: ESTES et al. 1984a; DESSELBERGER 1989). The basic genome pattern of group A rotaviruses consists of four different size classes containing RNA segments 1–4, 5–6, 7–9 and 10–11. Rotaviruses with different RNA patterns are either nongroup A rotaviruses or group A rotaviruses with genome rearrangements (see Sect. 7 below). Therefore, RNA profiles alone cannot be used to classify rotaviruses into different groups.

Table 3. Gene-protein assignments of group A rotaviruses

RNA segment	Noncoding sequence (bp) 5′	Noncoding sequence (bp) 3′	Protein product Designation	Deduced molecular weight (daltons)	Posttranslational modification	Remarks
1	18	17	VP1	125 005	–	Inner core protein RNA polymerase?
2	16	31	VP2	102 431	Myristylation	Inner core protein RNA binding Leucine zipper
3	49	38	VP3	88 000	–	Inner core protein RNA polymerase?
4	9	22	VP4	86 751	Proteolytic cleavage (VP5* + VP8*)	Surface protein (dimer) Hemagglutinin Neutralization antigen Fusogenic protein Virulence Pathogenicity
5	32	73	VP5 (NS53)	58 654	–	Nonstructural (?) Zinc fingers; assembly
6	23	139	VP6	44 816	Myristylation	Inner capsid protein (trimer) Subgroup antigen
7[a]	46	62	VP8 (NS35)	36 700	–	Nonstructural (?) RNA replication?
8[a]	48 –135	36	VP7(1) VP7(2)	37 368 33 919	Cleavage of signal sequence glycosylation	Surface glycoprotein Neutralisation antigen (serotype specific) Ca++ binding site?
9[a]	25	131	VP9 (NS34)	34 600	–	Nonstructural RNA binding
10	41	185	VP12 (NS20)	20 290	Glycosylation VP10 (NS29)	Nonstructural Morphogenesis
11	20	52	VP11 (NS26)	21 725	Phosphorylation	Nonstructural

[a] This gene-protein assignment is for A/bovine/UK rotavirus; for gene protein assignment of some other strains see Table 4.

Furthermore, as corresponding segments of different genome composition can comigrate (CLARKE and MCCRAE 1982; DESSELBERGER et al. 1986), and as comigrating RNAs of different strains can code for different serotype-specific proteins (BEARDS 1982), RNA profiles cannot be taken to classify viruses into different subgroups and serotypes.

5 Gene-Protein Assignments

Gene-protein coding assignments, and in a number of cases gene-function coding assignments, have been completed for several strains of rotaviruses using in vitro translation techniques or reassortant analysis (MASON et al. 1980, 1983; SMITH et al. 1980; DYALL-SMITH and HOLMES 1981; KALICA et al. 1981, 1983; MCCRAE and MCCORQUODALE 1982a; GREENBERG et al. 1981, 1983; LIU et al. 1988). A synopsis of this work is presented in Table 3.

Gene-protein-function assignments have been greatly advanced by careful analysis of collections of temperature-sensitive (ts) mutants of different rotavirus strains (e.g., GOMBOLD and RAMIG 1987). Further details are outlined in the chapters on rotavirus genetics and rotavirus proteins.

Gene products are either named according to the order of migration of rotavirus-infected cell proteins on gels (VP1–VP12; MCCRAE and MCCORQUODALE 1982a), or as viral proteins (VP) followed by numbers for structural proteins, and NS followed by a number indicating the molecular weight on gels (in thousands) for nonstructural proteins (ARIAS et al. 1982; MASON et al. 1983; LIU et al. 1988). The cleavage products of VP4 were named VP5* and VP8* (ESTES et al. 1981).

Table 4. Protein assignments for RNA segments 7, 8, and 9 of different rotavirus strains

Strain	RNA segments coding for		
	VP7	VP8/NS35	VP9/NS34
A/bovine/UK	8	7	9
A/bovine/NCDV	9	–[a]	–
A/simian/SA11	9	8	7
A/simian/RRV	7	9	8
A/human/Wa	9	8	7
A/human/HV-5	9	7	8
A/human/S2	9	–	–
A/human/ST3	9	–	–
A/porcine/OSU	9	–	–
A/porcine/Gottfried	9	–	–

[a] Not assigned

As the order of migration of RNA segments coding for related proteins changes, especially for RNA segments 7–9 but also 10 and 11, gene-protein assignments cannot be solely based on RNA profiles. Table 4 relates corresponding proteins of different rotavirus strains to their RNA order number. It is now established that the protein products VP1, VP2, VP3, VP6, VP7, and VP4 are structural proteins, whereas the products VP5/NS53, VP8/NS34, VP9/NS34, VP10/NS29, and VP11/NS26 are thought to be nonstructural and to have various functions in infected cells.

6 The Structure of Individual Rotavirus Genes and of Their Encoded Proteins

In the following, aspects of the structure of individual rotavirus genes and of encoded proteins will be discussed. A more detailed description of the protein functions is found in the chapter by G. Both et al.

6.1 Genes of Core Proteins

6.1.1 VP1

VP1 is coded for by RNA segment 1 and represents only 2% of all virion proteins (LIU et al. 1988). Its sequence (determined from the bovine rotavirus RF strain by COHEN et al. 1989) reveals a long open reading frame (ORF) coding for a 125 kDa a protein which has an excess of basic sequences. There is some similarity with sequences of RNA-dependent RNA polymerases of other dsRNA viruses and of picornaviruses (COHEN et al. 1989).

6.1.2 VP2

VP2 is encoded by RNA segment 2 and is the core protein present in highest concentration ($>$ 85% of all core proteins, 12%–15% of total virion proteins; LIU et al. 1988; KUMAR et al. 1989). VP2 seems to be partially exposed on the outside of single-shelled capsids (NOVO and ESPARZA 1981; TANIGUCHI et al. 1986). The gene codes for one ORF of 880 amino acids (KUMAR et al. 1989). The NH_2-terminal half of the protein exhibits two stretches of sequences containing leucines in every seventh position ("leucine zipper", KUMAR et al. 1989). This part of the protein is thought to bind to DNA or dsRNA (KUMAR et al. 1989), and nonspecific RNA binding has indeed been demonstrated for VP2 (BOYLE and HOLMES 1986). VP2 is myristylated which may provide a scaffolding function in the particle (CLARK and DESSELBERGER 1988). The VP2 gene of the human rotavirus Wa has been found to be closely related to that of the bovine RF strain (ERNST and DUHL 1989). VP2 sequences of other strains are also closely related (TIAN and McCRAE, in preparation; G. Both, personal communication).

6.1.3 VP3

VP3 is encoded for by RNA segment 3 and is only a minor component of the core (3% of core proteins, 0.5% of total virion proteins; BICAN et al. 1982; LIU et al. 1988). The first sequence of the gene was obtained from the SA11 strain (LIU and ESTES 1989) and reveals a basic protein, parts of which show some similarity to RNA polymerases. Similar sequences were obtained from VP3 genes of other rotaviruses (TIAN and McCRAE, in preparation; G. Both, personal communication).

6.2 Gene of the Inner Capsid Protein

6.2.1 VP6

VP6 is coded for by RNA segment 6 and is the most abundant protein, providing 50% of the weight of virions (LIU et al. 1988; PRASAD et al. 1988). Naturally VP6 occurs as a trimer (GORZIGLIA et al. 1985; ESTES et al. 1987; SABARA et al. 1987) and is myristylated in an amide bond (CLARK and DESSELBERGER 1988). VP6 contains epitopes which are common for all group A rotaviruses and, in addition, epitopes which specify subgroups I or II (GREENBERG et al. 1983). Some group A rotaviruses have neither subgroups I or II antigens (SVENSSON et al. 1988) or carry both (HOSHINO et al. 1987). Sequences of VP6 genes have been determined for a number of strains (Tables 1 and 2). Comparative sequence analysis of rotaviruses with different subgroup specificities (GORZIGLIA et al. 1988b) has not yet allowed clear definition of the subgroup specifying regions. Use of panels of monoclonal antibodies (mAbs) indicated that there are at least five nonoverlapping epitopes (POTHIER et al. 1987).

6.3 Genes of Outer Capsid Proteins

6.3.1 VP7

This is the major outer capsid protein (30% of virion weight; LIU et al. 1988) and is glycosylated. It is the main antigen for neutralizing antibodies, and serotypes have been defined according to the reactivity of VP7-specific mAbs (HOSHINO et al. 1984; SHAW et al. 1985; COULSON et al. 1987; TANIGUCHI et al. 1985; BIRCH et al. 1988; UNICOMB et al. 1989). So far 11 different serotypes have been discriminated (ESTES and COHEN 1989; and chapter on rotavirus proteins by G. Bотн et al.). The genes coding for VP7 are either RNA segment 7 (rhesus rotavirus (RRV), serotype 3), RNA segment 8 (bovine rotavirus, serotype 6) or RNA segment 9 (simian rotavirus SA11, serotype 3). Differences in protein coding assignment for segments 7, 8 and 9 are found for numerous other strains (Table 4) and are accompanied by differences in the protein coding assignments for the other two proteins coded by these segments (VP8 and VP9) which are probably nonstructural (see below). Therefore comparisons of genes are related to cognate corresponding sequences and not to RNA segments according to the order of migration on gels (Table 4).

The nucleotide sequences of VP7 genes of most serotypes have been determined (for detailed references see Table 1) and the following common features arise (Fig. 2): The nucleotide sequence reveals two ORFs of 326 and 286 amino acids length, respectively. The ORFs are in frame and are both used in SA11-infected cells (CHAN et al. 1986). They start with hydrophobic sequences which could be signal sequences directing the nascent VP7 to the endoplasmic reticulum. Both signal sequences seem to be cleaved at amino acid 51 (Gln) (STIRZAKER et al. 1987). The potential glycosylation site at amino acid 69 is present in almost all VP7 molecules (except for the calf rotavirus strains NCDV and RF) and is used. Other potential glycosylation sites are in amino acid positions 146, 238 and 318. Conserved cysteine residues are found in positions 82, 135, 165, 191, 196, 207, 244 and 249; conserved proline residues in positions 58, 86, 112, 131, 167, 197, 254, 275, and 279. Conserved cysteine residues allowing formation of disulfide bonds seem to be important for constituting neutralizing epitopes, as VP7 obtained from denaturing gels and contiguous peptides did not produce neutralizing antibody (BASTARDO et al. 1981; GUNN et al. 1985) and neutralizing mAbs failed to react with VP7 on western blots (COULSON et al. 1985; HEATH et al. 1986; TANIGUCHI et al. 1985). Areas of high sequence variation are shown in Fig. 2.

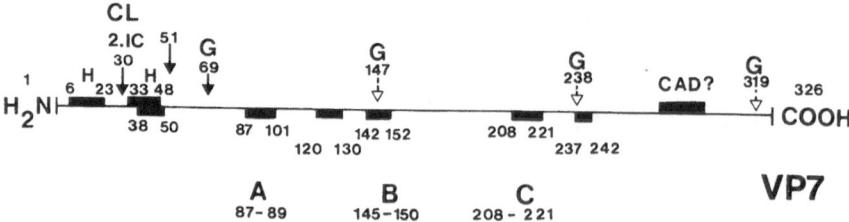

Fig. 2. Characteristics of VP7 on the basis of sequence, protein, and biological data. The protein is 326 amino acids long. *Above the line* the two hydrophobic (*H*) sequences, the second initiation codon (*2.IC*), the signal peptide cleavage site (51), four potential glycosylation sites (*G*), of which G69 is normally used, and the suggested cell attachment domain (*CAD*) are indicated. The *blocks below the line* delineate variable regions, and the three areas of neutralization epitopes (*A, B, C*) are indicated. Site A binds serotype-specific and cross-reactive monoclonal antibodies, site C only serotype-specific monoclonal antibodies

Competition experiments between soluble VP7 and virus particles and the use of VP7-specific mAbs seemed to indicate that VP7 is the viral cell attachment protein (SABARA et al. 1985; FUKUHARA et al. 1988). A peptide containing amino acids 275–295 blocked absorption of virus to cells as did mAbs directed against this peptide (SABARA et al. 1985; FRENCHICK et al. 1988). However, antibodies against this peptide gave conflicting results regarding protection (protection: FRENCHICK et al. 1988; no protection: GUNN et al. 1985).

Divergent sequences between different serotypes are found in six regions (Fig. 2), but only three of them (A, B, C, spanning amino acids 87–101, 142–152, and 208–221, respectively; Fig. 2) have been confirmed as neutralization-specific epitopes based on partial sequencing of mAb escape mutants (DYALL-SMITH et al. 1986; MACKOW et al. 1988a; TANIGUCHI et al. 1988a). Regions A and C seem to be in close proximity on the native protein (DYALL-SMITH et al. 1986). Only when the 3D structure of VP7 is established will it be possible to fully explain the relative significance of the different immunodominant sites.

The degree of glycosylation of the VP7 protein product may influence its antigenicity (CAUST et al. 1987; and chapter on rotavirus proteins by G. BOTH et al.) and also its capacity to bind to cells (FUKUHARA et al. 1988).

6.3.2 VP4

VP4 is coded for by RNA segment 4. In many papers before 1988 it was called VP3 (MASON et al. 1983), but is now generally called VP4 (MCCRAE and MCCORQUODALE 1982a; LIU et al. 1988). VP4 is a minor component of the virion (1.5% of virion proteins; LIU et al. 1988). However, it constitutes an important outer capsid protein with various functions: it has hemagglutinating activity (KALICA et al. 1983; MACKOW et al. 1989), carries neutralization-specific epitopes (HOSHINO et al. 1985; OFFIT et al. 1986; OFFIT and BLAVAT 1986), is posttranslationally cleaved by trypsin into products VP5* (approximately 60 kDa) and VP8* (approximately 28 kDa) which enhances the virion's infectivity (GRAHAM and ESTES 1980; ESTES et al. 1981; LOPEZ et al. 1985, 1986), determines growth in vitro and plaque formation (GREENBERG et al. 1983; KALICA et al. 1983), directs virulence in mice (OFFIT et al. 1986) and influences pathogenicity/attenuation in humans (FLORES et al. 1986; GORZIGLIA et al. 1986b, 1988a). It is also the likely protein to interact with the cellular receptor via its VP8* component (RUGGERI and GREENBERG 1991). However, the phenotype of plaquing efficiency and growth in vitro does not entirely depend on the segment 4 product, but is also influenced by the genetic background in which it is found, e.g., in reassortants (CHEN et al. 1989). VP4 is not glycosylated.

VP4 genes of various rotavirus strains have been sequenced (for references see Tables 1 and 2), and sequence comparisons have allowed the general features of the protein product to be deduced (Fig. 3).

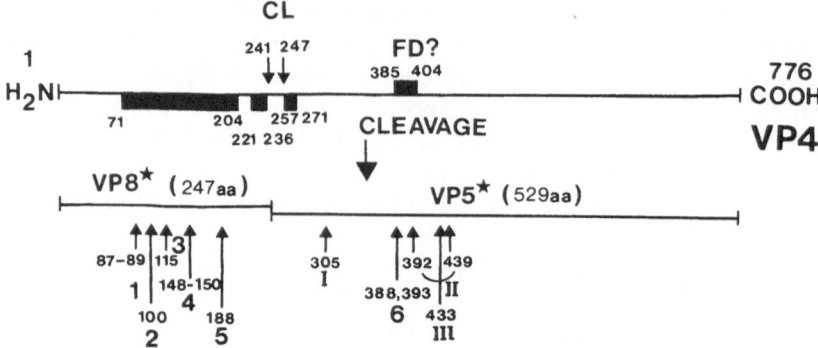

Fig. 3. Characteristics of VP4 on the basis of sequence, protein and biological data. The protein precursor is 776 amino acids long. The two proteolytic cleavage sites (*CL*) and the putative fusion domain (*FD*) are indicated *above the line*. The boxes *below the line* show the positions of a highly variable region (amino acids 71–204) and of two conserved regions around the cleavage sites. The cleavage products VP5* and VP8* are delineated as are the positions of neutralization epitopes 1–6 (MACKOW et al. 1988b) and I–III (TANIGUCHI et al. 1987, 1988b)

There is a long ORF coding for 776 amino acids in animal rotaviruses and for 775 amino acids in human rotaviruses. The latter have lost amino acid 136 of the animal strains, and their VP8* is therefore one amino acid shorter (KANTHARIDIS et al. 1987). Proteolytic cleavage occurs at arginine residues 241 and (preferentially) 247 (LOPEZ et al. 1985) which are conserved in all rotavirus VP4 sequences obtained so far (LOPEZ et al. 1986). Among various VP4s, 91 amino acids are conserved in human strains which are asymptomatic (M37, ST3), whereas symptomatic strains (Wa, RV-5, P, VA70) have a different amino acid conserved in all these positions (GORZIGLIA et al. 1988a). Three of these 91 amino acids are located within the six amino acid connecting peptide of the two cleavage products (Table 5). Some human strains may have additional cleavage sites after amino acid position 246 which has been suggested to be correlated with virulence (GORZIGLIA et al. 1988a). It is not clear whether the peptide spanning amino acids 242–247 is lost by the cleavage process. For further functional analysis of VP4, the reader is referred to the chapter on rotavirus proteins by G. BOTH et al.

There are conserved cysteine residues in positions 216, 318, 380, and 774 of all rotavirus strains (GORZIGLIA et al. 1988a; KANTHARIDIS et al. 1987, 1988; MACKOW et al. 1988b; NISHIKAWA et al. 1988). Most animal strains contain an additional cysteine residue at position 203 which might be important for stabilization of protein structure via disulfide bondage (MACKOW et al. 1988b). Various positions contain conserved prolines (amino acids 71, 76, 77, 110, 225, 226, 235, 334, 390, 395, 435, 451, 455, 482, 524, 669, 716, 749, 761). Analysis of mAb escape mutants has identified one antigenic site in VP5* and five antigenic sites in VP8* (amino acids 88/89, 114, 148/150, 188; MACKOW et al. 1988b; SHAW et al. 1985) which in general shows a high degree of variability between amino acids 71–204 (Fig. 3). Cross-reactive mAbs mostly bind to areas of VP5* (amino acids 306, 388, 393, 434,

Table 5. Amino acid sequences of the trypsin cleavage sites of different animal and human rotaviruses

Strain	Amino Acid Position											
	239	240	241	↓[a] 242	243	244	245	246	247	↓[a] 248	249	250
SA11	T	A	R	D	V	I	H	Y	R	A	Q	A
SA114F	V	S	–	N	I	V	Y	T	–	–	–	P
Bo486	V	S	–	N	I	V	Y	T	–	–	–	P
RRV	S	–	–	N	I	–	S	H	–	–	–	–
RV-5	S	S	–	S	I	Q	Y	R[b]	–	–	–	V
Wa	S	S	–	S	I	Q	Y	K[b]	–	–	–	V
P	S	S	–	S	I	Q	Y	K[b]	–	–	–	V
VA70	S	S	–	S	I	Q	Y	K[b]	–	–	–	V
M37	S	S	–	S	–	T	Y	Q	–	–	–	V
ST3	S	S	–	S	–	T	Y	Q	–	–	–	V

The single letter code is used.
Hyphens (-) indicate identity with corresponding aminoacid of the SA11 strain.
The sources for the sequences are: SA11 (NISHIKAWA et al. 1988); SA11 4F (LOPEZ et al. 1985); bovine 486 (POTTER et al. 1987); rhesus RRV M18006 (MACKOW et al. 1988a, b); human RV-5 (KANTHARIDIS et al. 1987); human M37, ST3, Wa, P, VA70 (GORZIGLIA et al. 1988a).
[a] Arrows indicate trypsin cleavage sites.
[b] Potential additional cleavage sites.

440; MACKOW et al. 1988b; TANIGUCHI et al. 1988b), and three epitopes (around amino acid 306, 393/440 (conformational) and 433) have been identified (TANIGUCHI et al. 1987, 1988b). A mAb which binds at amino acid 393 neutralizes rotaviruses of (VP7-determined) serotypes 3, 5 and 6 (TANIGUCHI et al. 1988b).

The region of VP4 spanning amino acids 384–401 (VP5*) has been found to be similar to hydrophobic and putative fusion promoting regions of glycoproteins of Sindbis and Semliki Forest viruses (MACKOW et al. 1988b). This sequence is relatively conserved in its hydrophobicity. For further functional analysis of VP4, the reader is referred to the chapter on rotavirus proteins by G. BOTH et al.

6.4 Genes Coding for Nonstructural Proteins

6.4.1 VP5/NS53

VP5/NS53 is encoded by RNA segment 5. It is produced early in infection in relatively low concentrations (ERICSON et al. 1982). There is some controversy as to whether it is only found in infected cells (ARIAS et al. 1982; ERICSON et al. 1982; HOLMES 1983) or also as a minor component of virus particles (MCCRAE and MCCORQUODALE 1982a; possibly BRUESSOW et al. 1987).

The nucleotide sequence of RNA 5 from bovine (RF) and simian (SA11) rotavirus strains (BREMONT et al. 1987; MITCHELL and BOTH 1989) revealed that there is one long ORF spanning 491 amino acids. Gene 5 is the only gene in which a sequence complementary to the 3' terminal sequence has been found internally (in positions 1513–1519; BREMONT et al. 1987). Sequences of amino acids 54–66 and 314–327 predict zinc binding and are called "zinc fingers" (EVANS and HOLLENBERG 1988). Such sequences have also been detected in the outer capsid protein s3 of reoviruses (SCHIFF et al. 1988). Besides conservation in an area around amino acids 37–81 (containing one of the "zinc fingers"), there is only 49% overall conservation at the nucleotide level and 36% at the amino acid level between the two sequenced genes (MITCHELL and BOTH 1989). A low degree of conservation had already been suggested earlier on the basis of cross-hybridization data (SCHROEDER et al. 1982). This surprising finding is in marked contrast to much higher degrees of conservation between corresponding other nonstructural proteins of different rotavirus isolates (MITCHELL and BOTH 1989). It will be interesting to see if this level of divergence of segment 5 is confirmed by further work.

Despite the presence of zinc fingers, RNA binding has not been shown for NS53 (BOYLE and HOLMES 1986). NS53 has, however, been demonstrated in "precore" RNA-protein complexes during early assembly stages in infected cells (PATTON 1986; GALLEGOS and PATTON 1989). RNA 5 has been shown to undergo rearrangements in vitro (HUNDLEY et al. 1985; see Sect. 7), and changes in the ORF have occurred (HUNDLEY et al. 1985; TIAN et al., in preparation), resulting in a novel enlarged VP5 or a loss of production of VP5 (HUNDLEY et al. 1985). Interestingly, growth kinetics and plaque formation of these gene 5 mutants of bovine rotavirus also differ significantly (TIAN et al., in preparation). It will be interesting to investigate how these gene 5 mutants form subviral particles. In this context it should be mentioned that reassortment of RNA 5 in cells or animals infected with two rotavirus strains is nonrandom (ALLEN and DESSELBERGER 1985; GOMBOLD and RAMIG 1986; GRAHAM et al. 1987).

6.4.2 VP8/NS35

This protein is encoded by RNA segment 7 (UK bovine rotavirus: MCCRAE and MCCORQUODALE 1982a), 8 (SA11 rotavirus: MASON et al. 1983), or 9 (RRV: GOMBOLD and RAMIG 1986) (Table 4). It has been detected in large amounts associated with core particles during later stages of assembly. Whether it is removed during intracellular assembly of double-shelled particles (HELMBERGER-JONES and PATTON 1986; GALLEGOS and PATTON 1989) or becomes a structural protein mediating attachment to cells (BASS et al. 1989) is controversial.

6.4.3 VP9/NS34

This protein is encoded by RNA segment 7 (SA11), 8 (RRV) or 9 (UK bovine rotavirus) (Table 4). It is a nonstructural protein found in replicase-active subviral particles (HELMBERGER-JONES and PATTON 1986; PATTON and

GALLEGOS 1988; GALLEGOS and PATTON 1989) and binds to nucleic acid (RNA or DNA, BOYLE and HOLMES 1986).

6.4.4 VP10/NS28

This protein is coded for by RNA 10. The gene of several strains has been sequenced (Tables 1 and 2). The protein has a predicted length of 175 amino acids and a deduced molecular weight of 20 290 and is found as VP12/NS20 in infected cells. By post-translational glycosylation it is modified to become VP10/NS28 (ERICSON et al. 1982; MCCRAE and MCCORQUODALE 1982a). Three amino-terminal hydrophobic regions (amino acids 7–21, 28–47, 67–85) have been identified which seem to anchor the protein in the endoplasmic reticulum (ER) (CHAN et al. 1988; BERGMANN et al. 1989). The two glycosylation sites are both in the hydrophobic region closest to the amino-terminal (amino acids 7–21), which makes it unlikely that this region is posttranslationally cleaved like a signal peptide (BOTH et al. 1983b). The cytoplasmic domain (i.e. the carboxy-terminal half of the molecule) is hydrophobic and seems to be involved in the processing of subviral particles, as purified single-shelled particles bind to ER membranes containing NS28 (AU et al. 1989; MEYER et al. 1989). Glycosylation of NS28 seems to be necessary for the particle processing (PETRIE et al. 1983). It is of interest that gene 10 possesses a relatively long untranslated 3′ terminal region (Table 3). At present it is unclear whether this has any functional significance.

6.4.5 VP11/NS26

This protein is coded for by the smallest genomic segment, RNA 11. There is a large ORF starting at nucleotide 22 and predicting a protein of 198 amino acids and an approximate size of 21.5 kDa (strain Wa: IMAI et al. 1983b; WARD et al. 1985; strain SA11: MITCHELL and BOTH 1988; WELCH et al. 1989). There was some controversy over the question whether the protein is structural (MATSUNO et al. 1980; WARD et al. 1985) or nonstructural (ARIAS et al. 1982; WELCH et al. 1989). The protein is very rich in serine and threonine. A second ORF out of phase with the first one has been found which can code for a protein of 92 amino acids (size 11 kDa). This protein has so far not been identified (MITCHELL and BOTH 1988; WELCH et al. 1989).

Most genome rearrangements have been found in segment 11 (see Sect. 7).

7 Genome Rearrangements

In 1984 PEDLEY et al. described "atypical" RNA profiles of rotaviruses obtained from multiple isolates of chronically infected children with severe combined immunodeficiency (SCID): Some RNA segments were decreased in concentration or missing from their normal position, and additional dsRNA bands of various sizes were found (PEDLEY et al. 1984). Hybridization to northern blots of genomic RNA using segment-specific cDNA probes showed that the additional dsRNA bands contained sequences of smaller genomic RNA segments (PEDLEY et al. 1984; ALLEN and DESSELBERGER 1985; HUNDLEY et al. 1987). The new sequences were covalently linked concatemers of segment-specific RNA and were called "genome rearrangements" (PEDLEY et al. 1984; HUNDLEY et al. 1985). Hybridization of cDNA probes synthesized on rearranged RNA bands always hybridized to the single RNA segment they seemed derived from (Hundley and Desselberger, unpublished results; SCOTT et al. 1989; GONZALEZ et al. 1989); i.e., genome rearrangements did not appear to be mosaic structures made up of sequences from more than one segment. Viruses with such rearranged genomes were also found in feces obtained from SCID children by others (DOLAN et al. 1985; WOOD et al. 1988). but – more importantly – in immunocompetent hosts as well (asymptomatically infected children: BESSELAAR et al. 1986; calves: POCOCK 1987; pigs: BELLINZONI et al. 1987, MATTION et al. 1988; rabbits: THOULESS et al. 1986, TANAKA et al. 1988). Recently, genome rearrangements have also been demonstrated in orbiviruses, another genus of the *Reoviridae* family (EATON and GOULD 1987). It is likely that an additional genomic band in the RNA profile of bluetongue virus strain 13V, as shown in Fig. 1 of RAMIG et al. (1985), is also a genome rearrangement. Therefore – besides point mutations and reassortment events – genome rearrangements are likely to be a further general and novel mechanism of genome variation and evolution of members of the *Reoviridae* family (DESSELBERGER 1989).

Serial passage in vitro of tissue culture-adapted bovine rotavirus (brv) at high multiplicity of infection has also led to the isolation of brv variants with genome rearrangements (HUNDLEY et al. 1985). Rearrangements have been observed of RNA segments 5, 6, 8, 10 and 11 (PEDLEY et al. 1984; HUNDLEY et al. 1985, 1987; THOULESS et al. 1986; POCOCK 1987; MATTION et al. 1988; TANAKA et al. 1988), i.e., mainly of genes coding for nonstructural proteins (Table 2) and of those mainly in segment 11.

Several of the rearranged genes have now been sequenced (GONZALEZ et al. 1989; GORZIGLIA et al. 1989; SCOTT et al. 1989; Tian et al., in preparation). In the case of segment 11 rearrangement (GONZALEZ et al. 1989; GORZIGLIA et al. 1989; SCOTT et al. 1989), the variant gene consists of a partial duplication of segment 11-specific sequences in which the normal reading frame is maintained; partially duplicated sequences are covalently linked to the 3' end after the termination codon but lack the initiation codon for the long ORF (see below). The normal conserved 5' and 3' terminal sequences are present only once at the ends of the rearranged segments. This silent duplication in the 3' half can be very conserved (SCOTT et al. 1989), but can also show a significant number of point mutations (GONZALEZ et al. 1989). The number of mutations can be higher than the nucleotide difference between the regular gene in the 5' half and the cogent regular genes of other strains (GONZALEZ et al. 1989). By contrast, the rearranged segment 5 of brv variant E (HUNDLEY et al. 1985) presented an extension of the normal open reading frame (TIAN et al., in preparation) with the consequence of an extended protein being made (HUNDLEY et al. 1985). The brv variant A showed a loss of the protein product of the rearranged segment 5 (HUNDLEY et al. 1985); a sequence of this gene is not available yet. A mechanism for all these rearrangements which is compatible with the data could be that during transcription the virion-associated RNA-dependent RNA polymerase falls back on the template at various stages, with the result of duplicating part of it (HUNDLEY et al. 1985, 1987). This mechanism would allow for conservation, extension, and abolition/shortage of the normal ORF.

Characterization of rotaviruses with genome rearrangements in vitro has shown that they are not defective (HUNDLEY et al. 1985, 1987) and can reassort (ALLEN and DESSELBERGER 1985; GRAHAM et al. 1987; BIRYAHWAHO et al. 1987). Rearranged genes can replace and be replaced by normal RNA segments, structurally as well as functionally (ALLEN and DESSELBERGER 1985; BIRYAHWAHO et al. 1987). Biophysical characterization of rotavirus particles with genome rearrangements has shown that up to 1800 additional bp (approximately 9.7% of the total genome size) were packaged without measurable change in the diameter or apparent S value of particles; however, the density of particles changed in direct proportion to the number of additionally packaged bp (MCINTYRE et al. 1987). Whereas in all cases 11 segments of RNA are packaged, there seems to be a considerable "reserve" capacity for packaging with regard to the length of RNA segments, and one wonders what the upper limit of this capacity might be.

As RNA segment 11 of the "long electropherotype" rotaviruses codes for a product related to that of segment 10 of "short electropherotype" viruses (DYALL-SMITH and HOLMES 1981), it has been speculated that the latter could have arisen from the former by a rearrangement event of RNA segment 11 (ALLEN and DESSELBERGER 1985; TANAKA et al. 1988). Sequencing of RNA segment 10 of short electropherotype rotaviruses, however, has shown that there is a long untranslated region in the 3' half of the gene which in some cases is and in others is not related by duplication to sequences of the 5' half (including the ORF; NUTTALL et al. 1989; MATSUI et al. 1990). The origin of these sequences is not clear at present.

8 Outlook

Much has been learned from extensive sequencing of the genes of group A rotaviruses, and the data have been extremely useful for biochemical and serological work with virus-specific proteins. Many of the data of the structural proteins will be enriched once a 3D structures of single- and double-shelled particles become available.

There is a need for sequence information of nongroup A rotaviruses in order to advance molecular and epidemiological work in this field. At present, (Autumn 1990), three nongroup A rotavirus nucleotide sequences have been published (SATO et al. 1989; BREMONT et al. 1990; CHEN et al. 1990; see Table 2). Two of them (segment 6 of a porcine group C rotavirus and segment 11 of a human group B rotavirus) revealed higher than expected sequence homologies with the corresponding genes of group A rotaviruses. However, additional sequence data are required to allow more far reaching conclusions. (See *Note added in proof* below.) The availability of procedures allowing reintroduction of genetically engineered rotavirus nucleic acid into viable particles will revolutionize wide aspects of the molecular biology of rotaviruses.

Acknowledgements. The authors' work has been supported by grants from the Medical Research Council, The Welcome Trust, the Scottish Home and Health Department, the European Economic Community and The British Council. MAM is a Lister Institute Fellow.

The authors thank Mary K. ESTES and J. COHEN for a preprint of their review on rotaviruses (Microbiological Reviews 1989). The help of Mrs. K. Rogers with typing of the manuscript is gratefully acknowledged.

References

Allen AM, Desselberger U (1985) Reassortment of human rotaviruses carrying rearranged genomes with bovine rotavirus. J Gen Virol 66: 2703–2714

Andrew ME, Boyle DB, Coupar BEH, Whitfeld PL, Both GW, Bellamy AR (1987) Vaccinia virus recombinants expressing the SA11 rotavirus VP7 glycoprotein gene induce serotype-specific neutralizing antibodies. J Virol 61: 1054–1060

Arias CF, Lopez S, Espejo RT (1982) Gene protein products of SA11 simian rotavirus genome. J Virol 41: 42–50

Arias CF, Lopez S, Bell JR, Strauss JH (1984) Primary structure of the neutralization antigen of simian rotavirus SA11 as deduced from cDNA sequence. J Virol 50: 657–661

Arias CF, Lizano M, Lopez S (1987) Synthesis in Escherichia coli and immunological characterization of a polypeptide containing cleavage sites associated with trypsin enhancement of rotavirus SA11 infectivity. J Gen Virol 68: 633–642

Au K-S, Chan W-K, Estes MK (1989) Rotavirus morphogenesis involves an endoplasmic reticulum transmembrane glycoprotein. In: Compans R, Helenius A, Oldstone M (eds) Cell biology of virus entry, replication and pathogenesis. Liss, New York, pp 257–267

Bass DM, Mackow ER, Greenberg HB (1989) The rotavirus gene 8 product, NS35 adheres to the surface of intestinal epithelial and MA104 cells and is a structural viral protein. Gastroenterology [Suppl] 96 (5,2): A30

Bastardo JW, McKimm-Breschkin JL, Souza S, Mercer LD, Holmes IH (1981) Preparation and characterization of antisera to electrophoretically purified SA11 virus polypeptides. Infect Immun 34: 641–647

Baybutt HN, McCrae MA (1984) The molecular biology of rotaviruses. VII. Detailed structural analysis of gene 10 of bovine rotavirus. Virus Res 1: 533–541

Beards GM (1982) Polymorphism of genomic RNAs within rotavirus serotypes and subgroups. Arch Virol 74: 65–70

Bellinzoni RC, Mattion NM, Burrone O, Gonzalez A, LaTorre JL, Scodeller EA (1987) Isolation of group A swine rotaviruses displaying atypical electropherotypes. J Clin Microbiol 25: 952–954

Bergmann CC, Mass D, Poruchynsky MS, Atkinson PH, Bellamy AR (1989) Topology of the non-structural receptor glycoprotein NS28 in the rough endoplasmic reticulum. EMBO J 8: 1695–1703

Besselaar TG, Rosenblatt A, Kidd AH (1986) Atypical rotavirus from South African neonates. Arch Virol 87: 327–330

Bican P, Cohen J, Charpilienne A, Scherrer R (1982) Purification and characterization of bovine rotavirus cores. J Virol 43: 1113–1117

Birch CJ, Heath RL, Gust ID (1988) Use of serotype-specific monoclonal antibodies to study the epidemiology of rotavirus infection. J Med Virol 24: 45–53

Biryahwaho B, Hundley F, Desselberger U (1987) Bovine rotavirus with rearranged genome reassorts with human rotavirus. Arch Virol 96: 257–264

Both GW, Bellamy AR, Street JE, Siegman LJ (1982) A general strategy for cloning double-stranded RNA: nucleotide sequence of the simian-11 rotavirus gene 8. Nucleic Acids Res 10: 7075–7088

Both GW, Mattick JS, Bellamy AR (1983a) Serotype-specific glycoprotein of simian 11 rotavirus: coding assignment and gene sequence. Proc Natl Acad Sci USA 80: 3091–3095

Both GW, Siegman LJ, Bellamy R, Atkinson PH (1983b) Coding assignment and nucleotide sequence of simian rotavirus SA11 gene segment 10: location of glycosylation sites suggests that the signal peptide is not cleaved. J Virol 48: 335–339

Both GW, Bellamy AR, Siegman LJ (1984a) Nucleotide sequence of the dsRNA genomic segment 7 of simian 11 rotavirus. Nucleic Acids Res 12: 1621–1626

Both GW, Siegman LJ, Bellamy AR, Ikegami N, Shatkin AJ, Furuichi Y (1984b) Comparative sequence analysis of rotavirus genomic segment 6—the gene specifying viral subgroups 1 and 2. J Virol 51: 97–101

Boyle JF, Holmes KV (1986) RNA-binding proteins of bovine rotavirus. J Virol 58: 561–568

Brandt CD, Kim HW, Rodriguez WJ, Arrobio JO, Jeffries BC, Stallings EP, Lewis C, Miles AJ, Chanock RM, Kapikian AZ, Parrot RH (1983) Pediatric viral gastroenteritis during eight years of study. J Clin Microbiol 18: 71–78

Bremont M, Charpilienne A, Chabanne D, Cohen J (1987) Nucleotide sequence and expression in Escherichia coli of the gene encoding the non-structural protein NCVP2 of bovine rotavirus. Virology 161: 138–144

Bremont M, Chabanne-Vantherot D, Vannier P, McCrae MA, Cohen J (1990) Sequence analysis of the gene (6) encoding the major capsid protein (VP6) of group C rotavirus: higher than expected homology to the corresponding protein from group A virus. Virology 178: 579–583

Bruessow H, Marc-Martin S, Eichhorn W, Sidoti J, Fryder V (1987) Characterization of a second bovine rotavirus serotype. Arch Virol 94: 29–41

Caust J, Dyall-Smith ML, Lazdins I, Holmes IH (1987) Glycosylation, an important modifier of rotavirus antigenicity. Arch Virol 96: 123–134

Chan W-K, Penaranda ME, Crawford SE, Estes MK (1986) Two glycoproteins are produced from the rotavirus neutralization gene. Virology 151: 243–252

Chan W-K, Au K-S, Estes MK (1988) Topography of the simian rotavirus non-structural glycoprotein (NS28) in the endoplasmic reticulum membrane. Virology 164: 435–442

Charpilienne A, Borras F, D'Auriol L, Galibert F, Cohen J (1986) Sequence of the gene encoding the outer glycoprotein of the bovine rotavirus (RF strain) and comparison with homologous genes from four bovine, simian and bovine rotaviruses. Ann Inst Pasteur/Virol 137E: 71–77

Chen D, Burns JW, Estes MK, Ramig RF (1989) The phenotypes of rotavirus reassortants depend upon the recipient genetic background. Proc Natl Acad Sci USA 86: 3743–3747

Chen GM, Hung T, Mackow ER (1990) cDNA cloning of each genomic segment of the group B rotavirus ADRV: molecular characterisation of the 11th RNA segment. Virology 175: 605–609

Christensen ML (1989) Human viral gastroenteritis. Clin Microbiol Rev 2: 51–89

Clark B, Desselberger U (1988) Myristylation of rotavirus proteins. J Gen Virol 69: 2681–2686

Clarke IN, McCrae MA (1982) Structural analysis of electrophoretic variation in the genome profiles of rotavirus field isolates. Infect Immun 36: 492–497

Cohen J, Leferre F, Estes MK, Bremont M (1984) Cloning of bovine rotavirus (RF strain): nucleotide sequence of the gene coding for the major capsid protein. Virology 138: 178–182

Cohen J, Charpilienne A, Chilmonczyk S, Estes MK (1989) Nucleotide sequence of bovine rotavirus gene 1 and expression of the gene product in baculovirus. Virology 171: 131–140

Coulson BS, Fowler KJ, Bishop RF, Cotton RGH (1985) Neutralizing monoclonal antibodies to human rotavirus and indications of antigenic drift among strains from neonates. J Virol 54: 14–20

Coulson BS, Fowler KJ, White JR, Cotton RGH (1987) Non-neutralizing monoclonal antibodies to a trypsin sensitive site on the major glycoprotein of rotavirus which discriminate
   between virus serotypes. Arch Virol 93: 199–211
Desselberger U (1989) Molecular epidemiology of rotaviruses. In: Farthing MJG (ed) Viruses
   and the gut. Swan, London, pp 55–69
Desselberger U, Hung T, Follett EAC (1986) Genome analysis of human rotaviruses by
   oligonucleotide mapping of isolated RNA segments. Virus Res 4: 357–368
Dolan KT, Twist EM, Horton-Slight P, Forrer C, Bell Jr LM, Plotkin SA, Clark HF (1985).
   Epidemiology of rotavirus electropherotypes determined by a simplified diagnostic technique
   with RNA analysis. J Clin Microbiol 21: 753–758
Dyall-Smith ML, Holmes IH (1984) Sequence homology between human and animal rotavirus
   RNA segments 10 and 11. J Virol 38: 1099–1103
Dyall-Smith ML, Holmes IH (1984) Sequence homology between human and animal rotavirus
   serotype-specific glycoproteins. Nucleic Acids Res 12: 3973–3982
Dyall-Smith ML, Elleman TC, Hoyne PA, Holmes IH, Azad AA (1983) Cloning and sequence
   of UK bovine rotavirus gene segment 7: marked sequence homology with simian rotavirus
   gene segment 8. Nucleic Acids Res 11: 3351–3362
Dyall-Smith ML, Lazdins I, Tregear GW, Holmes IH (1986) Location of the major antigenic
   sites involved in rotavirus serotype-specific neutralization. Proc Natl Acad Sci USA
   83: 3465–3468
Eaton BT, Gould AR (1987) Isolation and characterization of orbivirus genotypic variants. Virus
   Res 6: 363–382
Eiden J, Vonderfecht S, Theil K, Torres-Medina A, Yolken RH (1983) Genetic and antigenic
   relatedness of human and animal strains of antigenically distinct rotaviruses. J Infect Dis
   154: 972–982
Elleman TC, Hoyne PA, Dyall-Smith ML, Holmes IH, Azad AA (1983) Nucleotide sequence of
   the gene encoding the serotype-specific glycoprotein of UK bovine rotavirus. Nucleic Acid
   Res 11: 4689–4701
Ericson BL, Graham DY, Mason BB, Estes MK (1982) Identification, synthesis and modifications
   of simian rotavirus SA11 polypeptides in infected cells. J Virol 42: 825–839
Ernst H, Duhl JA (1989) Nucleotide sequence of genomic segment 2 of the human rotavirus
   Wa. Nucleic Acids Res 17: 4382
Estes MK, Cohen J (1989) Rotavirus gene structure and function. Microbiol Rev 53: 410–449
Estes MK, Graham DY, Mason BB (1981) Proteolytic enhancement of rotavirus infectivity:
   molecular mechanisms. J Virol 39: 879–888
Estes MK, Graham DY, Dimitrov DH (1984a) The molecular epidemiology of rotavirus
   gastroenteritis. In: Melnick JL (ed) Progress in medical virology. Karger, Basel, pp 1–22
Estes MK, Mason BB, Crawford S, Cohen J (1984b) Cloning and nucleotide sequence of the
   simian rotavirus gene 6 that codes for the major inner capsid protein. Nucleic Acids Res
   12: 1875–1887
Estes MK, Crawford SE, Penaranda ME, Petrie BL, Burns JW, Chan W-K, Ericson B, Smith GE,
   Summers MD (1987) Synthesis and immunogenicity of the rotavirus major capsid antigen
   using a baculovirus expression system. J Virol 61: 1488–1494
Evans RM, Hollenberg SM (1988) Zinc fingers. Gilt by association. Cell 52: 1–3
Flores J, Midthun K, Hoshino Y, Green K, Gorziglia M, Kapikian AZ, Chanock RM (1986).
   Conservation of the fourth gene among rotaviruses recovered from asymptomatic newborn
   infants and its possible role in attenuation. J Virol 60: 972–979
Follett EAC, Sanders RC, Beards GM, Hundley F, Desselberger U (1984) Molecular epidemiology
   of human rotaviruses. Analysis of outbreaks of acute gastroenteritis in Glasgow and the West
   of Scotland 1981/82 and 1982/83. J Hyg (Camb) 92: 209–222
Francavilla M, Miranda P, Di Matteo A, Sarasini A, Gerna G, Milanesi G (1987). Expression of
   bovine rotavirus neutralization antigen in Escherichia coli. J Gen Virol 68: 2975–2980
Frenchick P, Sabara MI, Ijaz MK, Babiuk LA (1988) Immune responses to synthetic peptide
   vaccines of veterinary importance. In: Kurstak E, Marusyk RG, Murphy FA, Van Regenmortel
   MHV (eds) Applied virus research vol. 1. Plenum, New York, pp 141–151
Fukuhara N, Yoshie D, Kitaoka S, Konno T (1988) Role of VP3 in human rotavirus internalization
   after target cell attachment via VP7. J Virol 62: 2209–2218
Fukuhara N, Nishikawa K, Gorziglia M, Kapikian AZ (1989) Nucleotide sequence of gene
   segment 1 of a porcine rotavirus strain. Virology 173: 743–749

Furuichi Y, Morgan MA, Muthukrishnan S, Shatkin AJ (1975) Reovirus messenger RNA contains a methylated, blocked 5' terminal structure: M^7 GpppG^{m}C. Proc Natl Acad Sci USA 72: 362–366

Gallegos CO, Patton JT (1989) Characterization of rotavirus intermediates: A model for the assembly of single shelled particles. Virology 172: 616–627

Glass RI, Keith J, Nakagomi O, Nakagomi T, Askaa J, Kapikian AZ, Chanock RM, Flores J (1985) Nucleotide sequence of the structural glycoprotein VP7 gene of Nebraska Calf Diarrhea Virus rotavirus: comparison with homologous genes from four strains of human and animal rotavirus. Virology 141: 292–298

Gombold JL, Ramig RF (1986) Analysis of reassortment of genome segments in mice mixedly infected with rotaviruses SA11 and RRV. J Virol 57: 110–116

Gombold JL, Ramig RF (1987) Assignment of simian rotavirus SA11 temperature-sensitive mutant groups A, C, F and G to genome segments. Virology 161: 463–473

Gonzalez SA, Burrone OR (1989) Porcine OSU rotavirus segment 11 shows common features with the viral gene of human origin. Nucleic Acids Res 17: 6402

Gonzalez SA, Mattion NM, Bellinzoni R, Scodeller A, Burrone OR (1989) Structure of rearranged genome segment 11 in two different rotavirus strains generated by a similar mechanism. J Gen Virol 70: 1329–1339

Gorziglia M, Cashdollar W, Hudson GR, Esparza J (1983) Molecular cloning of a human rotavirus genome. J Gen Virol 64: 2585–2595

Gorziglia M, Lanea C, Liprandi F, Esparza J (1985) Biochemical evidence for the oligomeric (possibly trimeric) structure of the major inner capsid polypeptide (45K) of rotavirus. J Gen Virol 66: 1889–1900

Gorziglia M, Aguirre Y, Hoshino Y, Esparza J, Blumentals I, Askaa J, Thompson M, Glass RI, Kapikian AZ, Chanock RM (1986a) Serotype-specific glycoprotein of OSU porcine rotavirus: coding assignment and gene sequence. J Gen Virol 67: 2445–2454

Gorziglia M, Hoshino Y, Buckler-White A, Blumentals I, Glass RI, Flores J, Kapikian AZ, Chanock RM (1986b) Conservation of amino acid sequence of VP8 and cleavage region of 84-kDa outer capsid protein among rotaviruses recovered from asymptomatic neonatal infection. Proc Natl Acad Sci USA 83: 7039–7043

Gorziglia M, Green K, Nishikawa K, Taniguchi K, Jones R, Kapikian AZ, Chanock RM (1988a) Sequence of the fourth gene of human rotaviruses recovered from asymptomatic or symptomatic infections. J Virol 62: 2978–2984

Gorziglia M, Hoshino Y. Nishikawa K, Maloy WL, Jones RW, Kapikian AZ, Chanock RM (1988b) Comparative sequence analysis of the genomic segment 6 of four rotaviruses each with a different subgroup specificity. J Gen Virol 69: 1659–1669

Gorziglia M, Nishikawa K, Green K, Taniguchi K (1988c) Gene sequence of the VP7 serotype specific glycoprotein of Gottfried porcine rotavirus. Nucleic Acid Res 16: 775

Gorziglia M, Nishikawa K, Fukuhara N (1989) Evidence of duplication and deletion in supershort segment 11 of rabbit rotavirus Alabama strains. Virology 170: 587–590

Gorziglia M, Nishikawa K, Hoshino Y, Taniguchi K (1990) Similarity of the outer capsid protein VP4 of the Gottfried strain of porcine rotavirus to that of asymptomatic human rotavirus strains. J Virol 64: 414–418

Graham DY, Estes MK (1980) Proteolytic enhancement of rotavirus infectivity: biologic mechanisms. Virology 101: 432–439

Graham A, Kudesia G, Allen AM, Desselberger U (1987) Reassortment of human rotavirus possessing genome rearrangements with bovine rotavirus: Evidence for host cell selection. J Gen Virol 68: 115–122

Green KY, Midthun K, Gorziglia M, Hoshino Y, Kapikian AZ, Chanock RM, Flores J (1987) Comparison of the amino acid sequences of the major neutralization protein of four human rotavirus serotypes. Virology 161: 153–159

Green KY, Hoshino Y, Ikegami N (1989) Sequence analysis of the gene encoding the serotype-specific glycoprotein (VP7) of two new human rotavirus serotypes. Virology 168: 429–433

Greenberg HB, Kalica AR, Wyatt RW, Jones RD, Kapikian AZ, Chanock RM (1981) Rescue of non-cultivatable human rotavirus by gene reassortment during mixed infection with its mutants of a cultivatable bovine rotavirus. Proc Natl Acad Sci USA 78: 420–424

Greenberg HB, Flores J, Kalica AR, Wyatt RG, Jones R (1983) Gene coding assignments of growth restriction, neutralization and subgroup specificities of the W and DS-1 strains of human rotavirus. J Gen Virol 64: 313–320

Gunn PG, Sato F, Powell KFH, Bellamy AR, Napier JR, Harding DRK, Hancock WS, Siegman LJ, Both GW (1985) Rotavirus neutralizing protein VP7 : Antigenic determinants investigated by sequence analysis and peptide synthesis. J Virol 54: 791–797

Heath R, Birch C, Gust I (1986) Antigenic analysis of rotavirus isolates using monoclonal antibodies specific for human serotypes 1, 2, 3 and 4 and SA11. J Gen Virol 67: 2455–2466

Helmberger-Jones M, Patton JT (1986) Characterization of subviral particles in cells infected with simian rotavirus SA11. Virology 155: 655–665

Herring AJ, Inglis NF, Ojeh CK, Snodgrass DR, Menzies JD (1982) Rapid diagnosis of rotavirus infection by the direct detection of viral nucleic acid in silver-stained polyacrylamide gels. J Clin Microbiol 16: 473–477

Hofer JMI, Sato F, Street JE, Bellamy AR (1987) Nucleotide sequence for gene 6 of rotavirus strain S2. Nucleic Acids Res 15: 7175

Holmes IH (1983) Rotaviruses. In: Joklik WK (ed) The Reoviridae. Plenum, New York, pp 359–423

Hoshino Y, Wyatt RG, Greenberg HB, Flores J, Kapikian AZ (1984) Serotypic similarity and diversity of rotaviruses of mammalian and avian origin as studied by plaque-reduction neutralization. J Infect Dis 149: 694–702

Hoshino Y, Sereno MM, Midthun K, Flores J, Kapikian AZ, Chanock RM (1985) Independent segregation of two antigenic specificities (VP3 and VP7) involved in neutralization of rotavirus infectivity. Proc Natl Acad Sci USA 82: 8701–8704

Hoshino Y, Gorziglia M, Valdesuso J, Askaa J, Glass RI, Kapikian AZ (1987) An equine rotavirus (FI-14) which bears both subgroup I and subgroup II specificities on its VP6. Virology 157: 488–496

Hum CP, Dyall-Smith ML, Holmes IH (1989) The VP7 gene of a new G serotype of human rotavirus (B37) is similar to G3 proteins in the antigenic C region. Virology 170: 55–61

Hundley F, Biryahwaho B, Gow M, Desselberger U (1985) Genome rearrangements of bovine rotavirus after serial passage at high multiplicity of infection. Virology 143: 88–103

Hundley F, McIntyre M, Clark B, Beards G, Wood D, Chrystie I, Desselberger U (1987) Heterogeneity of genome rearrangements in rotaviruses isolated from a chronically infected, immunodeficient child. J Virol 61: 3365–3372

Imai M, Akatani K, Ikegami N, Furuichi Y (1983a) Capped and conserved terminal structures in human rotavirus genome double-stranded RNA segments. J Virol 47: 125–136

Imai M, Richardson MA, Ikegami N, Shatkin AJ, Furuichi Y (1983b) Molecular cloning of double-stranded RNA virus genomes. Proc Natl Acad Sci USA 80: 373–377

Kalica AR, Greenberg HB, Wyatt RG, Flores J, Sereno MM, Kapikian AZ, Chanock RM (1981) Genes of human (strain Wa) and bovine (strain UK) rotaviruses that code for neutralization and subgroup antigens. Virology 112: 385–390

Kalica AR, Flores J, Greenberg HB (1983) Identification of the rotaviral gene that codes for the hemagglutinin and protease-enhanced plaque formation. Virology 125: 194–205

Kantharidis P, Dyall-Smith ML, Holmes IH (1987) Marked sequence variation between segment 4 genes of human RV-5 and simian SA11 rotaviruses. Arch Virol 93: 111–121

Kantharidis P, Dyall-Smith ML, Tregear GW, Holmes IH (1988) Nucleotide sequence of UK bovine rotavirus segment 4 : possible host restriction of VP3 genes. Virology 166: 308–315

Kapahnke R, Rappold W, Desselberger U, Riesner D (1986) The stiffness of dsRNA: Hydrodynamic studies on fluorescence-labelled RNA segments of bovine rotavirus. Nucleic Acids Res 14: 3215–3228

Kuchino Y, Nishimura S, Smith RE, Furuichi Y (1982) Homologous terminal sequences in the double-stranded RNA genome segments of cytoplasmic polyhedrosis virus of silkworm Bombyx mori. J Virol 44: 538–543

Kumar A, Charpilienne A, Cohen J (1989) Nucleotide sequence of the gene encoding for the RNA binding protein (VP2) of RF bovine rotavirus. Nucleic Acids Res 17: 2126

Lamb RA (1983) The influenza virus RNA segments and their encoded proteins In: Palese P, Kingsbury DW (eds) Genetics of influenzaviruses. Springer, Vienna, New York, pp 21–69

Liu M, Estes MK (1989) Nucleotide sequence of the simian rotavirus SA11 genome segment 3. Nucleic Acids Res 17: 7991

Liu M, Offit PA, Estes MK (1988) Identification of the simian rotavirus SA11 genome segment 3 product. Virology 163: 26–32

Lopez S, Arias CF (1987) The nucleotide sequence of the 5' and 3' ends of rotavirus SA11 gene 4. Nucleic Acids Res 15: 4691

Lopez S, Arias CF, Bell JR, Strauss JH, Espejo RT (1985) Primary structure of the cleavage site associated with trypsin enhancement of rotavirus SA11 infectivity. Virology 144: 11–19

Lopez S, Arias CF, Mendez E, Espejo RT (1986) Conservation in rotaviruses of the protein region containing the two sites associated with trypsin enhancement of infectivity. Virology 154: 224–227

Mackow ER, Shaw RD, Matsui SM, Vo PT, Benfield DA, Greenberg HB (1988a) Characterization of homotypic and heterotypic VP7 neutralization sites of rhesus rotavirus. Virology 165: 511–517

Mackow ER, Shaw RD, Matsui SM, Vo PT, Dang M-N, Greenberg HB (1988b) Characterization of the rhesus rotavirus gene encoding protein VP3: location of amino acids involved in homologous and heterologous rotavirus neutralization and identification of a putative fusion region. Proc Natl Acad Sci USA 85: 645–649

Mackow ER, Barnett JW, Chan H, Greenberg HB (1989) The rhesus rotavirus outer capsid protein VP4 functions as a hemagglutinin and is antigenically conserved when expressed by a baculovirus recombinant. J Virol 63: 1661–1668

Mason BB, Graham DY, Estes MK (1980) In vitro transcription and translation of simian rotavirus SA11 gene products. J Virol 33: 1111–1121

Mason BB, Graham DY, Estes MK (1983) Biochemical mapping of the simian rotavirus SA11 genome. J Virol 46: 413–423

Mason BB, Dheer SK, Hsiao CL, Zandle G, Kostek B, Rosanoff EI, Hung PP, Davis AR (1985) Sequence of the serotype-specific glycoprotein of the human rotavirus Wa strain and comparison with other human rotavirus serotypes. Virus Res 2: 291–299

Matsui SM, Mackow ER, Matsuno S, Paul PS, Greenberg HB (1990) Sequence analysis of gene 11 equivalents from short and supershort strains of rotavirus. J Virol 64: 120–124

Matsuno S, Hasegawa A, Kalica AR, Kono R (1980) Isolation of a recombinant between simian and bovine rotavirus. J Gen Virol 48: 253–256

Mattion N, Gonzalez SA, Burrone O, Bellinzoni R, LaTorre JL, Scodeller EA (1988) Rearrangement of genomic segment 11 in two swine rotavirus strains. J Gen Virol 69: 695–698

McCrae MA (1987) Nucleic acid analyses of non-group A rotaviruses. In: Bock G, Whelan JJ, Novel diarrhoea viruses. Ciba foundation symposium, vol 128. Wiley, Chichester, pp 24–36.

McCrae MA, McCorquodale JG (1982a) Molecular biology of rotaviruses. II. Identification of the protein-coding assignments of calf rotavirus genome RNA species. Virology 117: 435–443

McCrae MA, McCorquodale JG (1982b) Molecular biology of rotavirus. IV. Molecurlar cloning of the bovine rotavirus genome. J Virol 44: 1076–1079

McCrae MA, McCorquodale JG (1983) Molecular biology of rotaviruses. V. Terminal structure of viral RNA species. Virology 126: 204–212

McIntyre M, Rosenbaum V, Rappold W, Desselberger M, Wood D, Desselberger U (1987) Biophysical characterization of rotavirus particles containing rearranged genomes. J Gen Virol 68: 2961–2966

Mertens PPC, Sangar DV (1985) Analysis of the terminal sequence of the genome segments of four orbiviruses. In Barger TL, Yochim MM (eds) Biuetongue and related orbiviruses. Liss, New York, pp 371–387

Meyer JC, Bergmann CL, Bellamy AR (1989) Interaction of rotavirus cores with the non-structural glycoprotein NS28. Virology 171: 98–107

Mitchell DB, Both GW (1988) Simian rotavirus SA11 segment 11 contains overlapping reading frames. Nucleic Acids Res 16: 6244

Mitchell DB, Both GW (1989) Complete nucleotide sequence of the simian rotavirus SAll VP4 gene. Nucleic Acids Res 17: 2122

Mitchell DB, Both GW (1990) Conservation of a potential metal binding motif despite extensive sequence diversity in the rotavirus non-structural protein NS53. Virology 174: 618–621

Mueller H, Nitschke R (1987) Molecular weight determination of the two segments of double stranded RNA of infectious bursal disease virus, a member of the birnavirus group. Med Microbiol Immunol 176: 113–121

Nishikawa K, Gorziglia M (1988) The nucleotide sequence of the VP3 gene of porcine rotavirus OSU. Nucleic Acids Res 16: 11847

Nishikawa K, Taniguchi K, Torres A, Hoshino Y, Green K, Kapikian AZ, Chanock RM, Gorziglia M (1988) Comparative analysis of the VP3 gene of divergent strains of the rotaviruses simian SA11 and bovine Nebraska calf diarrhea virus. J Virol 62: 4022–4026

Novo E, Esparza J (1981) Composition and topography of structural polypeptides of bovine rotavirus. J Gen Virol 56: 325–335

Nuttall SD, Hunn CP, Holmes IH, Dyall-Smith ML (1989) Sequences of VP9 genes from short and supershort rotavirus strains. Virology 171: 453–457

Offit PA, Blavat G (1986) Identification of the two rotavirus genes determining neutralization specificities. J Virol 57: 376–378

Offit PA, Shaw RD, Greenberg HB (1986) Passive protection against rotavirus-induced diarrhea by monoclonal antibodies to surface proteins VP3 and VP7. J Virol 58: 700–703

Okada Y, Richardson MA, Ikegami N, Nomoto A, Furuichi Y (1984) Nucleotide sequence of human rotavirus genome segment 10, an RNA encoding a glycosylated virus protein. J Virol 51: 856–859

Patton JT (1986) Synthesis of simian rotavirus SA11 double-stranded RNA in a cell-free system. Virus Res 6: 217–233

Patton JT, Gallegos CO (1988) Structure and protein composition of the rotavirus replicase particle. Virology 166: 358–365

Pedley S, Bridger JC, Brown JF, McCrae MA (1983) Molecular characterization of rotaviruses with distinct group antigens. J Gen Virol 64: 2093–2101

Pedley S, Hundley F, Chrystie I, McCrae MA, Desselberger U (1984) The genomes of rotaviruses isolated from chronically infected immunodeficient children. J Gen Virol 64: 1141–1150

Pedley S, Bridger JC, Chasey D, McCrae MA (1986) Definition of two new groups of atypical rotaviruses. J Gen Virol 67: 131–137

Petrie BL, Estes MK, Graham DY (1983) Effect of tunicamycin on morphogenesis and infectivity. J Virol 46: 270–274

Pocock DH (1987) Isolation and characterization of two group A rotaviruses with unusual genome profiles. J Gen Virol 68: 653–660

Pothier P, Kohli E, Drouet E, Ghim S (1987) Analysis of the antigenic sites on the major inner capsid protein (VP6) of rotaviruses using monoclonal antibodies. Ann Inst Pasteur/Virol 138: 285–295

Potter AA, Cox G, Parker M, Babiuk LA (1987) The complete nucleotide sequence of bovine rotavirus C486 gene 4 cDNA. Nucleic Acids Res 15: 4361

Powell KFH, Gunn PR, Bellamy AR (1988) Nucleotide sequence of bovine rotavirus genomic segment 10: an RNA encoding the viral non-structural glycoprotein. Nucleic Acids Res 16: 763

Prasad BVV, Wang GJ, Clerx JPM, Chiu W (1988) Three-dimensional structure of rotavirus. J Mol Biol 199: 269–275

Ramig RF, Samal SK, McConnell S (1985) Genome RNAs of virulent and attenuated strains of bluetongue virus serotypes 10, 11, 13 and 17. In: Barger TL, Yochim MM (eds) Bluetongue and related orbiviruses. Liss, New York, pp 389–396

Reddy DA, Greenberg HB, Bellamy AR (1989) Rotavirus serotype 4: nucleotide sequence of genomic segment 9 of the St Thomas 3 strain. Nucleic Acids Res 17: 449

Richardson MA, Iwamoto A, Ikegami N, Nomoto A, Furuichi Y (1984) Nucleotide sequence of the gene encoding the serotype-specific antigen of human (Wa) rotavirus: comparison with the homologous genes from simian SA11 and UK bovine rotaviruses. J Virol 51: 860–862

Rixon F, Taylor P, Desselberger U (1984) Rotavirus RNA segments sized by electron microscopy. J Gen Virol 65: 233–239

Rodger SM, Bishop RF, Birch L, McLean B, Holmes IH (1981) Molecular epidemiology of human rotaviruses in Melbourne, Australia, from 1973–1979 as determined by electrophoresis of genomic ribonucleic acid. J Clin Microbiol 13: 272–278

Ruggeri FM, Greenberg HB (1991) Antibodies to the trypsin cleavage peptide VP8* neutralize rotavirus by inhibiting binding of virions to target cells in cell culture. J Virol 65: 2211–2219

Ruiz AM, Lopez IV, Lopez S, Espejo RT, Arias CF (1988) Molecular and antigenic characterization of porcine rotavirus YM, a possible new rotavirus serotype. J Virol 62: 4331–4336

Rushlow K, McNab A, Olson K, Maxwell F, Maxwell I, Stiegler G (1988) Nucleotide sequence of porcine rotavirus (OSU strain) gene segments 7, 8 and 9. Nucleic Acids Res 16: 367–368

Sabara M, Gilchrist JE, Hudson GR, Babiuk LA (1985) Preliminary characterization of an epitope involved in neutralization and cell attachment that is located on the major bovine rotavirus glycoprotein. J Virol 53: 58–66

Sabara M, Ready KFM, Frenchick PJ, Babiuk LA (1987) Biochemical evidence for the oligomeric arrangement of bovine rotavirus nucleocapsid protein and its possible significance in the immunogenicity of this protein. J Gen Virol 68: 123–133

Sato S, Yolken RM, Eiden JJ (1989) The complete nucleic acid sequence of gene segment 3 of the IDIR strain of group B rotavirus. Nucleic Acids Res 17: 10113

Schiff LA, Nibert ML, Co MS, Brown EG, Fields BN (1988) Distinct binding sites for zinc and double–stranded RNA in the reovirus outer capsid protein s3. Mol Cell Biol 8: 273–283

Schnagl RD, Rodger SM, Holmes IH (1981) Variation in human rotavirus electropherotypes occurring between rotavirus gastroenteritis epidemics in Central Australia. Infect Immun 33: 17–21

Schroeder BA, Street JE, Kalmakoff J, Bellamy AR (1982) Sequence relationships between the genome segments of human and animal rotavirus strains. J Virol 43: 379–385

Scott GE, Tarlow O, McCrae MA (1989) Detailed structural analysis of a genome rearrangement in bovine rotavirus. Virus Res 14: 119–128

Shaw RD, Stoner-Ma DL, Estes MK, Greenberg HB (1985) Specific enzyme-linked immuno-assay for rotavirus serotypes 1 and 3, J Clin Microbiol 22: 286–291

Smith ML, Lazdins I, Holmes IH (1980) Coding assignments of double-stranded RNA segments of SA11 rotavirus established by in vitro translation. J Virol 33: 976–982

Smith RE, Kister SE, Carozzi NB (1989) Cloning and expression of the major inner capsid protein of SA11 simian rotavirus in Escherichia coli. Gene 79: 239–248

Stirzaker SC, Whitfeld PL, Christie DL, Bellamy AR, Both GW (1987) Processing of rotavirus glycoprotein VP7: implications for the retention of the protein in the endoplasmic reticulum. J Cell Biol 105: 2897–2903

Svensson L, Grahnquist L, Pettersson CA, Grandien M, Stintzing G, Greenberg HB (1988) Detection of human rotaviruses which do not react with subgroup I- and II-specific monoclonal antibodies. J Clin Microbiol 26: 1238–1240

Tanaka TN, Conner ME, Graham DY, Estes MK (1988) Molecular characterization of three rabbit rotavirus strains. Arch Virol 98: 253–265

Taniguchi K, Urasawa S, Urasawa T (1985) Preparation and characterization of neutralizing monoclonal antibodies with different reactivity patterns to human rotaviruses. J Gen Virol 66: 1045–1053

Taniguchi K, Urasawa T, Urasawa S (1986) Reactivity patterns to human rotavirus strains of a monoclonal antibody against VP2, a component of the inner capsid of rotavirus. Arch Virol 87: 135–141

Taniguchi K, Morita Y, Urasawa T, Urasawa S (1987) Cross-reactive neutralization epitopes on VP3 of human rotavirus: Analysis with monoclonal antibodies and antigenic variants. J Virol 61: 1726–1730

Taniguchi K, Hoshino Y, Nishikawa K, Green KY, Maloy WL, Morita Y, Urasawa S, Kapikian AZ, Chanock RM, Gorziglia M (1988a) Cross-reactive and serotype-specific neutralization epitopes on VP7 of human rotavirus: Nucleotide sequence analysis of antigenic mutants selected with monoclonal antibodies. J Virol 62: 1870–1874

Taniguchi K, Maloy WL, Nishikawa K, Green KY, Hoshino Y, Urasawa S, Kapikian AZ, Chanock RM, Gorziglia M (1988b) Identification of cross-reactive and serotype 2-specific neutral-ization epitopes on VP3 of human rotavirus. J Virol 62: 2421–2426

Taniguchi K, Pongsuwanna Y, Choonthanon M, Urusawa S (1990) Nucleotide sequence of the VP7 gene of a bovine rotavirus (strain 61A) with different serotype specificity from serotype 6. Nucleic Acids Res 18: 4613

Tarlow O, McCrae MA (1990) Nucleotide sequence of the group antigen VP6 of the UK tissue culture adapted bovine rotavirus. Nucleic Acids Res 18: 4921

Taylor P, Rixon F, Desselberger U (1985) Rise per base pair in helices of double-stranded RNA determined by electron microscopy. Virus Res 2: 175–182

Thouless ME, DiGiacomo RF, Neumann DS (1986) Isolation of two lapine rotaviruses: characterization of their subgroup and serotype and RNA electropherotype. Arch Virol 89: 161–170

Tian Y, Tarlow O, McCrae MA (1990) Nucleotide sequence of gene 2 of the UK tissue culture adapted strain of bovine rotavirus. Nucleic Acids Res 18: 4015

Unicomb LE, Coulson BS, Bishop RF (1989) Experience with an enzyme immunoassay for serotyping human group A rotaviruses. J Clin Microbial 27: 586–588

Ward CW, Elleman TC, Azad AA, Dyall-Smith ML (1984) Nucleotide sequence of gene segment 9 encoding a nonstructural protein of UK bovine rotavirus. Virology 134: 249–253

Ward CW, Azad AA, Dyall-Smith ML (1985) Structural homologies between RNA gene segments 10 and 11 from UK bovine, simian SA11 and human Wa rotaviruses. Virology 144: 328–336

Welch SKW, Crawford SE, Estes MK (1989) The rotavirus SA11 genome segment 11 protein is a non-structural phosphoprotein. J Virol 63: 3974–3982

Whitton JL, Hundley F, O'Donnell B, Desselberger U (1983) Silver staining of nucleic acids. Application in virus research and in diagnostic virology. J Virol Methods 7: 185–198

Wood DJ, David TJ, Chrystie IL, Totterdell B (1988) Chronic enteric virus infection in two T-cell immunodeficient children. J Med Virol 24: 435–444

Xu L, Harbour D, McCrae MA (1991) Sequence of the gene encoding the major neutralization antigen of serotype 10 rotavirus. J Gen Virol 72: 177–180

Notes Added in Proof (Autumn 1993)

As considerable times has elapsed between submission and printing and as many more sequence data, particularly of VP4 and VP7 genes have accumulated since then, we have prepared *Table 6* which is a continuation of Table 2 and contains accession and identification numbers of rotavirus gene entries into the EMBL and Genbank databases between late 1990 and mid 1993. References were added where available. A detailed discussion of the data, particularly of non group A rotaviruses, has to be left to a future review.

Table 6. Accession and identification numbers of rotavirus gene entries in the EMBL and Genbank databases (late 1990-mid 1993)

RNA segment number	Accession number	ID Line	Accession number	Locus ID	Virus strain	Reference
1	M74216	PRVVP1A	M74216	PRVVP1A	C/porcine/Cowden	Bremont et al. (1992)
	X16830	ROSIVP1	X16830	ROSIVP1	A/simian/SA11	Mitchell and Both (1990)
	M97203	ROTROTAB	M97203	ROTROTAB	B/rodent/IDIR	Eiden and Hirshon (1993)
2	M74217	PRVVP2A	M74217	PRVVP2A	C/porcine/Cowden	Bremont et al. (1992)
	M91433	ROHCSP	M91433	ROHCSP	B/human/ADRV	Mitchell and Both (1990)
	X16831	ROSIVP2	X16831	ROSIVP2	A/simian/SA11	Mitchell and Both (1990)
3	M74219	PRVVP3A	M74219	PRVVP3A	C/porcine/Cowden	Bremont et al. (1992)
	X16387	ROSIVP3	X16387	ROSIVP3	A/simian/SA11	Mitchell and Both (1990)
4	S42279	S42279	S42279	S42279	A/human/AU-1	Isegawa et al. (1992)
	M28349	M28349	M28349	M28349	A/human/K8	Taniguchi et al. (1989)
	D10971	FRVVP4	D10971	FRVVP4	A/feline/FRV-1	
	L10359	PRVVAR4FA	L10359	PRVVAR4FA	A/porcine/	
	L10358	PRVVAR4SA	L10358	PRVVAR4SA	A/porcine/	
	M74218	PRVVP4A	M74218	PRVVPP4A	C/porcine/Cowden	Bremont et al. (1992)
	M63231	PRVYMVP	M63231	PRVYMVP	A/porcine/YM	Lopez et al. (1991)
	M96825	RO14GN	M96825	RO14GN	A/human/	
	L04638	RO1SP4	L04638	RO1SP4	A/equine/H-2	Hardy et al. (1993)
	L07657	RO1VP4A	L07657	RO1VP4A	A/human/I321	
	D14723	RO1VP4FRV	D14723	RO1VP4FRV	A/human/FRV64	
	D14725	RO1VP4K9	D14725	RO1VP4K9	A/human/K9	
	D14726	RO1VP4RO	D14726	RO1VP4RO	A/human/RO1845	
	D14367	ROBVP4	D14367	ROBVP4	A/bovine/	
	M92986	ROBVP4A	M92986	ROBVP4A	A/bovine/B223	Hardy et al. (1992)
	M63267	ROBVP4G	M63267	ROBVP4G	A/bovine/	Hardy et al. (1991)
	D90260	ROHK8	D90260	ROHK8	A/human/	Taniguchi et al. (1989)
	M58292	ROHVP4	M36397	ROHVP4	A/human/	Taniguchi et al. (1990)
	D10970	ROHVP41	D10970	ROHVP41	A/human/	
	M60600	ROHVP4A	M60600	ROHVP4A	A/human/69M	Qian and Green (1991)
	M88480	ROHVP4AA	M88480	ROHVP4AA	A/human/	Gorziglia et al. (1988)
	M91434	ROHVP4B	M91434	ROHVP4B	B/human/ADRV	
	L11599	ROTVP4X	L11599	ROTVP4X	A/ovine/LP14	
	X57319	RVSA114F	X57319	RVS114F	A/simian/SA11-4F	Mattion and Estes (1992)

Table 6. (Continued)

RNA segment number	Accession number	ID Line	Accession number	Locus ID	Virus strain	Reference
5	Z12108	BRUKVP5R	Z12108	BRUKVP5R	A/bovine/UKtc	
	Z12105	BRVP5RNA	Z12105	BRVP5RNA	A/bovine/B223	
	X59297	HSIGV803	X59297	HSIGV803	A/human/IGV-80-3	
	X60546	PRNS53	X60546	PRNS53	C/porcine/Cowden	
	Z12107	PRVP5RNA	Z12107	PRVP5RNA	A/porcine/OSU	
	M91435	ROHNS	M91435	ROHNS	B/human/ADRV	
	M55982	ROTMICP	M55982	ROTMICP	B/human/ADRV	
	Z12106	SRVP5RNA	Z12106	SRVP5RNA	A/human/Hochi	Chen et al. (1991)
6	S44813	S44813	S44813	S44813	C/porcine/Cowden	Cooke et al. (1992)
	S44818	S44818	S44818	S44818	C/human/Belem	Cooke et al. (1992)
	X57943	HRVP6A	X57943	HRVP6A	A/human/	
	X59843	HSRVP6	X59843	HSRVP6	C/human/	Cooke et al. (1991)
	M94157	PRVVP6C	M94157	PRVVP6C	C/porcine/Cowden	Cooke et al. (1992)
	X53667	REBRVP6	X53667	REBRVP6	A/bovine/UKtc	Tarlow and McCrae (1990)
	L12390	ROBNS34BS	L12390	ROBNS34BS	C/bovine/Shintoku	
	S56173	ROHVP6	S56173	ROHVP6	C/human/Bristol	Cooke et al. (1991)
	M94155	ROHVP6A	M94155	ROHVP6A	C/human/	Cooke et al. (1992)
	M94156	ROHVP6B	M94156	ROHVP6B	C/human/	Cooke et al. (1992)
	M84456	ROTBMICP	M84456	ROTBMICP	B/rodent/IDIR	Eiden et al. (1992)
	L11585	ROTDIRRPT	L11585	ROTDIRRPT	A/ovine/	
	L11596	ROTDORRPT	L11596	ROTDORRPT	A/ovine/	
	M88768	ROTVP6A	M88768	ROTVP6A	C/bovine/	Jiang et al. (1992)
	X69487	RVV6AC	X69487	RVV6AC	A/porcine/YM	
7	X57944	HRSEG7A	X57944	HRSEG7A	A/human/	Qian et al. (1991a)
	M69115	PRVNS34	M69115	PRVNS34	C/porcine/Cowden	
	M87502	RO1NS34A	M87502	RO1NS34A	A/simian/SA114F	Mattion et al. (1992)
	M91436	ROHNONST1	M91436	ROHNONST1	B/human/ADRV	
8	S83903	S83903	S83903	S83903	A/human/1407	Bessarab et al. (1991)
	S54574	S54574	D12710	S54574	A/bovine/KN-4	Matsuda et al. (1993)
	I02127	I02127	I02127	I02127		

Accession	Gene	Accession	Name	Strain	Reference
A01028	A01028	A01028	A01028	A/human/HU5	NISHIKAWA et al. (1991)
A01321	A01321	A01321	A01321	A/human/	
X56784	BRRNAVP7	X56784	BRRNAVP	A/avian/Ch2	
X57852	BRRVP7A	X57852	BRRVP7A	A/bovine/B223	BEARDS et al. (1993)
X63156	HRA64SP8	X63156	HRA64SP8	A/human/A64	QIAN et al. (1991b)
M61101	PRVGP7A	M61101	PRVGP7A	C/porcine/Cowden	AKOPIAN et al. (1992)
X58439	PRVP7G	X58439	PRVP7G	A/porcine/K	
L10360	PRVVAR4FB	L10360	PRVVAR4FB	A/porcine/	
L10361	PRVVAR4SB	L10361	PRVVAR4SB	A/porcine/	
X53403	REBRVP7G	X53403	REBRVP7G	A/bovine/61A	TANIGUCHI et al. (1990)
M61876	RO1G14	M61876	RO1G14	A/equine/F123	
D13549	RO1VP7	D00843	RO1VP7	A/equine/L338	BROWNING et al. (1991)
L01098	RO1VP7A	L01098	RO1VP7A	A/avian/Ty-1	
L07658	RO1VP7B	L07658	RO1VP7B	A/human/I321	
L11605	RO1VP7C	L11605	RO1VP7C	A/human/	
M64666	RO4UP7	M64666	RO4UP7	A/human/	
D01055	ROBA44VP7	D01055	ROBA44VP7	A/bovine/A44	TANIGUCHI et al. (1991)
D01054	ROBA5VP7	D01054	ROBA5VP7	A/bovine/A5	TANIGUCHI et al. (1991)
D12710	ROBD12710	D12710	ROBD12710	A/bovine/	MATSUDA et al. (1993)
D01056	ROBKK3VP7	D01056	ROBKK3VP7	A/bovine/KK3	TANIGUCHI et al. (1991)
M92651	ROBVP7A	M92651	ROBVP7A	A/bovine/	BLACKHALL et al. (1992)
M64679	ROBVP7B11	M64679	ROBVP7B11	A/bovine/B-11	HUANG et al. (1992)
M64680	ROBVP7B60	M64680	ROBVP7B60	A/bovine/B60	HUANG et al. (1992)
M63266	ROBVP7G	M63266	ROBVP7G	A/bovine/	HARDY et al. (1991)
M61100	ROHGP8A	M61100	ROHGP8A	C/human/88–220	QIAN et al. (1991b)
M91437	ROHNONSTR	M91437	ROHNONSTR	B/human/ADRV	TANIGUCHI et al. (1990)
M58290	ROHVP7	M36396	ROHVP7	A/human/	HUM et al. (1989)
J04334	ROHVP7B37	J04334	ROHVP7B37	A/human/B37	
L14072	ROTOUTCAP	L14072	ROTOUTCAP	A/human/116E	BROWNING et al. (1991)
D00843	ROTVP7	X57785	ROTVP7	A/equine/L338	GREEN et al. (1992)
M86490	ROTVP7A	M86490	ROTVP7A	A/human/	
M93006	ROTVP7AA	M93006	ROTVP7AA	A/human/	GREEN et al. (1992)
M86832	ROTVP7B	M86832	ROTVP7B	A/human/	GREEN et al. (1992)
M86833	ROTVP7C	M86833	ROTVP7C	A/human/	GREEN et al. (1992)
M86834	ROTVP7D	M86834	ROTVP7D	A/human/	
L11602	ROTVP7X	L11602	ROTVP7X	A/ovine/	
X65940	RVRFGP	X65940	RVRFGP	A/bovine/RF	

Table 6. (Continued)

RNA segment number	Accession number	ID Line	Accession number	Locus ID	Virus strain	Reference
9	M33872	ARRSEG9	M33872	ARRSEG9	B/human/ADRV	CHEN et al. (1990)
	D14613	RO1AU1115	D14613	RO1AU1115	A/human/	
	D14614	RO1AU125B	D14614	RO1AU125B	A/human/	
	D14615	RO1AU228C	D14615	RO1AU228C	A/human/	
	D14616	RO1AU379D	D14616	RO1AU379D	A/human/	
	D14617	RO1AU387E	D14617	RO1AU387E	A/human/	
	D14618	RO1AU720F	D14618	RO1AU720F	A/human/	
	D14619	RO1AU785G	D14619	RO1AU785G	A/human/	
	D14620	RO1AU938H	D14620	RO1AU938H	A/human/	
	D14621	RO1CAT21	D14621	RO1CAT21	A/human/	
	D14622	RO1MZ58J	D14622	RO1MZ58J	A/human/	
	L04529	RO1NS35A	L04529	RO1NS35A	A/human/DS-1	PATTON et al. (1993)
	L04530	RO1NS35B	L04530	RO1NS35B	A/bovine/NCDV	PATTON et al. (1993)
	L04531	RO1NS35C	L04531	RO1NS35C	A/simian/SA11-P	PATTON et al. (1993)
	L04532	RO1NS35D	L04532	RO1NS35D	A/simian/SA11-R	PATTON et al. (1993)
	L04533	RO1NS35E	L04533	RO1NS35E	A/avian/Ty-1	PATTON et al. (1993)
	L04534	RO1NS35F	L04534	RO1NS35F	A/human/Wa	PATTON et al. (1993)
	D14623	RO1PA151K	D14623	RO1PA151K	A/human/	
	D14624	RO1PCP5L	D14624	RO1PCP5L	A/human/	
	D00911	ROTIDIR9	D00911	ROTIDIR9	B/rodent/IDIR	PETRIC et al. (1991)
	X65939	RVCNS5	X65939	RVCNS5	C/porcine/Cowden	
10	L12391	ROBNS26MR	L12391	ROBNS26MR	C/bovine/Shintoku	
	D10771	ROHSG10	D01145	ROHSG10	A/human/	BALLARD et al. (1992)
	D10772	ROHSG10R	D01146	ROHSG10R	A/human/	BALLARD et al. (1992)
	M81488	ROTNONSTP	M81488	ROTNONSTP	A/human/	LAMBDEN et al. (1992)
	X69485	RVNS28A	X69485	RVNS28A	A/porcine/YM	
11	D00912	ROTIDIR11	D00912	ROTIDIR11	B/rodent/IDIR	PETRIC et al. (1991)
	X65938	RVCNS26	X65938	RVCNS26	C/porcine/Cowden	
	X69486	RVNS26B	X69486	RVNS26B	A/porcine/YM	LOPEZ and ARIAS (1993)
	M28377	M28377	M28377	M28377	A/human/B37	NUTTALL et al. (1989)
	M28378	M28378	M28378	M28378	A/human/RV5	NUTTALL et al. (1989)

Note: The gene sequences of group B and C rotaviruses are listed under segment numbers as published but are not always the genome equivalents of group A virus segments of the same number.

Additional References Relating to Database
Entries Listed in Table 6

Akopian TA, Lunin VG, Kruglyak VA, Ruchadze GG, Bakhutashvili VI, Naroditsky BS, Tichonenko TI (1992) Nucleotide sequence of the cDNA for porcine rotavirus VP7 gene (strain K). Virus Genes 6: 393–396

Ballard A, McCrae MA, Desselberger U (1992) Nucleotide sequences of normal and rearranged RNA segments 10 of human rotaviruses. J Gen Virol 73: 633–638

Beards G, Xu L, Ballard A, Desselberger U, McCrae MA (1993) A serotype 10 human rotavirus. J Clin Microbiol 30: 1432–1435

Bessarab IN, Epifanova NV, Novikova NA, Borodin AM (1991) An analysis of the gene coding for the basic neutralizing antigen VP7 of human rotavirus isolate 1407. Vopr Virusol 36: 480–483

Blackhall J, Bellinzoni R, Mattion N, Estes MK, La Torre J, Magnusson G (1992) A bovine rotavirus serotype 1: serologic characterization of the virus and nucleotide sequence determination of the structural glycoprotein VP7 gene. Virology 189: 833–837

Bremont M, Juste-Lesage P, Chabanne-Vautherot D, Charpilienne A, Cohen J (1992) Sequences of the four larger proteins of a porcine group C rotavirus and comparison with the equivalent group A rotavirus proteins. Virology 186: 684–692

Browning GF, Chalmers RM, Fitzgerald TA, Snodgrass DR (1991) Serological and genomic characterization of L338, a novel equine group A rotavirus G serotype. J Gen Virol 72: 1059–1064

Chen GM, Hung T, Mackow ER (1990) Identification of the gene encoding the group B rotavirus VP7 equivalent: Primary characterization of the ADRV segment 9 RNA. Virology 178: 311–315

Chen GM, Werner-Eckert R, Mackow ER, Tao H (1991) Expression of the major inner capsid protein of the group B rotavirus ADRV: primary characterization of genome segment 5. Virology 182: 820–829

Cooke SJ, Lambden PR, Caul EO, Clarke IN (1991) Molecular cloning, sequence analysis and coding assignment of the major inner capsid protein gene of human group C rotavirus. Virology 184: 781–785

Cooke SJ, Clarke IN, Freilas RB, Gabbay YB, Lambden PR (1992) The correct sequence of the porcine group C/Cowden rotavirus major inner capsid protein shows close homology with human isolates from Brazil and the UK. Virology 190: 531–537

Eiden JJ, Nataro J, Vonderfecht S, Petric M (1992) Molecular cloning, sequence analysis, in vitro expression, and immunoprecipitation of the major inner capsid protein of the IDIR strain of group B rotavirus (GBR). Virology 188: 580–589

Eiden JJ, Hirshon C (1993) Sequence analysis of group B rotavirus gene 1 and definition of a rotavirus-specific sequence motif within the RNA polymerase gene. Virology 192: 154–160

Green KY, Sarasini A, Qian Y, Gerna G (1992) Genetic variation in rotavirus serotype 4 subtypes. Virology 188: 362–368

Hardy ME, Woode GN, Xu Z, Gorziglia M (1991) Comparative amino acid sequence analysis of VP4 for VP7 serotype 6 bovine rotavirus strains NCDV, B641, and UK. J Virol 65: 5535–5538

Hardy ME, Gorziglia M, Woode GN (1992) Amino acid sequence analysis of bovine rotavirus B223 reveals a unique outer capsid protein VP4 and confirms a third bovine VP4 type. Virology 191: 291–300

Hardy ME, Gorziglia M, Woode GN (1993) The outer capsid protein VP4 of equine rotavirus H-2 represents a unique VP4 type by amino acid sequence analysis. Virology 193: 492–497

Huang JA, Nagesha HS, Snodgrass DR, Holmes IH (1992) Molecular and serological analyses of two bovine rotaviruses (B-11 and B-60) causing calf scours in Australia. J Clin Microbiol 30: 85–92

Isegawa Y, Nakagomi O, Nakagomi T, Ueda S (1992) A VP4 sequence highly conserved in human rotavirus strain AU-1 and feline rotavirus strain FRV-1. J Gen Virol 73: 1939–1946

Jiang B, Tsunemitsu H, Gentsch JR, Glass RI, Green KY, Qian Y, Saif LJ (1992) Nucleotide sequence of gene 5 encoding the inner capsid protein (VP6) of bovine group C rotavirus: Comparison with corresponding genes of group C, A, and B rotaviruses. Virology 190: 542–547

Lambden PR, Cooke SJ, Caul EO, Clarke IN (1992) Cloning of noncultivatable human rotavirus by single primer amplification. J Virol 66: 1817–1822

Lopez S, Lopez I, Romero P, Mendez E, Soberon X, Arias CF (1991) Rotavirus YM gene 4: Analysis of its deduced amino acid sequence and prediction of the secondary structure of the VP4 protein. J Virol 65: 3738–3745

Lopez S, Arias CF (1993) Protein NS26 is highly conserved among porcine rotavirus strains. Nucleic Acids Res 21: 1042

Matsuda Y, Isegawa Y, Woode GN, Zheng S, Kaga E, Nakagomi T, Ueda S, Nakagomi O (1993) Two-way cross-neutralization mediated by a shared P (VP4) serotype between bovine rotavirus strains with distinct G (VP7) serotypes. J Clin Microbiol 31: 354–358

Mattion N, Estes MK (1992) Nucleotide sequence of a rotavirus gene 4 associated with unique biologic properties. Arch Virol 120: 109–113

Mattion N, Cohen J, Aponte C, Estes M (1992) Characterization of an oligomerization domain and RNA properties on rotavirus nonstructural protein NS34. Virology 190: 68–83

Mitchell DB, Both GW (1990) Completion of the genomic sequence of the simian rotavirus SA11: nucleotide sequences of segments 1, 2 and 3. Virology 177: 324–331

Nishikawa K, Hoshino Y, Gorziglia M (1991) Sequence of the VP7 gene of chicken rotavirus Ch2 strain of serotype 7 rotavirus. Virology 185: 853–857

Nuttall SD, Hum CP, Holmes IH, Dyall-Smith ML (1989) Sequences of VP9 genes from short and supershort rotavirus strains. Virology 171: 453–457

Patton JT, Salter-Cid L, Kalbach A, Mansell EA, Kattoura M (1993) Nucleotide and amino acid sequence analysis of the rotavirus non-structural RNA-binding protein NS35. Virology 192: 438–446

Petric M, Mayur K, Vonderfecht S, Eiden JJ (1991) Comparison of group B rotavirus genes 9 and 11. J Gen Virol 72: 2801–2804

Qian Y, Green KY (1991) Human rotavirus strain 69M has a unique VP4 as determined by amino acid sequence analysis. Virology 182: 407–412

Qian Y, Jiang B, Saif LJ, Kang SY, Ojeh CK, Green KY (1991a) Molecular analysis of the gene 6 from a porcine group C rotavirus that encodes the NS34 equivalent of group A rotaviruses. Virology 184: 752–757

Qian Y, Jiang B, Saif LJ, Kang SY, Ishimaru Y, Yamashita Y, Oseto M, Green KY (1991b) Sequence conservation of gene 8 between human and porcine group C rotaviruses and its relationship to the VP7 gene of group A rotaviruses. Virology 182: 562–569

Taniguchi K, Nishikawa K, Urasawa T, Urasawa S, Midthun K, Kapikian AZ, Gorziglia M (1989) Complete nucleotide sequence of the gene encoding VP4 of a human rotavirus (strain K8) which has unique VP4 neutralization epitopes. J Virol 63: 4101–4106

Taniguchi K, Urasawa T, Kobayashi N, Gorziglia M, Urasawa S (1990) Nucleotide sequence of VP4 and VP7 genes of human rotaviruses with subgroup I specificity and long RNA pattern: Implication for new G serotype specificity. J Virol 64: 5640–5644

Taniguchi K, Pongsuwanna Y, Choonthanom M, Urasawa S (1990) Nucleotide sequence of the VP7 gene of a bovine rotavirus (strain 61A) with different serotype specificity from serotype 6, Nucleic Acids Res 18: 4613

Rotavirus Protein Structure and Function

G.W. Both[1], A.R. Bellamy[2], and D.B. Mitchell[1]

[1] CSIRO Division of Biomolecular Engineering, P.O. Box 184, North Ryde, NSW 2113, Australia
[2] Centre for Gene Technology, School of Biological Sciences, University of Auckland, Private Bag 92019, Auckland, New Zealand

1 Introduction

It is now some 30 years since simian rotavirus particles were first described (MALHERBE et al. 1963) and 20 years since rotaviruses were first observed in humans and identified as a possible cause of acute diarrheal disease (BISHOP et al. 1973). Since that time we have gradually come to appreciate the staggering amount of morbidity and mortality caused by these viruses, especially in young children (KAPIKIAN and CHANOCK 1985). The recognition of the clinical importance of rotaviruses has stimulated basic research into their structure, gene regulation, and mode of replication.

The development of the techniques of genetic engineering, DNA sequencing, and monoclonal antibody production have also ensured that our knowledge of rotavirus has expanded rapidly. The genome sequence of the simian rotavirus SA11 has been completed and there are multiple sequences available for many gene segments, especially those coding for proteins of immunological significance. In addition, the structure of the virion has been determined to 40 Å resolution (PRASAD et al. 1988). Amongst eukaryotic viruses, perhaps only for HIV-1 has such an extensive basic knowledge been acquired so quickly.

In some cases the protein sequences inferred from the cDNA have yielded important clues as to the structure and function of the rotavirus proteins and, for some of these proteins, biochemical confirmation of their role has been obtained. However, in other cases, our knowledge still does not extend beyond the predicted sequence of the polypeptide. In this chapter, we summarize the information available for each gene product in relation to its role in virus structure, replication, and morphogenesis. A brief overview of these areas is presented to allow the discussion which follows to be read in context; for more detailed descriptions the reader is referred to Chaps. 2 and 5 of this volume.

2 Components of the Virus

Rotaviruses have a genome which consists of 11 segments of double-stranded-RNA (RODGER et al. 1975; NEWMAN et al. 1975) which code for six nonstructural and six structural polypeptides (SMITH et al. 1980; KALICA et al. 1981; McCRAE and McCORQUODALE, 1982; MASON et al. 1983; BOTH et al. 1983a; GREENBERG et al. 1983a; LIU et al. 1988) (Table 1). The arrangement of the structural proteins that comprise the outer layer of the single-stranded (ss) and double-stranded (ds) particles was recently determined by PRASAD et al. (1988). In this chapter we will refer to virions as double shelled particles (dsps); particles lacking the outer capsid layer will

Table 1. Rotavirus SA11 genes coding assignments[a]

Genomic segment	Length (bp)	Protein	ORF (aa)	Protein MW[b]	Function	Comments/function
1	3302	VP1	1088	125 128	Central core	Probable RNA polymerase component
2	2690	VP2	881	102 698	Central core	Binds RNA. Nucleocapsid protein?
3	2591	VP3	835	98 142	Central core	Exhibits guanylyltransferase activity
4	2362	VP4	776	86 775	Outer capsid	Hemagglutinin. Cleaved to yield VP5* VP8*
5	1611	NS53	495	58 484	Nonstructural	Possible metal binding domain
6	1356	VP6	397	44 903	Major protein in ssps	Subgroup specific antigen. Ligand for NS28
7	1104	NS34	312	36 072	Nonstructural	ORF starts at second AUG
8	1059	NS35	317	36 629	Nonstructural	–
9	1062	VP7 (GP)	326	37 198	Outer capsid or ER membrane	ORF starts at first AUG. Type-specific antigen
10	751	NS28 (GP)	175	20 309	Nonstructural glycoprotein in ER membrane	Acts as receptor for ssps
11	667	NS26	198	21 772	Nonstructural	Phosphorylated
		NS12	92	11 010	Nonstructural	

bp, base pairs; aa, amino acids; MW, molecular weight; ER, endoplasmic reticulum; ORF, open reading frame; ssps, single-shelled particles.
[a] Based on data derived from cloned SA11 rotavirus genes. Sequence references are: genes 1, 2 and 3, MITCHELL and BOTH (1990b); gene 4, MITCHELL and BOTH (1989); gene 5, MITCHELL and BOTH (1990a); gene 6, BOTH et al. (1984b); gene 7, BOTH et al. (1984a); gene 8, BOTH et al. (1982); gene 9, BOTH et al. (1983a); gene 10, BOTH et al. (1983c); gene 11, MITCHELL and BOTH (1988).
[b] For the unmodified polypeptide.

be referred to as single shelled particles (ssps). The inner particle, which also lacks VP6 (BICAN et al. 1982), will be called the central core. The latter particles are transcriptionally inactive, in contrast to intact ssp which will synthesize mRNA in vitro for long periods (COHEN 1977; MASON et al. 1980). Because the rotavirus mRNA and genomic dsRNAs are capped and methy-

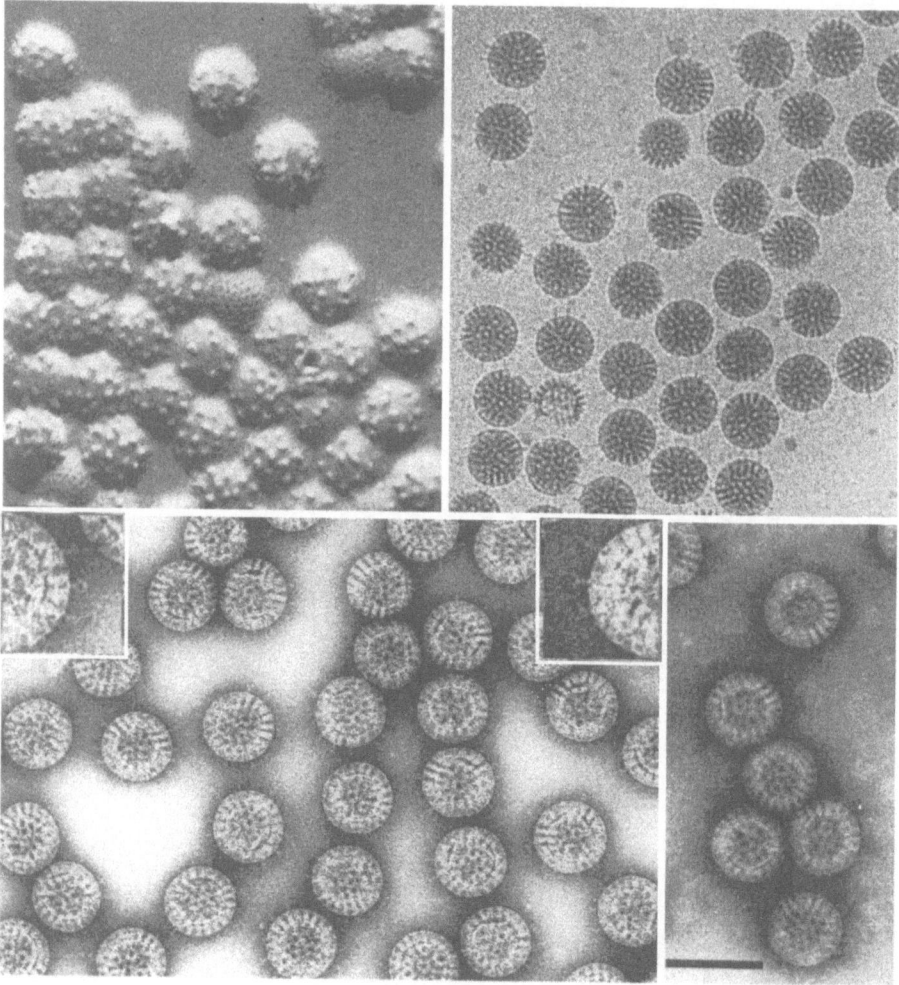

Fig. 1. Surface proteins of rotavirus observed using different preparative methods. *Top panels*: surface projections observed in a shadowed replica of a deep etched sample (*left*) and in the hydrated state with the virus in a thin layer of amorphous ice (*right*). *Bottom panels*: negatively stained preparations observed at low electron dose which reveals better detail of the spikes (*left*) and their characteristic "Y" and "lollipop"-shaped conformation (*inset*). Treatment at elevated pH removes the projections yielding smooth particles (*right*) and releases VP4 to the supernatant (see text). The *scale bar* is 1000 Å and *inserts* are 2x magnification. (Photomicrographs courtesy of C. Bailey and J. Berriman, Department of Cellular and Molecular Biology, University of Auckland)

lated (IMAI et al. 1983a; SPENCER and GARCIA 1984), the ssps are assumed to contain the enzymatic activities required to produce these modified mRNAs that is, the methylase and guanylate transferase activities, and the nucleotide triphosphohydrolases. These enzymes are present in reovirus cores (FURUICHI et al. 1975a, b), the prototype for the family *Reoviridae*, but their assignments to the ssps of rotavirus is largely inferential. It is widely assumed that these enzyme activities must be associated with the central core components VP1, VP2, or VP3 or perhaps with VP6, the protein which forms the outer capsomeres of the ssps.

The outer capsid of the dsp is composed of VP4 and VP7, the two major antigens of the virus. The dsp presents a relatively smooth surface, but when freeze-etched preparations of virus are shadowed with platinum or the virus is examined in vitreous ice, spikes are clearly visible on the surface (Fig. 1). These structures are also evident when the virus is examined by negative staining, provided that the conditions are carefully controlled. Under these conditions, the surface projections exhibit a "Y" and "lollipop" configuration (Fig. 1). PRASAD et al. (1988), when considering the volume occupied by the surface projections in their reconstruction, tentatively identified them as single molecules of VP4. More recently, further work by this group (PRASAD et al. 1990) and a separate investigation of Rhesus rotavirus (YEAGER et al. 1990) revealed a dimeric structure (see Sect. 7.1).

3 Replication and Morphogenesis

Members of the family *Reoviridae* multiply in the cytoplasm of the infected cell, but it is becoming increasingly clear that there are significant differences in the replication strategy adopted by the different members of the group. For example, rotaviruses probably differ from reoviruses since they appear to enter cells both by endocytosis and direct penetration of the membrane (SUZUKI et al. 1985, 1986); the latter may be the major route for productive infection (KALJOT et al. 1988). Both membrane penetration (KALJOT et al. 1988) and infectivity (ESTES et al.1981; CLARK et al. 1981; ESPEJO et al. 1981) are enhanced by cleavage of VP4. During entry, virions appear to be uncoated (FUKUHARA et al. 1988). This presumably activates the transcriptase which then produces ss mRNAs. Immunoelectron microscopy and immunofluorescence using monospecific polyclonal antisera (CHASEY 1980; PETRIE et al. 1982; WELCH et al. 1989) or monoclonal antibodies (PETRIE et al. 1984; PORUCHYNSKY et al. 1985; RICHARDSON et al. 1986; KABCENELL et al. 1988) has shown that proteins produced by translation of the viral mRNAs accumulate in viroplasmic inclusion bodies in the cytoplasm. However, the two glycoproteins, VP7 and NS28, are directed to the membrane of the endoplasmic reticulum (ER). The location of VP4 is less certain but it appears to accumulate near the ER membrane (KABCENELL et al. 1988). Electron microscopy has also revealed the presence of ssps in viroplasmic inclusions (PORUCHYNSKY et al. 1985; KABCENELL et al. 1988) and the initial stages of virus assembly probably occur at that site. However, the virus matures in the lumen of the ER and the process by which ssps leave the cytoplasm and enter the ER is a distinctive and unusual feature of rotavirus morphogenesis.

The ssps enter the lumen of the ER by budding across the membrane, a step which is mediated by the glycoprotein NS28, which acts as a receptor, and VP6, which serves as the ligand on the outer surface of ssps (AU et al. 1989 a, b; BERGMANN et al. 1989; MEYER et al. 1989). The budding particles acquire an ER membrane envelope during this process. However, mature virus is not enveloped and the mecahnisms by which the envelope is lost and VP7 is acquired are not understood. Since the two rotavirus glycoproteins VP7 and NS28 have been studied intensively at the molecular level, a large part of the following discussion will focus on their biology. Nevertheless, for clarity of presentation, we will deal in turn with the products of the 11 genome segments, beginning with the largest.

4 Gene Segment 1 (VP1)

VP1, encoded by gene segment 1, is a minor component of the central core and comprises about 2% of the mass of the virion (ESTES and COHEN 1990). This gene has now been sequenced for the bovine RF (COHEN et al. 1989) and SA11 strains (MITCHELL and BOTH 1990b). The sequences are highly conserved at the nucleotide and protein levels (86% and 96%, respectively) and for the SA11 gene a polypeptide of 1088 residues with a molecular weight of 125 128 is predicted (Table 1). This is in agreement with the apparent size of the protein as judged by SDS-PAGE (ERICSON et al. 1982). Rotavirus ssps contain only VP1, VP2, VP3, and VP6, but nevertheless are active both in transcription (COHEN 1977) and RNA capping and methylation (SPENCER and GARCIA 1984). COHEN et al. (1989) noted that the arrangement of certain amino acid residues in RNA-dependent RNA poly-merases of several plant and animal viruses (KAMER and ARGOS 1984; GORBALENYA and KOONIN 1988) are also conserved in the region between amino acids 517 and 636 of VP1. The equivalent proteins from reovirus (λ3) and bluetongue viruses (VP1) also share these homologies. A lysate from cells infected with a recombinant baculovirus expressing bluetongue virus gene segment 1 was shown to exhibit a (new?) poly-U-dependent ATP polymerase activity (URAKAWA et al. 1989). The lesion in a temperature-sensitive mutant with an RNA-negative phenotype also maps to gene segment 1 (GOMBOLD and RAMIG 1987). Thus, it seems likely that VP1 (and its equivalent in other members of the *Reoviridae*) may constitute at least part of the viral RNA polymerase complex. VP1 from the bovine RF strain has also been expressed as a full-length, non-fusion protein using a recombinant baculovirus (COHEN et al. 1989), but the protein was not tested for RNA polymerase activity. Expressed VP1 induced antibodies in guinea pigs but these possessed no neutralizing activity (COHEN et al. 1989).

5 Gene Segment 2 (VP2)

VP2 is encoded by gene segment 2 and is the most abundant protein in the rotavirus central core (BICAN et al. 1982). It may be exposed at the surface of the ssp since antibodies to VP2 were able to react with these particles (TANIGUCHI et al. 1986). The gene has now been sequenced for the bovine RF (KUMAR et al. 1989), human Wa (ERNST and DUHL 1989), and simian SA11 strains (MITCHELL and BOTH 1990b). The sequences are greater than 81% and 91% homologous at the nucleotide and protein levels, respectively and two separate "leucine zipper" motifs, hypothesized to be involved in the dimerization of nucleic acid binding proteins (LANDSCHULZ et al. 1988), are conserved between SA11 residues 535 and 556 and 666 and 687. These generally contain leucine at every seventh residue over eight turns of a hypothetical α-helix. However, the probability that these sequences actually form an α-helix is low (EISENBERG et al. 1984) and the existence of the helices is therefore doubtful. Nevertheless, VP2, is the only protein in the central core with demonstrated nucleic acid binding activity, although this was not sequence-specific (BOYLE and HOLMES 1985) and it has been suggested as a nucleocapsid protein (KUMAR et al. 1989). For the SA11 gene, a prominent α-helical region betwen residues 55 and 89, which could potentially be stabilized by ion-pair formation, may be involved in RNA binding or in protein–protein interactions (ERNST and DUHL 1989). VP2 is also thought to be myristylated (CLARK and DESSELBERGER 1988), although probably not NH2-terminally, since it lacks the NH2-terminal consensus sequence Gly-X-X-X-Ser/Thr present in many other myristylated proteins (CHOW et al. 1987). Many other myristylated proteins associate with membranes (HENDERSON et al. 1983; STREULI and GRIFFIN 1987) and modified VP2 might therefore provide a membrane anchor point for virus assembly (MUSALEM and ESPEJO 1985).

6 Gene Segment 3 (VP3)

VP3, the product of gene segment 3, was described only recently because it proved difficult to resolve from other viral proteins by PAGE (LIU et al. 1988). An earlier study suggested that segment 3 encoded a protein of molecular weight 91 000 which was a component of the central core (DYALL-SMITH and HOLMES 1981) and this has now proved to be the case. VP3 is a minor component of the central core, and is approximately one third as abundant as VP1 (LIU et al. 1988). VP3 is synthesized at low levels in the infected cell and poorly translated in vitro. Segment 3 of SA11 has now been sequenced (LIU and ESTES 1989; MITCHELL and BOTH 1990b), completing the sequence of the entire SA11 genome. The amino acid sequence shows scattered short regions of homology with several viral polymerases (LIU and ESTES 1989) and the arrangement of key residues in a "Polymerase module" (POCH et al. 1989) described for RNA polymerases from diverse sources is also found in VP3 (MITCHELL and BOTH 1990b). Temperature-sensitive mutants that map to gene 3 are defective in RNA synthesis (GOMBOLD and RAMIG 1987) and VP3 can detected in early replication complexes (GALLEGOS and PATTON 1989). Further suggesting that the protein may be involved in RNA synthesis and replication.

7 Gene Segment 4 (VP4): The Viral Hemagglutinin

7.1 Structure, Sequence, and Cleavage Site

VP4, the product of gene segment 4 was formerly known as VP3 (ESTES et al. 1983) but was renamed following the identification of the product of gene segment 3 as a structural protein (LIU et al. 1988). VP4 was defined as the viral hemagglutinin when the ability of some rotavirus strains to agglutinate red cells was shown to segregate with this gene segment (KALICA et al. 1983; ESTES et al. 1983). Although it is the minor component of the outer capsid of the virus, VP4 is an important determinant of virulence and growth in cell culture (GREENBERG et al. 1983a; OFFIT et al. 1986) and an important neutralizing antigen (HOSHINO et al. 1985; reviewed in MATSUI et al. 1989) (see also Chap. 7). The protein also contributes in large part to the altered growth characteristics of SA11 strain 4F. This is a more stable variant of the virus which produces larger plaques and grows to higher titer than the original SA11 isolate (PEREIRA et al. 1984; BURNS et al. 1989).

Considerable effort has been devoted to determining the structure of the VP4 gene and protein. This has now been determined for ten human (KANTHARIDIS et al. 1987; GORZIGLIA et al. 1988a; TANIGUCHI et al. 1988, 1989), three bovine (POTTER et al. 1987; KANTHARIDIS et al. 1988; NISHIKAWA et al. 1988), one porcine (NISHIKAWA and GORZIGLIA 1988), and three monkey isolates (LOPEZ et al. 1985; LOPEZ and ARIAS 1987; NISHIKAWA et al. 1988; MACKOW et al. 1988b; MITCHELL and BOTH 1988). In most cases the protein from human isolates is 775 residues in length while VP4 from animal-derived strains is one amino acid longer. However, VP4 from the human strain K8 differs in that it has a single amino acid insertion after residue

135 and a single deletion at residue 575, relative to the other human strains. The K8 isolate also shares only 60%–70% amino acid homology with other human rotaviruses and appears to represent a separate VP4 type (TANIGUCHI et al. 1989). There is no signal peptide at the NH2-terminal of VP4 which would direct it into the ER and, consistent with this observation, the protein is not glycosylated. It is presumed that VP4 is incorporated into the virus prior to budding and is carried into the ER in association with the ssp.

The arrangement of VP4 on the surface of the virus has been clarified somewhat by electron microscopic studies undertaken on virus embedded in amorphous ice (PRASAD et al. 1988, 1990; YEAGER et al. 1990). The initial reconstructions showed projections 35 Å in width which extended 45 Å beyond the surface of the virus. On the basis of molecular volume calcutions, it was concluded that a single VP4 molecule constitutes each of the 60 "spikes" (PRASAD et al. 1988). Recent work, using noise reduction methods, has produced reconstructions of higher resolution. They show that the projections extend to between 100 and 120 Å and have a distinct bilobed distal portion (Fig. 2a). Fab fragments from a monoclonal antibody directed against VP4 were used to decorate the virus (PRASAD et al. 1990). The reconstruction of the Fab-complexed virus showed that each projection exposed two epitopes, implying a dimeric structure. The larger molecular volume of the spike was also consistent with the projections being composed of a dimer of VP4.

Direct identification of the spike as VP4 is confirmed by the results obtained when the dsp is treated with alkali (Fig. 1). This treatment removes the projections and yields a smooth particle which retains VP7 but lacks VP4 (ANTHONY et al. 1991). Since alkali treatment removes the projections and yields an intact particle that lacks VP4 and VP5* it is reasonable to conclude that the surface projections indeed are composed of VP4.

Negatively stained images of the virus (Fig. 1) indicate that the surface projections are often in a Y or lollipop configuration, a finding which could possibly be due to the dimeric nature of VP4. There are some similarities between the appearance of these structures in the electron microscope and the appearance of the reovirus hemagglutinin which is thought to be a tetramer composed of two dimers of protein σ1 (LEE et al. 1981; BASSEL-DUBY et al. 1987). The bifurcated appearance of the rotavirus VP4 projections therefore probably is due to the dimeric nature of the structure: The appearance of the spike may also be related in some way to the proteolytic cleavage event which converts VP4 to VP5* and VP8* (see below). The identification of VP4 as the protein forming the rotavirus spike is consistent with the situation found for other viruses possessing surface projections (e.g., reovirus and influenza virus): in these viruses, oligomeric surface projections also function in both virus attachment and hemagglutination (LEE et al. 1981; MASRI et al. 1986; FURLONG et al. 1988; WEBSTER et al. 1982).

Proteolytic cleavage of the viral form of VP4 (88 kDa) yields VP5* (60 kDa) and VP8* (28 kDa) and this cleavage event greatly enhances

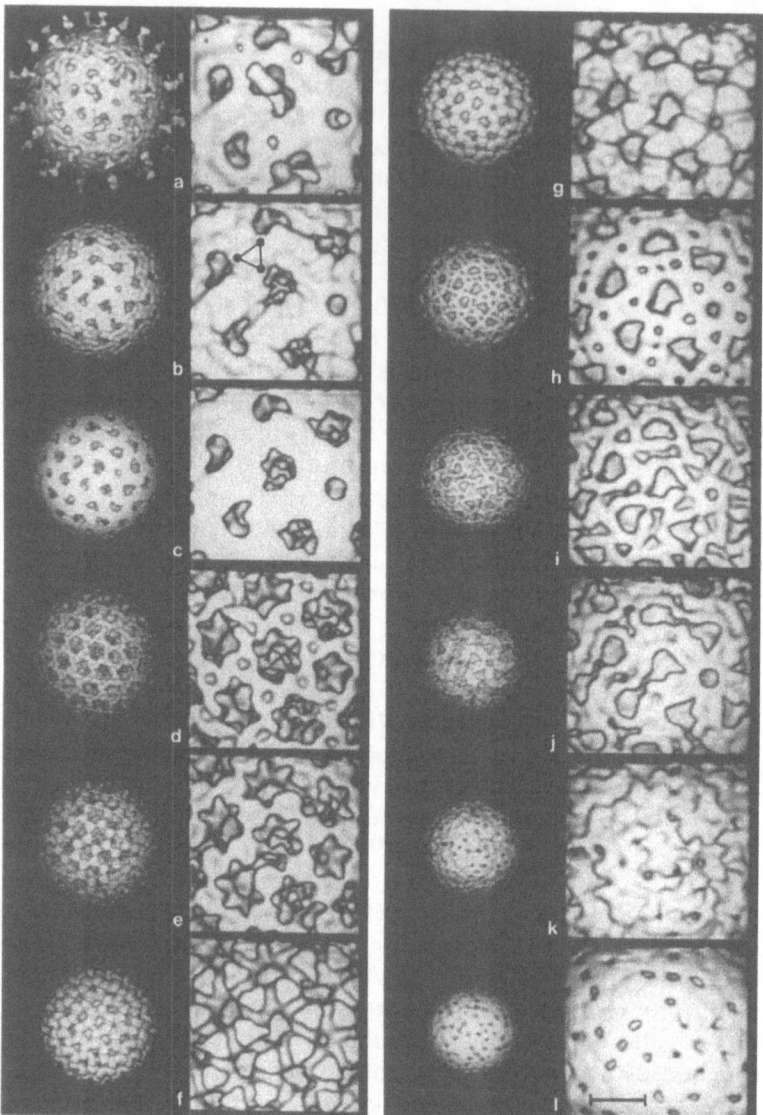

Fig. 2a–l. Surface-shaded representations of Rhesus rotavirus reconstructions viewed at successively smaller radii, obtained by truncating the three-dimensional maps with spherical envelopes to reveal the internal structure. All surface views are from an equatorial direction, close to the axis of a larger (type II) hole and have been calculated at the same threshold density. All views except (**a**) are from a 26Å resolution map. Fine details are revealed in the enlarged central region of each full view. One *triangular motif*, presumed to be a trimeric aggregate of VP7 molecules situated in the outer capsid, is schematically labeled in (**b**). The radii corresponding to the different levels are : **b**, 393Å; **c**, 377Å; **d**, 365Å; **e**, 353Å; **f**, 335Å; **g**, 324Å; **h**, 312Å; **i**, 300Å; **j**, 288Å; **k**, 271 Å; and **l**, 259Å. *Scale bar* for close-up views [shown in l] is 100Å. (Plate kindly supplied by Dr M. YEAGER (Fig. 10 from YEAGER et al. 1990) and reproduced with permission of The Journal of Cell Biology.)

infectivity in vitro (CLARK et al. 1981; ESPEJO et al. 1981; ESTES et al. 1981). However, the NH_2-terminal of viral VP4 and its cleavage product VP8* was found to be blocked and not amenable to sequencing by Edman degradation, indicating that VP8* constituted the NH_2-terminal portion of the primary translation product and VP5* the COOH-terminal part (LOPEZ et al. 1985). Partial NH_2-terminal amino acid sequences were determined for two species of VP5* and these were in almost complete agreement with the amino acid sequence deduced for the gene 4 sequence from isolate SA11fm (LOPEZ et al. 1985). This suggested that trypsin cleaves the protein at residues Arg-241 and Arg-247, i.e., six residues are probably removed during trypsin activated cleavage (LOPEZ et al. 1985) Arg-241 is conserved in all VP4 sequences determined to date and Arg-247 varies in only three bovine isolates. In these cases, however, an alternative Arg occurs at residue 246 (LOPEZ et al. 1985; KANTHARIDIS et al. 1988). Considerable amino acid variation occurs between and adjacent to the cleavage sites (LOPEZ et al. 1986; GORZIGLIA et al. 1986b, 1988a; NISHIKAWA et al. 1988; KANTHARIDIS et al. 1988; TANIGUCHI et al. 1989) and this perhaps influences rotavirus virulence in much the same manner as is known to occur for some strains of influenza virus (KLENK and ROTT 1988).

7.2 VP4 as the Hemagglutinin

It appears that rotaviruses can be taken up by two routes, one involving endocytosis, the other utilizing direct penetration of the membrance (PETRIE et al. 1982; QUANN and DOANE 1983; SUZUKI et al. 1985, 1986; FUKUHARA et al. 1987, 1988; KALJOT et al. 1988) and these pathways have very different kinetics (KALJOT et al. 1988). However, the addition of lysosomotropic agents had little effect on the yield of virus (FUKUHARA et al. 1987; LUDERT et al. 1987; KALJOT et al. 1988), suggesting that entry via endocytosis and the subsequent acidification of endosomes is not an important route for virus infection. The interpretation of experiments aimed at investigating mechanisms of viral entry is fraught with difficulties, but these results raise the possibility that if there are two routes of entry for the virus, there may also be two modes of cell attachment. VP7 has been shown to bind to the surface of MA 104 cells under conditions which promote virus adsorption and, on this basis, it has been proposed as the cell attachment protein (SABARA et al. 1985; FUKUHARA et al. 1988). However, these observations do not necessarily preclude a role for VP4 in virus attachment and productive infection; a role for VP4 in cell attachment would be consistent with its identification as the hemagglutinin. The involvement of VP4 in virus attachment leading to productive infection would also be consistent with the observed restriction of growth of certain rotavirus strains in tissue culture and in mice, a property which segregates with gene 4 (GREENBERG et al. 1983a; OFFIT and BLAVATT 1986; OFFIT et al. 1986) and with the location of several neutralizing epitopes on VP8* (MACKOW et al. 1988b; TANIGUCHI et al. 1988), especially if it were to be shown that antibodies to these sites could prevent virus adsorption. In addition, it appears that similar sialic acid structures are required for hemagglutination of erythrocytes by rotavirus strains SA11 and NCDV and for the adsorption of these viruses to cells, a finding consistent with the involvement of VP4 in both processes (FUKUDOME et al. 1989).

Penetration of rotavirus into cells, but not adsorption, is enhanced by cleavage of VP4 (CLARK et al. 1981; FUKUHARA et al. 1988; KALJOT et al. 1988). Similarly, cleavage of the cell attachment protein enhances infectivity of the myxo- and paramyxoviruses, but for these membrane-enveloped viruses cleavage generates a hydrophobic NH_2-terminal on the COOH-terminal cleavage fragment. This region mediates fusion between the viral envelope and the endosomal membrane following virus uptake by endocytosis (GETHING et al. 1976; DANIELS et al. 1985). Cleavage of VP4 does not produce a hydrophobic region at the NH_2-terminal of VP5*, and this is consistent with the observation that uptake via endocytosis does not lead to a productive rotavirus infection. Thus, rotavirus particles may penetrate the cell membrane directly (KALJOT et al. 1988) and sequences other than the NH_2-terminal of VP5* may be involved. Indeed, a sequence showing homology with the proposed fusion domain of Sindbis and Semliki Forest viruses is present between residues 137 and 154 of VP5* and a neutralizing epitope also maps to this region (MACKOW et al. 1988b). FUKUHARA et al. (1988) have also shown that components of the inner capsid are selectively taken up following cleavage of VP4 with trypsin, whereas without trypsin activation all the viral proteins enter the cell. Thus, rotavirus particles may enter the cells by a mechanism involving direct penetration of the membrane and shedding of the outer capsid layer, but the mechanism is yet to be detailed.

Clones of VP4 genes from several isolates are now available (POTTER et al. 1987; MACKOW et al. 1989; MITCHELL and BOTH 1989) and these will facilitate studies of the biology and antigenicity of VP4. For example, VP4 from rhesus rotavirus has been expressed in insect cells using a recombinant baculovirus (MACKOW et al. 1989). About 5% of the total protein in the cells was VP4 and this was antigenically identical to viral VP4, as judged by its reactivity with a panel of neutralizing monoclonal antibodies. Expressed VP4 was also able to bind erythrocytes, confirming its role as the hemagglutinin. Thus, VP4 expressed in the baculovirus system appears to maintain its structural and functional integrity, although it has not been determined whether the protein can form dimers in vitro. Hemagglutination was inhibited by hyperimmune serum, VP4-specific monoclonal antibodies, and by glycophorin (MACKOW et al. 1989). Glycophorin A also inhibited hemagglutination of human erythrocytes by SA11 and NCDV virus particles (FUKUDOME et al. 1989). These findings therefore suggest that this protein might constitute the erythrocyte receptor for rotaviruses. Large-scale expression of this gene has the potential to yield amounts of biologically-active protein sufficient for structural studies to be undertaken.

8 Gene Segment 5 (NS53)

The nonstructural protein NS53, coded for by gene segment 5, is produced at low levels in the cell (ERICSON et al. 1982) but can be detected early during infection in particles which have replicase activity and appear to be precursors of the ssps (PATTON et al. 1986; GALLEGOS and PATTON 1989). Segment 5 from the bovine RF strain (BREMONT et al. 1987) and simian SA11 strains (MITCHELL and BOTH 1990a) has now been sequenced. In agreement with earlier studies, which revealed great genetic diversity between gene 5 segments from certain rotavirus strains (SCHROEDER et al. 1982), the RF and SA11 genes are only 49% and 36% homologous at the gene and protein levels, respectively. This represents a very low level of conservation compared with other rotavirus nonstructural proteins (MITCHELL and BOTH 1990a). Despite this diversity, amino acids 37–79 are largely conserved between the bovine and simian proteins, suggesting that they may be functionally important. The region contains numerous cysteine and histidine residues which can be made to fit the general motif for a metal binding domain (BERG 1986) and, depending on precisely which residues are used, one or two "zinc fingers" could be formed. Cysteine residues 325 and 328 are also conserved and could be involved in metal ion binding if brought into juxtaposition by protein folding. Based on preliminary experiments, NS53 appears capable of binding zinc (ESTES and COHEN 1990) and proteins containing zinc fingers are known to bind nucleic acids (MILLER et al. 1985). This suggests a role for NS53 in binding RNA at some stage during infection, perhaps during selection or packaging of RNA segments. However, an earlier study which examined the RNA binding ability of rotavirus proteins failed to detect binding with NS53 (BOYLE and HOLMES 1985), but this may have been due to the low level of expression of the protein. NS53 has now been expressed in *E. coli* (BREMONT et al. 1987) and in the baculovirus system (ESTES and COHEN 1990) and this may assist studies to assess the RNA binding ability of the protein. Given the appearance of the protein in early replication intermediates (GALLEGOS and PATTON 1989), demonstration of a capacity of the protein to bind ss- or dsRNA would provide an important clue as to its function.

9 Gene Segment 6 (VP6)

VP6, a 42 kDa polypeptide, is the major structural protein of ssp (PRASAD et al. 1988) and the subgroup-specific antigen (GREENBERG et al. 1983b). In addition, VP6 is the protein which acts as the ligand in the interaction with NS28 during the early events that precede the transfer of ssps across the rough ER (RER) membrane (AU et al. 1989a; MEYER et al. 1989). Thus, VP6 plays important structural, immunological and morphogenic roles in the infectious process. These will be considered in turn.

9.1 VP6 Structure

The use of the term rota (Latin), meaning wheel, to describe this group of viruses (FLEWETT et al. 1974) derived from the characteristic arrangement of the capsomeres that comprise the surface lattice of the ssp (PRASAD et al. 1988). There are several lines of evidence which support the conclusion that the prominent surface projections of ssps are composed of VP6, but perhaps the most compelling is the ability of purified forms of VP6 to form tubular and sheet structures that exhibit an almost identical arrangement to that observed in the ssp.

CHASEY and LABRAM (1983) were among the first to describe the hexagonal packing present in the sheet and tubular forms of rotavirus proteins found in fecal material. The presence of these structures in this intractable material is something of a tribute to the stability and protease-resistance of the trimeric form of VP6. The protein composition of the tubular forms of VP6 was clarified by the observation that VP6 could be prepared from ssps by treatment with calcium chloride (BICAN et al. 1982) or lithium chloride (READY and SABARA 1987) and that the protein in this form also showed a tendency to form extended tubular forms when the high concentrations of $CaCl_2$ used in preparation were removed by dialysis. Similar observations were reported by ESTES et al. (1987), who expressed a cloned copy of the VP6 gene in *Spodoptera frugiperda* cells using a recombinant baculovirus. Tubular forms of VP6 were observed that were indistinguishable from those formed when the viral form of the protein was allowed to reassemble in vitro.

READY et al. (1988) processed images of the tubular forms of VP6. This yielded a hexagonal lattice but the resolution achieved in the optical reconstruction was insufficient to yield details of the arrangement of the trimers of VP6. However these authors calculated the volume of the trimer and deduced that it probably was an elongated structure with a diameter of the order of 55 Å. This dimension was consistent with the first description of the inner capsid protein as a "trimeric truncated cone" (ESPARZA and GIL 1978).

The two-dimensional paracrystalline arrays found in preparations of purified VP6 enable the trimeric arrangement of VP6 to be observed when the images are further processed (Fig. 3). This negatively stained image has sufficient resolution to reveal the right-handed asymmetric arrangement of VP6 molecules in the trimer. This view is directly comparable with the reconstruction of the virus at the level of the inner capsid, where the trimers of VP6 are also evident but at lower resolution (Fig. 2g, YEAGER et al. 1990). In the reconstruction of the hexagonal array (Fig. 2) the trimers enclose a 20 Å channel. This feature can also be seen deeper within the virus in Fig. 2h.

There is also good biochemical evidence that the basic assembly unit of VP6 is the trimer. GORZIGLIA et al. (1985) and SABARA et al. (1987) demonstrated that treatment of the ssp with SDS at a reduced temperature (65 °C) prior to electrophoresis on polyacrylamide gels yielded the trimeric form of

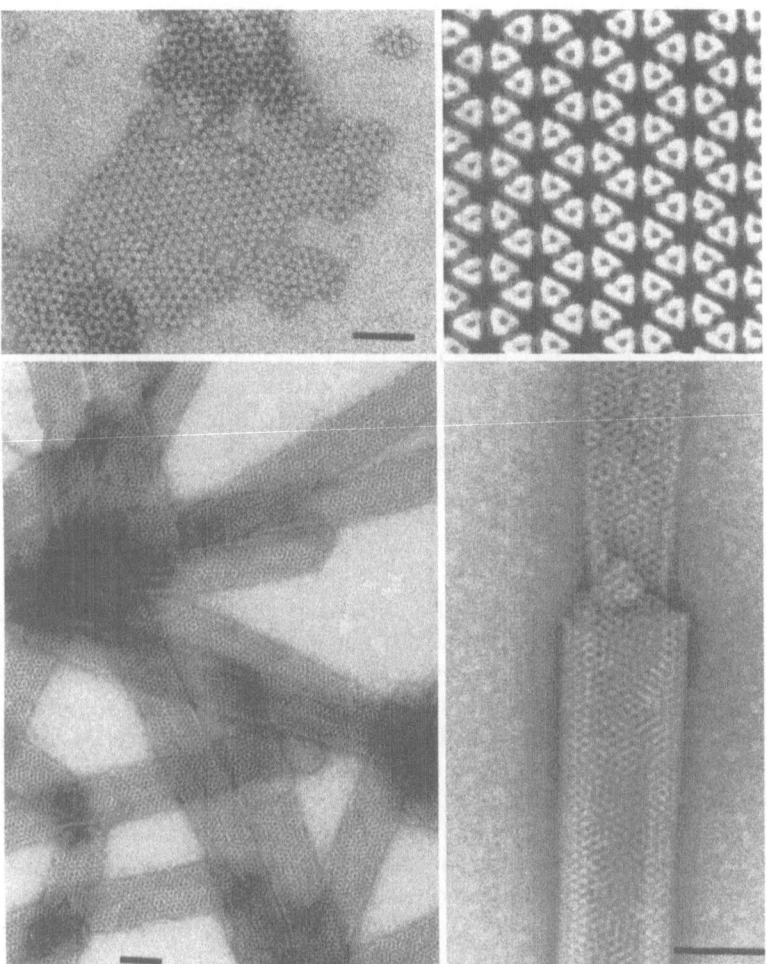

Fig. 3. Two-dimensional crystalline tubes of VP6. Negatively stained preparations show some regions where the 600 Å diameter tubes have broken (*top left*) to yield small areas suitable for image analysis (*top right*). The 100 Å hexagonal lattice (protein shown in *white*) shows centers of mass in which the individual molecules of VP6 can be seen asymmetrically distributed around the basic trimer motif. Tubes at *lower left* are from VP6 purified from virus and display a pattern that is a moiré of the two sides of the tube. VP6 expressed in the baculovirus system forms similar tubes (*lower right*), in some cases multilayered. The *scale bar* is 500 Å. (Photomicrographs courtesy of Dr. J. Berriman, School of Biological Sciences, University of Auckland).

the protein. Indeed it has also been demonstrated (GORZIGLIA et al. 1988b) that a subgroup-specific monoclonal antibody, 255/60 (GREENBERG et al. 1983b), specific for subgroup 1 viruses, recognizes only the trimeric form of the protein, implying that the epitope recognized by this monoclonal antibody is composed of more than one polypeptide chain.

Thus, with respect to the construction of the surface lattice of the ssp, the basic unit is the trimeric form of VP6, and these in turn are arranged as hexamers in the form dictated by the principles of icosahedral symmetry (CASPER and KLUG 1962). The central channel enclosed by the trimeric cluster is roughly 20 Å in diameter (Fig. 2) and the larger channels enclosed by the hexameric arrangement of the trimers are approximately 60 Å in diameter. In the three-dimensional reconstruction of the ssp (PRASAD et al. 1988, and Chap. 2, this volume) the packing of the hexameric units of VP6 leaves large channels between the inner and outer layers of capsomeres. These channels penetrating the ssp provide a potential route for the entry of nucleoside triphosphates and for the egress of mRNA following transcription. It is also attractive to speculate that the binding of VP6 to the NS28 receptor might involve the central 20 Å channel, which is roughly of the correct dimensions to accommodate a tetrameric aggregate of NS28 (see below).

9.2 VP6 as an Antigen

VP6 is also the protein involved in the subgroup-specific immunological response to rotaviruses (reviewed in MATSUI et al. 1989). The identification of this protein as the subgroup antigen was first established unequivocally by GREENBERG et al. (1983b), who demonstrated that subgroup specificity segregated with genomic segment 6. The sequence of VP6 has been determined for a number of different rotavirus strains representing both major subgroups (BOTH et al. 1984b; ESTES et al. 1984; HOFER et al. 1987; GORZIGLIA et al. 1988b). Although amino acid comparisons reveal a number of amino acid differences, these cannot be interpreted at present in any predictive fashion due to the absence of high resolution structural information.

VP6 is a highly immunogenic protein and an epitope on VP6 enables most (GREENBERG et al. 1983b), but not all (SVENSSON et al. 1988), group A rotaviruses to be separated into one of two subgroups. The traditional method of mapping epitopes on viral proteins, propagation of the virus in the presence of neutralizing monoclonals followed by sequence analysis of variants that escape neutralization (WEBSTER et al. 1982; DYALL-SMITH et al. 1986), is difficult to apply to VP6 because it is an inner protein that is not readily accessible to antibody. Monoclonal antibodies directed against VP6 therefore are nonneutralizing and of little value for selecting escape mutants. The recent success in achieving crystallization of the ssp (B. Harris, I. Anthony, A.R. Bellamy and S. Harrison, unpublished results) may eventually provide crystallographic information on which a high resolution structure could be based. Alternatively, the relative ease with which VP6 may be expressed and purified may lead to the crystallization and determination of the three-dimensional structure for this protein.

9.3 VP6 as a Ligand

The third important feature of VP6 is its role as the ligand in the interaction that occurs between the ssp and NS28, the receptor protein. This interaction was first described by Au et al. (1989a) and the essential feature of the receptor: ligand interaction was clarified in greater detail by MEYER et al. (1989) and Au et al. (1989b). The nature of the interaction that occurs between the ligand, VP6, and the receptor, NS 28, will be discussed in greater detail below. However it is relevant to consider here what is known of the binding site on the ligand. The only substantive piece of information available at present is that this binding site appears to be conserved across different rotavirus strains. MEYER et al. (1989) demonstrated that ssps of strain Wa (subgroup 2) were able to interact with NS 28 of strain SA11 (subgroup 1), and it has been shown subsequently that the NS28 receptor of a bovine strain (NCDV) is also able to interact with VP6 of strain SA11 (Meyer, Street, Bergman and Bellamy, unpublished observation). The ssps of these different strains also are able to competitively inhibit the binding of VP6 and NS28. Taken collectively, this information suggests that despite the significant sequence variation found between different VP6 proteins, they probably share a conserved binding site. Since a monoclonal antibody specific for the major epitope on VP6 is able to block the binding to ssps to membranes that contain NS28, the ligand binding site is probably in close proximity to the epitope recognized by this antibody.

VP6 is presented on the ssp as a trimeric aggregate, which in turn is arranged into hexamers (PRASAD et al. 1988; Fig. 1) and it is attractive to speculate that the binding of VP6 to the receptor involves the central 2 nm channel. This channel is roughly of the correct dimensions to accommodate a tetrameric aggregate of NS28 (see below). The purified form of VP6 also is able to competitively inhibit the binding of the ssp to NS28 (MEYER et al. 1989) indicating that the trimeric form of VP6 retains the binding site. However, further definition of the ligand binding site on VP6 awaits the determination of the tertiary and quaternary structure of the protein.

10 Gene Segment 7 (NS34)

NS34 is a nonstructural polypeptide encoded by gene segment 7 of SA11, but the segment number can vary due to electropherotype variation between strains (DYALL-SMITH et al. 1983a). Little is known about this protein or its function, although the gene sequence from several strains has been determined (BOTH et al. 1984a; WARD et al. 1984; RUSHLOW et al. 1988). No temperature-sensitive mutations have so far been mapped to this gene segment (GOMBOLD and RAMIG 1987). The equivalent protein from a bovine isolate was found to bind ss- and dsRNA equally well in a manner which was neither nucleotide sequence-nor rotavirus RNA-specific (BOYLE and HOLMES 1985). NS34 has also been found in the infected cell in association with complexes containing replicase activity (HELMBERGER-JONES and PATTON 1986; PATTON and GALLEGOS 1988). However, it precise role in replication remains to be determined.

11 Gene Segment 8 (NS35)

Segment 8 of SA11 codes for the nonstructural protein NS35, which, from imunocytochemical studies, is located in the viroplasms of infected cells (PETRIE et al. 1984). The sequence of the gene and protein, which has been determined for three strains (BOTH et al. 1982; DYALL-SMITH et al. 1983b; RUSHLOW et al. 1988), has yielded little information concerning the role of NS35 during infection. Viruses carrying a temperature-sensitive mutation in the gene have an RNA-negative phenotype (RAMIG 1983; RAMIG and FIELDS 1983; GOMBOLD et al. 1985) and produce a large proportion of empty particles (RAMIG and PETRIE 1984), suggesting a role for the protein in the replication of ssRNA or its packaging into subviral particles. NS35, together with NS34, has also been found in the infected cell in association with complexes containing replicase activity (HELMBERGER-JONES and PATTON 1986; PATTON and GALLEGOS 1988), but no RNA binding activity was detected for the protein (BOYLE and HOLMES 1985).

12 Gene Segment 9 (VP7): The Major Glycoprotein

Segment 9 of the SA11 genome codes for VP7 (BOTH et al. 1983a; MASON et al. 1983), but the electrophoretic mobility and hence the segment number can vary depending on the virus strain (DYALL-SMITH et al. 1983a). The structure of the VP7 gene was first determined for the SA11 strain (BOTH et al. 1983a) but the sequences of the equivalent genes from many other isolates are now known (ELLEMAN et al. 1983; ARIAS et al. 1984; DYALL-SMITH and HOLMES 1984; RICHARDSON et al. 1984; GUNN et al. 1985; GLASS et al. 1985; MASON et al. 1985; RUIZ et al. 1988; MACKOW et al. 1988a; GORZIGLIA et al. 1986a, 1988c; GREEN et al. 1989; REDDY et al. 1989. HUM et al. 1989; NISHIKAWA et al. 1989). These confirm the original gene and protein structure and show that, at the 3' end, there is complete conservation of eight bases between the various strains sequenced. Even more extensive sequence conservation exists at the 5' end. Among eight isolates, representing six serotypes, there are no changes in the first ten bases and only nine changes from the SA11 gene sequence are found in the different serotypes prior to position 72. However, this conserved RNA sequence overlaps the first of two, potential, in-frame initiation codons in the gene which are also conserved in all isolates (summarized in NISHIKAWA et al. 1989). The 5' proximal codon has a weak and the second codon a strong consensus initiation sequence (KOZAK 1984, 1986). Therefore, the conserved RNA sequence may not necessarily be maintained for the initiation of protein synthesis. Rather, it may contain signals involved in the packaging of the segment. The longest open reading frame of the gene (bases 49-1026) contains 326 codons.

12.1 Nature of the Carbohydrate

VP7 is the only structural glycoprotein present in the rotavirus particle and the site(s) at which it is glycosylated varies. For the SA11 and OSU strains, there is only a single glycosylation site, Asn-Ser-Thr at residues 69–71, and this site is common to all strains except NCDV (GUNN et al. 1985; GLASS et al. 1985) and the human McN strain (NISHIKAWA et al. 1989). For other strains there are alternative Asn-X-Ser/Thr sites at residues 145–147 (equine serotype III) and 146–148 (human serotype II), 238–140 (serotypes I and II, serotype III–mouse, feline and human strains only – serotypes IV, VI and VIII) and 318–320 (serotype VI); two sites are filled in the Wa (serotype I) and the bovine strain BDV 486 strains (KOUVELOS et al. 1984a, b).

In early investigations it was found that VP7 from SA11 could be radiolabeled with ^{3}H-mannose or ^{3}H-glucosamine (ARIAS et al. 1982; ERICSON et al. 1983). The carbohydrate was also sensitive to digestion with endoglycosidase H, indicating that it was of the high mannose type, a finding which was later confirmed by direct analysis of the oligosaccharides released from VP7 isolated from the bovine BDV 486 strain (KOUVELOS et al. 1984b) and the simian SA11 virus (BOTH et al. 1983b; KABCENELL and ATKINSON 1985; KABCENELL et al. 1988). For the latter, the original Glc3Man9GlcNac2 moiety was found to be processed predominantly to the Man8 and Man6 forms, but no complex carbohydrate was detected. This provided an important clue to the fate of VP7 with respect to protein transport in the infected cell. Since complex carbohydrate is only added to proteins in transit through the Golgi apparatus (HUBBARD and IVATT 1981), this result clearly indicated that VP7 never reaches this organnelle in the cell, but remains confined to the ER. Such a cellular location is entirely consistent with the site of assembly of the virus as deduced from electron microscopy.

12.2 Location of the VP7 Signal Peptide

Entry of proteins into the ER is mediated by a hydrophobic signal peptide which is often, but not always, cleaved from the nascent polypeptide during translocation (BLOBEL and DOBBERSTEIN 1975; WALTER and LINGAPPA 1986).

It was first observed by ERICSON et al. (1983) that VP7 carried a cleavable signal peptide and when the sequence became available (BOTH et al. 1983a) a hydrophobicity plot of the protein showed that two prominent regions of hydrophobic amino acids were present near the NH_2-terminal. Both regions had the characterisitics of signal peptides and each was preceded by an initiation codon (BOTH et al. 1983a; GUNN et al. 1985). These regions, referred to as the H1 and H2 domains (WHITFELD et al. 1987), are conserved in all VP7 sequences determined to date (NISHIKAWA et al. 1989) and their presence raises questions as to which region actually functions as the signal sequence and therefore where cleavage occurs. These questions were addressed by introducing mutations in the VP7 gene (PORUCHYNSKY et al. 1985; WHITFELD et al. 1987) Several mutations were constructed in which the regions coding for either the first, the second, or both hydrophobic domains were deleted and these genes were expressed in COS cells using an SV40-based expression vector. Signal peptide acitivity was monitored by whether or not the expressed protein was glycosylated. It was clear that the presence of the H1 or the H2 domain was sufficient to direct glycosylation of the protein, i.e., signal peptide function was present in both regions (PORUCHYNSKY et al. 1985; WHITFELD et al. 1987), but it was later shown that the H2 domain alone was sufficient to direct synthesis of authentic VP7 (STIRZAKER et al. 1987; PORUCHYNSKY and ATKINSON 1988). It is still not clear whether or not translation begins in vivo at the first (weak) or second (strong) initiation codon. Conceivably, a minor amount of a protein with a subtly different function could be made if initiation occurred at the first AUG.

12.3 The VP7 Signal Peptide Cleavage Site

VP7 purified from virus contains a blocked NH_2-terminal (ARIAS et al. 1984) which made it difficult to determine the site of signal peptide cleavage. Therefore, STIRZAKER et al. (1987) constructed modified forms of the VP7 gene in which initiation began specifically at the first or second AUG codons. The mutated genes were transcribed and translated in vitro in a reticulocyte lysate system in the presence of canine pancreatic microsomes to determine whether or not cleavage of the signal peptide occurred. The processed products produced in each case where identical in size, suggesting that processing occurred downstream of the second signal peptide domain, irrespective of whether one or both domains were present. A likely cleavage site between Ala-50 and Gln-51 was identified from empirical rules (VON HEIJNE 1986) and indirect evidence that this site was used was obtained by introducing a mutation designed to prevent signal peptide cleavage (STIRZAKER et al. 1987). The cleavage site was identified directly by obtaining partial amino acid sequence data from radiolabeled viral protein or VP7 translated in vitro. Both proteins were processed in vitro to produce Gln-51 as the major NH_2-terminal residue (STIRZAKER et al. 1987). These results are not consistent with an earlier study (CHEN et al. 1986) which suggested that the VP7 precursor was cleaved between the H1 and H2 domains. However, both studies showed that multiple (at least two) forms of VP7 were present in purified virus, although their origin remains to be determined.

Other studies have also shown that VP7, translated in vitro in the presence of microsomes, becomes resistant to digestion with trypsin, indicating that it is translocated completely into the lumen of the ER (KABCENELL and ATKINSON 1985; STIRZAKER et al. 1987). The processed form of the protein is also resistant to extraction with high salt and sodium carbonate and by this criterion VP7 becomes an integral membrane protein after entering the ER (KABCENELL and ATKINSON 1985). However, processing of the signal peptide removes the most prominent hydrophobic domains in the protein and therefore it is not obvious how the protein is anchored in the membrane (STIRZAKER et al. 1987). The need to recover VP7 from the membrane for assembly into virus particles may impose special requirements on the membrane-anchoring mechanism of the protein. This requirement might explain why there appears to be no obvious transmembrane anchor domain in the mature protein. The membrane bound form of VP7 can also be distinguished immunologically and biochemically from the polypeptide already incorporated into virus particles, indicating that there are two pools of the protein present in the cell during infection (KABCENELL et al. 1988). Based on the kinetics of processing of oligosaccharides attached to membrane bound vs viral VP7, it was suggested that the former is the precursor of the latter. Recent evidence also suggests that VP7 may form oligomeric structures both with itself and with NS28 and VP4 (MAASS and ATKINSON 1990).

12.4 Retention of VP7 in the Endoplasmic Reticulum

Proteins are synthesized in the cytoplasm of the cell and remain there by default unless they carry a signal peptide. Depending on its type, the signal peptide may direct the protein into one of several organelles. Most proteins which are directed into the ER by a signal peptide pass through that organelle en route to other intra- or extracellular locations, including the Golgi, the plasma membrane, and the medium. Evidence derived from the study of its attached Carbohydrate (BOTH et al. 1983b; KABCENELL and ATKINSON 1985; KABCENELL et al. 1988) and from immuno electron microscopy (PETRIE et al. 1984; PORUCHYNSKY et al. 1985; KABCENELL et al. 1988) indicates that VP7 does not leave the ER. Thus, the localization of VP7 is unusual and the protein is one of only a small number of membrane-associated proteins which localize to the ER. NS28 (see below) is another.

Considerable effort has been devoted to determining how VP7 is retained in the ER. Deletion studies with the cloned gene showed that removing amino acids 47–61 in the protein converted VP7 from a resident ER protein into a secreted polypeptide (PORUCHYNSKY et al. 1985). Because VP7 never normally traverses the secretory pathway, it seemed unlikely that the protein would carry a signal necessary for transport. Thus, this result suggested that the protein carries a positive signal for retention in the ER.

A comparison between two deletion mutants, 47–61 (the numbers identify the amino acids deleted), which was secreted, and 51–61, which was not, allowed the sequences involved in retention to be partially characterized. With the identification of the signal peptide cleavage site, it became clear that in mutant 51–61 the H2 signal peptide remained intact, while in deletion 47–61 the last four residues of the H2 region were removed. This suggested that the signal peptide had two functions: (1) targeting VP7 to the ER and (2) retaining it in that organelle. This was tested by replacing the H2 signal peptide with sequences coding for the signal peptide from influenza hemagglutinin, a membrane protein which normally is transported to the cell surface (GETHING and SAMBROOK 1981). This hybrid precursor gave rise to authentic VP7 which was secreted from the cells, indicating that the H2 signal peptide was required for both targeting and retention (STIRZAKER and BOTH 1989). Since the uncleaved VP7 precursor could not be detected in the cells, the signal peptide exerted its effect despite being rapidly cleaved from the protein.

Although the H2 signal peptide is required for retention of the protein in the ER it is not sufficient on its own. When the H2 sequences were spliced onto another, secreted, reporter molecule (the malaria S antigen) the modified molecule was still secreted (STIRZAKER and BOTH 1989), i.e., sequences in mature VP7 were also required for retention of the protein. In another study, in which α-amylase was used as a reporter molecule, it was concluded that two regions of VP7 (one of which included residues 62–111) were necessary, but each alone was insufficient for retention (PORUCHYNSKY and ATKINSON 1988). However, both of these were present in the mature secreted form of VP7 produced form a hybrid precursor with a foreign signal peptide (STIRZAKER and BOTH 1989). Collectively therefore the results indicate that the H2 signal peptide together with residues 62–111 is required for retention of VP7 in the ER.

12.5 Is VP7 the Cell Attachment Protein ?

On the basis that VP7 adheres to the surface of MA104 cells, it has been suggested that the protein is the cell attachment polypeptide of the virus (SABARA et al. 1985; FUKUHARA et al. 1988). Radiolabeled lysates were prepared from rotavirus infected cells. These were clarified by high speed centrifugation and layered onto intact MA104 cells under conditions normally used for virus adsorption in vitro. The free form of VP7 (presumably released from the membrane at some stage during infection) bound to an unidentified cell surface component (SABARA et al. 1985; FUKUHARA et al.

1988). Binding of VP7, or adsorption of virus particles to cells, was inhibited by the addition of antiviral serum (SABARA et al. 1985) or VP7-specific neutralizing monoclonal antibodies (SABARA et al. 1985; FUKUHARA et al. 1988). Binding of the protein was also competitively inhibited by the addition of increasing amounts of unlabeled, intact, homologous virus particles (SABARA et al. 1985), whether treated with trypsin or not (FUKUHARA et al. 1988). A monoclonal antibody to VP4 also inhibited virus adsorption to a lesser extent, but a vp6-specific antibody and a non-neutralizing VP7 monoclonal had no effect (SABARA et al. 1985). Thus, VP7 was proposed as the cell attachment protein of the virus. However this conclusion is difficult to reconcile with the fact that, for many viruses, the hemagglutinating protein is also the cell attachment protein and VP4 would therefore be expected to perform this role (KALICA et al. 1983). In addition, for other viruses (e.g., reovirus and influenza) the cell attachment protein forms a protruding spike (LEE et al. 1981; FURLONG et al. 1988; WEBSTER et al. 1982) and VP4 almost certainly constitutes the spike protein on the surface of the rotaviruses (see discussion above and PRASAD et al. 1990). If it is confirmed that infection occurs following attachment via VP4, how then might the predominance of neutralizing antibodies directed against VP7 be explained? At least some of these inhibit hemagglutinating activity of the virus and therefore they might act by interfering indirectly with VP4-mediated cell attachment (GREENBERG et al. 1983c). Alternatively, it is conceivable that VP7 neutralizing antibodies might interfere predominantly with the removal of the outer capsid which appears to occur during membrane penetration (FUKUHARA et al. 1988), or that the two proteins form a combined domain on the surface of the virus.

13 Gene Segment 10 (NS28)

Assignment of the NS28 glycoprotein as the product of gene segment 10 was achieved by direct translation of the denatured genome segment (McCRAE and McCORQUODALE 1982) or mRNA (MASON et al. 1983) or by using a cloned copy of the gene to hybrid select transcripts able to direct the synthesis of NS28 in vitro (BOTH et al. 1983c). In the same study, it was also demonstrated that, in the presence of canine pancreatic microsomes, the protein was glycosylated with high mannose carbohydrate. However, the signal peptide was not cleaved, since the glycosylation sites, located at residues 8 and 18, were retained and filled.

The amino acid sequence for a wide range of NS28 proteins has been determined from cloned cDNA transcripts of segment 10 (BOTH et al. 1983c; OKADA et al. 1984; BAYBUTT and McCRAE 1984; WARD et al. 1985; POWELL et al. 1988). All proteins are 175 amino acids long and initiate from the first of three in-frame AUG's which possesses a strong consensus sequence for initiation (KOZAK 1984, 1986). Three hydrophobic regions (H1–H3) are located at the NH$_2$-terminal of the molecule. Since the carbohydrate present on the molecule is exclusively of the high mannose type (BOTH et al. 1983c; KABCENELL and ATKINSON 1985), the protein clearly does not reach the Golgi apparatus but rather must be localized to the RER, as is the case for the other rotavirus glycoprotein, VP7.

13.1 Membrane Insertion

The insertion of NS28 in the ER membrane was investigated by ERICSON et al. (1983), who found that the protein inserted into microsomes but was not cleaved and remained sensitive to trypsin digestion. KABCENELL and ATKINSON (1985) confirmed that the protein had a transmembrane orientation with N-linked glycosylation occurring at the luminal face of the membrane while the bulk of the protein remained protease-sensitive and therefore on the cytoplasmic side.

No information is available concerning the nature of the signal(s) which enables NS28 to be retained in the RER. However several lines of evidence indicate that the second hydrophobic domain spans the membrane (BERGMANN et al. 1989). Deletion of H2 abolished the ability of the protein to be correctly inserted and glycosylated and the size of the fragment resistant to protease digestion indicated that the protein emerged from the membrane near amino acid 42 (BERGMANN et al. 1989). Therefore, approximately 132 amino acids are exposed to the cytoplasmic side of the membrane, with the H3 region (amino acids 63–80) partially protected in some way from proteolytic degradation. CHAN et al. (1988) attributed this protection to the integral membrane status of the H3 domain of the protein, but it is clear that the trypsin sites between amino acids 48 and 67 are accessible and therefore they must be exposed on the cytoplasmic side of the membrane (BERGMANN et al. 1989).

13.2 Receptor Function

The receptor function of NS28 was first described by AU et al. (1989a), who demonstrated that membranes prepared from insect cells expressing the NS28 gene under the direction of the baculovirus promoter were able to specifically interact with the ssp. Membranes from cells not expressing the gene exhibited no receptor activity. This observation was confirmed by MEYER et al. (1989), who used a series of recombinant vaccinia viruses to deliver a range of rotavirus genes to cells. Membranes from these cells were used to further characterize the interaction between NS28 and its ligand (VP6; see above). Other rotavirus proteins are not required for the interaction to occur, but in the infected cell their involvement is not necessarily excluded. The interaction between NS28 and VP6 in vitro is highly specific and exhibits an affinity constant of approximately 5×10^{-11} M. The interaction therefore is of roughly the same affinity as the interaction of icosahedral viruses with their receptors in the plasma membrane (e.g., human rhinovirus at 5×10^{-11} M; COLONNO et al. 1988). Essentially similar results were obtained by AU et al. (1989b) but these authors inferred the presence of both a high affinity (12×10^{-11} M) and low affinity component (4×10^{-11} M) in the binding reaction.

It is likely that the relatively high affinity constant found for the VP6: NS28 interaction reflects a cooperative interaction between many receptors in the membrane and a correspondingly large number of VP6 molecules on the surface of the ssp. It is known that the ligand (VP6) exists as a trimer (GORZIGLIA et al. 1985; SABARA et al. 1987), and there is evidence that NS28 is tetrameric (MAASS and ATKINSON 1990). Analysis of the region of NS28 available to form the receptor domain (133 amino acids) by techniques such as site-directed mutagenesis, coupled with the ability to measure the receptor activity of engineered variants of NS28 (Bergmann et al., unpublished results), should soon provide further information on the domains of the protein which are essential for receptor activity.

The interaction of the rotavirus ssp with NS28 also provides an interesting model for the analysis of the events which precede the transfer of the icosahedral core across the RER membrane. The transfer of the ssp across the RER membrane is the topological equivalent of the transfer of viral components across the plasma membrane, from the cytoplasm to the exterior of the cell. Budding is a process utilized by many viruses that mature at the cell surface and the NS28:VP6 interaction therefore may be well placed to make a contribution to our understanding of the molecular processes involved in the maturation of membranous viruses.

14 Gene Segment 11 (NS26)

NS26, the smallest known product of the SA11 genome, is coded for by gene segment 11, although the electrophoretic migration of segment 11 may differ for strains with a "super short" electropherotype (NUTTALL et al. 1989). The protein is serine- and threonine-rich and has an apparent molecular weight of 26000, but its predicted size is only 21 500 (ERICSON et al. 1982; IMAI et al. 1983b; WARD et al. 1985; MITCHELL and BOTH 1988; NUTTALL et al. 1989; GONZALEZ and BURRONE 1989). NS26 undergoes posttranslational modification following synthesis (ERICSON et al. 1982) and is phosphorylated (WELCH et al. 1989), but this modification does not account for the discrepancy between the observed and predicted sizes; some other modification to the protein must occur. No O-linked glycosylation was detected (WELCH et al. 1989).

Discussion as to whether NS26 is a minor structural protein or a non-structural polypeptide (ERICSON et al. 1982; MCCRAE and MCCORQUODALE 1982; ARIAS et al. 1982) was resolved following expression of the protein in the baculovirus system and the production of a high-titer antiserum (WELCH et al. 1989). Antibody to the expressed protein failed to neutralize virus particles and did not react with them by immunoelectron microscopy, immunoblotting, or immunoprecipitation.

Gene 11 is also the only segment in the SA11 genome containing a substantial, alternative open reading frame (coding for 92 amino acids) and this was also conserved in several other strains, suggesting that it may be used (MITCHELL and BOTH 1988). However, in sequences recently determined for this gene from the lapine Alabama (GORZIGLIA et al. 1989) and porcine OSU strains of rotavirus (GONZALEZ and BURRONE 1989) this reading frame is modified or truncated, making it less likely that it gives rise to a gene product.

15 Conclusions

Techniques for the cloning and sequencing of rotavirus genes have led to the rapid accumulation of information concerning the primary sequence of rotavirus proteins in the absence of information on their functions. Thus, for several rotavirus nonstructural proteins, complete sequence data are now available in the almost total absence of information concerning the role they play during infection. Only for NS28 is there now a substantial body of information concerning its role in the infected cell. Indeed the role of NS28 as a receptor is one of the most interesting aspects to emerge from the study of rotavirus genes. The interaction of this protein with its ligand VP6 promises to provide important insights into the fundamental processes involved in the transfer of icosahedral particles across membranes.

For a number of the structural rotavirus proteins, VP4, VP6, and VP7 in particular, our knowledge is now quite extensive, and although all the features of these proteins have yet to be fully characterized, a remarkable amount of information has accumulated in a very short period of time. However, for the central core proteins VP1, VP2, and VP3, the information is fragmentary, at best.

To advance our knowledge of the rotavirus proteins significantly, the primary amino acid sequences of the polypeptides need to be related to the tertiary and quaternary structures of the proteins. This approach has provided a wealth of information when applied to other viral proteins such as the influenza virus hemagglutinin (WILSON et al. 1981; WILEY et al. 1981) and neuraminidase (VARGHESE et al. 1983; COLMAN et al. 1983). The availability of appropriate expression systems capable of yielding sufficient quantities of individual proteins for use in structural studies should greatly accelerate our progress in this area. It is therefore encouraging that authentic rotavirus protein appears to be produced in the baculovirus system (ESTES et al. 1987; MACKOW et al. 1989). The ssp itself appears to be amenable to crystallographic analysis (I. Anthony, B. Harris, A.R. Bellamy and S. Harrison, unpublished observations) and determination of its structure may help us to understand how the genome is transcribed. Thus, the next few years should see an expansion in our knowledge of rotavirus structure and function, although, we suspect, probably not at the pace experienced since 1973, when the medical significance of these viruses was first recognized.

Acknowledgement. DBM was supported by CSIRO and a postgraduate studentship from the Dairy Research Council. We thank Dr. J. Berriman for helpful comments on the manuscript and for the electron micrographs in Figs. 1 and 3. Dr. M. Yeager (Scripps Clinic and Research Foundation, California, U.S.A.) supplied us with a copy of his manuscript (YEAGER et al. 1990) prior to publication, from which Fig. 2 is taken. We thank Anne McGill for assistance in preparing the manuscript. Supported by a programme grant (to ARB) from the Medical Research Council of New Zealand.

References

Arias CF, Lopez S, Espejo RT (1982) Gene protein products of SA11 simian rotavirus genome. J Virol 41: 42–50

Arias CF, Lopez S, Bell JR, Strauss JH (1984) Primary structure of the neutralization gene of simian rotavirus SA11 as deduced from cDNA sequence. J Virol 50: 657–661

Au K-S, Chan W-K, Burns JW, Estes MK (1989a) Receptor activity of rotavirus nonstructural glycoprotein NS28. J Virol 63; 4553-4562

Au K-S, Chan W-K, Estes MK (1989b) Compans R, Helenius A, Oldstone M (eds), UCLA symposium on cell biology of viral entry, replication and pathogenicity. Liss, New York, pp 257–267

Bassel-Duby R, Nibert ML, Homcy CJ, Fields BN, Sawutz DG (1987) Evidence that the σ 1 protein of reovirus serotype 3 is a multimer. J Virol 61: 1834–1841

Baybutt HN, McCrae MA (1984) The molecular biology of rotaviruses 7. Detailed structural analysis of gene 10 of bovine rotaviruses. Virus Res 1: 533–542

Berg JM (1986) Potential metal-binding domains in nucleic acid binding proteins. Science 232: 485–487

Bergmann CC, Maass D, Poruchynsky MS, Atkinson PH, Bellamy AR (1989) Topology of the non-structural rotavirus receptor glycoprotein NS28 in the rough endoplasmic reticulum. EMBO J 8: 1695–1703

Bican P, Cohen J, Charpilienne A, Scherrer R (1982) Purification and characterization of bovine rotavirus cores. J Virol 43: 1113–1117

Bishop RF, Davidson GP, Holmes IH, Ruck BJ (1973) Virus particles in epithelial cells of duodenal mucosa from children with acute non-bacterial gastroenteritis. Lancet 2: 1281–1283

Blobel G, Dobberstein B (1975) Transfer of proteins across membranes. I. Presence of proteolytically processed and unprocessed immunoglobulin light chains on membrane-bound ribosomes of murine myeloma. J Cell Biol 67: 835–851

Both GW, Bellamy AR, Street JE, Siegman LJ (1982) A general strategy for cloning double-stranded RNA: nucleotide sequence of the simian-11 rotavirus gene 8. Nucleic Acids Res 10: 7075–7088

Both GW, Mattick JS, Bellamy AR (1983a) Serotype-specific glycoprotein of simian-11 rotavirus: coding assignment and gene sequence. Proc Natl Acad Sci USA 80: 3091–3095

Both GW, Mattick J, Siegman L, Atkinson PH, Weiss S, Bellamy AR, Street JE, Metcalf P (1983b) Compans RW, Bishop DHL (eds) Double stranded RNA viruses, Elsevier, New York, pp 73–82

Both GW, Siegman LJ, Bellamy AR, Atkinson PH (1983c) Coding assignment and nucleotide sequence of simian rotavirus SA11 gene segment 10: location of glycosylation sites suggests that the signal peptide is not cleaved. J Virol 48: 335–339

Both GW, Bellamy AR, Siegman LJ (1984a) Nucleotide sequence of the dsRNA genomic segment 7 of simian-11 rotavirus. Nucleic Acids Res 12: 1621–1626

Both GW, Siegman LJ, Bellamy AR, Ikegami N, Shatkin AJ, Furuichi Y (1984b) Comparative sequence analysis of rotavirus genomic segment 6 – the gene specifying viral subgroups 1 and 2. J Virol 51: 97–101

Both GW, Stirzaker SC, Bergmann CC, Andrew ME, Boyle DB, Bellamy AR (1990) Kurstak E, Marusyk RG, Murphy FA, van Regenmortel MHV (eds), Applied virology research. Plenum, New York, pp 267–290

Boyle JF, Holmes KV (1985) RNA-binding proteins of bovine rotavirus. J Virol 58: 561–568

Bremont M, Charpilienne D, Chabanne D, Cohen J (1987) Nucleotide sequence and expression in Escherichia coli of the gene encoding the nonstructural protein NCVP2 of bovine rotavirus. Virology 160: 138–144

Burns JW, Chen D, Estes MK, Ramig RF (1989) Biological and immunological characterization of a simian rotavirus SA11 variant with an altered genome segment 4. Virology 169: 427–435

Caspar DLD, Klug A (1962) Physical principles in the construction of regular viruses. Cold Spring Harbor Symp Quant Biol 27: 1–24

Chan W-K, Penaranda ME. Crawford SE, Estes MK (1986) Two glycoproteins are produced from the rotavirus neutralization gene. Virology 151: 243–252

Chan W-K, Au K-S, Estes MK (1988) Topography of the simian rotavirus nonstructural glycoprotein (NS28) in the endoplasmic reticulum membrane. Virology 164: 435–442

Chasey D (1980) Investigation of immunoperoxidase-labeled rotavirus in tissue culture by light and electron microscopy. J Gen Virol 50: 195–200

Chasey D, Labram J (1983) Electron microscopy of tubular assemblies associated with naturally occurring bovine rotavirus. J Gen Virol 64: 863–872

Chow M, Newman JFE, Filman D, Hogle JM, Rowlands DJ, Brown F (1987) Myristylation of picornavirus capsid protein VP4 and its structural significance. Nature 327: 482–486

Clark B, Desselberger U (1988) Myristylation of rotavirus proteins. J Gen Virol 69: 2681–2686

Clark SM, Roth JR, Clark ML, Barnett BB, Spendlove RS (1981) Trypsin enhancement of rotavirus infectivity:mechanisms of enhancement. J Virol 39: 816–822

Cohen J (1977) Ribonucleic acid polymerase associated with purified calf rotavirus. J Gen Virol 36: 395–402

Cohen J, Charpilienne A, Chilmonczyk S, Estes MK (1989) Nucleotide sequence of bovine rotavirus gene 1 and expression of the gene in baculovirus. Virology 171: 131–140

Colman PM, Varghese JN, Laver WG (1983) Structure of the catalytic and antigenic sites in influenza virus neuraminidase. Nature 303: 41–44

Colonno RJ, Condra JH, Mitzutani S, Callahan PL, Davies ME, Murcko MA (1988) Evidence of the direct involvement of the rhinovirus canyon in receptor binding. Proc Natl Acad Sci USA 85: 5449–5453

Daniels RS, Downie JC, Hay AL, Knossow M, Skehel JJ, Wang ML, Wiley DC (1985) Fusion mutants of the influenza virus haemagglutinin glycoprotein. Cell 40: 431–439

Dyall-Smith MI, Holmes IH (1981) Comparisons of rotavirus polypeptides by limited proteolysis: close similarity of certain polypeptides of different strains. J Virol 40: 720–728

Dyall-Smith ML, Holmes IH (1984) Sequence homology between human and animal rotavirus serotype-specific glycoproteins. Nucleic Acids Res 12: 3973–3982

Dyall-Smith ML, Azad AA, Holmes IH (1983a) Gene mapping of rotavirus double-stranded RNA segments by northern blot hybridization: application to segments 7, 8 and 9. J Virol 46: 317–320

Dyall-Smith ML, Elleman TC, Hoyne PA, Holmes IH, Azad AA (1983b) Cloning and sequence of UK bovine rotavirus gene segment 7: marked sequence homology with simian rotavirus gene segment 8. Nucleic Acids Res 11: 3351–3362

Dyall-Smith ML, Lazdins I, Tregear GW, Holmes IH (1986) Location of the major antigenic sites involved in rotavirus serotype-specific neutralization. Proc Natl Acad Sci USA 83: 3465–3468

Eisenberg D, Schwarz E, Komaromy M, Wall R (1984) Analysis of membrane and surface protein sequences with the hydrophobic moment plot. J Mol Biol 179: 125–142

Elleman TL, Hoyne PA, Dyall-Smith ML, Holmes IH, Azad AA (1983) Nucleotide sequence of the gene encoding the serotype-specific glycoprotein of UK bovine rotavirus. Nucleic Acids Res 11: 4689–4701

Ericson BL, Graham DY, Mason BB, Estes MK (1982) Identification, synthesis, and modifications of simian rotavirus SA11 polypeptides in infected cells. J Virol 42: 825–839

Ericson BL, Graham DY, Mason BB, Hansenn H, Estes MK (1983) Two types of glycoprotein precursors are produced by the simian rotavirus SA11. Virology 127: 320–332

Ernst H, Duhl JA (1989) Nucleotide sequence of genomic segment 2 of human rotavirus Wa. Nucleic Acids Res 17: 4382

Esparza J, Gil F (1978) A study on the ultrastructure of human rotavirus. Virology 91: 141–150

Espejo RT, Lopez S, Arias C (1981) Structural polypeptides of simian rotavirus SA11 and the effect of trypsin. J Virol 37: 156–160

Estes MK, Cohen J (1990) Rotavirus gene structure and function. Microbiol Rev 53: 410–449

Estes MK, Graham DY, Mason BB (1981) Proteolytic enhancement of rotavirus infectivity: molecular mechanisms. J Virol 39: 879–888

Estes MK, Palmer EL, Obijeski JF (1983) Rotaviruses: a review. In: Compans RW, Cooper M, Koprowski H et al. (eds) Current topics in microbiology and immunology, vol 105. Springer, Berlin Heidelberg, New York, pp 123–184

Estes MK, Mason BB, Crawford S, Cohen J (1984) Cloning and nucleotide sequence of the simian rotavirus gene 6 that codes for the major inner capsid protein. Nucleic Acids Res 12: 1875–1887

Estes MK, Crawford SE, Penaranda ME, Petrie BL, Burns J, Chan W-K, Ericson B, Smith GE, Summers MD (1987) Synthesis and immunogenicity of the rotavirus major capsid antigen using a baculovirus expression system. J Virol 61: 1488–1494

Flewett TH, Davies H, Bryden AS, Robertson MJ (1974) Diagnostic electron microscopy of feces. II. Acute gastroenteritis associated with reovirus-like particles. J Clin Pathol 27: 608–614

Fukudome K, Yoshie O, Konno T (1989) Comparison of human, simian, and bovine rotaviruses for requirement of sialic acid in haemagglutination and cell adsorption. Virology 172: 196–205

Fukuhara N, Yoshie O, Kitaoka S, Konno T, Ishida N (1987) Evidence for endocytosis-independent infection by human rotavirus. Arch Virol 97: 93–99

Fukuhara N, Yoshie O, Kitaoka S, Konno T (1988) Role of VP3 in human rotavirus internalization after target cell attachment via VP7. J Virol 62: 2209–2218

Furlong DB, Nibert ML, Fields BN (1988) σ1 protein of mammalian reoviruses extends from the surfaces of viral particles. J Virol 2: 246–256

Furuichi Y, Morgan M, Muthukrishnan S, Shatkin AJ (1975a) Reovirus messenger RNA contains a methylated, blocked 5'-terminal structure: m7G(5')ppp(5')GmpCp-. Proc Natl Acad Sci USA 72: 362–366

Furuichi Y, Muthukrishnan S, Shatkin AJ (1975b) 5'-terminal m7G(5')ppp(5')Gmp in vivo: identification in reovirus genome RNA. Proc Natl Acad Sci USA 72: 742–745

Gallegos CO, Patton JT (1989) Characterization of rotavirus replication intermediates: a model for the assembly of single-shelled particles. Virology 172: 616–627

Gething MJ, Sambrook J (1981) Cell-surface expression of influenza haemagglutinin from a cloned DNA copy of the RNA gene. Nature 293: 620–625

Gething M-J, White JM, Waterfield M (1976) Purification of the fusion protein of Sendai virus: analysis of the NH2-terminal sequence generated during precursor activation. Proc Natl Acad Sci USA 75: 2737–2740

Glass RI, Keith J, Nakagomi O, Nakagomi T, Askaa J, Kapikian AZ, Chanock RM, Flores J (1985) Nucleotide sequence of the structural glycoprotein VP7 gene of nebraska calf diarrhoea virus rotavirus: comparison with homologous genes from four strains of human and animal rotaviruses. Virology 141: 292–298

Gombold JI, Ramig RF (1987) Assignment of simian rotavirus SA11 temperature-sensitive mutant groups A, C, F and G to genome segments. Virology 161: 463–473

Gombold JL, Estes ML, Ramig RF (1985) Assignment of simian rotavirus SA11 temperature-sensitive mutant groups B and E to genome segments. Virology 143: 309–320

Gonzalez SA, Burrone OR (1989) Porcine rotavirus segment 11 sequence shows common features with the viral gene of human origin. Nucleic Acids Res 17: 6402

Gorbalenya AE, Koonin EV (1988) Birnavirus RNA polymerase is related to polymerases of positive strand RNA viruses. Nucleic Acids Res 15: 7735

Gorziglia M, Larrea C, Liprandi F, Esparza J (1985) Biochemical evidence for the oligomeric (possibly trimeric) structure of the major inner capsid polypeptide (45K) of rotaviruses. J Gen Virol 66: 1889–1900

Gorziglia M, Aguirre Y, Hoshino Y, Esparza J, Blumentals I, Askaa J, Thompson M, Glass R, Kapikian AZ, Chanock RM (1986a) VP7 serotype-specific glycoprotein of OSU porcine rotavirus: coding assignment and gene sequence. J Gen Virol 67; 2445–2454

Gorziglia M, Hoshino Y, Buckler-White A, Blumentals I, Glass R, Flores J, Kapikian AZ, Chanock Rm (1986b) Conservation of amino acid sequence of VP8 and cleavage region of 84-kd outer capsid protein among rotaviruses recovered from asymptomatic neonatal infection. Proc Natl Acad Sci USA 83: 7039–7043

Gorziglia M, Green K, Nishikawa K, Taniguchi K, Jones R, Kapikian AZ, Chanock RM (1988a) Sequence of the fourth gene of human rotavirus recovered from asymptomatic or symptomatic infections. J Virol 62: 2978–2984

Gorziglia M, Hoshino Y, Nishikawa K, Maloy WL, Jones RW, Kapikian AZ, Chanock RM (1988b) Comparative sequence analysis of the genomic segment 6 of four rotaviruses each with a different subgroup specificity. J Gen Virol 69: 1659–1669

Gorziglia M, Nishikawa K, Green K, Taniguchi K (1988c) Gene sequence of the VP7 serotype specific glycoprotein of Gottfried porcine rotavirus. Nucleic Acids Res 16: 775

Gorziglia M, Nishikawa K, Fukuhara N (1989) Evidence of duplication and deletion in super short segment 11 of rabbit rotavirus Alabama strain. Virology 170: 587–590

Green KY, Hoshino Y, Ikegami N (1989) Sequence analysis of the gene encoding the serotype-specific glycoprotein (VP7) of two new human rotavirus serotypes. Virology 168: 429–433

Greenberg HB, Flores J, Kalica AR, Wyatt RG, Jones R (1983a) Gene coding assignments for growth restriction, neutralization and subgroup specificities of the W and DS-1 strains of human rotavirus. J Gen Virol 64: 313–20

Greenberg HB, McAuliffe V, Valdesuso J, Wyatt R, Flores J, Kalica A, Hoshino Y, Singh NH (1983b) Serological analysis of the subgroup protein of rotavirus, using monoclonal antibodies. Infect Immun 39: 91–99

Greenberg HB, Valdesuso J, van Wyke K, Midthun K, Walsh M, McAuliffe V, Wyatt RG, Kalica AR, Flores J, Hoshino Y (1983c) Production and preliminary characterization of monoclonal antibodies directed at two surface proteins of rhesus rotavirus. J Virol 47: 267–275

Gunn PG, Sato F, Powell KFH, Bellamy AR, Napier JR, Harding DRK, Hancock WS, Siegman LJ, Both GW (1985) Rotavirus neutralizing protein VP7: antigenic determinants investigated by sequence analysis and peptide synthesis. J Virol 54: 791–797

Helmberger-Jones M, Patton JT (1986) Characterization of subviral particles in cells infected with simian rotavirus SA11. Virology 155: 655–665

Henderson LE, Krutzsch HC, Oroszlan S (1983) Myristyl amino-terminal acylation of murine retrovirus proteins: an unusual post-translational protein modification. Proc Natl Acad Sci USA 80: 339–343

Hofer JMI, Street JE, Bellamy AR (1987) Nucleotide sequence for gene 6 of rotavirus strain S2. Nucleic Acids Res 15: 7175

Hoshino Y, Sereno MM, Midthun K, Flores J, Kapikian AZ, Chanock RM (1985) Independent segregation of two antigenic specificities (VP3 and VP7) involved in neutralization of rotavirus infectivity. Proc Natl Acad Sci USA 82: 8701–8704

Hubbard SC, Ivatt J (1981) Synthesis and processing of asparagine-linked oligosaccharides. Annu Rev Biochem 50: 555–583

Hum CP, Dyall-Smith ML, Holmes IH (1989) The VP7 gene of a new G serotype of human rotavirus (B37) is similar to G3 proteins in the antigenic C region. Virology 170: 55–61

Imai M, Akatani K, Ikegami N, Furuichi Y (1983a) Capped and conserved terminal structures in human rotavirus genome double-stranded RNA segments. J Virol 47: 125–136

Imai M, Richardson MA, Ikegami N, Shatkin AJ, Furuichi Y (1983b) Molecular cloning of double-stranded RNA virus genomes. Proc Natl Acad Sci USA 80: 373–377

Kabcenell AK, Atkinson PA (1985) Processing of the rough endoplasmic reticulum membrane glycoproteins of rotavirus SA11. J Cell Biol 101: 1270–1280

Kabcenell AK, Poruchynsky MS, Bellamy AR, Greenberg HB, Atkinson PA (1988) Two forms of VP7 are involved in the assembly of SA11 rotavirus in the endoplasmic reticulum. J Virol 62: 2929–2941

Kalica AR, Greenberg HB, Wyatt RG, Flores J, Sereno MM, Kapikian AZ, Chanock RM (1981) Genes of human (strain Wa) and bovine (strain UK) rotaviruses that code for neutralization and subgroup antigens. Virology 112: 385–390

Kalica AR, Flores J, Greenberg HB (1983) Identification of the rotaviral gene that codes for haemagglutination and protease-enhanced plaque formation. Virology 125: 194–205

Kaljot KT, Shaw RD, Rubin DH, Greenberg HB (1988) Infectious rotavirus enters cells by direct membrane penetration, not by endocytosis. J Virol 62: 1136–1144

Kamer G, Argos P (1984) Primary structural comparison of RNA-dependent polymerases from plant, animal and bacterial viruses. Nucleic Acids Res 12: 7269–7282

Kantharidis P, Dyall-Smith ML, Holmes IH (1987) Marked sequence variation between segment 4 genes of human RV-5 and simian SA11 rotaviruses. Arch Virol 93: 111–121

Kantharidis P, Dyall-Smith M, Tregear GW, Holmes IH (1988) Nucleotide sequence of UK bovine rotavirus segment 4: possible host restriction of VP3 genes. Virology 166: 308–315

Kapikian AZ, Chanock RM (1985) Rotaviruses, In: Fields BN (ed) Virology, Raven, New York, pp 863–906

Klenk H-D, Rott R (1988) The molecular biology of influenza virus pathogenicity. Adv Virus Res 34: 247–275

Kouvelos K, Petric M, Middleton PJ (1984a) Comparison of bovine, simian, and human rotavirus structural glycoproteins. J Gen Virol 65: 1211–1214

Kouvelos K, Petric M, Middleton PJ (1984b) Oligosaccharide composition of calf rotavirus. J Gen Virol 65: 1159–1164

Kozak M (1984) Selection of initiation sites by eukaryotic ribosomes: effect of inserting AUG triplets upstream from the coding sequence for preproinsulin. Nucleic Acids Res 12: 3873–3893

Kozak M (1986) Point mutations that define a sequence flanking the AUG initiator codon that modulates translation by eukaryotic ribosomes. Cell 44: 283–292

Kumar A, Charpilienne A, Cohen J (1989) Nucleotide sequence of the gene for the RNA binding protein (VP2) of RF bovine rotavirus. Nucleic Acids Res 17: 2126

Landschulz WH, Johnson PF, McKnight SL (1988) The leucine zipper: a hypothetical structure common to a new class of DNA, binding proteins. Science 240: 1759–1764

Lee PWK, Hayes EC, Joklik WK (1981) Protein 1 is the reovirus cell attachment protein. Virology 108: 156–163

Liu M, Estes MK (1989) Nucleotide sequence of the simian rotavirus SA11 genome segment 3. Nucleic Acids Res 17: 7991

Liu M, Offit PA, Estes MK (1988) Identification of the simian rotavirus SA11 genome segment 3 product. Virology 163: 26–32

Lopez S, Arias CF (1987) The nucleotide sequence of the 5′ and 3′ ends of rotavirus SA11 gene 4. Nucleic Acids Res 15: 4691

Lopez S, Arias CF Bell JR, Strauss JH, Espejo RT (1985) Primary structure of the cleavage site associated with trypsin enhancement of rotavirus SA11 infectivity. Virology 144: 11–19

Lopez S, Arias CF, Mendez E, Espejo R (1986) Conservation in rotaviruses of the protein region containing the two sites associated with trypsin enhancement of infectivity. Virology 154: 224–227

Ludert JE, Michelangeli F, Gil F, Liprandi F, Esparza J (1987) Penetration and uncoating of rotaviruses in cultured cells, Intervirol 27; 95–101

Maass DR, Atkinson PH (1990) Rotavirus proteins VP7, NS28 and VP4 form oligomeric structures. J Virol 64: 2632–2641

Mackow ER, Shaw RD, Matsui SM, Vo PT, Benfield DA, Greenberg HB (1988a) Characterization of homotypic and heterotypic VP7 neutralization sites of rhesus rotavirus. Virology 165: 511–517

Mackow ER, Shaw RD, Matsui SM, Vo PT, Dang M-N, Greenberg HB (1988b) Characterization of the rhesus rotavirus VP3 gene: localization of amino acids involved in homologous and heterologous rotavirus neutralization and identification of a putative fusion region. Proc Natl Acad Sci USA 85: 645–649

Mackow ER, Barnett JW, Chan H, Greenberg HB (1989) The rhesus rotavirus outer capsid protein VP4 functions as a haemagglutinin and is antigenically conserved when expressed by a baculovirus recombinant. J Virol 63: 1661–1668

Malherbe HH, Harwin R, Ulrich M (1963) The cytopathic effect of vervet monkey viruses. S Afr Med J 37: 407–411

Mason BB, Graham DY, Estes MK (1980) In vitro transcription and translation of simian rotavirus SA11 gene products. J Virol 33: 1111–1121

Mason BB, Graham DY, Estes MK (1983) Biochemical mapping of the simian rotavirus SA11 genome. J Virol 46: 413–423

Mason BB, Dheer SK, Hsaio C-L, Zandle G, Kostek B, Rosanoff EI, Hung PP, Davis AR (1985) Sequence of the serotype-specific glycoprotein of the human rotavirus Wa strain and comparison with other human rotavirus serotypes. Virus Res 2: 328–336

Masri SA, Nagata L, Mah DCW, Lee PWK (1986) Functional expression in Escherichia coli of cloned reovirus S1 gene encoding the viral attachment polypeptide $\sigma 1$. Virology 149: 83–90

Matsui SM, Mackow E, Greenberg HB (1989) Molecular determinants of rotavirus neutralization and protection. In: Maramorosch K, Murphy FA, Shatkin AJ (eds) Advances in virus research. Academic, San Diego, pp 181–214

McCrae MA, McCorquodale JG (1982) The molecular biology of rotaviruses. II. Identification of the protein-coding assignments of calf rotavirus genome RNA species. Virology 117: 435–443

Metcalf P (1982) The symmetry of reovirus. J Ultrastruct Res 78: 292–301

Meyer JC, Bergmann CC, Bellamy AR (1989) Interaction of rotavirus cores with the nonstructural glycoprotein NS28. Virology 170: 98–107

Miller J, McLachlan AD, Klug A (1985) Reptitive zinc-binding domains in the protein transcription factor IIIA from Xenopus oocytes. EMBO J 4: 1609–1614

Mitchell DB, Both GW (1988) Simian rotavirus SA11 segment 11 contains overlapping reading frames. Nucleic Acids Res 16: 6244

Mitchell DB, Both GW (1989) Complete nucleotide sequence of the simian rotavirus SA11 VP4 gene. Nucleic Acids Res 17: 2122

Mitchell DB, Both GW (1990a) Conservation of a potential metal binding motif despite extensive sequency diversity in the rotavirus nonstructural protein NS53. Virology 174: 618–621

Musalem C, Espejo RT (1985) Release of progeny virus from cells infected with simian rotavirus SA11. J Gen Virol 66: 2715–2724

Newman JFE, Brown F, Bridger JC, Woode GN (1975) Characterization of a rotavirus. Nature 258: 631–633

Nishikawa K, Gorziglia M (1988) The nucleotide sequence of the VP3 gene of porcine rotavirus OSU. Nucleic Acids Res 24: 11847

Nishikawa K, Taniguchi K, Torres A, Hoshino Y, Green K, Kapikian AZ, Chanock RM, Gorziglia M (1988) Comparative analysis of the VP3 gene of divergent strains of the rotaviruses simian SA11 and bovine Nebraska calf diarrhoea virus. J Virol 62: 4022–4026

Nishikawa K, Hoshino Y, Taniguchi K, Green KY, Greenberg HB, Kapikian AZ, Chanock RM, Gorziglia M (1989) Rotavirus VP7 neutralization epitopes of serotype 3 strains. Virology 171: 503–515

Nuttall SD, Hum CP, Holmes IH, Dyall-Smith ML (1989) Sequences of VP9 genes from short and supershort rotavirus strains. J Virol 171: 453–457

Offit PA, Blavat G (1986) Identification of the two rotavirus genes determining neutralization specificities. J Virol 57: 376–378

Offit PA, Blavat G, Greenberg HB, Clark CF (1986) Molecular basis of rotavirus virulence: role of gene segment 4. J Virol 57: 46–49

Okada Y, Richardson MA, Ikegami N, Nomoto A, Furuichi Y (1984) Nucleotide sequence of human rotavirus genome segment 10, an RNA encoding a glycosylated virus protein. J Virol 51: 856–859

Patton JT (1986) Synthesis of simian rotavirus SA11 double-stranded RNA in a cell-free system. Virus Res 6: 217–233

Patton JT, Gallegos CO (1988) Structure and protein composition of the rotavirus replicase particle. Virology 166: 358–365

Pereira HG, Azeredo RS, Fiahlo AM, Vidal MNP (1984) Genomic heterogeneity of simian rotavirus SA11. J Gen Virol 65: 815–818

Petrie BL, Graham DY, Hanssen H, Estes MK (1982) Localization of rotavirus antigens in infected cells by ultrastructural immunocytochemistry. J Gen Virol 63: 457–467

Petrie BL, Greenberg HB, Graham DY, Estes MK (1984) Ultrastructural localization of rotavirus antigens using colloidal gold. Virus Res 1: 133–152

Poch O, Sauvaget I, Delarue M, Tordo N (1989) Identification of four conserved motifs among RNA-dependent polymerase encoding elements. EMBO J 8: 3867–3874

Poruchynsky MS, Atkinson PH (1988) Primary sequence domains required for the retention of rotavirus VP7 in the endoplasmic reticulum. J Cell Biol 107: 1697–1706

Poruchynsky MS, Tyndall C, Both GW, Sato F, Bellamy AR, Atkinson PA (1985) Deletions into an NH_2-terminal hydrophobic domain result in secretion of rotavirus VP7, a resident endoplasmic reticulum membrane glycoprotein. J Cell Biol 101: 2199–2209

Potter AA, Cox G, Parker M, Babiuk LA (1987) The complete sequence of bovine rotavirus C486 gene 4 cDNA. Nucleic Acids Res 15; 4361

Powell KFH, Gunn PR, Bellamy AR (1988) Nucleotide sequence of bovine rotavirus genomic segment 10: an RNA encoding the viral non-structural glycoprotein. Nucleic Acids Res 16: 763

Prasad BVV, Wang GJ, Clerx JPM, Chiu W (1988) Three-dimensional structure of rotavirus. J Mol Biol 199: 269–275

Prasad BVV, Burns JW, Marietta E, Estes MK Chiu W (1990) Localisation of VP4 neutralisation sites in rotavirus by three-dimensional cryo-electron microscopy. Nature 343: 476–479

Quann CM, Doane FW (1983) Ultrastructural evidence for the celllar uptake of rotavirus by endocytosis. Intervirology 20: 223–231

Ramig RF (1983) Isolation and genetic characterization of temperature-sensitive mutants that define five additional recombination groups in simian rotavirus SA11. Virology 46: 464–473

Ramig RF, Fields BN (1983) Genetics of reoviruses. in: Joklik WK (ed) The reoviridae, Plenum, New York, pp 197–228

Ramig RF, Petrie BL (1984) Characterization of temperature-sensitive mutants of simian rotavirus SA11: protein synthesis and morphogenesis. J Virol 49: 665–673

Ready KFM, Sabara M (1987) In vitro assembly of bovine rotavirus nucleocapsid protein. Virology 157; 189–198

Ready KFM, Buko KMA, Whippey PW, Alford WP, Bancroft JB (1988) The structure of tubes of bovine rotavirus nucleocapsid protein (VP6) assembled in vitro. Virology 167: 50–55

Reddy DA, Greenberg HB, Bellamy AR (1989) Rotavirus serotype IV: nucleotide sequence of genomic segment nine of St Thomas 3 strain. Nucleic Acids Res 17: 449

Richardson MA, Iwamoto A, Ikegami N, Nomoto A, Furuichi Y (1984) Nucleotide sequence of the gene encoding the serotype-specific antigen of human (Wa) rotavirus: comparison with the homologous genes from simian SA11 and UK bovine rotaviruses. J Virol 51: 860–862

Richardson SC, Mercer LE, Sonza S, Holmes IH (1986) Intracellular location of rotaviral proteins. Arch Virol 88: 251–264

Rodger SM, Schnagl RD, Holmes IH (1975) Biochemical and biophysical characterization of diarrhoea viruses of human and calf origin. J Virol 16: 1229–1235

Roseto A, Esciag J, Delain E, Cohen J, Scherrer R (1979) Structure of rotaviruses as studied by the freeze-drying technique. Virology 98: 471–475

Ruiz AM, Lopez IV, Lopez S, Espejo RT, Arias CF (1988) Molecular and antigenic characterization of porcine rotavirus YM, a possible new rotavirus serotype. J Virol 62: 4331–4336

Rushlow K, McNab A, Olson K, Maxwell F, Maxwell I, Stiegler G (1988) Nucleotide sequence of porcine rotavirus (OSU strain) gene segments 7, 8 and 9. Nucleic Acids Res 16: 367–368

Sabara M, Gilchrist JE, Hudson GR, Babiuk LA (1985) Preliminary characterization of an epitope involved in neutralization and cell attachment that is located on the major bovine rotavirus glycoprotein. J Virol 53: 58–66

Sabara M, Ready KFM, Frenchick PJ, Babiuk LA (1987) Biochemical evidence for the oligomeric arrangement of bovine rotavirus nucleocapsid protein and its possible significance in the immunogenicity of this protein. J Gen Virol 68: 123–133

Schroeder BA, Street JE, Kalmakoff J, Bellamy AR (1982) Sequence relationships between the genome segments of human and animal rotavirus strains. J Virol 43: 379–385

Smith ML, Lazdins I, Holmes IH (1980) Coding assignments of double-stranded RNA segments of SA11 rotavirus established by in vitro translation. J Virol 33: 976–982

Spencer E, Garcia BI (1984) Effect of S-adenosylmethionine on human rotavirus RNA synthesis. J Virol 52: 188–197

Stirzaker SC, Both GW (1989) The signal peptide of rotavirus glycoprotein VP7 is essential for its retention in the ER as an integral membrane protein. Cell 56: 741–747

Stirzaker SC, Whitfeld PL, Christie DL, Bellamy AR, Both GW (1987) Processing of rotavirus glycoprotein VP7: implications for the retention of the protein in the endoplasmic reticulum. J Cell Biol 105: 2897–2903

Streuli CH, Griffin BE (1987) Myristic acid is coupled to a structural protein of polyoma virus and SV40. Nature 326: 619–622

Suzuki H, Kitaoka S, Konno T, Sato T, Ishida N (1985) Two modes of human rotavirus entry into MA104 cells. Arch Virol 82: 25–43

Suzuki H, Kitaoka S, Sato T, Konno Y, Iwasaki Y, Numazaki Y, Ishida N (1986) Further investigation on the mode of entry of human rotavirus into cells. Arch Virol 91: 135–144

Svensson L, Grahnquist L, Pettersson C-A, Grandein M, Stintzing G, Greenberg HB (1988) Detection of human rotaviruses which do not react with subgroup I-and subgroup II-specific monoclonal antibodies. J Clin Microbiol 26: 1238–1240

Taniguchi K, Urasawa T, Ursawa S (1986) Reactivity patterns to human rotavirus strains of a monoclonal antibody against VP2, a component of the inner capsid of rotavirus. Arch Virol 87: 135–141

Taniguchi K, Maloy WL, Nishikawa K, Green KY, Hoshino Y, Urasawa S, Kapikian AZ, Chanock RM (1988) Identification of cross-reactive and serotype 2-specific neutralization epitopes on VP3 of human rotavirus. J Virol 62: 2421–2426

Taniguchi K, Nishikawa K, Urasawa T, Urasawa S, Midthun K, Kapikian AZ, Gorziglia M (1989) Complete nucleotide sequence of the gene encoding VP4 of a human rotavirus (strain K8) which has unique VP4 neutralization epitopes. J Virol 63: 4101–4106

Urakawa T, Ritter DG, Roy P (1989) Expression of the largest RNA segment and synthesis of VP1 protein of bluetongue virus in insect cells by recombinant baculovirus: association of VP1 protein with RNA polymerase activity. Nucleic Acids Res 17: 7395–7401

Varghese JN, Laver WG, Colman PM (1983) Structure of the influenza virus glycoprotein antigen neuraminidase at 2.9Å resolution. Nature 303: 35–40

von Heijne G (1986) A new method for predicting signal sequence cleavage sites. Nucleic Acids Res 14: 4683–4690

Walter P, Lingappa VR (1986) Mechanism of protein translocation across the endoplasmic reticulum membrane. Annu Rev Cell Biol 2; 499–516

Ward CW, Elleman TC, Azad AA, Dyall-Smith ML (1984) Nucleotide sequence of gene segment 9 encoding a nonstructural protein of UK bovine rotavirus. Virology 134: 249–253

Ward CW, Azad AA, Dyall-Smith ML (1985) Structural homologies between RNA gene segments 10 and 11 from UK bovine, simian SA11 and human Wa rotaviruses. Virology 144: 328–336

Webster RG, Laver WG, Air GB, Schild GC (1982) Molecular mechanisms of variation in influenza viruses. Nature 296: 115–221

Welch S-KW, Crawford SE, Estes MK (1989) Rotavirus SA11 genome segment 11 is a nonstructural phosphoprotein. J Virol 63: 3974–3982

Whitfield PL, Tyndall C, Stirzaker SC, Bellamy AR, Both GW (1987) Location of signal sequences within the rotavirus SA11 glycoprotein VP7 which direct it to the endoplasmic reticulum. Mol Cell Biol 7: 2491–2497

Wiley DC, Wilson IA, Skehel JJ (1981) Structural identification of the antibody-binding sites of Hong Kong influenza haemagglutinin and their involvement in antigenic variation. Nature 289: 373–378

Wilson IA, Skehel JJ, Wiley DC (1981) Structure of the haemagglutinin membrane glycoprotein of the influenza virus at 3Å resolution. Nature 289: 366–373

Yeager M, Dryden KA, Olson NH, Greenberg HB, Baker TS (1990) Three-dimensional structure of Rhesus rotavirus by cryo-electron microscopy and image reconstruction. J Cell Biol 110: 2133–2144

Notes Added in Proof

Since the completion of this review, other papers which impact on these discussions have been published and the sequence of the rotavirus SA11 genome has been completed (MITCHELL and BOTH 1990b). Several of the predictions concerning protein function which emerged from these and other gene sequences now have biochemical support.

VP1, VP2 and VP3

Gene segment one from the porcine Gottfried (FUKUHARA et al. 1989) and UK bovine strains (TARLOW and MCCRAE 1990) has now also been sequenced. When single-shelled particles were incubated with a photoreactable compound, [^{32}P]azido-ATP and then exposed to uv light, VP1 became radiolabeled. Transcriptional activity was also decreased, identifying this protein as a probable component of the RNA polymerase complex (VALENZUELA et al. 1991). When VP2 was expressed using a recombinant baculovirus system, empty core-like particles were recovered from cell lysates by sucrose gradient centrifugation (LABBE et al. 1991). In the presence of VP6 these formed ssps. Studies with temperature-sensitive mutants have now implicated VP2 as an essential component of enzymatically active replicase particles (MANSELL and PATTON 1990). VP6 was not required. Thus, VP2 is confirmed as an important structural and functional protein of the inner core of rotavirus particles. Similarly, when double- and transcriptionally active, single-shelled particles were incubated with [α-^{32}P] GTP, only VP3 bound the nucleotide covalently, identifying the protein as the rotavirus guanylyltransferase (LIU et al. 1992; PIZARRO et al. 1991). VP3 expressed by a recombinant baculovirus also demonstrated the same property, suggesting that this protein alone in the virus particle possesses capping activity (LIU et al. 1992).

VP4

New data available for VP4 confirm this protein as the viral haemagglutinin and suggest additional structural and functional features. Baculovirus-expressed rhesus rotavirus VP4 bound to high molecular weight glycoproteins on murine enterocytes and competed with whole virus particles for those binding sites (BASS et al. 1991). Expression of the N-terminal VP8* fragment in a baculovirus system produced a protein capable of agglutinating human erythrocytes (FIORE et al. 1991). Certain monoclonal antibodies to VP8* have also been shown to neutralize infectivity by blocking virus attachment (RUGGERI and GREENBERG 1991). These properties are consistent with a role for VP4 in cell attachment mediated by VP8* which presumably, therefore has a surface location. Although the crystallographic structure of VP4 has not been solved, from studies with the electron microscope it appears that the viral protein may exist as a dimer (see Sect. 7.1 above). However, VP4 spikes, which appear to be dimeric, can be decorated on the virion with a monoclonal antibody, 2G4 (PRASAD et al. 1990), the binding epitope for which maps near residue 393 of VP5* (MACKOW et al. 1988), implying that this portion of the protein is also exposed at the tip of the spike. From a comparison between the porcine rotavirus YM and other sequences, it has been suggested that VP4 may be composed of a globular domain at the N-terminus, while the carboxy-terminal domain may contain heptad repeats characteristic of α-helical coiled-coil structures (LOPEZ et al. 1991). Such a general structure is reminiscent of other spike proteins from reovirus and influenza which also function as cellular binding proteins (see Sect. 7.1 above).

NS53

The conservation of cystine-rich motifs within NS53 proteins which otherwise showed great amino acid sequence diversity suggested the presence of a metal binding domain (zinc finger) near the N-terminus of the protein (MITCHELL and BOTH 1990a). It has now been demonstrated that NS53 expressed using the beaculovirus system was able to bind Zn^{++} as well as ssRNA in vitro (BROTTIER et al. 1992). The protein was also confirmed as non-structural. Such properties are consistent with a role for the protein in the early stages of virus assembly as discussed elsewhere in this volume.

VP6

Rotavirus VP6, which normally associates as trimers to form the outer layer of the ssps can also be found as paracrystalline arrays in infected cells. The VP6 gene was modified by adding sequences specifying an N-terminal signal peptide and a C-terminal membrane-spanning anchor domain on the protein. This polypeptide, expressed by a recombinant vaccinia virus, was exported from the cytoplasm and became anchored in the plasma membrane. Surprisingly, these modifications did not affect the ability of the protein to trimerise and form arrays which could be seen on the surface of the cells. The antigenicity of the protein in mice was also enhanced by this mode of presentation (REDDY et al. 1992).

NS34 and NS35

Further analysis of the amino acid sequence predicted for NS34, the product of gene seven, indicated that the protein was slightly acidic and rich in sequences with an alpha-helical conformation. The protein expressed from gene 7 was detected as oligomers in insect cells. Similar oligomers were detected in SA11-infected cells. Oligomer formation may be attributable to two regions of heptad repeats present in the carboxy-terminal half of the molecule. NS34 was also associated with the cytoskeleton and bound RNA as reported previously (BOYLE and HOLMES 1985; MATTION et al. 1992). More recently it was shown that multimeric and monomeric forms of NS34 recognize and bind specifically to rotavirus mRNAs *via* the conserved sequence present at their 3' ends (PONCET et al. 1993).

NS35 has also been identified as an RNA binding protein in SA11-infected cells by using an anti-NS35 monoclonal antibody to recover uv-cross-linked protein/RNA complexes. However, this protein bound both double- and single-standard RNA and lacked sequence specificity (KATTOURA et al. 1992). By comparing the sequences for NS35 obtained from distantly related rotavirus isolates, a highly conserved sequence of 37 amino acids with a nett basic charge was identified and suggested as an RNA binding domain. A stem-loop structure formed by base-pairing of conserved nucleotides from both ends of the NS35 mRNA was also predicted and suggested as a packaging signal (PATTON et al. 1993). Both the NS34 and NS35 proteins are present in early replication complexes, but presumably they have separate functions. NS34 may assist in assembly of the segments for packaging while NS35 appears to be essential for replication of the RNA segments.

VP7

Further evidence has been obtained to show that the sequences responsible for retention of VP7 in the ER reside wholly in the N-terminal one third of the protein since these residues confer the property of ER retention on the malaria S antigen, a protein which is normally secreted (STIRZAKER et al. 1990).

VP7 forms oligomeric structures with itself and with NS28 and VP4 during virus particle budding and assembly (MAASS and ATKINSON, 1990). The enveloped viral particles which are believed to be intermediates in the assembly process have now been isolated (PORUCHYNSKY and ATKINSON 1991). Further evidence for the formation of hetero-oligomers was also presented. It was suggested that during budding, VP7 may be repositioned from its lumenal location in the ER back across the membrane. In a further study the effect of calcium ionophores on oligomerisation and virus assembly was also examined (PORUCHYNSKY et al. 1991). In the absence of calcium, membrane-enveloped particles accumulated, VP7 was excluded from oligomeric structures and glycosylation of both VP7 and NS28 was altered. A major role for calcium in glycosylation and the correct conformation of proteins for virus assembly was implied.

NS28

The NS28 sequences which are important for its function as a receptor for the budding ssps have been examined (AU et al. 1993; TAYLOR et al. 1992; TAYLOR et al. 1993). A variety of mutant proteins were constructed and tested for receptor activity using a transient expression assay. Surprisingly, the N-terminal region (53 aas) of NS28 could be completely removed without causing the loss of function or the ability to tetramerise, but removal of just the C-terminal residue was sufficient to abolish all receptor activity (TAYLOR et al. 1992). In other experiments portions of NS28 were synthesized as glutathione-S-transferase fusion proteins which could be cleaved to generate soluble C-terminal receptor domains (TAYLOR et al. 1993). These retained receptor function when linked to a solid matrix, demonstrating that integral membrane status was not required for function. In a separate study involving truncated NS28 proteins it was found that deletion of residues 110–155 or the removal of the three C-terminal residues (aas 173–175) reduced, but did not abolish receptor activity. However, deletion of residues 161–175 did abolish ssp binding activity (AU et al. 1993). Synthetic peptides which mimicked the C-terminus of NS28 were not able to compete for binding to ssps. It was concluded that the conformational integrity of the C-terminal region of NS28 was most important for ssp binding (AU et al. 1993). Evidence that NS28 may contain a binding site for VP4 was also presented.

NS26

The non-structural NS26 protein is now known to be modified with O-linked N-acetylglucosamine, which accounts for its maturation from a 26kd to a 28kd species (GONZALEZ and BURRONE 1991). The alternative reading frame of gene segment 11 (MITCHELL and BOTH 1988) which is conserved in some form in most rotavirus isolates has been shown to encode a 12kd non-structural protein which is expressed in infected cells (MATTION et al. 1991).

Additional References

Anthony ID, Bullivant S, Dayal S, Bellamy AR, Berriman JA (1991) Rotavirus spike structure and polypeptide composition. J Virol 65: 4334–4340

Au KS, Mattion NM, Estes MK (1993) A subviral particle binding domain on the rotavirus nonstructural glycoprotein-NS28. Virology 194: 665–673

Bass DM, Mackow ER, Greenberg HB (1991) Identification and partial characterization of a rhesus rotavirus binding glycoprotein on murine enterocytes. Virology 183: 602–610

Brottier P, Nandi P, Bremont M, Cohen J (1992) Bovine Rotavirus Segment-5 Protein Expressed in the Baculovirus System Interacts with Zinc and RNA. J Gen Virol 73: 1931–1938

Fiore L, Greenberg HB, Mackow ER (1991) The VP8 fragment of VP4 is the rhesus rotavirus hemagglutinin. Virology 181: 553–563

Fukuhara N, Nishikawa K, Gorziglia M, Kapikian AZ (1989) Nucleotide sequence of gene segment 1 of a porcine rotavirus strain. Virology 173: 743–749

Gonzalez SA, Burrone OR (1991) Rotavirus NS26 is modified by addition of single O-linked residues of N-acetylglucosamine. Virology 182: 8–16

Kattoura MD, Clapp LL, Patton JT (1992) The rotavirus nonstructural protein, NS35, possesses RNA-binding activity in vitro and in vivo. Virology 191: 698–708

Labbe M, Charpilienne A, Crawford SE, Estes MK, Cohen J (1991) Expression of rotavirus VP2 produces empty corelike particles. J Virol 65: 2946–2952

Liu M, Mattion NM, Estes MK (1992) Rotavirus VP3 Expressed in insect cells possesses guanylyltransferase activity. Virology 188: 77–84

Lopez S, Lopez I, Romero P, Mendez E, Soberon X, Arias CF (1991) Rotavirus YM gene 4: analysis of its deduced amino acid sequence and prediction of the secondary structure of the VP4 protein. J Virol 65: 3738–3745

Mansell EA, Patton JT (1990) Rotavirus RNA replication: VP2, but not VP6, is necessary for viral replicase activity. J Virol 64: 4988–4996

Mattion NM, Cohen J, Aponte C, Estes MK (1992) Characterization of an oligomerization domain and RNA-binding properties on rotavirus nonstructural protein NS34. Virology 190: 68–83

Mattion NM, Mitchell DB, Both GW, Estes MK (1991) Expression of rotavirus proteins encoded by alternative open reading frames of genome segment 11. Virology 181: 295–304

Mitchell DB, Both GW (1990b) Completion of the genomic sequence of the simian rotavirus SA11: Nucleotide sequences of segments 1, 2 and 3. Virology 177: 324–331

Patton JT, Saltercid L, Kalbach A, Mansell EA, Kattoura M (1993) Nucleotide and amino acid sequence analysis of the rotavirus nonstructural RNA-binding protein NS35. Virology 192: 438–446

Pizarro JL, Sandino AM, Pizarro JM, Fernandez J, Spencer E (1991) Characterization of rotavirus guanylyltransferase activity associated with polypeptide VP3. J Gen Virol 72: 325–332

Poncet D, Aponte C, Cohen J (1993) Rotavirus protein NSP3 (NS34) is bound to the 3' consensus sequence of viral mRNAs in infected cells. J Virol 67: 3159–3165

Poruchynsky MS, Atkinson PH (1991) Rotavirus protein rearrangements in purified membrane-enveloped intermediate particles. J Virol 65: 4720–4727

Poruchynsky MS, Maass DR, Atkinson PH (1991) Calcium depletion blocks the maturation of rotavirus by altering the oligomerization of virus-encoded proteins in the ER. J Cell Biol 114: 651–661

Prasad BVV, Burns JW, Marietta E, Estes MK, Chiu W (1990) Localization of VP4 neutralization sites in rotavirus by three-dimensional cryo-electron microscopy. Nature 343: 476–479

Reddy DA, Bergmann CC, Meyer JC, Berriman J, Both GW, Coupar BEH, Boyle DB, Andrew ME, Bellamy AR (1992) Rotavirus VP6 modified for expression on the plasma membrane forms arrays and exhibits enhanced immunogenicity. Virology 189: 423–434

Ruggeri FM, Greenberg HB (1991) Antibodies to the trypsin cleavage peptide VP8 neutralize rotavirus by inhibiting binding of virions to target cells in culture. J Virol 65: 2211–2219

Stirzaker SC, Poncet D, Both GW (1990) Sequences in the rotavirus glycoprotein VP7 which mediate delayed translocation and retention of the protein in the ER. J Cell Biol 111: 1343–1350

Tarlow O, McCrae MA (1990) Nucleotide sequence of gene 1 of the UK tissue culture adapted strain of bovine rotavirus. Nucleic Acids Res. 18: 7150

Taylor JA, Meyer JC, Legge MA, O'Brien JA, Street JE, Lord VJ, Bergmann CC, Bellamy AR (1992) Transient expression and mutational analysis of the rotavirus intracellular receptor— The C-terminal methionine residue is essential for ligand binding. J Virol 66: 3566–3572

Taylor JA, O'Brien JA, Lord VJ, Meyer JC, Bellamy AR (1993) The RER-localized rotavirus intracellular receptor—a truncated purified soluble form is multivalent and binds virus particles. Virology 194: 807–814

Valenzuela S, Pizarro J, Sandino AM, Vasquez M, Fernandez J, Hernandez O, Patton J, Spencer E (1991) Photoaffinity labeling of rotavirus VP1 with 8-azido-ATP: identification of the viral RNA polymerase. J Virol 65: 3964–3967

Rotavirus Replication

J.T. PATTON

1 Introduction

The development of methods for the propagation of rotaviruses in cell culture approximately 10 years ago represented an important milestone in the study of the rotavirus replication cycle. Although a general picture of most aspects of rotavirus replication has emerged since that time, a detailed description of the molecular biology of the replication cycle remains to be accomplished. By comparison with other RNA viruses, particularly those not of the *Reoviridae* family, it appears that the rotavirus replication cycle is unusually complex and will require considerable effort to fully elucidate. The fact that rotavirions contain two complete shells of protein made up of a total of six different proteins, bud through the endoplasmic reticulum as a step in morphogenesis, and have a genome that consists of 11 unique molecules of double-stranded RNA (dsRNA) indicate that the number of distinct intermediates formed

Department of Microbiology and Immunology, University of Miami School of Medicine, P.O. Box 016960 (R-138), Miami, FL 33101, USA

Current Topics in Microbiology and Immunology, Vol. 185
© Springer-Verlag Berlin · Heidelberg 1994

during the replication cycle are considerable. Perhaps the greatest gap in our knowledge of the replicative cycle concerns the mechanism of genome assortment. Understanding this process is made more difficult by our lack of information concerning the mechanism of assortment for any viruses containing segmented genomes, especially those which have fewer genome segments than the rotaviruses. Advances in our understanding of the replication cycle of the rotaviruses are allowing comparisons to be made with the reoviruses, the prototypic members of the family *Reoviridae*. Interestingly, although the mechanisms used in replicating the dsRNA genome of the rotaviruses and reoviruses appear to be quite similar, processes used by these viruses to assemble their outer shells are quite distinct. Indeed, one of the more unique features of the replication cycle of the rotaviruses is the role played by the endoplasmic reticulum in viral morphogenesis.

2 Overview of the Replication Cycle

As an introduction to rotavirus replication, an overview of this process is presented below. Details concerning the replication cycle will be discussed in later sections. Double-shelled rotaviruses bind to the host cell via the outer shell glycoprotein, VP7. As a result of cleavage of the outer shell hemagglutinin protein VP4 by a trypsin-like protease, the virus is then able to enter the cell by passing directly through the cytoplasmic membrane. In the cytoplasm, the outer shell of the virion is disrupted stimulating the RNA-dependent RNA polymerase (transcriptase) that is associated with the inner shell to synthesize viral messenger RNAs (mRNAs). The mRNAs are subsequently translated giving rise to all the viral proteins necessary in the replication cycle. The accumulation of viral proteins in the cytoplasm results in the formation of large inclusions termed viroplasms that are proposed to be the sites of genome replication and assembly of progeny single-shelled particles. In a process that requires genome assortment, subviral particles containing newly made protein and 11 different viral mRNAs, each representing a different genome segment, are formed. An RNA polymerase (replicase) activity associated with the subviral particles uses the mRNAs as templates for the synthesis of minus-strand RNA resulting in the formation of dsRNA. After the maturation of replicase particles into single-shelled particles, some will function as transcriptase particles and synthesize additional mRNAs, thus leading to an amplification of the level of RNA replication in the cell. Other newly formed single-shelled particles associate with VP4 and bud through regions of the endoplasmic reticulum containing NS28 and VP7. As a result of this process, single-shelled particles may be transiently surrounded by a membrane. During the final step in morphogenesis, the membrane covering the single-shelled particles is lost, while VP4 and VP7 condense around the particles producing the outer shell of protein found on rotavirions.

3 Adsorption and Entry

Two lines of evidence indicate that the outer shell glycoprotein VP7 is responsible for attachment of rotavirus to cellular receptors: (a) Antibodies directed against VP7 block the adsorption of rotaviruses to host cells (MATSUNO and INOUYE 1983; SABARA et al. 1985; FUKUHARA et al. 1988) and VP7 purified from virions has the capacity by itself to bind to cells under conditions used to adsorb intact virions (SABARA et al. 1985; FUKUHARA et al. 1988). Although little is known about the cellular receptors for rotavirus, components of the receptors for simian and bovine rotaviruses include sialic acids (YOLKEN et al. 1987; KELJO and SMITH 1988; FUKUDOME et al. 1989). For the embryonic monkey kidney cell line MA014, KELJO and SMITH (1988) have estimated the number of cellular receptors for rotavirus SA11 to be 13000 units per cell. After virus attachment, the outer shell hemagglutinin protein VP4 catalyzes entry of rotavirus into the cell. The role of VP4 in virus entry was perhaps best shown by FUKUHARA et al. (1988), who demonstrated that anti-VP4 antibodies were capable of blocking infection of cells containing preadsorbed, but not internalized, human rotavirus. Using electron microscopy to examine viral morphogenesis, early studies concluded that rotavirus gained entry into the cell through endocytosis (QUANN and DOANE 1983; PETRIE et al. 1981; ESTES et al. 1983). About the same time these studies were being carried out, investigators became aware that treatment of rotavirus with trypsin enhanced the infectivity of simian and bovine rotaviruses and was essential for propagation of human rotaviruses in cell culture (CLARK et al. 1979, 1981; GRAHAM and ESTES 1980). Later studies determined that trypsin-activated rotavirus did not penetrate the host via endocytosis but rather by direct penetration of the cell membrane (SUZUKI et al. 1985,1986; KALJOT et al. 1988). Trypsin treatment of rotavirus results in the cleavage of the outer shell hemagglutinin protein VP4, producing the two polypeptides VP5 (60 kDa and VP8 (28 kDa) (CLARK et al. 1981; ESPEJO et al. 1981; ESTES et al. 1981). The larger cleavage fragment VP5 contains a hydrophobic region similar to those found in the fusion protein of Sindbis and Semliki Forest viruses (MACKOW et al.1988). Thus a fusion-related activity of VP5 may play a role in the entry of rotavirus into cells by direct penetration. Further information on the structure and function of VP4 and VP7 in rotavirus replication is presented elsewhere in this volume.

Trypsin-activation of rotavirus significantly increases the kinetics of virus entry into the cell. Trypsin-activated rhesus rotavirus enters MA104 cells with a half-time of 3–5 min, but if non-activated the half-time of entry is 30–50 min (KALJOT et al. 1988). Studies by SUZUKI et al. (1986) indicate that during penetration through the cell membrane the outer shell of rotavirions is disrupted. The removal of the outer shell would presumably result in activation of the transcriptase associated with the inner shell of the virus (COHEN et al. 1979). As a result, perhaps it is not surprising that viral mRNAs and protein are made soon after cells are infected with trypsin-activated virus.

4 Viral Gene Expression

Although studies have examined temporal features of rotavirus replication, the results of these studies have not always precisely agreed. Given that these studies often varied with respect to the virus strain, type of host cell, multiplicity of infection, and usage of trypsin-activated virus, differences in results might be expected as all of these factors can have an affect on the kinetics of the replication cycle (ALMEIDA et al. 1978; MCCRAE and FAULKNER-VALLE 1981; ESTES et al. 1979, 1983). Newly made proteins have been detected in cells infected with trypsin-activated simian rotavirus SA11 as early as 70 min postinfection indicating that viral transcription begins soon after entry of the virus into the cell (LIU et al. 1988). Rotavirus RNA replication can be detected by 3 h postinfection, approximately the same time that single- and double-shelled particles initially appear in the infected cell (MCCRAE and FAULKNER-VALLE 1981; STACY-PHIPPS and PATTON 1987). Electrophoretic analysis of the synthesis of SA11 RNA during the replication cycle indicates that the level of transcription (plus-strand RNA synthesis) becomes maximum at 9–12 h postinfection while the level of RNA replication (minus-strand RNA synthesis) reaches maximum significantly earlier, at 6–9 h postinfection (STACY-PHIPPS and PATTON 1987). The fact that the level of replication does not directly parallel that of transcription suggests that the size of mRNA pools in the cell is not the only factor affecting the level of dsRNA synthesis. In contrast to reports on the mechanism of reovirus replication (ACS et al. 1971), it is clear that mRNAs made both at early and late times of infection serve as templates for rotavirus replication (STACY-PHIPPS and PATTON 1987). An interesting feature of the rotavirus replication cycle is that the levels of viral protein synthesis are greatest at 3–5 h postinfection, significantly earlier than periods of peak transcription (MCCRAE and FAULKNER-VALLE 1981; ERICSON et al. 1982). This suggests that a large amount of the viral mRNA made after 5 h postinfection is somehow sequestered, perhaps by movement into viroplasmic inclusions, such that the RNA is no longer available for translation. Depending on several factors including the multiplicity of infection and strain of virus, the rotavirus replication cycle is completed anywhere from 8.5 to 15 h postinfection (ESTES et al. 1979; MCCRAE and FAULKNER-VALLE 1981).

4.1 Regulation

No qualitative differences have been noted between the early and late products of rotavirus transcription or translation (MCCRAE and FAULKNER-VALLE 1981; ERICSON et al. 1982; STACY-PHIPPS and PATTON 1987; JOHNSON and MCCRAE 1989). Thus, unlike the DNA viruses, rotavirus gene expression cannot be divided easily into early and late phases. Rotavirus gene expression is regulated at the level of viral transcription, however, as viral mRNAs are not produced in equimolar concentrations in the cell despite the presence of equimolar concentrations of template dsRNAs in particles responsible for mRNA synthesis (STACY-PHIPPS and PATTON 1987; JOHNSON and McCRAE 1989; PATTON 1990). Viral gene expression is also regulated at the level of translation, as viral proteins are synthesized in the cell in relative concentrations significantly different from those of viral mRNAs (JOHNSON and McCRAE 1989). The result of regulation is that some proteins, such as VP6, the major inner shell protein, are produced at concentrations that exceed other proteins, such as the minor core protein, VP1, by greater than 100-fold. With nucleotide sequences known for all 11 of the rotavirus genome segments (MITCHELL and BOTH 1990), it should be possible to use site-specific mutagenesis to identify sequences in rotaviral mRNAs that affect the efficiency of their translation.

4.2 Transcriptase Particles

Single-shelled particles have been isolated from the cytoplasm of rotavirus-infected cells (HELMBERGER-JONES and PATTON 1986). When assayed for associated RNA polymerase activity in vitro in the presence of the four ribonucleotides, single-shelled particles are able to synthesize viral mRNAs indicating that such particles are responsible for viral transcription in the cell (COHEN 1977; SPENCER and ARIAS 1981; MASON et al. 1983). Particles with transcriptase activity exhibit optimal activity at a pH of 8–8.5 and in the presence of magnesium ions (COHEN 1977; SPENCER and ARIAS 1981). Single-shelled particles of simian rotavirus SA11 have a density of 1.38 g/cm^3 in CsCl and contain the viral proteins VP1, VP2, VP3, and VP6 (ERICSON et al. 1982; LIU et al. 1988). The molar ratio of these proteins in single-shelled particles is estimated to be 1:10:0.35:78, respectively (LIU et al. 1988). The identity of the viral transcriptase has not been directly demonstrated. However, sequencing of gene 1 of the simian, porcine, and bovine strains of rotavirus have indicated that VP1 shares structural similarities with known and putative RNA-dependent RNA polymerases (COHEN et al. 1989; FUKUHARA et al. 1989; MITCHELL and BOTH 1990). These finding have led to the suggestion that, in fact, VP1 is the rotavirus RNA polymerase and therefore would operate as the viral transcriptase. Because VP3 also shares some sequence similarity with RNA polymerases (LIU and ESTES 1989; MITCHELL and BOTH 1990), it remains possible that this protein functions as the rotavirus polymerase or is one of the components of a multicomponent polymerase complex. The fact that both VP1 and VP3 are in low copy number in single-shelled particles is consistent with the idea that one or both of these proteins are constituents of the viral RNA polymerase.

   A yet unexplained feature of rotavirus transcription is why this process requires the major inner shell protein VP6. BICAN et al. (1982) and SANDINO et al. (1986) have shown that chaotropic agents can be used to selectively remove VP6 from single-shelled particles. When assayed in vitro, such particles lack transcriptase activity but regain activity when reconstituted with purified VP6. Given that VP6 has no reported ability to bind nucleotides or nucleic acids (BOYLE and HOLMES 1986) and has no sequence homology with RNA polymerases, it will be interesting to determine how VP6 interacts with other proteins in single-shelled particles to stimulate the viral transcriptase.

Like the reoviruses, transcription of the rotavirus genome is a conservative process that uses as template the dsRNAs located within single-shelled particles (BANERJEE and SHATKIN 1970; PATTON 1986–1987). The mRNAs synthesized by single-shelled particles differ significantly from most cellular mRNAs as they contain 5'-cap structures but lack 3'-polyadenylate sequences (IMAI et al. 1983; MCCRAE and MCCORQUODALE 1983). SPENCER and GARCIA (1984) have shown that the enzymes necessary for capping of the viral transcripts, e.g., the guanylyl transferase, are components of the transcriptase particle. These same investigators have found that efficient elongation of transcripts requires the presence of 5'-terminal cap structures on the RNA molecules. The guanylyl transferase activity may be associated with the minor core component, VP3, as a protein with a similar molecular weight (88 kDa) present in rotavirions, is able to bind GTP specifically (FUKUHARA et al. 1989; Spencer, unpublished results).

Single-shelled particles that synthesize mRNA contain ATPase activity which appears to be necessary for the elongation of plus-strand RNA (SPENCER and ARIAS 1981; SPENCER and GARCIA 1984). Possibly, the energy produced from the hydrolysis of ATP is used during transcription by a helicase activity to disrupt basepairing in the dsRNA template. Alternatively, the energy derived from hydrolysis may be required to displace the progeny transcript from the minus-strand template during transcription.

Double-shelled particles also contain the viral transcriptase but that activity is masked by the presence of the outer shell. Exposure of double-shelled particles to low concentrations of Ca^{2+} chelating agents or to brief heat shock disrupts the outer shell giving rise to particles able to synthesize mRNAs in vitro (COHEN et al. 1979; SPENCER and ARIAS 1981). The affect of chelating agents on double-shelled particles suggests that Ca^{2+} may play an important role in virion stability and in the activation of the virion-associated transcriptase in the cell (COHEN et al. 1979).

Little in known about the nature of the protein-protein interactions that exist in single-shelled particles. Using cross-linking agents to examine the structure of single-shelled particles, SANDINO and SPENCER (unpublished results) have found that VP6 is in direct contact with VP2 and VP3. Given that VP3 may be the guanylyl transferase, these results suggest that VP3, located in channels extending into the core of single-shelled particles spans the core protein, VP2, and the surrounding inner shell protein, VP6 (Fig. 1). Perhaps the interaction of VP6 with VP3 stimulates capping activity and thus

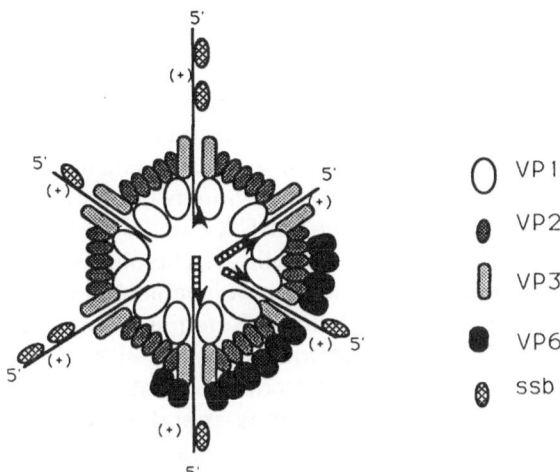

Fig. 1. Proposed structure of a rotavirus replicase particle. VP1, VP2, and VP3 are required components of an enzymatically active replicase particle. During replication, the plus-strand RNA template, which initially extends from the particle, moves to the interior of the particle via channels located at the vertices. In the interior, the RNA template is replicated by the putative RNA polymerase, VP1, to produce double-stranded RNA. VP3, as the putative guanylyl transferase, is positioned in channels of the vertices such that it is able to interact directly with VP2 and VP6. During RNA replication, VP6 associates with the replicase particle allowing the particle to simultaneously undergo assembly into a single-shelled particle. The role of viral single-stranded RNA binding proteins (ssb) in RNA replication is unknown

promotes VP1 to synthesize viral mRNA. This would provide one possible explanation for the requirement of VP6 in transcription.

Despite the many uncertainties concerning the structure of single-shelled particles, the structure must allow movement of the transcriptase relative to the dsRNA template during the synthesis of mRNA. Transcriptase particles are able to carry out mRNA synthesis in vitro continuously over a period of hours indicating an ability of the RNA polymerase within the particles to reinitiate plus-strand RNA synthesis (COHEN 1977; SPENCER and GARCIA 1984). This ability implies that the RNA polymerase is recycled during transcription such that the same molecule of polymerase can repeatedly transcribe at least one of the genome segments. YAZAKI et al. (1986) have provided evidence that each of the genome segments of cytoplasmic polyhedrosis virus, a member of the family *Reoviridae*, are held in circular configuration by a protein that binds specifically to the 5'- and 3'-termini of the genome segments. If the genome segments of the rotaviruses are similarly circularized, then during transcription it may be possible that the RNA polymerase could move directly from the 3'-end to the 5'-end of the segment without releasing from the template. Presumably that would significantly increase the rate of reinitiation as compared to the rate that would be produced by the polymerase dropping off the template after elongation and then having to reassociate with the template to initiate transcription.

5 Rotavirus RNA Replication

Although the rotaviruses differ significantly from the reoviruses in several aspects, including gene number and morphogenesis, their mechanisms for genome replication are quite similar. Both types of viruses replicate their genomes asymmetrically with viral plus-strand RNA serving as the template for the synthesis of minus-strand RNA to produce dsRNA (ACS et al. 1971; SAKUMA and WATANABE 1971, 1972; PATTON 1986–1987). The plus-strand RNA template used in rotavirus replication is indistinguishable from viral mRNA containing a 5'-terminal cap but lacking a 3'-terminal poly(A) sequence (IMAI et al. 1983; MCCRAE and MCCOROUODALE 1983). Viral mRNA, therefore, serves dual functions in the cell: to direct the synthesis of viral protein and to template the synthesis of minus-strand RNA. The mechanism used to regulate the relative amounts of viral mRNA that participate in RNA replication versus translation remains unknown. Naked viral dsRNA cannot be found in rotavirus-infected cells nor is it released from subviral particles following synthesis in vitro. These findings indicate that each replicase particle is responsible for only a single cycle of RNA replication. The viral replicase presumably remains in the replicase particle following the synthesis of dsRNA serving ultimately as the viral transcriptase.

Electrophoretic analyses of the synthesis of rotavirus plus- and minus-strand RNAs during the replication cycle has shown that, although the genome segments are transcribed with unequal frequencies, they are replicated with equal frequencies (STACY-PHIPPS and PATTON 1987; PATTON 1990). How replication is regulated to give equimolar synthesis of the genome segments is not known. However, one possibility is that a replicase particle cannot initiate synthesis of dsRNA until the particle via assortment contains 11 different viral mRNAs, each corresponding to a different genome segment. Since replicase particles can carry out only a single-cycle of replication, replicase activity in such particles would, under normal situations, lead to equal frequencies of genome replication. Studies by ACS et al. (1971) indicate that the reoviruses also replicate their genome segments with equal frequencies. The results of another study, in which the synthesis of rotavirus RNAs during the replication cycle was monitored by hybridization assay, suggested that differences may exist in the replication frequencies of some genome segments (JOHNSON and MCCRAE 1989). The apparent conflict in the results of this study with those described above remains unexplained but may be due to differences in technical approaches used to assay for minus-strand RNA. Another possibility is that due to the use of high multiplicities of infection in experiments by JOHNSON and MCCRAE (1989), defective virus particles were generated leading to the appearance of non-equimolar synthesis of genomic dsRNAs in the cell.

5.1 Genome Assortment

A property common to the rotaviruses and reoviruses is their ability to form reassortants upon coinfection of cells with two different strains of viruses (RAMIG and FIELDS 1983; KAPIKIAN and CHANOCK 1985). The process of assortment for the rotaviruses and other viruses containing segmented genomes is not understood. However, certain features of rotavirus replication provide some insight into the mechanism of this process. For instance, replicase particles isolated from infected cells have the capacity to synthesize all 11 genomic dsRNAs (HELMBERGER-JONES and PATTON 1986; PATTON and GALLEGOS 1988). To date, it has not been possible to separate replicase particles into subsets, each responsible for replication of a different genome segment (unpublished results). Thus replicase particles apparently do not initiate replication until assortment has occurred. Since replication follows assortment, it must be viral mRNAs that undergo assortment as opposed to dsRNAs. From this it follows that the recognition signals for assortment must exist on the viral mRNAs. The fact that reassortant viruses are readily formed in coinfected cells suggests that, during genome assortment and the assembly of replicase particles, the viral mRNAs that are destined to serve as templates for replication are assorted independently. Successful reassortment between different strains of viruses requires that they share common recognition signals.

Little progress has been made in defining the recognition signals present on viral mRNAs that are used for genome assortment. The assortment process is likely to require two signals, one that allows viral mRNAs and host mRNAs to be distinguished from each other and another that allows the 11 rotavirus mRNAs to be individually recognized. Sequence analysis has shown that the 5'- and 3'-ends of the rotavirus dsRNA genome segments are conserved containing identical 5'–10 base terminal sequences (IMAI et al. 1983; McCRAE and McCORQUODALE 1983). These conserved sequences and the lack of a 3'-terminal poly(A) tail may help differentiate viral from host mRNAs during the assortment process. The 3' conserved sequence on viral mRNAs can also be expected to function as a recognition signal for binding of the viral replicase. However, as the terminal sequences are identical on all 11 types of viral mRNAs produced in infected cells, these sequences cannot alone make up the recognition signals ("sorting signals") necessary for assortment of viral mRNAs (ANZOLA et al. 1987). The sorting signals may be located near the 5'- and 3'-ends of viral mRNAs, as comparison of the sequence of the gene for VP9 of Wa (segment 11), SA11 (11), UK (11) and short (RV-5, segment 10), and supershort (B37, 10) strains of rotaviruses demonstrated that the first 21 and last 42 bases of all the genes are highly conserved (NUTTAL et al. 1989). Length of the mRNA probably does not constitute part of the sorting signals as rotavirus strains have been isolated that, as a result of rearrangements, contain genome segments of dramatically altered size which do not appear to affect their ability to undergo assortment

(GONZALEZ et al. 1989; HUNDLEY et al. 1985, 1987; MATTION et al. 1988). Studies on wound tumor virus (WTV), a member of the family *Reoviridae* with a genome of 12 segments of dsRNA, indicate that this virus contains segment-specific inverted repeats adjacent to conserved terminal sequences (ANZOLA et al. 1987). The inverted repeats and terminal conserved sequences have been proposed to allow stable, segment-specific basepairing between the 5′ and 3′-terminal regions of WTV mRNAs. Xu et al. (1989) have suggested that 5′-terminal-3′-terminal interactions may form the sorting signals necessary for genome assortment for members of the family *Reoviridae*. Indeed, this may be the case for the rotaviruses, as OKADO et al. (1984) reported that, based on sequence analysis, rotavirus mRNAs may be able to form "panhandle" structures involving their 5′- and 3′-terminal regions. The possibility that 5′-terminal-3′-terminal interactions may constitute part of the sorting signals for *Reoviridae* is far from certain, however, as such terminal interactions do not appear to be possible in the case of some reovirus mRNAs (RICHARDSON and FURUICHI 1983).

5.2 Replicase Particles

The development of cell-free systems that support the synthesis of rotavirus dsRNA has proven useful for the study of the structure and function of replicase particles. Treatment of rotavirus and reovirus replicase particles with single-strand-specific RNase abolishes their capacity to synthesize full-length dsRNA in vitro (ACS et al. 1971; PATTON 1986–1987). This phenomenon suggests that the 11 plus-strand RNAs that serve as templates for minus-strand synthesis are located in replicase particles such that they are accessible to the nuclease. Therefore, replicase particles can be viewed as having plus-strand RNA near to or extending from their surface (Fig. 1). In contrast to the plus-strand RNAs, the dsRNA products of replication are resistant to digestion by double-strand-specific RNase and, hence, are likely sequestered within the protein shell surrounding the replicase particle (PATTON and GALLEGOS 1988). Comparison of the nuclease sensitivities of the plus-stand RNA templates and the dsRNA products supports a hypothesis that, during RNA replication, the RNA templates move from areas extending out of the replicase particle to areas within the particle. Electrophoretic analyses of the overall size of replicase particles during RNA replication have demonstrated that replicase particles undergo a reduction in size as they synthesize the 11 genomic dsRNAs (PATTON and GALLEGOS 1990). As determined by migration in nondenaturing agarose gels, newly assembled replicase particles have an overall size of 100 nm or more but after completing replication their size is reduced to 45–75 nm. The reduction in size during replication correlates directly with a decrease in the amount of plus-strand RNA templates associated with the particles. Treatment of SA11 replicase particles with single-strand-specific RNase, besides interfering with their ability to synthesize full-length dsRNAs, reduces their size to that of untreated replicase particles which have completed genome replication (45–75 nm; PATTON and GALLEGOS 1990). These results are in agreement with the hypothesis that during replication the plus-strand RNA templates move into the replicase particle during dsRNA synthesis. The fact that the size of replicase particles changes during RNA replication indicates that the plus-strand templates must physically extend from the surface of newly formed replicase particles (Fig. 1). If the templates were located only within channels near the surface of the replicase particle, a change in the overall size of the particle during the synthesis of dsRNA would not be expected. The decrease in size of the particle during replication may also be affected by the loss of single-strand RNA binding proteins as the plus-strand RNA template moves into the particle and is converted to dsRNA. ZWEERINK (1974) previously reported that reovirus replicase particles may similarly undergo a change in size during RNA replication.

Although the data indicate that plus-strand templates for dsRNA synthesis extend from replicase particles, it is not certain which regions of the templates this actually includes. As the 3'-terminal regions of the RNA templates should contain the promoter for the viral replicase, the 3'-termini can be expected to be located within the interior of the particle. In contrast, the 5'-terminal regions of the template RNAs may extend from the replicase particle or loop back into the replicase particle due to basepairing with its 3'-terminal regions. In the latter case, significant portions of the looped out regions of the templates must extend from the replicase particle to account for the extreme sensitivity of these particles to enzymatic inactivation by RNase treatment.

Analysis of the dsRNA product of SA11 replicase particles as they carry out genome replication in a cell-free system suggests that the synthesis of viral dsRNA is an ordered process, with the smallest genome segments replicated to completion first and the largest segment completed last (PATTON and GALLEGOS 1990). This result is in agreement with those obtained for the reovirus replicase particle which suggests that the reovirus genome is also sequentially replicated beginning with the smallest segments and ending with the largest segments (ZWEERINK 1974). Although replication of the rotavirus genome segments may proceed sequentially, this idea does not necessarily reflect the pattern of initiation of minus-strand synthesis in the replicase particle. Minus-strand synthesis may be initiated simultaneously for all 11 genome segments and, because of differences in the length of the RNA templates, the smallest segments could be replicated to completion first and the largest segments replicated to completion last. Alternatively, minus-strand RNA synthesis may be initiated in a sequential manner, beginning with the smallest genome segments and ending with the largest genome segments. Given that these results indicate a very ordered process of genome replication in the replicase particle, the synthesis of dsRNA in the infected cell would be expected to be equimolar. For nonequimolar synthesis of dsRNA to occur would appear to require a mechanism to bypass any requirement that might exist for smaller segments to be replicated prior to larger segments.

5.3 Replication Intermediates

To help define the role of viral proteins in rotavirus replication, several studies have attempted to identify the protein components of subviral particles that have associated polymerase activity. In initial studies, subviral particles prepared from the cytoplasm of rotavirus SA11-infected cells were resolved by centrifugation on CsCl gradients and those fractions of the gradients containing replicase and transcriptase particles were identified by in vitro polymerase assay (HELMBERGER-JONES and PATTON 1986; PATTON and GALLEGOS 1988). Analysis of the protein composition of these particles showed that, as a population, replicase particles differed from transcriptase paricles in density and protein composition. Transcriptase particles were shown to be like single-shelled virions in density (approximately 1.38 g/cm^3) and in protein content containing VP1, VP2, and VP6. In contrast, replicase particles were found to be extremely heterogenous in density, with most denser than transcriptase particles. In some very early experiments, SAKUMA and WATANABE (1971) similarly showed that reovirus replicase particles, in contrast to transcriptase particles, varied considerably in density. The heterogeneity of replicase particles indicates that they vary structurally. One source of the heterogeneity might be due to the changes in the spatial arrangement of RNA in the replicase particle as the plus-strand RNA templates move from outside into the interior of the particle where they are converted to dsRNA. Protein analysis showed that SA11 replicase particles recovered from CsCl gradients contained cores of VP1 and VP2 at a molar ratio similar to those found for single-and double-shelled virions (HELMBERGER-JONES and PATTON 1986). These results provided evidence that VP1 and VP2 are components of the viral replicase. However, protein analysis of replicase particles recovered from CsCl gradients showed that they contained, on average, significantly less VP6 per particle than those with transcriptase activity, i.e., single-shelled particles. The low molar ratio of VP6 to VP2 indicated that some replicase particles lack a complete shell of VP6 protein and thus supports the hypothesis that, in contrast to transcription, VP6 is not required for SA11 RNA replication (PATTON and GALLEGOS 1988). NS34 and NS35 were present in fractions of CsCl gradients that contained replicase particles and co-sedimented with replicase particles on glycerol gradients providing evidence that these nonstructural proteins may play a role in genome replication. Recent studies have shown that NS35 is an RNA binding protein with no detectable sequence specificity (M. KATTOURA and J. PATTON, unpublished results). Due to this property, NS35 may be associated with replicase particles because of its affinity for the mRNAs that serve as templates for replication in such particles.

More recent studies have used nondenaturing electrophoretic gel systems to characterize intracellular subviral particles involved with the replication of SA11 rotavirus dsRNA (GALLEGOS and PATTON 1989). Such gel systems can fully resolve, without disruption, rotavirus double-shelled, single-shelled, and core particles. When SA11 intracellular subviral particles were assayed for polymerase activity in vitro and then subjected to electrophoresis on nondenaturing agarose gels, those particles containing newly made genome-length dsRNAs migrated at positions between virion-derived cores and intermediate of single- and double-shelled virions. As an indication of the protein composition of particles containing newly made dsRNA nonassayed subviral particles that comigrated on the nondenaturing gels with particles containing newly made dsRNA were examined for protein content. The results indicated that rotavirus-infected cells contain three unique types of replication intermediates (RIs). The largest intermediate (75 nm, 320–390 S) contains the inner shell proteins VP1, VP2, VP3 and VP6 and was therefore referred to as the single-shelled RI (GALLEGOS and PATTON 1989). The similarity of single-shelled RIs and single-shelled virions in protein composition and in rates of migration on agarose gels suggests that the overall structure of these particles is similar. The nonstructural proteins NS35, NS34 and NS36 (gene 11 product, previously referred to as VP9, WELCH et al. 1989) were also detected in regions of the gel that contained the single-shelled RI suggesting that they are components of these structure. The second type of RI (core RI) that was identified contains the same proteins found in virion-derived cores, i.e., VP1, VP2, and VP3. The core RI lacks the major inner shell protein VP6 and is significantly greater in size than virion-derived cores (60 nm vs 45 nm). The size of core RIs may exceed that of virion-derived cores because of the presence of associated nonstructural proteins (NS35, NS34, and NS26). Like the results obtained by analyzing the protein composition of replicase particles recovered from CsCl gradients, the presence of newly made dsRNA in core RIs indicates that VP6 is not required for replicase activity. Characterization of nonassayed subviral particles that comigrated on nondenaturing gels with the smallest paticles containing newly made dsRNA suggested that an RI (precore RI, 45 nm, 160 S) consisting of only the structural proteins VP1 and VP3 may have replicase activity. In addition to the nonstructural proteins NS35, NS34, and NS26, in some instances a protein of similar molecular weight to the SA11 gene 5 product, NS53, was detected in preparations of the precore RI. The size of the precore RI (45 nm) was similar to that of virion-derived cores although the latter structure contains the major core protein VP2. The lack of VP2 in subviral particles that comigrated with particles that contained newly made dsRNA suggests that VP2 may not be necessary for replicase activity. However, given that the size of replicase particles decreases during dsRNA synthesis (see Sect. 5.2), the protein composition of nonassayed subviral particles may not fairly represent the protein composition of comigrating replicase particles that have assayed for polymerase activity. It may be

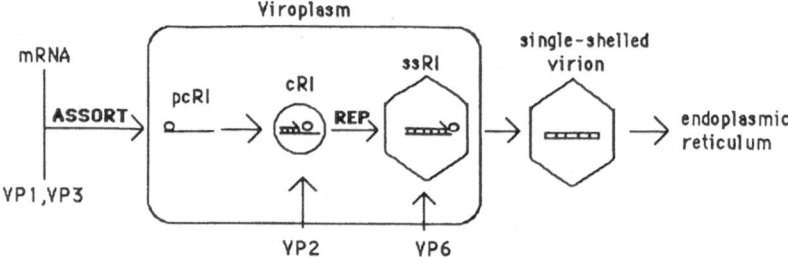

Fig. 2. Concurrent RNA replication and the assembly of single-shelled particles. Viral mRNA associates with the structural proteins VP1 and VP3 and other nonstructural proteins to form mRNA-protein complexes (mRNPs). Via assortment (Assort), the mRNPs are assembled into precore RIs (pcRIs) which contain 11 viral RNAs, each representing a different genome segment. In the viroplasm, VP2 interacts with the precore RI producing the core RI (cRI) and stimulating the replicase in the particle to initiate RNA replication (*REP*). As RNA replication continues, VP6 binds to the core RI giving rise to the single-shelled RI (ssRI). Following replication, the single-shelled RI moves out the viroplasm and, as a single-shelled virion, either functions as a transcriptase particle or associates with the endoplasmic reticulum where it undergoes maturation into a double-shelled virion

that some replicase particles containing VP2, although initially having sizes significantly greater than 45 nm, once assayed decrease in size such that they comigrate with particles identified as the precore RI.

As discussed more fully in a later section (Sect. 6.2), single-shelled particles are proposed to assemble by the sequential addition of first VP2 and then VP6 to the precore RIs (GALLEGOS and PATTON 1989). Thus, precore RIs may be the direct precursors of core RIs which, in turn, may be the direct precursors of single-shelled RIs and single-shelled particles (Fig. 2). As precore and core RIs assemble into single-shelled particles, replicase activity associated with the RIs is proposed to concurrently synthesis minus-strand RNA leading to the production of dsRNA (GALLEGOS and PATTON 1989). This model for concurrent morphogenesis and RNA replication provides an explanation for why multiple structurally distinct types of subviral particles with replicase activity can be detected in infected cells. The heterogeneity of replicase particles with respect to density in CsCl is probably due, in no small part, to their variability in protein composition stemming from replicase particles undergoing maturation into single-shelled particles as they synthesize dsRNA.

5.4 Proteins Required for Replicase Activity

Temperature-sensitive (ts) mutants have been developed which map to most of the rotavirus genome segments (RAMIG 1982, 1983; GOMBOLD and RAMIG 1987; GOMBOLD et al. 1987). As described in detail elsewhere in this volume, these mutants have been used to genetically define the function of viral proteins in the rotavirus replication cycle (RAMIG and PETRIE 1984). Two SA11 mutants, tsF and tsG, possessing temperature sensitive lesions that map to genes encoding VP2 and VP6, respectively, have recently been used to examine the role of these inner shell proteins in rotavirus RNA replication at the molecular level (MANSELL and PATTON 1990). Cells infected with tsF and maintained at temperatures permissive for replication (31°C) for several hours and then shifted to nonpermissive temperatures (39°C) lose the ability to form subviral particles that contain replicase activity. In contrast, cells infected with tsG, maintained at permissive temperatures and then shifted to nonpermissive temperatures, continue to form particles that have associated replicase activity. Examination of particles assembled in tsG-infected cells at nonpermissive temperatures revealed that particles with replicase activity lacked VP6 and were similar in protein composition to the core RIs described by GALLEGOS and PATTON (1989). These results provide compelling evidence that VP6 is not a necessary component of the viral replicase. Thus the viral replicase and transcriptase can be distinguished based upon their need for VP6 for polymerase activity (MANSELL and PATTON 1990). The fact that tsF was unable to assemble replicase particles at nonpermissive temperatures indicates that functional VP2 is required for the assembly of replicase particles with polymerase activity. This result is consistent with studies examining tsF-infected cells by electron microscopy which showed that morphogenic intermediates were not formed at 39°C (RAMIG and PETRIE 1984). The study of the morphogenic intermediates and polymerase activities present in cells infected with ts mutants other than tsF and tsG should provide additional information on the function of viral proteins in rotavirus replication.

6 Virion Morphogenesis

6.1 Assembly of the Inner Shell

Several studies have used electron microscopy to define steps in rotavirus morphogenesis. A study by ESPARZA et al. (1980) described three types of particles in the cytoplasm of cells infected with human rotavirus: the nucleoid (37 nm), core (50 nm), and single-capsid (70 nm) particles. The nucleoid and core particles were suggested to be the morphological subunits of single-shelled particles. Other studies have also detected nucleoid- and core-like structures in rotavirus-infected cells and similarly suggested that they serve as precursors of single-shelled particles (CHASEY 1977; PETRIE et al. 1981; HOLMES 1983). Core particles are proposed to assemble in viroplasmic inclusions and then undergo additional morphogenesis into single-shelled particles near the periphery of the inclusions (ALTENBURG et al. 1980; PETRIE et al. 1981; KABCENELL et al. 1988). Further support for the hypothesis that single-shelled particles are assembled in viroplasmic inclusions comes from studies using immunoelectron microscopy, which have shown that VP6 and VP2, the major components of single-shelled and core particles, respectively, both accumulate in the inclusions (PETRIE et al. 1984). In addition to containing the nonstructural proteins NS35 and NS26, the inclusions may also contain viral mRNA and serve as the sites of genome assortment and RNA replication (PETRIE et al. 1984). The assembly of single-shelled particles is a rapid process, as KABCENELL et al. (1988) have found that particles containing newly synthesized VP2 and VP6 can be detected within 5 min of chase after a 5 min pulse-labeling.

Based on a comparison of sizes, the nucleoid, core, and single-capsid particles described by ESPARZA et al. (1980) are probably counterparts of the precore (45 nm), core (60 nm), and single-shelled RIs (75 nm) identified by GALLEGOS and PATTON (1989). Characterization of rotavirus RIs suggests that there are at least three intermediates in the assembly of single-shelled particles (GALLEGOS and PATTON 1989). These intermediates, the precore, core, and single-shelled RIs, have common nonstructural proteins (NS35, NS34, NS26) but differ significantly in structural protein composition. Based on a comparison of their protein composition, the initial intermediate formed in the assembly of single-shelled particles is probably the precore RI which contains the structural proteins VP1 and VP3 (Fig. 2). From the nucleotide sequence of the genes encoding VP1 and VP3, one of these two proteins, most likely VP1, is the viral RNA-dependent RNA polymerase (COHEN et al. 1989; FUKUHARA et al. 1989; LIU and ESTES 1989; MITCHELL and BOTH 1990). This suggests that in the formation of single-shelled particles the first structural proteins to associate with the viral mRNA template for replication is the viral polymerase. Comparison of the protein composition of the precore and core RIs indicates that the core RI may be formed in vivo by the addition of VP2 to the precore RI. Based on studies of the SA11 gene 2 ts mutant tsF, association of VP2 with the precore RI to form the core RI likely activates the viral polymerase, i.e., VP1 and /or VP3, to initiate the synthesis of minus-strand RNA (MANSELL and PATTON 1990). Studies on the RNA binding properties of rotavirus proteins suggest that VP2 may associate with the dsRNA product as it is synthesized by the core RI (BOYLE and HOLMES 1986).

As replication proceeds, the core RI interacts with VP6 to form the single-shelled RI which then contains the complete complement of structural proteins that are found in single-shelled virions. VP6 has been shown to exist in infected cells and purified virions as a homotrimer (GORZIGLIA et al. 1985; SABARA et al. 1987; PRASAD et al. 1988). SABARA et al. (1987) have proposed that VP6 trimers or multimers of these trimers associate with cores to form the single-shelled particles. Cell-free systems containing truncated molecules of VP6 have been used to identify domains in VP6 required for assembly into single-shelled particles and for trimerization (L. Clapp and J. Patton, unpublished results). The results indicate that the domain essential for binding of VP6 to cores resides at the carboxyl end of the protein located between amino acid residues 251 and 397. In contrast, the domain essential for trimerization resides near the center of VP6 located between residues 105 and 328. The fact that some truncated species of VP6, although able to assemble into single-shelled particles in vitro , are unable to form trimers suggests that trimerization of VP6 is not necessary for the assembly of single-shelled particles.

Completion of minus-strand RNA synthesis in the single-shelled RI and the loss of associated nonstructural proteins would constitute the final step in the assembly of single-shelled particles. A complete understanding of the rotavirus replication cycle will require additional information on the types of protein-protein interactions that are required in the assembly of single-shelled particles and the possible function of nonstructural proteins in this process. A direct demonstration that the precore and core RIs are precursors of single-shelled particles will also be necessary to verify the steps in the assembly pathway.

Once formed, a single-shelled particle can presumably serve either of two functions: (1) as an intermediate in the assembly of a double-shelled particle or (2) as a transcriptase particle to synthesize viral mRNAs. It is not known if a transcriptase particle can ultimately act as an intermediate in the assembly of a double-shelled particle or is destined to synthesize viral message throughout the replication cycle. Transcriptase particles are probably produced through-out the replication cycle, i.e., both at early and late times postinfection, as the level of viral transcription continually increases in the infected cell until 9–12 h postinfection (STACY-PHIPPS and PATTON 1987).

6.2 Assembly of the Outer Shell

The final stages of rotavirus morphogenesis occur in the rough endoplasmic reticulum (RER) where the hemagglutinin protein VP4 and the glycoprotein VP7 condense around single-shelled particles to form the outer shell of protein present on rotavirions (ESTES et al. 1983; DUBOIS-DALQ et al. 1984). During virion maturation, single-shelled particles are proposed to migrate from viroplasmic inclusions to regions of the RER that contain the rotavirus glycoproteins VP7 and NS28 (PETRIE et al. 1982; CHAN et al. 1988). Single-shelled particles then bind to the RER because of their affinity for the viral nonstructural receptor protein NS28 (AU et al. 1988; MEYER et al. 1989). MEYER et al. (1989) have shown that VP6 is the ligand that allows binding of single-shelled particles to NS28. These investigators also demonstrated that efficient binding of single-shelled particles requires Ca^{2+}. This result perhaps explains the observation by SHAHRABADI and LEE (1986) that single-shelled particles, but not double-shelled, are assembled in cells maintained in the absence of Ca^{2+}. After binding to the RER, single-shelled particles bud into the lumen of the RER resulting in particles that are bound by membranes which presumably contain VP7 and NS28 (DUBOIS-DALQ et al. 1984; KABCENELL et al. 1988). How VP4 becomes associated with single-shelled particles during morphogenesis is not clear but probably occurs prior to the entry of particles into the RER for the following reasons: (1) Analysis of the amino acid sequence of VP4 indicates that this protein lacks an NH_2 terminal signal peptide and thus is not likely to be able to pass unassisted into the RER (LOPEZ et al. 1985, MACKOW et al. 1988). (2) Immunofluorescence studies suggest that unassembled VP4 may accumulate in the cytoplasm on or near the RER (KABCENELL et al. 1988). Taken together, these results imply that what passes into the RER via budding are single-shelled particles which already contain the necessary VP4 that will be required to assemble the outer shell. As a result of budding, single-shelled particles are membrane-bound but only transiently (DUBOIS-DALQ et al. 1984). In the RER, through an unexplained mechanism, the membrane that surrounds the particles disintegrates releasing NS28 and allowing VP7 to associate with VP4 and condense around the particle to form the outer shell. The time required to assemble the outer shell beginning with a single-shelled particle that has yet to enter the RER is as little as 10–15 min (KABCENELL et al. 1988). Maturing double-shelled virions continue to accumulate in the lumen of the RER until cell lysis (CHASEY 1977; HOLMES 1983). Additional details concerning the structure and function of VP4 and VP7 in rotavirus morphogenesis is presented elsewhere in this volume.

References

Acs G, Schonberg H, Christman J, Levin DH, Silverstein SC (1971) Mechanism of reovirus double-stranded ribonucleic acid synthesis in vivo and in vitro. J Virol 8: 684–689

Almeida JD, Hall T, Banatvala JE, Totterdell BM, Chrystie IL (1978) The effect of trypsin on the growth of rotavirus. J Gen Virol 40: 213–218

Altenburg BC, Graham DY, Estes MK (1980) Ultrastructural study of rotavirus replication in cultures cells. J Gen Virol 46: 75–85

Anzola JV, Xu Z, Asamizu T, Nuss DL (1987) Segment-specific inverted repeats found adjacent to conserved terminal sequences in wound tumor virus genome and defective interfering RNAs. Proc Natl Acad Sci USA 84: 8301–8305

Au K-S, Chan W-K, Estes MK (1988) Rotavirus morphogenesis involves an endoplasmic reticulum transmembrane glycoprotein. In: Compans R, Helenius A, Oldstone M (eds) UCLA symposium on cell biology of viral entry, replication and pathogenesis. Lis, New York, pp 257–267

Banerjee AK, Shatkin AJ (1970) Transcription in vitro by reovirus-associated ribonucleic acid dependent polymerase. J Virol 6: 1–11

Bican P, Cohen J, Charplilienne A, Scherrer R (1982) Purification and characterization of bovine rotavirus cores. J Virol 43: 1113–1117

Boyle JF, Holmes KV (1986) RNA-binding proteins of bovine rotavirus. J Virol 58: 561–568

Chan W-K, Au K-S, Estes MK (1988) Topography of the simian rotavirus nonstructural glycoprotein (NS28) in the endoplasmic reticulum membrane. Virology 164: 435–442

Chasey D (1977) Different particle types in tissue culture and intestinal epithelium infected with rotavirus. J Gen Virol 37: 443–451

Clark SM, Barnett BB, Spendlove RS (1979) Production of high-titer bovine rotavirus with trypsin. J Clin Microbiol 9: 413–417

Clark SM, Roth JR, Clark ML, Barnett BB, Spendlove RS (1981) Trypsin enhancement of rotavirus infectivity: mechanism of enhancement. J Virol 39: 816–822

Cohen J (1977) Ribonucleic acid polymerase activity associated with purified calf rotavirus. J Gen Virol 36: 395–402

Cohen J, Laporte J, Charpilienne A, Scherrer R (1979) Activation of rotavirus RNA polymerase by calcium chelation. Arch Virol 60: 177–186

Cohen J, Charpilienne A, Chilmonczyk S, Estes MK (1989) Nucleotide sequence of bovine rotavirus gene 1 and expression of the gene in baculovirus. Virology 171: 431

Dubois-Dalq M, Holmes KV, Rentier B (1984) Assembly of rotaviruses. In: Dubois-Dalq M, Holmes KV, Rentier B (eds) Assembly of enveloped RNA viruses. Springer, Vienna, New York, pp 171–184

Ericson BL, Graham DY, Mason BB, Estes MK (1982) Identification, synthesis, and modifications of simian rotavirus SA11 polypeptides in infected cells. J Virol 42: 825–839

Esparza J, Gorziglia M, Gil F, Romer H (1980) Multiplication of human rotaviruses in cultured cells: an electron microscopic study. J Gen Virol 47: 461–472

Espejo RR, Lopez S, Arias C (1981) Structural polypeptides of simian rotavirus and the effect of trypsin. J Virol 37: 156–160

Estes MK, Graham DY, Gerba CP, Smith EM (1979) Simian rotavirus SA11 replication in cell cultures. J Virol 31: 810–815

Estes MK, Graham DY, Mason BB (1981) Protelytic enhancement of rotavirus infectivity: molecular mechanisms. J Virol 39: 879–888

Estes MK, Palmer EL, Obijeski JF (1983) Rotaviruses: a review. In: Compans RW, Cooper M, Koprowski JF et al. (eds) Current topics in microbiology and immunology, vol 105. Springer, Berlin Heidelberg New York, pp 123–184

Fukudome K, Yoshie O, Konno T (1989) Comparison of human, simian, and bovine rotaviruses for requirement of sialic acid in hemagglutination and cell adsorption. Virology 172: 196–205

Fukuhara N, Yoshie O, Kitaoka S, Konno T (1988) Role of VP3 in human rotavirus internalization after target cell attachment via VP7. J Virol 62: 2209–2218

Fukuhara N, Nishikawa K, Gorziglia M, Kapikian AZ (1989) Nucleotide sequence of gene segment 1 of a porcine rotavirus strain. Virology 173: 743–749

Gallegos CO, Patton JT (1989) Characterization of rotavirus replication intermediates: A model for the assembly of single-shelled particles. Virology 172: 616–627

Gombold JL, Ramig RF (1987) Assignment of simian rotavirus SA11 temperature-sensitive mutant groups A,C,F and G to genome segments. Virology 161: 463–473

Gombold JL, Estes MK, Ramig RF (1985) Assignment of simian rotavirus SA11 temperature-sensitive mutant groups B and E to genome segments. Virology 143: 309–320

Gonzalez SA, Mattion NM, Bellinzoni R, Burrone OR (1989) Structure of rearranged genome segment 11 in two different rotavirus strains generated by a similar mechanism. J Gen Virol 70: 1329–1336

Gorziglia M, Larrea C, Liprandi F, Esparza J (1985) Biochemical evidence for the oligomeric (possibly trimeric) structure of the major inner capsid polypeptide (45K) of rotaviruses. J Gen Virol 66: 1889–1900

Graham DY, Estes MK (1980) Proteolytic enhancement of rotavirus infectivity: biologic mechanisms. Virology 101: 432–439

Helmberger-Jones M, Patton JT (1986) Characterization of subviral particles in cells infected with simian rotavirus SA11. Virology 155: 655–665

Holmes IH (1983) Rotavirus. In: Joklik WK (ed) The reoviridae. Plenum, New York, pp 359–423

Hundley F, Biryahwaho B, Gow M, Desselberger U (1985) Genome rearrangements of bovine rotavirus after serial passage at high multiplicity of infection. Virology 143: 88–103

Hundley F, McIntyre M, Clark B, Beards G, Wood D, Chrystie I, Desselberger U (1987) Heterogeneity of genome rearrangements in rotaviruses isolated from a chronically infected immunodeficient child. J Virol 61: 3365–3372

Imai M, Akatani K, Ikegami N, Furuchi Y (1983) Capped and conserved terminal structures in human rotavirus genome double-stranded RNA segments. J Virol 47: 125–136

Johnson MA, McCrae, MA (1989) Molecular biology of rotaviruses. VIII. Quantitative analysis of regulation of gene expression during virus replication. J Virol 63: 2048–2055

Kabcenell AK, Poruchynsky MS, Bellamy R, Greenberg HB, Atkinson PH (1988) Two forms of VP7 are involved in assembly of SA11 rotavirus in endoplasmic reticulum. J Virol 62: 2929–2941

Kaljot KT, Shaw RD, Rubin DH, Greenberg HB (1988) Infectious rotavirus enter cells by direct cell membrane penetration, not by endocytosis. J Virol 62: 1136–1144

Kapikian AZ, Chanock RM (1985) Rotaviruses. In: Fields BN (ed) Virology. Raven, New York, pp 863–906

Keljo DJ, Smith AK (1988) Characterization of binding of simian rotavirus SA-11 to cultured epithelial cells, J Ped Gastroent Nutr 7: 257–263

Liu M, Estes MK (1989) Nucleotide sequence of the simian rotavirus SA11 genome segment 3. Nucleic Acids Res 17: 7991

Liu M, Offit PA, Estes MK (1988) Identification of the simian rotavirus SA11 genome segment 3 product. Virology 163: 26–32

Lopez S, Arias CF, Bell JR, Strauss JH, Espejo RT (1985) Primary structure of the cleavage site associated with trypsin enhancement of rotavirus SA11 infectivity. Virology 144: 11–19

Mackow ER, Shaw RD, Matsui SM, Vo PT, Dang M-N, Greenberg, HB (1988) The rhesus rotavirus gene encoding protein VP3: Location of amino acids involved in homologous and heterologous rotavirus neutralization and identification of a putative fusion region. Proc Natl Acad Sci USA 85: 645–649

Mansell EA, Patton JT (1990) Rotavirus RNA replication: VP2, but not VP6, is necessary for viral replicase activity. J Virol 64: 4988–4996

Mason BB, Graham DY, Estes MK (1983) In vitro transcription and translation of simian rotavirus SA11 gene products. J Virol 33: 1111–1121

Matsuno S, Inouye S (1983) Purification of an outer capsid glycoprotein of neonatal calf diarrhea virus and preparation of its antisera. Infect Immun 39: 155–158

Mattion N, Gonzalez SA, Burrone O, Bellinzoni R, LaTorre JL, Scodeller EA (1988) Rearrangement of genomic segment 11 in two swine rotavirus strains. J Gen Virol 69: 695–698

McCrae MA, Faulkner-Valle GP (1981) Molecular biology of rotaviruses: I. Characterization of basic growth parameters and pattern of macromolecular synthesis. J Virol 39: 490–496

McCrae MA, McCorquodale JG (1983) Molecular biology of rotaviruses. V. Terminal structure of viral RNA species. Virology 126: 204–212

Mitchell DB, Both GW (1990) Completion of the genomic sequence of the simian rotavirus SA11: nucleotide sequences of segments 1,2, and 3. Virology 177: 324–331.

Meyer JC, Bergmann CC, Bellamy AR (1989) Interaction of rotavirus cores with the nonstructural glycoprotein NS28. Virology 170: 98–107

Nuttal SD, Hum CP, Holmes IH, Dyall-Smith ML (1989) Sequences of VP9 from short and supershort rotavirus strains. Virology 171: 453–457

Okada Y, Richardson MA, Ikegami N, Nomoto A, Furuichi Y (1984) Nucleotide sequence of human rotavirus genome segment 10, an RNA encoding a glycosylated virus protein. J Virol 51: 856–859

Patton JT (1986–1987) Synthesis of simian rotavirus SA11 double-stranded RNA in a cell-free system. Virus Res 6: 217–233

Patton JT (1990) Evidence for equimolar synthesis of double-stranded RNA and minus-strand RNA in rotavirus-infected cells. Virus Res 17: 199–208

Patton JT, Gallegos CO (1988) Structure and protein composition of the rotavirus replicase particle. Virology 166: 358–365

Patton JT, Gallegos CO (1990) Rotavirus RNA replication: single-stranded RNA extends from the replicase particle. J Gen Virol 71: 1087–1094

Petrie BL, Graham DY, Estes MK (1981) Identification of rotavirus particle types. Intervirology 16: 20–28

Petrie BL, Graham DY, Hanssen H, Estes MK (1982) Localization of rotavirus antigens in infected cells by ultrastructural immunocytochemistry. J Gen Virol 63: 457–467

Petrie BL, Greenberg HB, Graham DY, Estes MK (1984) Ultrastructural localization of rotavirus antigens using colloidal gold. Virus Res 1: 133–152

Prasad BVV, Wang GJ, Clerx JPM, Chiu W (1988) Three dimensional structure of rotavirus. J Mol Biol 199: 269–275

Quann CM, Doane FW (1983) Ultrastructural evidence for the cellular uptake of rotavirus by endocytosis. Intervirology 20: 223–231

Ramig RF (1982) Isolation and genetic characterization of temperature-sensitive mutants of simian rotavirus SA11. Virology 12: 93–105

Ramig RF (1983) Isolation and genetic characterization of temperature-sensitive mutants that define five additional recombination groups in simian rotavirus SA11. Virology 130: 465–473

Ramig RF, Fields BN (1983) Genetics of reoviruses. In: Joklik WK (ed) The reoviridae. Plenum, New York, pp 197–228

Ramig RF, Petrie BL (1984) Characterization of temperature-sensitive mutants of simian rotavirus SA11: Protein synthesis and morphogenesis. J Virol 49: 665–673

Richardson MA, Furuichi Y (1983) Nucleotide sequence of reovirus genome segment S3, encoding non-structural protein sigma NS. Nucleic Acids Res 11: 6399–6408

Sabara M, Gilchrist JE, Hudson GR, Babuik LA (1985) Preliminary characterization of an epitope involved in neutralization and cell attachment that is located on the major bovine rotavirus glycoprotein. J Virol 53: 58–66

Sabara M, Ready KFM, Frenchick PJ, Babuik LA (1987) Biochemical evidence for the oligomeric arrangement of bovine rotavirus nucleocapsid protein and its possible significance in the immunogenicity of this protein. J Gen Virol 68: 123–133

Sakuma SV, Watanabe Y (1971) Unilateral synthesis of reovirus double-stranded ribonucleic acid by a cell-free replicase system. J Virol 8: 190–196

Sakuma SV, Watanabe Y (1972) Reovirus replicase-directed synthesis of double-stranded ribonucleic acid. J Virol 10: 628–638

Sandino AM, Jashes M, Faundez G, Spencer E (1986) Role of the inner protein capsid on in vitro human rotavirus transcription. J Virol 60: 797–802

Shahrabadi MS, Lee PWK (1986) Bovine rotavirus maturation is a calcium-dependent process. Virology 152: 298–307

Spencer E, Arias MI (1981) In vitro transcription catalyzed by heat-treated human rotavirus. J Virol 40: 1–10

Spencer E, Garcia BI (1984) Effect of S-adenosylmethionine on human rotavirus RNA synthesis. J Virol 52: 188–197

Stacy-Phipps S, Patton JT (1987) Synthesis of plus- and minus-strand RNA in rotavirus-infected cells. J Virol 61: 3479–3484

Suzuki H, Kitaoka S, Konno T, Sato T, Ishida N (1985) Two modes of human rotavirus entry into MA104 cells. Arch Virol 85: 25–34

Suzuki H, Kitaoka S, Sato T, Konno T, Iwasaki Y, Numazaki Y, Ishida N (1986) Further investigation on the mode of entry of human rotavirus into cells. Arch Virol 91: 135–144

Welch SW, Crawford SE, Estes MK (1989) Rotavirus SA11 genome segment 11 protein is a nonstructural phosphoprotein. J Virol 63: 3974–3982

Xu Z, Anzola JV, Nalin CM, Nuss DL (1989) The 3'-terminal sequence of a wound tumor virus transcript can influence conformational and functional properties associated with the 5'-terminus. Virology 170: 511–522

Yazaki K, Mizuno A, Sano T, Fujii H, Miura K (1986) A new method for extracting circular and supercoiled genome segments from cytoplasmic polyhedrosis virus. J Virol Methods 14: 275–283

Yolken RH, Willoughby R, Wee S-B, Miskuff R, Vonderfecht S (1987) Sialic acid glycoproteins inhibit in vitro and in vivo replication of rotaviruses. J Clin Invest 79: 148–154

Zweerink HJ (1974) Multiple forms of SS → DS RNA polymerase activity in reovirus infected cells. Nature 247: 313–315

Genetics of the Rotaviruses

J.L. GOMBOLD[1] and R.F. RAMIG[2]

[1] Department of Microbiology, University of Pennsylvania School of Medicine, Philadelphia, PA 19104, USA
[2] Division of Molecular Virology, Baylor College of Medicine, One Baylor Plaza, Houston, TX 77030, USA

Current Topics in Microbiology and Immunology, Vol. 185
© Springer-Verlag Berlin · Heidelberg 1994

1 Introduction

In the brief period since rotaviruses were first identified as human pathogens (BISHOP et al. 1973; FLEWETT et al. 1973) and the genus *Rotavirus* was created (MATHEWS 1979), our understanding of these viruses has increased dramatically. Current research on the structure and function of rotavirus proteins had its beginnings in the characterization of viral phenotypes and the identification of the polypeptides responsible for those properties. The ability to quickly map phenotypes to viral proteins is responsible for much of our understanding of protein function. In many cases, the association of phenotypes with specific protein species was accomplished using genetic techniques. The ease and versatility of these techniques resulted in their use in studying virion structure, serology, and pathogenesis. Just as important has been the study of the genetic features of rotaviruses which has led to a better understanding of the fundamental properties of rotaviruses and the processes behind their evolution.

This chapter will review briefly the structural components of rotaviruses that are important in understanding their genetic properties (see Chaps. 2 and 3 for more detailed information on viral structure). Next will follow discussions of conditional-lethal mutations and other genetic markers that have been used to study rotaviruses, the interactions between viruses during mixed infections, and the factors that modulate those interactions. The rotavirus genetic map and the significance of the genetic reassortment between rotaviruses will then be considered. Finally, the implications of genetic studies for the future of rotavirus research will be examined.

2 Features of the Rotaviruses Relevant to Genetic Studies

Rotaviruses comprise one of six genera in the family *Reoviridae* and are unified through morphological, biochemical, and serological characteristics. Virions are composed of six structural proteins arranged to form a core (proteins VP1, VP2, VP3) surrounded by an inner (VP6) and outer (VP4, VP7) capsid (ESTES et al. 1983). Most reports describe five nonstructural proteins in infected cells designated by their molecular weights: NS53, NS35, NS34, NS28, and NS26. Some evidence exists to suggest that NS53 may be a structural protein (MCCRAE and MCCORQUODALE 1982; BASS et al. 1989), although most groups have been unable to detect it in purified virus. Contained within the core of the particle are 11 double-stranded RNA (dsRNA) genome segments numbered 1–11 in order of increasing electrophoretic mobility on polyacrylamide gels (WELCH 1971; MUCH and ZAJAC 1972; WELCH and THOMPSON 1973; NEWMAN et al. 1975; RODGER et al. 1975). The pattern of segments after electrophoresis is a reproducible characteristic of a virus strain and is known as the viral electropherotype (Fig. 1). In general, each segment is monocistronic. However, segment 9 of rotavirus strain SA11, encoding the outer capsid protein VP7, has two conserved in-frame initiation codons and produces two polypeptides that differ by an NH_2-terminal extension of four or five amino acids (CHAN et al. 1986). Although these initiation codons are conserved in many rotavirus strains, it is not clear if all rotaviruses produce the two related polypeptides identified in SA11. It is also unclear if this genome segment is genetically bicistronic since mutants mapping to this segment have not been identified. However, the ability to engineer specific mutations in rotavirus coding sequences may make this issue available for study.

The dsRNA genome serves as template for the synthesis of mRNA. As demonstrated by COHEN et al. (1979), removal of the outer layer of the viral capsid activates a virion-associated RNA polymerase that transcribes mRNA from the dsRNA genome. The parental dsRNA genome is contained within the core throughout the infection and is never free within the cytoplasm of the infected cell. Likewise, progeny dsRNA genomes are synthesized in subviral particles and are never free in the cytoplasm. Thus, mRNA is the only species of viral RNA found free in the cytoplasm and is the only type of RNA available for genetic interactions during coinfection of cells with two viruses. It is also important to note that the fully conservative mode of replication dictates that mutations in the positive-sense strand of dsRNA will not be expressed.

Rotaviruses are divided into six broad groups (or serogroups) by immunologic tests that detect common determinants shared by all viruses within the group. By far the best characterized are the group A viruses. These viruses infect most mammalian and avian species and are a significant cause

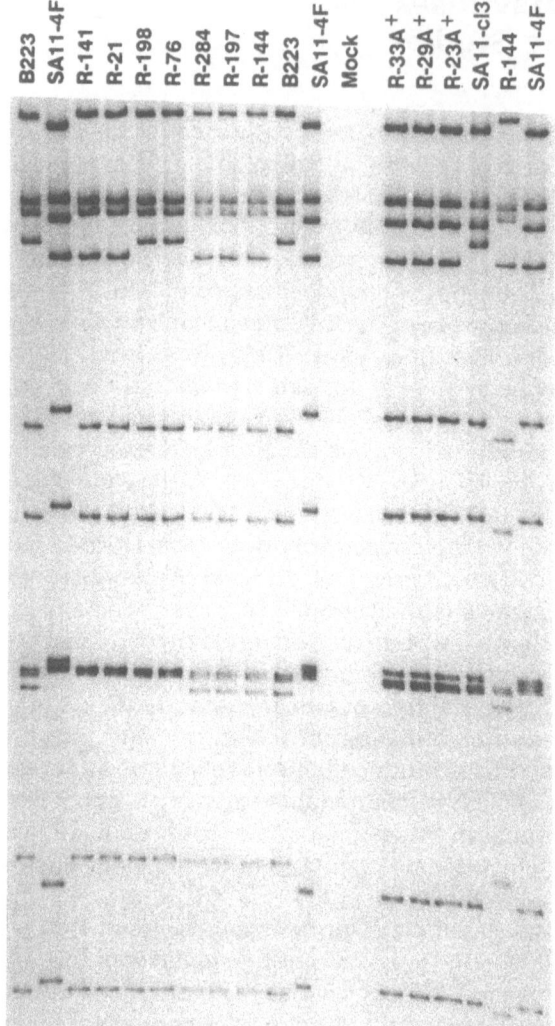

Fig. 1. Electropherotypes of rotavirus strains and reassortants. Shown are the electrophoretic profiles of ^{32}P-labeled genome RNAs of rotavirus strains and reassortants derived from those strains. Lanes labeled *B223, SA11-4F*, and *SA11-c/3* contain the RNA of the three virus strains. Note the polymorphic electrophoretic mobility of each genome segment of each of these strains, a feature particularly evident when comparing B223 and SA11-4F. Lanes labeled with an R-prefix contain reassortant viruses derived from crosses of parental varuses. The cluster on the *left* contains reassortants derived from a cross of B223 and SA11-4F. The cluster on the *right* contains reassortants derived from the cross of SA11-cl3 and R-144. Note in the reassortants that the parental origin of each genome segment can be determined by comparing segment mobility with the mobilities of the cognate segments of the parental viruses. The lane labeled *mock* contained a mock-infected cell RNA preparation. (From Chen et al. 1989)

of morbidity and mortality. However, it is becoming increasingly apparent that viruses in groups B–F are also important animal pathogens. Difficulties in cultivating nongroup A viruses have severely restricted the characterization of these "atypical" rotaviruses. The recent success in growing one group C virus, together with the advent of reliable diagnostic tests for the nongroup A viruses, may pave the way for a more complete understanding of rotaviruses (NAKATA et al. 1986; SAIF et al. 1988; YOLKEN et al. 1988b; BURNS et al. 1989b). Because of the paucity of information on the nongroup A viruses, this chapter with be restricted to genetics of group A rotaviruses only.

The group A viruses are further divided into subgroups and serotypes. Subgroups are defined by reactivities of antigenic determinants located on the inner capsid protein VP6 (KAPIKIAN et al. 1981; GREENBERG et al. 1983). Most virus isolates belong to either subgroup I (SG I) or subgroup II (SG II), but some strains possess determinants of both subgroups while others display neither (HOSHINO et al. 1987a; SVENSSON et al. 1988). Reactivities of the outer capsid proteins VP4 and VP7 determine serotype specificity (KALICA et al. 1981a; GREENBERG et al. 1983; HOSHINO et al. 1985). In contrast to anti-VP6 serum, both monoclonal and hyperimmune polyclonal antiserum specific for these proteins neutralize viral infectivity.

3 Rotavirus Genetic Markers

Genetic analyses require genetic markers with phenotypes that can be readily identified in the progeny of mixed infection. With rotaviruses, as for other viruses, a wide variety of phenotypes have served as genetic markers. Below, some of the markers used in studies of the rotaviruses are described, beginning with the extremely useful conditional-lethal class of mutations.

3.1 Conditional-Lethal Mutations

Conditional-lethal mutations have been useful markers in viral systems because the mutant phenotype is conditional, that is, mutants replicate well at the permissive condition but fail to replicate at the restrictive condition. The conditional-lethal phenotype allows for the production of mutant virus at permissive condition, but specific and strong selection against the mutant virus at restrictive condition. Temperature-sensitive (ts) mutants and non-sense (polypeptide chain termination or "amber") mutants are the two most common types of conditional-lethal mutants. With ts mutants, conditional lethality is accomplished with temperature, low temperature (e.g., 31°C) being permissive and high temperature (e.g., 39°C) nonpermissive. In contrast, nonsense mutants grow in a permissive cell, one that contains a host gene product (tRNA) capable of suppressing the mutation, but not in cells which lack the host gene encoding the suppressor tRNA. A measure of a mutant's combined reversion frequency and "leakiness" is its efficiency of plating (EOP), defined as the ratio of titers at the nonpermissive condition to the permissive condition. For rotavirus ts mutants, EOPs range from 10^{-3} to 10^{-6}.

Temperature-sensitive mutations are the only conditional-lethal mutation type that have been exploited for the study of rotaviruses and have been isolated from simian rotaviruses SA11 (RAMIG 1982, 1983a; GREENBERG et al. 1981) and RRV (KALICA et al. 1983) and the bovine rotavirus UK (GREENBERG et al. 1981; FAULKNER-VALLE et al. 1982). Mutants were placed into genetic groups based on their ability to reassort genome segments and generate wild-type progeny in crosses between pairs of mutants (see below). The maximum number of groups distinguishable by reassortment tests is equivalent to the number of segments in the genome, 11 in the case of rotaviruses.

The SA11 mutant collection is the most complete, defining ten of the 11 reassortment groups expected for rotaviruses (RAMIG 1982, 1983a; Table 1). The prototype mutants of groups A–J each contain a single ts lesion as determined by reassortment tests (see below). Only one mutant, defining group K, remains to be identified. Rotavirus segment 9 is bicistronic (CHAN et al. 1986). However it is unlikely that mutations in the two cistrons could be distinguished by reassortment tests. Several SA11 and RRV ts mutants

Table 1. Rotavirus SA11 temperature-sensitive mutants

Mutant group	Prototype mutant	Mutagen[a]	Genome segment[b]	Mutant protein product[c]	RNA phenotype[d] ssRNA	dsRNA	Protein phenotype[e]
A	tsA (778)	NA	4	VP4	+	+	+
B	tsB (339)	NA	3	VP3	−	−	+/−
C	tsC (606)	NA	1	VP1	−	−	+/−
D	tsD (975)	NA	?	?	+	+	+
E	tsE (1400)	HA	8	NS35	−	−	+/−
F	tsF (2124)	5-AC	2	VP2	+	+	+/−
G	tsG (2130)	5-AC	6	VP6	+	−	+/−
H	tsH (2384)	5-AC	?	?	+	+	+
I	tsI (2403)	5-AC	?	?	+	+	+
J	tsJ (2131)	5-AC	?	?	+	+	+

NA, nitrous acid; HA, hydroxylamine; 5-AC, 5-azacytidine.
[a] Compiled from RAMIG (1982, 1983a).
[b] Compiled from GOMBOLD et al. (1985) and GOMBOLD and RAMIG (1987).
[c] ESTES and COHEN (1989).
[d] Compiled from RAMIG (1983b) and CHEN et al. 1990.
[e] RAMIG and PETRIE (1984). All mutants synthesize protein at 30° but those indicated as +/− synthesize 20–25% the levels of wild-type.

isolated by GREENBERG et al. (1981) were grouped against the SA11 ts mutant collection of Ramig. All appeared to contain single, double, or triple mutations of previously established groups (Ramig, unpublished data; GOMBOLD and RAMIG 1987). It should be noted, however, that a group K mutation would not have been detected if present as a component of any of the multiple mutants. Isolation of the last mutant group will require that the K mutation is the only lesion present so that reassortment with the other ten groups can be demonstrated.

The seven UK ts mutants described by GREENBERG et al. (1981) define four genetic groups. FAULKNER-VALLE et al. (1982, 1983) found that their 51 mutants defined nine distinct mutant groups. Because the UK mutant groups have not been reconciled with each other or with the SA11 mutants, it is not known whether they overlap or if any represent the group K mutant absent in the SA11 collection.

Although ts mutations have been used extensively, they are often poor tools for studies of protein structure and function. Proteins containing ts defects are generally full length, and even at the nonpermissive temperature may provide a low level of function. This "leaky" nature of ts mutants, together with their often high reversion rates, is sufficient to obscure results in genetic experiments. In contrast, nonsense mutations result in truncated proteins and the functional defects are generally severe enough to prevent even minor levels of activity. Moreover, reversion frequencies are often substantially less than for ts mutants since reversion can occur only by back mutation of the original mutant codon. Consequently, nonsense mutants have proven useful in numerous studies of viruses, mostly bacteriophages. Their absence among animal viruses is due solely to the lack of suppressor cell lines in which to grow mutants. However, the recent construction of cell lines capable of suppressing nonsense mutations has made the use of such mutants a viable option (HUDZIAK et al. 1982; SEDIVY et al. 1987). Indeed, amber mutants of poliovirus (SEDIVY et al. 1987) and vesicular stomatitis virus (WHITE and McGEOCH 1987) have been reported. The development of a collection of nonsense mutants of rotaviruses would undoubtedly help to further our knowledge of rotavirus genetics and physiology.

3.2 Neutralization-Resistant Mutants

One disadvantage of mutants obtained either spontaneously or following mutagenesis is that the site undergoing mutation is essentially random. The use of monoclonal antibodies to select for neutralization- or monoclonal antibody-resistant (MAR) mutants allows selection of mutations limited to regions encoding neutralizing epitopes. Many MAR mutations have been localized at the nucleotide level by sequencing. It should be noted, however, that conformational epitopes recognized by antibodies may select mutations at distant locations in the protein target. Obviously, the use of MAR mutants is limited to proteins involved in neutralization since mutants are selected by their ability to escape neutralization.

MAR mutants in both VP4 (MACKOW et al. 1988b; TANIGUCHI et al. 1987, 1988a; SHAW et al. 1986) and VP7 (DYALL-SMITH et al. 1986; MACKOW et al. 1988a; TANIGUCHI et al. 1988b) have been reported. Variants of VP4 have revealed domains involved in both homotypic and heterotypic neutralization (MACKOW et al. 1988b; TANIGUCHI et al. 1988b). Serotype specific sites are clustered in VP8*, the 60 kDa NH_2-terminal fragment of VP4 generated by digestion with trypsin (MACKOW et al. 1988b; SHAW et al. 1986). In contrast, the COOH-terminal fragment VP5* contains mostly cross-reactive sites.

Sequence analysis of genome segment 9 (encoding VP7 in SA11) of MAR variants selected with anti-VP7 monoclonal antibodies has identified three domains involved in neutralization (DYALL-SMITH et al. 1986; MACKOW et al. 1988a; TANIGUCHI et al. 1988a). Although one epitope (site C) appears to be immunodominant, each site alone is capable of mediating neutralization.

3.3 Genetically Engineered Mutants

Molecular biology has provided a powerful tool for the study of protein structure and function through the ability to generate site-specific mutations. These techniques are only now being applied to the study of rotaviruses. PORUCHYNSKY et al. (1985) generated specific deletions involving one or both hydrophobic domains in VP7. While deletions in the first domain had little effect, deletions in the second caused the protein to be transported from the endoplasmic reticulum (ER) to the Golgi and secreted. The aberrant processing observed with this second mutant led to the conclusion that the second domain acts to anchor the protein in the ER membrane and may function as a positive signal for ER localization.

As yet unavailable are techniques for introducing specific genetically engineered mutations into infectious virus. The determinants responsible for packaging the 11 genome segments have not been identified, limiting studies with engineered mutations. The identification of segment packaging signals and the advent of techniques to produce rotaviruses containing specific mutations will be a major contribution to the study of rotavirus genetics.

3.4 Polymorphisms and Rearrangements

A widely used marker in reassortment studies of the rotaviruses has been the electrophoretic polymorphism exhibited when the dsRNA genome segments are resolved by gel electrophoresis (MATSUNO et al. 1980; GREENBERG et al. 1981; ALLEN and DESSELBERGER 1985; GOMBOLD et al. 1985). These genomic polymorphisms between virus strains are commonly referred to as electropherotypes. Since cognate segments from different viral strains often migrate at different rates in gels, differences in viral electropherotypes provide markers for parental origin of genome segments in reassortant viruses. In some cases, viral proteins also migrate differently, allowing protein polymorphisms to be used as markers.

RNA segment polymorphisms have been used extensively as genetic markers in experiments designed to physically map specific phenotypic traits by correlating the segregation of parental genome segments with a given phenotype. This technique of segregation analysis has identified proteins confering growth properties (GREENBERG et al. 1983), hemagglutination (KALICA et al. 1983), Subgroup and serotype specificity (KALICA et al. 1981a; GREENBERG et al. 1983), and virulence (OFFIT et al. 1986a). Analysis of the segregation of genome segments in reassortants has also allowed known protein-coding assignments to be extended to several rotavirus strains and enabled the assignment of ts lesions to specific genome segments (GARBARG-CHENON et al. 1984; ALLEN and DESSELBERGER 1985; GOMBOLD et al. 1985; GOMBOLD and RAMIG 1987). These topics will be discussed more thoroughly below.

In contrast to other viruses with segmented genomes, notably reovirus and influenza, serial high-multiplicity passage of rotaviruses results in rearrangements of genome segments (HUNDLEY et al. 1985). The rearranged segments appear to be concatemers of smaller segments (ALLEN and DESSELBERGER 1985; MATTION et al. 1989; GONZALEZ et al. 1989) and are similar to rearrangements seen in viruses isolated from immunodeficient children (PEDLEY et al. 1984). Rearrangements appear to result from aberrant transcription of single segments; they do not contain sequences from more than one gene (McCrae, personal communication; GONZALEZ et al. 1989). Coding regions are maintained in all observed rearrangements, although some produce proteins of abnormal size. Viruses with rearranged genome segments are genetically stable and reassort segments in mixed infections (ALLEN and DESSELBERGER 1985; HUNDLEY et al. 1985). Therefore, rearrangements make excellent genetic markers in experiments employing two otherwise indistinguishable virus strains.

3.5 Biological Phenotypes as Genetic Markers

In addition to conventional genetic markers and electrophoretic polymorphisms, several biological properties of rotaviruses have been used as genetic markers. These include serological characteristics (subgroup and serotype specificity; GREENBERG et al. 1983; BURNS et al. 1989a), hemagglutination (KALICA et al. 1983), growth properties (GREENBERG et al. 1983; CHEN et al. 1989), plaque size (CHEN et al. 1989), protection from challenge (OFFIT et al. 1986b), and virulence (OFFIT et al. 1986a). As in all mapping experiments, these biological phenotypes have been used together with segment poly-morphisms as markers to identify the gene and protein responsible for a particular phenotype.

4 Genetic Interactions During Mixed Infections

4.1 Studies with Temperature-Sensitive Mutants

Temperature-sensitive mutants have been isolated from several different rota-virus strains, including the bovine UK (FAULKNER-VALLE et al. 1982, 1983; GREENBERG et al. 1981), the simian SA11 (RAMIG 1982, 1983a, b; Greenberg, personal communication), and the rhesus rotavirus RRV (Greenberg, per-sonal communication). The standard method of assigning the ts mutants to groups with ts lesions in the same gene, the complementation test, did not prove useful because the ts mutants interfered with the growth of wild-type virus and presumably with each other during mixed infection (RAMIG 1983a, c). The rotavirus ts mutants were capable of reassortment, and this method (reassortment or recombination test) was used to place the mutants into groups that contained lesions on different genome segments (FAULKNER-VALLE et al. 1982, 1983; GREENBERG et al. 1981; RAMIG 1982, 1983a–c). In addition to allowing the division of ts mutants into groups, studies of reassortment revealed a number of other features of the rotavirus life cycle. These features of rotavirus mixed infections with ts mutants are described below.

4.1.1 Complementation and Interference

Attempts to divide rotavirus ts mutants collections into groups of mutants having lesions in identical function by complementation revealed that comple-mentation between mutants was inefficient and nonreproducible (RAMIG 1982). In the reovirus system similar results had been obtained, and the failure of certain mutant pairs to complement was shown to correlate with the interference phenotype of the mutants (CHAKRABORTY et al. 1979). Mixed infections between individual SA11 ts mutants and wild-type SA11 revealed that all of the ts mutants tested had a strong interference phenotype, the phenotype being most prominent at restrictive temperature (RAMIG 1983a, c). Interference of the ts mutants with the growth of wild type was assumed (but not proven) to indicate that mutants could interfere with each other during mixed infection at restrictive temperature, leading to a failure of complementa-tion to produce an enhanced yield. However, this failure of complementation did not prevent grouping of mutants into groups with lesions in the same function. Since the rotavirus genome segments are generally monocistronic

(ESTES and COHEN 1989), assignment of mutations to specific genome segments served the same purpose. Reassortment tests to group the ts mutants are described below.

4.1.2 The "All or None" Nature of Reassortment

When mixed infections of permissive cells were performed with pairs of ts mutants and the yields of the infections were scored for the presence of wild-type (ts$^+$) recombinant (reassortant) progeny, one of two outcomes was noted (GREENBERG et al. 1981; FAULKNER-VALLE et al. 1982, 1983; RAMIG 1982, 1983a–c). Either high frequencies ($>$ 1.0%) or low frequencies ($<$ 0.2%) of ts$^+$ reassortant progeny were found, with no mutant pairs repeatedly giving intermediate values between these extremes. This all-or-none fashion of generating ts$^+$ recombinant progeny was consistent with predictions based on independent reassortment of genome segments during mixed infection. The failure to see a continuous gradient from low to high recombination frequencies (i.e., between 0 and 0.2%) among the different mutant pairs suggested that generation of ts$^+$ progeny did not occur by the mechanism of intramolecular recombination involving breakage and reformation of phosphodiester bonds. The all-or-none recombination seen with rotavirus ts mutants was similar to that noted for other viruses with segmented genomes (SIMPSON and HIRST 1968; FIELDS and JOKLIK 1969), in which generation of recombinant progeny had been shown to occur by reassortment by demonstration of progeny containing genome segments from both parental viruses (SHARPE et al. 1978). MATSUNO et al. (1980) had shown that reassortment could occur in rotaviruses, strengthening the conclusion that recombination frequencies observed between rotavirus ts mutants were indicative of recombination by mechanism of reassortment.

The all-or-none recombination (reassortment) frequencies were used to divide the ts mutant collections as follows. Mutants in pairs that failed to reassort were assigned to the same group, as they had mutations on the cognate genome segments of the two parental viruses and were unable to reassort ts$^+$ progeny. Mutants in pairs that reassorted at high frequency were assigned to different groups since they had mutations on noncognate segments that could reassort ts$^+$ progeny. Thus, mutants within a single reassortment-defined group contain lesions in the same segment and mutants in different reassortment-defined groups contain lesions on distinct reassorting segments. The most complete ts mutant collection, derived from SA11, contains mutations that define ten reassortment groups (RAMIG 1983a). Another ts mutant collection, derived from UK virus, contains nine reassortment groups (FAULKNER-VALLE et al. 1982, 1983). Other mutant collections define fewer groups. Since the rotavirus genome consists of 11 reassorting segments, it is clear that none of the ts mutant collections contains mutations in every gene. However, it is possible that reconciliation of the different mutant collections will reveal that all 11 segments have been marked with a ts lesion. This reconciliation remains to be done.

When discussing the generation of recombinants in segmented genome viruses, including rotaviruses, the more descriptive term "reassortment" is generally used to indicate the mechanism of recombinant generation.

4.1.3 Stability of the Reassortant Phenotype/Genotype

For reassortants to provide a useful tool for the analysis of rotaviruses, it was important to show that the reassortant genotype arose from the stable exchange of genome segments and not from some other phenomenon that would not breed true.

The possibility existed that progeny plaques scored as reassortant following mixed infection of two ts mutants could arise from an aggregate of virus that contained two mutants capable of complementation, a phenomenon observed with certain other enveloped and nonenveloped viruses (WALLIS and MELNICK 1967; DAHLBERG and SIMON 1969). Apparent reassortant progeny plaques arising from such a mechanism would be expected to segregate the parental viruses upon subsequent passage and plaquing, an indication that they were not stable reassortants. Analysis of the temperature phenotype of ts$^+$ progeny plaques from mixed rotavirus infections revealed that they did not segregate ts virus on passage. Furthermore, ultrafiltration or sonication of yields of mixed infections to remove virus aggregates did not significantly reduce the frequency of ts$^+$ progeny following mixed infection. Taken together, these observations indicated that aggregates of complementing virus did not account for the formation of ts$^+$ progeny and that they indeed represented stable reassortants (RAMIG 1983c).

The possibility also existed that the ts$^+$ progeny among the yields of mixed infections with two ts mutants could represent heterozygous progeny particles. Heterozygotes had been shown to contribute significantly to the frequency of ts$^+$ progeny in the yield of mixed infections between ts mutants of other viruses (KINGSBURY and GRANOFF 1970; DAHLBERG and SIMON 1969). In the case of rotavirus, these heterozygotes would be partially diploid and contain two copies of a specific genome segment, one containing a ts lesion and the other derived from the second parent and representing a wild-type copy of the segment. Dominance of the wild phenotype would account for the ts$^+$ phenotype of the heterozygous particle. Such heterozygous particles should be of greater than average density due to the presence of at least 12 genome segments rather than the normal 11 segments. The yield of a cross of two rotavirus mutants was fractionated by density gradient centrifugation, and the frequency of reassortants was determined across the peak of infectious virus. The variation of reassortant frequency across the virus peak fell within the variation seen for replicate titrations of a single cross and centered on the value found for an unfractionated aliquot of the same yield. The failure to enrich for ts$^+$ progeny at higher density indicated that heterozygosis did not contribute significantly to the formation of ts$^+$ progeny and suggested that the ts$^+$ progeny represented stable reassortants (RAMIG 1983c).

4.1.4 Other Features of Rotavirus Reassortment

Several other features of reassortment among rotavirus ts mutants have been studied and have revealed, or confirmed, features of the rotavirus life cycle.

When replicate mixed infections were performed with ts mutant pairs, either within single experiments or on different days, a wide range of reassortment frequencies were observed for the cross of any given mutant pair (FAULKNER-VALLE et al. 1982; RAMIG 1983a). In some cases, certain mutant pairs were scored as reassortment-positive in some experiments and reassortment negative in others (FAULKNER-VALLEE et al. 1982). These variations were ascribed to variations in the efficiency of plating shown by wild type from infection to infection. This variation in EOP would also be observed in the ts$^+$ reassortant progeny (FAULKNER-VALLE et al. 1982). In addition, the variation was shown to be inherent in the genetic system as the standard deviation of the mean reassortment frequency was shown to be no different for replicate crosses performed in a single experiment when compared to crosses performed in different experiments on different days (RAMIG 1983c).

The reassortment frequencies observed between pairs of ts mutants are generally lower than the 25% that would be theoretically predicted for random reassortment and range from about 1% to about 20%. The range of recombination frequencies observed with different mutant pairs suggested that the genome segments might be loosely linked and that statistical analysis of a large amount of reassortment data might reveal such linkage. However, statistical analysis of reassortment data showed that the reassortment frequencies for different ts mutant pairs were not significantly different, indicating that linkage did not exist and that a linear map of the mutations could not be generated (RAMIG 1983c). It is not understood why observed reassortment frequencies are lower than theoretically predicted. However, in the reovirus system, this phenomenon has been shown to be related to the interference phenotype of the ts mutants. Mutant pairs that strongly interfered with the growth of wild-type virus also yielded characteristically low frequencies of ts$^+$ reassortant progeny (CHAKRABORTY et al. 1979). Analysis of mixed infections between rotavirus ts mutants and wild-type revealed that all mutants tested interfered with the growth of the coinfecting wild-type virus (RAMIG 1983a, c). It is thought that the interference of the ts mutants with wild type accounts for the lower than expected frequencies of reassortment since the ts mutants could also interfere with wild-type (ts$^+$) reassortants generated during mixed infection. Support for this notion has come from the higher frequencies of reassortment observed in crosses where two wild-type parents were utilized and no intentional selection pressures were placed on them either during the cross or selection of plaques for analysis (GOMBOLD and RAMIG 1986).

Kinetic analysis of reassortment between ts mutants of rotavirus revealed that reassortment was maximal of near maximal at the earliest time after infection that reassortant progeny could be detected: 16 h at 31°C (RAMIG 1983c). The early time of reassortment is typical of viruses with segmented genomes (FIELDS 1971; MACKENZIE 1970). It is thought to reflect the maturation of the first assembled viruses and, so far as is known, assembly and reassortment represent different aspects of the morphogenesis process. The fact that reassortment frequency is maximal at the earliest times indicates that each progeny genome undergoes only a single round of reassortment and does not participate in multiple genetic exchanges as do many of the DNA genome viruses. This observation is consistent with removal of the reassortant progeny genome from the mating pool by morphogenesis. However, this does not imply that genome segments are not available for reassortment late in the infectious cycle. Then, the proportion of reassortant progeny simply increases in parallel with the total yield (RAMIG 1983c). Recently it has been demonstrated that, during asynchronous mixed infections, a second virus infection 24 h after the first virus infection results in the formation of reassortants between the first and second viruses (Ramig, unpublished observation). This observation may have significance for the evolution of the rotaviruses (see below).

The effect of multiplicity of infection (MOI) on reassortment between rotavirus ts mutants has also been examined. The frequency of reassortment was found to increase in parallel with MOI until an MOI of 2.5–5.0 pfu/cell of each parent was attained. At higher MOI no further increase in reassortment frequency was noted (RAMIG 1983c). This observation most likely reflects the fact that two viruses must infect the same cell for reassortants to be formed. It is only at an MOI in the range of 2.5–5.0 of each virus that 100% of the cells in a culture are infected with both viruses (based on the Poisson distribution). However, additional information can be gained from this data. Since crude lysates of infected cells generally contain a vast excess of noninfectious to infectious paricles, the parallel between reassortment frequency and MOI (by definition a measure of infectious particles) indicates that the noninfectious particles present in the inoculum do not significantly participate in the genetic interactions in the mixed-infected cell (RAMIG 1983c). Trypsin activation of viruses before mixed infection has also been shown to increase the frequency of reassortment between rotavirus ts mutants (RAMIG 1982). This effect is most likely a result of trypsin treatment converting noninfectious particles to an infectious form, thus effectively increasing the MOI of the infection.

By definition, ts mutants do not grow at nonpermissive temperature. However, mixed infection of cells with pairs of ts mutants at the nonpermissive temperature resulted in the production of ts$^+$ reassortant progeny, often at frequencies greater than observed if the infection was performed at permissive temperature (RAMIG 1983a, c). The yields of the mixed infections performed at nonpermissive temperature were generally low, suggesting that the growth advantage of the ts^+ reassortants accounted for the high frequency of reassortment observed. The ten mutant groups of rotavirus SA11 were defined by reassortment analysis under standard conditions of permissive temperature. However, when mixed infections were performed at nonpermissive temperature, one mutant pair (tsG and tsJ) failed to yield reassortant progeny (RAMIG 1983a). These mutants reassorted normally at permissive temperature, and each reassorted with all other mutant groups at nonpermissive temperature. These observation suggest that tsG and tsJ contain lesions which render a protein complex involved in reassortment nonfunctional at the nonpermissive temperature. The tsG lesion has been mapped to genome segment 6 and VP6 (GOMBOLD and RAMIG 1987), but the tsJ lesion has not been mapped. Thus, the nature of this specific conditional block to reassortment with a single mutant pair remains unresolved.

4.2 Other Studies of Genetic Interaction During Mixed Infection

Studies with ts mutants indicated that reassortment was relatively unrestricted between mutants of the same rotavirus strain. However, indications that reassortment may be somewhat limited in nature came from a generally observed association of certain subgroup, serotype, and electropherotype markers (HOSHINO et al. 1984; KALICA et al. 1981b). For example, identifiable reassortants between subgroup I and subgroup II rotaviruses have been only rarely isolated. Crosses between two viruses representing either the same or different subgroups have confirmed and apparent restriction to intersubgroup reassortment (WARD et al. 1988c; WARD and KNOWLTON 1989). For a more complete discussion of this apparent restriction to reassortment between rotaviruses belonging to different subgroup, serotype and electropherotypic groups see Sect 7.2.1.

5 The Genetic Map of Rotavirus

5.1 Reassortment and Segregation Analysis

Segregation analysis using reassortant viruses has been the method of choice for mapping most phenotypes. Genome segments carrying specific markers can be identified by the cosegregation of the phenotype of interest and a single genome segment or combination of genome segments. Crosses between two viruses are used to generate reassortants and the parental origin of segments is determined. If a group of reassortants are examined that all possess the phenotype of interest, the genome segment responsible for that phenotype must be present in all progeny clones. Since reassortment is random in the absence of selective pressure, only the genome segment carrying the selected marker will segregate nonrandomly. Conversely, a group of reassortants lacking the phenotype will always derive one genome segment from the parental virus that did not display the phenotype.

Two requirements must be met if segregation analysis is to be successful. First, the phenotypes of the parental viruses must be distinguishable. To map the hemagglutinin, for example, one virus must agglutinate erythrocytes but the other must not. In the case of ts mutants, the mutations must belong to different genetic groups so that ts^+ reassortants can form. Second, the method requires that the electropherotypes of the parental viruses be different. Ideally, each of the 11 segments should migrate differently and thus be distinguishable. In practice this means that the parental viruses will be different strains.

In the specific case of mapping ts mutations, the use of two ts mutants in the cross provides an advantage not realized when wild-type viruses are used in mapping. To see this, it must be understood that a cross between two ts viruses (e.g., tsA and tsB) will yield four progeny types: (1) parental virus tsA, (2) parental virus tsB, (3) tsA/tsB reassortants (double mutants), and (4) ts^+ reassortants (and at some low level determined by reversion frequency, ts^+ revertants of the parents). If the yield of this cross is plaqued at the nonpermissive temperature, only ts^+ reassortant or revertant progeny will grow. This allows for selection of ts^+ reassortants and eliminates the tedious chore of randomly analyzing hundreds of clones for reassortants. It is possible, however, to use one wild-type parent in crosses if appropriate means of selection against it exist. GOMBOLD et al. (1985) used the human rotavirus Wa in crosses to map ts mutations in SA11. Wa grows poorly in culture, and the growth advantage of wild-type reassortants in the yield of the cross resulted in virtually no parental virus being recovered. It should be noted that the genome segment responsible for the fastidious growth of Wa (segment 4) would confer poor growth characteristics on reassortants possessing that segment. Therefore, mutations in SA11 segment 4 would be difficult to map using Wa as the second parent.

5.2 Physical Map of the Rotavirus SA11 Genome

5.2.1 Group A ts Mutations

The prototype SA11 group A mutant tsA (778) was mapped using two different RRV mutants, tsC/H(2) or tsC/F(4), as the second parent in segregation analyses. Wild-type (ts^+) progeny obtained from the crosses were titered at both the permissive and nonpermissive temperatures to verify the ts$^+$ phenotype, and the genome RNAs were resolved by gel electrophoresis to identify the parental origin of each genome segment of each progeny clone. A total of 28 wild-type clones from the cross between tsA (778) and tsC/H(2) were examined. Table 2 lists clone numbers, EOPs, and electropherotypes for 14 representative reassortants. Since the reassortants had ts$^+$ phenotype, the SA11 genome segment with the group A mutation was required to be replaced with the cognate segment from RRV. All clones derived segment 4 from the RRV parent, indicating segment 4 was the site of the group A lesion (GOMBOLD and RAMIG 1987). Six clones derived only segment 4 from RRV, providing strong support for this assignment. The cross between tsA(778) and tsC/F(4) yielded 26 clones that also suggested SA11 segment 4 as the site of the group A mutation: all 26, including seven single segment reassortants, derived segment 4 from RRV. It is evident from Table 2 that other segments, in addition to segment 4, segregated nonrandomly. These segments, rather than being derived uniformly from RRV, were of SA11 origin. Selection against the RRV C and F lesions in RRV ts4 explains the selection against RRV segment 1s and 2, as will be discussed below. Therefore, all nonrandom segregation could be explained and group A mutants were assigned to genome segment 4.

Segment 4 encodes the 88 KDa outer capsid protein VP4 (MASON et al. 1983). This protein is responsible for hemagglutination and, together with VP7, viral neutralization (KALICA et al. 1983; HOSHINO et al. 1985, 1987b; OFFIT and BLAVAT 1986). VP4 also accounts for trypsinenhanced plaque formation and infectivity (ESTES et al. 1981; KALICA et al. 1983), and the growth restriction of certain rotavirus strains in cell culture (GREENBERG et al. 1981,1983). Some evidence suggests that, in the presence of trypsin, rotavirus particles can penetrate cellular membranes directly (SUZUKI et al. 1985; FUKUHARA et al. 1988; KALJOT et al. 1988), arguing that VP4 mediates internalization of virions.

VP4 also confers at least some properties of virulence in the mouse model of rotavirus gastroenteritis (OFFIT et al. 1986a). This is unlikely to be due to cell tropism since most data suggest that VP7, the segment 9 gene product, mediates interactions with the host cell, at least in vitro (SABARA et al. 1985). Recently, however, BASS et al. (1989) reported that NS35 is actually a structural protein that binds to intestinal epithelia and fibroblasts in culture. Antibodies to VP4 neutralize virus in vitro in plaque-reduction assays (GREENBERG et al. 1983; COULSON et al. 1985; HOSHINO et al. 1985; TANIGUCHI et al. 1985; KITAOKA et al. 1986; BURNS et al. 1988) and protect suckling mice from disease (OFFIT et al. 1986a).

FLORES et al. (1986) observed that the deduced amino acid sequence adjacent to the trypsin cleavage site in VP4 is highly conserved among avirulent rotaviruses but different from the more virulent strains. This sequence conservation suggests that the region of VP4 surrounding the cleavage site is an important determinant of rotavirus pathogenicity. Although tsA(778) is virulent in mice (Gombold, unpublished observations), other

Table 2. Efficiency of plating and electropherotype of reassortant clones derived from crosses with SA11 tsA(778)

RRV mutant	Experiment	Clone	Efficiency of plating	Parental origin of indicated genome segment										
				1	2	3	4	5	6	7	8	9	10	11
tsC/H(2)	I	1196	0.31	S[a]	S	S	R[b]	R	R	S	S	S	R	R
		1197	0.35	S	R	R	R	S	S	R	S	S	R	S
		1198	0.72	S	S	S	R	S	R	S	S	S	S	S
		1199	1.35	S	S	S	R	S	R	S	S	S	R	S
		1200	0.43	S	S	S	R	S	R	S	S	S	S	R
		1201	3.44	S	S	S	R	S	S	S	S	S	S	S
		1202	0.70	S	R	S	R	S	R	S	R	S	R	R
		1203	0.21	S	S	S	R	S	S	S	S	R	R	S
		1204	0.53	S	S	S	R	S	S	S	S	R	S	S
		1205	1.43	S	S	S	R	S	S	S	S	S	S	S
		1206	1.40	S	S	S	R	S	S	S	S	S	S	S
		1207	0.76	S	S	S	R	S	S	S	S	S	S	S
		1208	1.02	S	S	S	R	S	S	S	S	S	R	S
		1210	0.77	S	S	S	R	S	S	S	S	S	R	S
		Ratio:	$\dfrac{\text{RRV}}{\text{SA11}}$	$\dfrac{0}{14}$	$\dfrac{2}{12}$	$\dfrac{1}{13}$	$\dfrac{14}{0}$	$\dfrac{1}{13}$	$\dfrac{5}{9}$	$\dfrac{1}{13}$	$\dfrac{1}{13}$	$\dfrac{2}{12}$	$\dfrac{6}{8}$	$\dfrac{3}{11}$
	II[c]													
		Ratio:	$\dfrac{\text{RRV}}{\text{SA11}}$	$\dfrac{0}{14}$	$\dfrac{5}{9}$	$\dfrac{2}{12}$	$\dfrac{14}{0}$	$\dfrac{3}{11}$	$\dfrac{2}{12}$	$\dfrac{8}{6}$	$\dfrac{4}{10}$	$\dfrac{6}{8}$	$\dfrac{4}{10}$	$\dfrac{2}{12}$

tsC/F(4)												
Clone	Ratio (RRV/SA11)											
1211	2.22	S	S	S	R	S	S	S	S	S	S	S
1212	0.91	S	S	S	R	S	S	R	S	S	S	S
1213	1.08	S	S	S	R	S	S	S	S	S	S	S
1214	1.81	S	S	S	R	R	R	S	S	S	S	R
1215	0.53	S	S	S	R	S	S	S	S	S	S	R
1216	7.89	S	S	S	R	S	S	S	S	S	S	S
1218	0.94	S	S	S	R	S	S	S	S	S	S	S
1219	2.28	S	S	S	R	S	R	S	R	S	S	S
1220	5.28	S	S	S	R	S	S	S	S	R	S	R
1221	2.08	S	S	S	R	S	S	R	S	S	R	S
1222	1.31	S	S	S	R	S	S	S	S	R	S	S
1223	5.23	S	S	S	R	S	R	S	S	R	S	R
1224	1.55	S	S	S	R	S	S	S	S	S	S	S
1225	1.52	S	S	S	R	S	S	S	S	S	S	S
Ratio (I):	RRV/SA11	0/14	0/14	0/14	14/0	1/13	3/11	2/12	1/13	3/11	1/13	4/10
Ratio (IIc):	RRV/SA11	0/12	0/12	2/10	12/0	1/11	5/7	3/9	7/5	10/2	2/10	2/10

RRV, rhesus rotavirus.

From GOMBOLD and RAMIG (1987).

[a] Segment derived from SA11.

[b] Segment derived from RRV.

[c] Segregation ratios from second independent but identical cross; electropherotypes of individual clones not shown. All clones had ts$^+$ phenotype.

group A mutants may display various degrees of virulence depending on the precise site of the ts mutation. Group A mutants altering the sequence or conformation of the trypsin cleavage site would be potentially useful in analyzing the role of VP4 in virulence. Group A mutants may also identify additional domains in VP4 that are involved in virulence.

Physiologic studies of tsA(788) have shown that neither RNA nor protein synthesis are defective at the nonpermissive temperature (RAMIG 1983b; RAMIG and PETRIE 1984; CHEN et al. 1990). This observation is consistent with the location of VP4 in the outer shell of the virion and its absence in transcriptionally active particles in vitro. Viral morphogenesis occurs normally at the nonpermissive temperature arguing that the ts defect does not interfere with viral assembly (RAMIG and PETRIE 1984). Given the importance of trypsin cleavage of VP4, it is tempting to speculate that the mutation in tsA(778) alters the structure of VP4 at the nonpermissive temperature rendering it less susceptible to cleavage.

5.2.2 Group B ts Mutations

The mapping of the SA11 group B prototype mutation tsB(339) is one of three cases in which a ts mutation was mapped using crosses with the wild-type human rotavirus Wa. Twenty ts[+] reassortants were isolated from two independent crosses between Wa and SA11 tsB(339). The parental origin of genome RNAs was determined by gel electrohporesis and, in the case of segment 2, by dot hybridization. Unambiguous identification of segment 2 required dot hybridization because of small differences in electro-phoretic mobility between the SA11 and Wa segment 2. All reassortants derived segments from both parents, although there was a clear bias towards SA11 as the parent of origin. This bias was likely due to the predominance of the SA11 parent in the crosses. However, segment 3 originated from Wa in every reassortant examined indicating that segment 3 was the location of the tsB lesion (GOMBOLD et al. 1985). In four ts[+] reassortants, segment 3 was the only segment derived from Wa. Wa segment 4 was selected against in all 20 reassortants, presumably because that segment imparts restricted growth in culture on any reassortant containing it (GREENBERG et al. 1981, 1983).

The product of genome segment 3, VP3, is a minor component of the virion core and has an apparent molecular weight of 88 kDa in SDS-polyacrylamide gels (LIU et al. 1988). The deduced amino acid sequence indicates that VP3 is basic and shares homology with viral RNA polymerases (LIU and ESTES 1989). The similarity between VP3 and other known RNA polymerases is interesting given that levels of both single-stranded (ss) RNA and dsRNA are reduced in cells infected with tsB(339) at the nonpermissive temperature (RAMIG 1983b; CHEN et al. 1990). It appears that the defect primarily alters the synthesis of ssRNA since the ratio of dsRNA to ssRNA is normal compared to wild-type-infected cells. The poor synthesis of ssRNA by tsB(339) at the nonpermissive temperature may reflect inefficient assembly and subsequent transcription of RNA by newly made subviral particles. This hypothesis is supported by the observation that cells infected with tsB(339) at the non-permissive temperature contain fewer viroplasms and larger proportions of the transiently enveloped intermediates and produce fewer mature virus particles (RAMIG and PETRIE 1984). Recently, GALLEGOS and PATTON (1989) identified VP3 as a component of replicase-competent particles in infected cells. However, it is not clear if VP3 has an active role in replicase activity or simply plays a structural role.

5.2.3 Group C ts Mutations

An RRV double mutant, tsA/I(5), was used as the second parent in crosses to map the lesion in SA11 tsC(606). Thirty-two ts$^+$ reassortants were isolated from this cross and characterized. Only segment 1 was derived from RRV exclusively, and four reassortants contained segment 1 as the only RRV segment. Consequently, the group C mutation was assigned to segment 1 (GOMBOLD and RAMIG 1987). Segment 4 originated from SA11 in every case, presumably due to the group A mutation in the RRV parent. Selection against a second segment that could be attributed to the group I mutation of tsA/I(5) was not seen (see below).

Structural protein VP1, the largest of the viral proteins, is a 125 kDa polypeptide encoded by genome segment 1 (MASON et al. 1983). It is a minor component of the virion core, an observation leading some to suggest that its role may be enzymatic rather than structural (ESTES and COHEN 1989). Indeed, it contains domains that share homology with RNA polymerases of other RNA viruses. VP1 has recently been identified as a component of replication intermediates in SA11-infected cells (GALLEGOS and PATTON 1989).

A role for VP1 in RNA synthesis is consistent with the defect in both transcription and replication by tsC(606) at the nonpermissive temperature (RAMIG 1983b; CHEN et al. 1990). VP1 of tsC(606) also displays a reduced mobility on polyacrylamide gels, clearly showing that it differs from wild-type VP1 (RAMIG and PETRIE 1984). Although the ts lesion is likely to be the cause of the altered mobility, there is no direct evidence of this, and it is clear that proteins can display altered migration rates without an association with ts mutations.

5.2.4 Group E ts Mutations

Mapping of the group E mutation was performed with 21 reassortants isolated from a cross between Wa and tsE(1400). Genome RNAs were resolved by gel electrophoresis to determine the parental origin of each RNA segment, and the origin of segment 2 was confirmed by dot hybridization. In every ts$^+$ reassortant, segment 8 was of Wa origin, and segment 8 was the only Wa segment in 16 of the 21 reassortants. These data led to the assignment of the group E mutation to SA11 segment 8 (GOMBOLD et al. 1985). With few exceptions, the remaining RNA segments were of SA11 origin. This was not unexpected because of the predominance of the SA11 parent in the cross. No reassortants contained Wa segment 4, presumably for the growth restriction it confers (GREENBERG et al. 1983).

SA11 genome segment 8 encodes a 35 kDa nonstructural protein, NS35 (MASON et al. 1983). Because of the similar sizes and small differences in electrophoretic mobility in segments 7, 8 and 9 the coding assignments for these segments vary among rotavirus strains. In RRV, NS35 is the product of segment 9 (GOMBOLD and RAMIG 1986), but in the bovine rotavirus UK it is encoded by segment 7 (MCCRAE and MCCORQUODALE 1982), indicating that coding assignments for segments 7–9 of each virus examined must be carefully determined. Sequence analysis suggests that NS35 is basic (BOTH et al. 1982) and has some secondary structural characteristics of nucleic acid binding proteins. However, NS35 fails to bind nucleic acids in a solid-phase RNA binding assay (BOYLE and HOLMES 1986), although this may result from the denaturing conditions of the assay.

NS35 is not detectable by gel electrophoresis of tsE-infected cell proteins labeled at either the permissive or the nonpermissive temperature (RAMIG and PETRIE 1984). It is not known whether the mutant protein is made at levels undetectable by autoradiography or if the protein has an altered mobility in gels and comigrates with another protein. Infected cell poly-peptides have not been examined by immunoprecipitation to address this latter possibility.

Synthesis of mRNA by tsE(1400) is defective at the nonpermissive temperature, producing about 2% of RNA made in wild-type-infected cells (RAMIG 1983b; CHEN et al. 1990). The small amount of ssRNA that is made at the restrictive temperature is converted to dsRNA with the same efficiency as wildtype. The lesion in tsE (1400) therefore affects ssRNA synthesis preferentially. NS35 has been shown to be present in subviral particles active in replication of the negative-sense genome strand (GALLEGOS and PATTON 1989). Given that the ts lesion in NS35 does not appear to inhibit RNA replication, the precise role of this nonstructural protein is not clear. NS35 may function as a scaffolding protein during morphogenesis or, alternatively, in the gathering of each of the 11 RNA segments that make up the genome. The latter hypothesis is attractive since a large proportion of empty particles accumulate at the nonpermissive temperature in tsE-infected cells (RAMIG and PETRIE 1984).

5.2.5 Group F ts Mutations

The prototype SA11 group F mutant tsF(2124) was mapped using crosses with RRV mutants tsC/H(2) or RRV tsA/I(5) as the second parent. Twenty-six ts[+] reassortants were isolated from the cross with tsC/H(2) and ten from the cross with tsA/I(5). Segment 2 was derived from RRV in all 36 reassortants, eight of which contained segment 2 as the only RRV segment. Therefore, the group F mutation was assigned to genome segment 2 (GOMBOLD and RAMIG 1987).

Three additional clones from the cross between SA11 tsF(2124) and RRV tsA/I(5) derived all 11 genome segments from SA11 yet had ts[+] phenotype. Since control infections with tsF(2124) alone exhibited significantly high reversion frequencies, the selection of revertants from the cross was not unexpected (GOMBOLD and RAMIG 1987).

Two genome segments were expected to originate solely from SA11 in each cross since the RRV parental mutants were double mutants. RRV segment 1, the site of the group C mutation in tsC/H(2), was selected against and was always of SA11 origin. However, selection against the group H lesion of tsC/H(2) was not observed. A similar observation was made in the case of clones derived from the cross with tsA/I(5). Segment 4 was always of SA11 origin, consistant with the assignment of the A lesion, but selection against a second RRV segment due to the group I lesion in tsA/I(5) was not seen. The reason selection against the RRV tsH and tsI lesions was not observed is unknown; however, lack of selection against these mutations was observed in other experiments (see below).

RNA segment 2 encodes the 94 kDa polypeptide VP2 (MASON et al. 1983), the major structural protein of the viral core (BICAN et al. 1982). VP2 is myristilated (CLARK and DESSELBERGER 1988) and binds RNA in a solid-phase assay (BOYLE and HOLMES 1986). The predicted amino acid sequence of VP2 contains domains rich in leucine which may catalyze dimerization through "leucine zippers", a characteristic of some DNA binding proteins (ESTES and COHEN 1989).

Consistent with the RNA binding properties of VP2, the group F ts lesion mapped to this protein results in increased synthesis of ssRNA at the nonpermissive temperature (RAMIG 1983b; CHEN et al. 1990). Conversion of ssRNA to dsRNA is about 6% of wild-type rates, but because of the elevated amounts of single-stranded template, the overall level of dsRNA in infected cells is normal. At the nonpermissive temperature, tsF-infected cells contain few viroplasms and are essentially devoid of virus particles. However, large tubules resembling aberrant virus are observed at the permissive temperature (RAMIG and PETRIE 1984), suggesting that the effects on the morphogenesis of tsF(2124) are not necessarily temperature-dependent.

5.2.6 Group G ts Mutations

The prototype SA11 group G mutant tsG(2130) was crossed with Wa and with two RRV mutants, tsC/H(2) or tsC/F(4), to derive mapping reassortants (GOMBOLD and RAMIG 1987). Nineteen ts[+] reassortants examined from the cross with Wa strongly indicated that the ts mutation in tsG was on segment 6: 18 of the reassortants derived segment 6 from Wa. Segment 6 in the one discordant clone migrated more slowly than either Wa or wild-type SA11 segment 6 and was probably the result of a second site reversion.

Fourteen clones obtained from the cross between tsG(2130) and tsC/H(2) were ts$^+$ reassortants and all derived genome segment 6 from the RRV parent. An additional ten reassortants, eight of which were ts$^+$ were isolated from the cross of tsG(2130) with tsC/F(4). The eight ts$^+$ reassortants all derived segment 6 from RRV, supporting the assignment of the group G mutation to genome segment 6 (GOMBOLD and RAMIG 1987). However, the two reassortants that maintained the ts phenotype also derived segment 6 from RRV. Backcrosses of the these ts clones to each of the ten prototype mutants indicated that none of the ts reassortants contained the group G mutation. One clone was a C/F double mutant, as shown by its failure to yield ts$^+$ reassortants in crosses with either SA11 tsC(606) or SA11 tsF(2124) (Gombold and Ramig, unpublished data). This result was consistent with the presence of RRV segments 1 and 2 in this clone and the assignments of the group C and F mutations to genome segments 1 and 2, respectively. The origin and identity of the ts lesion(s) in the other clone was not clear. Significant proportions of ts$^+$ reassortants were observed from crosses with each of the ten SA11 prototype mutants (Gombold and Ramig, unpublished data). Since neither of the ts reassortants contained the group G lesion, the ts reassortants along with the 41 ts$^+$ reassortants derived from the three mapping crosses were all consistent with the assignment of the group G mutation to genome segment 6 (GOMBOLD and RAMIG 1987).

The major structural protein of the virion, VP6, is the product of RNA segment 6 (MASON et al. 1983). VP6 is a 41 kDa protein and, like VP2, is myristilated (CLARK and DESSELBERGER 1988). In its native state, the protein forms noncovalent trimers that are bound through disulfide bonds into hexamers (NOVO and ESPARZA 1981; SABARA et al. 1987). For reasons that are not understood, VP6 is probably the most immunogenic of the viral structural proteins although neither polyclonal hyperimmune serum nor anti-VP6 monoclonal antibodies are capable of neutralizing viral infectivity. It also bears antigenic determinants responsible for both group and subgroup specificity.

VP6 is necessary for transcriptase activity in vitro. Its removal from single-shelled particles abolishes activity of the virion polymerase and will restore activity when added to cores (BICAN et al. 1982; SANDINO et al. 1986). Furthermore, VP6 is required for replication of ssRNA intermediates into the dsRNA genome (HELMBERGER-JONES and PATTON 1986). RNA synthesis by SA11 tsG(2130) supports these observations in that the amounts of both ssRNA and dsRNA are significantly reduced at the nonpermissive temperature compared to wild-type-infected cells. The group G mutant is thus classified as RNA-negative (CHEN et al. 1990).

Clearly, there are several mechanisms by which lesions in VP6 could result in the ts phenotype. It is not known whether mutations in VP6 disrupt an enzymatic or a structural function. However, the requirement for inter-actions between VP6 monomers and between VP6 and other viral poly-peptides suggests that interference of protein interactions is one possibility. The postulated interference with the higher order structure of the inner shell apparently would have no effect on viral assembly, given the absence of morphogenic defects in tsG-infected cells (RAMIG and PETRIE 1984), but would have to be severe enough to prevent proper functioning of the viral polymerase.

5.2.7 Unmapped SA11 ts Mutant Groups

Genome segment locations for four of the ten SA11 ts mutants, tsD(975), tsH(2384), tsI(2403), and tsJ(2131), have not been determined. The ts$^+$ reassortants derived from mapping crosses with these mutants segregated genome segments relatively randomly and prevented identification of map positions. Reversion of the ts lesions and/or leakiness of the lesions are the most likely reasons for the difficulty in mapping these mutant groups since all but tsH have high efficiencies of plating.

All four of these mutants synthesize normal or elevated amounts of both ss- and dsRNA at the nonpermissive temperature (RAMIG 1983b; CHEN et al. 1990). Except for tsH(2384), which produces a large proportion of empty particles at the restrictive temperature, none display aberrant morphogenesis (RAMIG and PETRIE 1984).

5.3 Mapping ts Mutations of Other Virus Strains

Mapping the SA11 ts mutant groups by crosses with RRV ts mutants allowed verification of some assignments since the SA11 and RRV mutant groups used in these crosses had been reconciled by reassortment grouping (GOMBOLD and RAMIG 1987). Confirmation of the assignments carries added significance because the RRV and SA11 mutants were isolated independently.

The RRV mutant tsA/I(5) was used in crosses with SA11 tsC(606) and SA11 tsF(2124) (GOMBOLD and RAMIG 1987). In total, 45 reassortant progeny were isolated and examined. All reassortants derived genome segment 4 from the SA11 parent, identifying segment 4 as the site of either the group A or I lesion in RRV ts(5). Because SA11 tsA(778) mapped unambiguously to segment 4 in several independent crosses, it was unlikely that the selection against RRV segment 4 in crosses with tsA/I(5) was due to the group I mutation. Although the data appeared to support the assignment of the SA11 group A mutant, this assignment should be held tentative until the site of the group I mutant is identified. It seems probable that the RRV tsI lesion could not be mapped in these crosses because the I lesion is likely to be leaky and of high reversion like the SA11 tsI mutant.

The group C mutation was present in both the RRV tsC/H(2) and RRV tsC/F(4) parents used in mapping crosses with SA11 tsA(778), SA11 tsF(2124), and SA11 tsG(2130) (GOMBOLD and RAMIG 1987). If these crosses are considered together, 103 of 104 reassortants contained SA11 segment 1, confirming the mapping of SA11 tsC to segment 1. The one discordant clone derived segment 1 (and 2) from RRV and has a ts phenotype. Recombination tests indicated that the ts reassortant was a C/F double mutant. Therefore, all mapping data indicated that segment 1 is the site of the RRV group C mutants.

The only other mutant group mapped independently was the group F mutant. Thirty-six clones obtained from crosses of RRV ts C/F(4) with SA11 tsA(778) and SA11 tsG(2130) clearly supported the assignment of group F mutants to segment 2. All but one clone derived genome segment 2 from the SA11 parents; the one reassortant containing RRV segment 2 was the ts reassortant described above, which contain mutations belonging to both groups C and F. These data strongly support the assignment of the RRV group F lesion to segment 2.

A single mutant from the UK ts mutant collection, the group 1 ts17 mutant, has been mapped by segregation analysis (FAULKNER-VALLE et al. 1983). ts$^+$ reassortants were derived from a cross of UK ts17 and the porcine rotavirus OSU. The parental origin of genome segments was determined by gel electrophoresis and, for certain segments, hybridization. All ts$^+$ reassortants contained segment 7 of OSU origin, indicating that the UK group 1 lesion mapped to segment 7. Since the UK and SA11 ts mutant collections have not been reconciled by reassortment testing, it is unclear which of the SA11 mutant groups maps to the same segment as UK group 1.

6 Other Genetic Studies with Rotaviruses

6.1 Reassortment as a Tool to Identify Rotavirus Gene Function

Reassortment has proven to be the primary tool for identification of gene function in rotaviruses. Although the use of reassortment to map gene function is conceptually similar to the mapping of ts mutations, there are significant differences in the approaches taken. Reassortant viruses are most easily selected if they result from the cross of parental viruses with selectable mutations such as ts mutations. However, because all rotavirus ts mutants have been isolated following mutagenesis of wild type, the potential exists for the presence of nontemperature-sensitive mutations in these mutants.

To avoid the problems associated with non-expected phenotypes associated with nontemperature-sensitive mutations (RUBIN and FIELDS 1980), reassortants for mapping gene function are generally derived from crosses of two wild-type parental viruses that differ in the phenotype of interest. Reassortant progeny are identified and characterized either by random "brute-force" screening or by selection for new combinations based on the phenotypes of the wild-type parental viruses (often using monoclonal antibodies to select for new combinations of neutralizing antigens). Once a collection of reassortants is generated and the parental original of the genome segments determined by electrophoresis, the reassortants are screened for expression of the phenotype of interest. The phenotype of interest is generally seen to segregate with one, or a specific constellation of, segment(s) derived from the parent that expresses that phenotype. Once the genome segment responsible for a phenotype has been identified, the protein responsible can be deduced, since the coding assignments for the rotavirus genome segments are known (ESTES and COHEN 1989). The identification of gene(s) and gene product(s) responsible for several rotavirus phenotypes are described below.

6.1.1 Identification of the Gene Responsible for Fastidious Growth of Rotaviruses in Cell Culture

Rotaviruses have generally proven to be fastidious agents, requiring significant effort to adapt to culture. Reassortment between culture-adapted viruses and noncultivable viruses provided a means of rescuing genes from the noncultivable viruses that encoded antigens not associated with the culture-adapted parent. Treating the progeny of such a cross with neutralizing antisera to the culture-adapted parent allowed selection of reassortants that expressed neutralization antigens characteristic of the noncultivable virus. This method was refined by the use of ts mutants of the culture-adapted parent, so that temperature selection could be applied to the progeny without restricting the segregation of the antigen-encoding genes (GREENBERG et al. 1981, 1982). Analysis of a large number of reassortants derived from crosses of culture-adapted and fastidious viruses revealed that genome segment 4 was always derived from the culture-adapted parental virus (GREENBERG et al. 1982, 1983), indicating that genome segment 4 of the fastidious virus encoded the function that restricted growth in culture.

6.1.2 Identification of Genes Encoding Rotavirus Antigenic Specificities

Rotaviruses were shown to contain several different antigenic specificities by a number of methods. The group antigen is shared by all group A rotaviruses and cannot be studied by reassortment since all group A viruses express that antigen. However, most rotavirus strains express one of two subgroup antigens that are detected by immune adherence hemagglutination assay and one of several serotype antigens that are detected by neutralization assay. Reassortment was used to demonstrate that subgroup specificity and neutralization specificity were encoded in different genome segments, since neutralization and subgroup antigens could be separated by reassortment (KAPIKIAN et al. 1981). Later analysis of genome segment segregation in reassortants that separated these antigenic specificities revealed that the subgroup specificity was encoded in segment 6 and neutralization specificity in segment 9 (KALICA et al. 1981a; GREENBERG et al. 1983). However, the studies showing subgroup and neutralization specificity to be encoded in different segments made use of reassortants generated by rescue of fastidious viruses with culture-adapted viruses. All of these reassortants contained genome segment 4 from the culture-adapted virus, so that the antigen encoded in this segment could not be studied. Subsequently, the question of genes encoding neutralization antigens was reexamined using reassortants derived from two culture-adapted parental viruses so that all genome segments could segregate freely. When these reassortants were analyzed for neutralization using antisera made to the parental viruses or antisera that bridged the serotypes of the two parents, neutralization specificity was found to segregate with two segments, segments 4 and 9 (HOSHINO et al. 1985; OFFIT and BLAVAT 1986).

6.1.3 Identification of Determinants of Rotavirus Virulence and Pathogenesis

The different in virulence seen among rotavirus strains in animal models suggests that viral genes encode this difference. In the suckling mouse model, the simian strain SA11 (virulent) was found to induce diarrhea in 50% of inoculated mice at a 50-fold lower dose than did the bovine strain NCDV (nonvirulent). Reassortment was utilized to identify the viral gene responsible for the difference in virulence. Analysis of the 50% diarrhea dose of a large number of SA11/NCDV reassortant revealed that virulence segregated with genome segment 4 of SA11 (OFFIT et al. 1986a). However, more recent studies that utilized different virus strains and either the suckling mouse or procine models of infection have not been able to confirm this assignment of virulence to genome segment 4 (Chen and Ramig, unpublished data; Hoshino and Saif, personal communication). The reason for this failure to confirm segment 4 as the determinant of virulence in rotaviruses is not understood, although it may relate to the virus strains chosen for study (see below, Sect. 6.3).

6.1.4 Identification of Rotavirus Antigens Protective Against Challenge

Reassortant rotaviruses have been used in two ways to identify the rotavirus antigens responsible for protection from challenge. Reassortment viruses expressing VP4 and VP7 of different parental specificity were used to induce immunity that was tested by challenge with the parental viruses, or immunity to parental viruses was tested by challenge with reassortant viruses.

In the suckling mouse model active immunity cannot be tested because of the short "window" of susceptibility to rotavirus infection and disease. However, protection from challenge by immunity passively transferred from the dam has been demonstrated (OFFIT and CLARK 1985a, b). In the mouse model of passive protection, reassortant viruses have been used both as immunogens and as challenge viruses. These studies showed that both of the independently segregating neutralization antigens (VP4 and VP7) were capable of inducing a protective immune response (OFFIT et al. 1986b). Passive protection to rotavirus challenge is also conferred by the oral administration of monoclonal or polyclonal antibody to suckling mice (OFFIT et al. 1986c). Recently, challenge of mice passively immunized with various anti-VP4 monoclonal antibodies with reassortant viruses has demonstrated that both VP4 and VP7 contain epitopes that confer protection against heterologous chellenge (MATSUI et al. 1989).

Immunization and challenge of gnotobiotic piglets with reassortant viruses that segregate the VP4 and VP7 neutralization antigens of serotypically distinct porcine rotaviruses has demonstrated that both VP4 and VP7 are capable of inducing active immune responses that are protective (HOSHINO et al. 1988).

The identification of two distinct neutralization antigens in rotavirus raised the question of the relative immunogenicity of each antigen as a question important for vaccine design. Two studies have addressed this question, both involving infection of human volunteers and measurement of VP4- and VP7-specific neutralizing responses by titration against reassortants expressing only VP4 or VP7 serologically homologous to the infecting virus. In one study of adult volunteers infected with a human virus it was found that greater than 80% of the neutralizing antibodies were reactive with VP4 (WARD et al. 1988a). However, in a second study of infants immunized with the vaccine strain bovine WC3, reassortants containing the homologous VP7 were neutralized to 20-fold higher titer than reassortants containing homologous VP4 (WARD et al. 1990). The basis for this difference in response in the two studies is not understood, but further analysis with reassortants will likely shed light on the question.

6.1.5 Other Studies Involving Reassortment

Reassortment has provided a useful tool for answering other more diverse questions of gene coding and function in the rotaviruses. Several of these are described briefly below.

The product of SA11 genome segment 3 could not be unequivocally identified by the standard methods of in vitro translation. Reassortants were constructed using SA11 and NCDV, or SA11 and RRV, as parental viruses, since both the genome segments and gene products of these virus pairs could be distinguished electrophoretically. Analysis of reassortants that segregated segment 3 of one virus on a background of segments from the other virus allowed the clear assignment of the SA11 segment 3 product as a protein of 88 kDa that had not been identified because it comigrated with the segment 4 product, VP4, which also was 88 kDa (LIU et al. 1988).

Reassortment was also used to identify the rotaviral gene that encoded the viral hemagglutinin and the requirement for protease for plaque formation. Reassortants were constructed using RRV (hemagglutination-positive, protease-requiring) and UK (hemagglutination-negative, protease-independent) as parents. Scoring the ability of reassortants to agglutinate human erythrocytes and form plaques in the absence of protease revealed that both of these phenotypes were encoded in genome segment 4 (KALICA et al. 1983). The protease-dependent phenotype has recently been reexamined utilizing a different pair of viruses, with the finding that in some virus pairs this phenotype segregated with segment 4, but with other virus pairs the phenotype segregated with a combination of two segments including segment 4 (CHEN et al. 1989). This unexpected segregation result obtained with some virus pairs will be discussed below (Sect. 6.2).

Occasionally, monoclonal antibodies are identified for which standard methods of determining protein specificity such as radioimmunoprecipitation or western blot analysis are not successful. In one such study, a neutralizing monoclone for which protein specificity could not be determined by standard means was determined to be specific for VP4 through the use of reassortants that segregated VP4 and VP7 of the virus which the monoclonal antibody neutralized (ZHENG et al. 1989).

Recently, some rotavirus strain have been shown to induce hepatitis in mice and to replicate antigen in HepG2 (human liver) cells, while other virus strains do not show these phenotypes (UHNOO et al. 1989; SCHWARZ et al. 1989). Examination of reassortants between parental viruses able and unable to grow in HepG2 cells revealed that the growth phenotype segregated with genome segment 4. In addition, it was shown that viruses unable to grow in HepG2 cells experienced a replication block prior to protein synthesis. The block to replication that occurred prior to protein synthesis was also determined by genome segment 4 (RAMIG and GALLE 1990).

6.2 Limitations and Potential Problems with Reassortment and Segregation Analysis for Mapping Rotavirus Gene Function

Analysis of cosegregation of genome segments with phenotypes in reassortants has been widely applied for mapping in viruses with segmented genomes (PALESE 1984). In these studies phenotypes have generally been seen to segregate cleanly with one or some constellation of genome segments. Most of these studies have segregated the genes of one parental virus onto the background of segments from the other parental virus. The mapping data obtained are clearly correct for the pair of parental viruses used to generate the reassortants. However, it has been recently shown that there can be exceptions to the generality of the mapping data obtained from reassortment experiments (CHEN et al. 1989). It is currently unclear if exceptions to the general applicability of mapping data will be frequent or rare. However, the possibility that expression of phenotypes may depend on the parental viruses chosen for reassortment means that the interpretation and application of the results of reassortment mapping studies must be approached with caution.

The general applicability of the results of reassortment mapping studies was called into question by the results described below. A variant of the simian rotavirus SA11 (SA11-4F) was isolated (PERIERA et al. 1984) that had unusual properties when compared to wild type (SA11-Cl3), including making large plaques in the presence of protease, making small clear plaques in the absence of protease, growth to significantly higher titer, and synthesis of a VP4 that was more susceptible to trypsin cleavage (BURNS et al. 1989a). The variant also synthesized a VP4 of higher apparent molecular weight than that of SA11-Cl3, suggesting that the unusual properties of the variant were associated with segment 4. To confirm that segment 4 was indeed responsible for the unusual phenotypes associated with the variant, SA11-4F was used to generate reassortants with the bovine virus B223 (CHEN et al. 1989). Reassortants were constructed that reassorted the outer capsid genes of SA11-4F in all combinations with the other genes derived entirely from B223. The reciprocal reassortants were also constructed. Analysis of the unusual phenotypes that were expected to segregate with segment 4 from SA11-4F revealed that formation of large plaques in the presence of protease, the formation of small clear plaques in the absence of protease, and growth to high titer all segregated with two segments (4 and 9) from SA11-4F and not with segment 4 alone as expected. Since protease-associated phenotypes had previously been shown to segregate with segment 4 (KALICA et al. 1983), the SA11-4F genes present on a B223 genetic background in reassortants were moved by a second reassortment experiment onto the genetic background of SA11-Cl3 segments. On this recipient genetic background, the same phenotypes segregated with genome segment 4 alone of SA11-4F. Thus, identical phenotypes segregated with two genes or a single gene, depending on the recipient genetic background (SA11-Cl3 or B223). The basis for this difference is not understood. However, it was clear that on a B223 background the two outer capsid proteins must be of homologous origin to express the phenotypes in question. When VP4 and VP7 were of heterologous origin the phenotypes were not expressed. This suggested that expression of the protease-associated phenotypes depended upon the physical interaction between the two outer capsid protein species (CHEN et al. 1989).

Further evidence for the interaction of proteins in the outer capsid of rotavirus affecting the expression of phenotypes comes from examination of the neutralization of the same SA11-4F/B223 reassortants with the VP4-specific monoclonal antibody 2G4 (CHEN et al. 1992). SA11-4F is neutralized by 2G4 and B223 is not. Among the SA11-4F/B223 reassortants, the expected neutralization is observed if both VP4 and VP7 are derived from the SA11-4F parent, and the expected lack of neutralization is observed if both VP4 and VP7 are from B223. However, if the outer capsid proteins VP4 and VP7 are of heterologous parental origin, unexpected neutralization results are obtained. Specifically, reassortants containing SA11-4F VP4 and B223 VP7 are not neutralized, and reassortants containing B223 VP4 and SA11-4F VP7 are neutralized, the converse of what would be predicted. This result indicates that the interaction of the outer capsid proteins of heterologous origin affects the presentation of epitopes. Interaction of SA11-4F VP4, which normally displays a 2G4 epitope, with B223 VP7 renders the 2G4 epitope on the SA11-4F VP4 cryptic and the virus is not neutralized. Likewise, interaction of B223 VP4, which normally does not react with 2G4, with SA11-4F VP7 exposes a 2G4 epitope on the B223 VP4 and the virus in neutralized (CHEN et al. 1992). Similar unexpected reactivities, or unexpected variations in titer, have been reported when the neutralization of reassortants with polyclonal sera was examined (HOSHINO et al. 1988; I.H. Holmes, personal communication). The basis for this unexpected presentation of neutralizing epitopes is not understood. However, it may provide a system in which to probe the complex interactions of VP4 and VP7 in the outer capsid of the virus. This result may also have implications for the use of reassortants as attenuated rotavirus vaccines, in which VP4 amd VP7 of heterologous origin has been suggested as a means of broadening the immune response (KAPIKIAN et al. 1986).

Finally, one other potential problem could be encountered during the analysis of reassortants, phenotypic mixing. Phenotypic mixing has been shown to occur at high frequency during mixed infection between serologically distinct viruses, with 40%–75% of the progeny particles being neutralization mosaics (WARD et al. 1988b). If the yield of a cross were examined directly for a phenotype, phenotypically mixed particles could confound the result. However, if single plaques were picked and passaged or the mass yield were passaged once at low MOI (< 1), the phenotype expressed on the surface would reflect the genotype of the virus. The occurrence of phenotypic mixing empasizes the necessity for biologically cloning and genotypically characterizing reassortants before they are used to score the segregation of phenotype.

7 Comparison of Rotavirus Genetics to the Genetics of Other Members of the *Reoviridae*

The majority of the genetic features of the rotaviruses are shared with other members of the family *Reoviridae*. However, some features are different from other *Reoviridae* in detail or substance, and some features in other *Reoviridae* have not yet been examined in rotaviruses. The similarities and differences in genetics between rotaviruses and other members of the *Reoviridae* are briefly discussed below.

7.1 Common Features of Genetics in the *Reoviridae*

Many of the features of rotavirus genetics described above are shared with reovirus, the most extensively studied of the remaining members of the *Reoviridae*. Most of the features are also shared with other members of the *Reoviridae* for which the appropriate studies have been made.

Among all of the *Reoviridae* reassortment occurs in an "all-or-none" fashion (FIELDS and JOKLIK 1969; SHIPHAM and DE LA REY 1976; RAMIG 1982) that indicates functional segmentation of the genome. These "all or none" reassortment frequencies have been used to divide the various mutant collections into groups in which reassortment does not occur (mutations on cognate segments) but between which reassortment occurs at high frequecy (mutations on reassorting segments). The absence of a gradient of a gradient of recombination between mutants that fail to reassort and those that do indicates that intramolecular recombination mediated by breakage and reunion of genome segments does not occur at detectable levels in this virus family. However, the detection of rotavirus genome segments in which rearrangements have occurred (PEDLEY et al. 1984; HUNDLEY et al. 1985) and the demonstration that these rearrangements involve duplications of information within a single segment (ALLEN and DESSELBERGER 1985; NUTTALL et al. 1989) suggest that an intramolecular recombination-like mechanism may function in rotavirus-infected cells (see Sect. 8.4).

Among all member of the *Reoviridae* in which reassortment kinetics have been examined, reassortment is an early event and maximal reassortment frequencies are attained early in the life cycle (FIELDS and JOKLIK 1969; RAMIG 1983c; RAMIG et al. 1989). The early nature of reassortment is consistent with the early maturation of infectious virus particles, since the earliest events in maturation and reassortment are thought to be identical. The identity of the maturation and reassortment events is also consistent with the early appearance of maximal frequencies, a phenomenon which indicates that a reassortant progeny genome, once generated, is unable to undergo further rounds of mating. Thus, genetic evidence supports the early formation of infectious progeny virus and the effective removal of these progeny genomes from the replicating pool.

In studies of the randomness of reassortment in the *Reoviridae*, cosegregation of segments or selection for or against specific genome segments was not observed unless specific selective pressures were placed on the reassortant progeny (CROSS and FIELDS 1976; GOMBOLD and RAMIG 1986). However, if experiments were performed under conditions in which selective pressures could occur, nonrandom segregation was expected and was observed (GRAHAM et al. 1987; WARD et al. 1988c; WARD and KNOWLTON 1989). This contrasts with the situation in bluetongue virus, in which nonrandom segregation was observed in the absence of international selective pressure and against unequal parental multiplicity gradients (RAMIG et al. 1989).

In all cases among the *Reoviridae* in which it has been examined, complementations is inefficient, apparently due to interference of ts mutants with the wild type (FIELDS 1971; RAMIG 1983c; SHIPHAM and DE LA REY 1976). The inefficient complementation observed with these viruses makes the use of reassortment tests necessary to place mutations into groups having lesions in the same function (cistron).

7.2 Features of Rotavirus Genetics Unique Among the *Reoviridae*

Some features of rotavirus genetics are currently considered unique among the *Reoviridae*. However, while the uniqueness of the observations for rotaviruses may be real, they may also reflect that fact that similar phenomena have not been examined in other members of the *Reoviridae*.

7.2.1 Restrictions on Reassortment Among Viruses of Different Serotype, Subgroup, or Group Antigens

Subgroup (SG)I rotavirus isolates have generally been serotype (ST)2, with the other serotypes falling into SG II (HOSHINO et al. 1984). In addition, SGI ST2 viruses generally have a characteristic "short" migration pattern for segment 11 (KALICA et al. 1981b). Identifiable reassortants between SGI and SGII rotaviruses have been only rarely isolated (NAKAGOMI et al. 1987; BROWN et al. 1988a; STEELE and ALEXANDER 1988; SETHI et al. 1988), suggesting that the genome segments encoding ST2 neutralization determinants, SGI determinants, and short migration of segment 11 remain closely associated (genetically linked?) in reassortment situations. Several studies have examined this question. Crossess between ST2/SGI/short electropherotype viruses and ST 1, 2, or 3/SG II/long electropherotype viruses have demonstrated that the association (linkage ?) between these determinants, while strong, can be broken by reassortment. Specifically, from crosses of SGII/ST1/ long and SGI/ST2/ short viruses, GARBARG-CHENON et al. (1984, 1986) were able to isolate at low frequency viruses that were SGII/ST1/ short. This demonstrated that the short electropherotype could be separated from the SGI determinants with which it almost always naturally occurred. Similar results were reported for other crosses of this type, and it was also shown with the isolation of progeny that were SGI/ST1 and SGII/ST2 that the association of SGI with ST2 could be broken (URASAWA et al. 1986).

In the studies described above, reassortants that contained new combinations of SG and ST antigens were isolated from the progeny of mixed infection only at relatively low frequency. This result was surprising as recombination frequencies between different SGII viruses had been shown to be relatively high (33%, WARD et al. 1988c; 38%, GOMBOLD and RAMIG 1986). To determine the basis for the low reassortment frequencies between SGI and SGII viruses, two different crosses between SGI and SGII viruses were examined and only about 14% of progeny virus were found to be reassortant (WARD and KNOWLTON 1989). Genome segments from the SGII parent were overrepresented among the individual progeny picked after single rounds of infection, and subsequent passage of the mass yield of the infection led to virtually complete domination of SGII segments among plaque picked progeny after multiple passages (WARD and KNOWLTON 1989). Thus, the potential for formation of reassortants between SGI and SGII viruses in the natural situation in which multiple rounds of infection and replication occur would appear to be severely limited. The dominance of SGII segments after several rounds of infection did not appear to result from some specific selection based on increased growth potential, but rather appeared to result from the adaptation of the SGII segments for optimal function as a group (WARD and KNOWLTON 1989). Thus, the adaptation of the SGII segments for optimal function as a group, rather than a specific block to SGI/SGII reassortment, would appear to account for the relative paucity of SGI/SGII reassortants in nature.

In addition to the group A rotaviruses on which the discussion has focussed to this point, five other groups (B–F) of rotaviruses have been identified . These viruses, while having the morphology of rotaviruses, show no antigenic reactions with each other or with group A rotaviruses. The similarity between group A and nongroup A rotavirus genomes suggested that they may be able to reassort, an event that could have potential evolutionary significance. In one experiment designed to address this question, mixed infections of suckling rats were performed with group A and nongroup A viruses that could infect the rat. Examination of the progeny virus from such an infection revealed no reassortants among 323 progeny plaques examined suggesting that rotaviruses of the different groups were genetically isolated and incapable of reassortment (YOLKEN et al. 1988a).

The genetic isolation of the group A and nongroup A rotavirus also has been shown to extend to viruses that are even more distantly related: rotaviruses and reoviruses. Crosses between a wild-type rotavirus and ts mutants of reovirus did not yield any ts$^+$ progeny in which a rotavirus genome segment had rescued reovirus to ts$^+$ phenotype (HRDY 1982). Under the conditions of these crosses rescue by reassortment would have been detected if it occurred with a frequency of 10^{-5} or greater. Similar crosses between rotaviruses and reoviruses or orbiviruses have been performed in which both of the parental viruses contained ts mutations. Examination of approximately 200 progeny ts$^+$ progeny plaques from each cross revealed none of the ts$^+$ progeny to be reassortant. It was calculated that reassortment between these viruses must occur at a frequency of $<10^{-7}$. Revertants of both parental viruses were detected, indicating that both viruses had replicated in the mixed-infected cell without reassorting (Ramig, unpublished data; Samal and Ramig, unpublished data).

The failure of reassortment between antigenically unrelated rotaviruses, or rotaviruses and viruses from other genera in the *Reoviridae*, is not surprising in light of recent studies of genetic isolation among the antigenically complex orbivirus genus. In the orbiviruses, reassortment studies between viruses from within a serogroup or from different serogroups showed that the antigenic relationships generally predicted whether reassortment would occur. An even better correlation was found between the ability of genomes of the viruses to hybridize; those having significant sequence similarity were able to reassort, while those lacking sequence homology did not form reassortants (GONZALEZ 1987; BROWN et al. 1988b).

7.2.2 Host Effects on Rotavirus Reassortment

Specific host effects on the reassortant genotypes isolated from mixed infection of cells in vitro have been observed in rotaviruses (GRAHAM et al. 1987). Mixed infections were performed with a human rotavirus that carried a rearranged genome and a normal bovine rotavirus. The genotypes present among the progeny were examined after plaque formation on MA104 cells or BSC-1 cells. It was found that the cell line used to isolate the reassortants from the yield of the mixed infection had an effect on the genotypes isolated. Some reassortant genotypes could be isolated on either indicator cell line, although the frequency of isolation of a given genotype was often quite different between the two cell lines. In other cases, reassortant genotypes were isolated at relatively high frequency on one indicator cell line and not isolated at all on the other indicator cell line. Statistical analysis showed significant differences in the isolation frequency of specific genotypes from cell line to cell line and that different genotypes was favored on one cell line over the other. This result emphasizes the importance of the indicator cell line, especially when data from different experiments are to be compared.

No specific in vivo host effects have been observed in rotaviruses, although this may be due to the relatively small number of reassortment experiments performed in vivo. It is unclear if such differences should be expected. However, in the orbivirus system, the different mammalian host species and the vector have been shown to be permissive for reassortment to significantly different degrees. Hosts such as cattle or the vector *Culicoides variipennis*, experience either long-term viremia or life-long infection, produce reassortant viruses at high frequency (STOTT et al. 1987; SAMAL et al. 1987A), whereas hosts that experience an acute infection that is rapidly cleared (sheep) produce only a very low frequency of reassortants (SAMAL et al. 1987b). In the reovirus system, no specific effects of different host species have been reported. However in the mouse, in which reovirus infection is systemic with many different organ systems infected, different reassortant constellations were obtained from different tissues (WENSKE et al. 1985). In a similar fashion, mixed infection of mice with rotaviruses led to the isolation of different reassortant constellations at different times after infection (GOMBOLD and RAMIG 1986). It is not clear if this observation was significant since all virus replication occurred in the local environment of the small intestine.

7.2.3 Reassortment and Superinfection Exclusion

Since reassortment is clearly important to the evolution of the rotaviruses, one pertinent question is the frequency of mixed infection in individuals so that the potential exists for the production of reassortant progeny. The incidence of mixed infection is clearly high, given that the frequency of isolation of viruses with mixed electropherotypes from infected individuals is approximately 10% (LOURENCO et al. 1981; SPENCER et al. 1983; NICOLAS et al. 1984), and based on studies in mice, simultaneous infections with two viruses results in high frequencies of reassortment (GOMBOLD and RAMIG 1986). However, if the mixed infection is not simultaneous, the temporal separation of the two infections may not permit reassortment because of superinfection exclusion. In one attempt to examine this possibility, asynchronous mixed infections were performed in vitro using ts mutants and the production of ts$^+$ reassortants determined. Simultaneous infection with a pair of mutants yielded 17% reassortants, and reassortment occurred if infection with the first and second viruses was separated by as much as 24 h (RAMIG 1990). Reassortment frequency was reduced as the time between the viruses increased, but it did not fall below 3%, a value constituting a significant proportion of the progeny. Thus, at the level of the individual cell, a significant degree of asynchrony in the infection (about the time of the one-step growth cycle at 31 °C) of two viruses still allows genetic interaction at significant levels. In the infected individual organism the time of asynchrony could possibly be much longer, provided the two viruses were ultimately able to find their way to the same cell.

The failure of rotaviruses to establish superinfection exclusion contrasts with the rapid onset of exclusion in bluetongue virus-infected vero cells, in which only 4–8 h are required to establish the exclusion and prevent reassortment (RAMIG et al. 1989). In bluetongue virus, the in vitro system mimics what is seen in vivo. Superinfection exclusion is also seen in the vector species, although the time to establish the exclusion state is days rather than hours (EL-HUSSEIN et al. 1990). Thus, in the orbiviruses, superinfection exclusion would appear to limit the reassortment potential of the virus whereas the absence of superinfection exclusion in rotaviruses increases their potential for reassortment.

7.3 Genetics of Non-selectable Phenotypes or Mutations with Complex Phenotype

It is clear from the above discussions that phenotypes can be mapped by reassortment without resorting to selection on the basis of the phenotype of interest. A reassortant collection is generated, and individual reassortants are scored for expression of the phenotype without any selection being applied. However, in some cases the phenotype of a mutation may be identifiable only through the use of reassortment. No such cases have yet been identified in rotaviruses, but the phenomenon of suppression of mutant phenotype was identified by reassortment analysis in reovirus.

A revertant of a reovirus ts mutant had unusual properties which suggested that it still contained the original ts lesion in suppressed form. To demonstrate that the revertant did indeed contain the original ts lesion it was necessary to satisfy two requirements: (1) that ts progeny could be rescued from the revertant following backcross of the revertant to wild type, and (2) that any rescued ts mutations were identical to the parental ts mutant from which the revertant was isolated. These two requirements were satisfied, the first by separating by reassortment the parental ts lesion from the mutation that was suppressing its phenotype, and the second by using reassortment testing to show the identity of the rescued ts mutation and the parental ts mutation (RAMIG et al. 1977). The suppressor mutation that was present in the revertant has no identifiable phenotype other than the ability to suppress the ts mutation, making it difficult to work with. However, through application of reassortment tests it was possible both to determine that suppression of ts lesions was a common phenomenon in reovirus (RAMIG and FIELDS 1979) and to map the genome segment locations of some of the suppressor mutations (MCPHILLIPS and RAMIG 1984).

Similar analyses of revertants of rotavirus ts mutants have not revealed the presence of suppression of ts phenotype in the rotavirus system (Ramig, unpublished data). However, this example emphasizes the utility of reassortment to dissect complex phenotypes in segmented genome viruses.

8 Implications of Genetics for Future Studies on the Rotaviruses and *Reoviridae*

Genetics in general, and reassortment in particular, have provided a powerful tool for the investigation of problems relating to the rotaviruses and other segmented genome viruses. It seems likely that reassortment will occupy a central place in the armamentarium of the rotavirologist in the future. Some future problems in which reassortment is likely to play a central role are briefly discussed below.

8.1 Mapping of Viral Phenotypes and Host Responses

Mapping viral phenotypes by the creation of hybrids between viruses having different phenotypes is experiencing wide application in virology. In the rotaviruses and other segmented genome viruses, the technique is particularly useful for two reasons: (1) the high frequency of reassortment makes construction of reassortants a relatively simple task and (2) the monocistronic nature of the rotavirus genome segments ensures that entire genes are transferred and hybrid genes are not created (although such hybrid genes might be useful).

Reassortment will continue to play a role in the analysis of rotavirus infection, particularly in identifying viral genes that control pathogenesis and viral gene products to which the host immunologically responds in biologically meaningful ways.

However, while reassortment will continue to provide a powerful tool for analysis, the researcher must be aware of the limitations of the method. Reassortment experiments will not map phenotypes with any precision greater than assigning them to a segment(s). The methods of molecular biology will be required for fine structure mapping within segments. Furthermore, mapping of certain phenotypes will be severely limited untill the problems of rescue are solved (see Sect. 8.2). It will also be important to bear in mind the possibility that the results of a reassortment mapping experiment will apply only to the virus strains examined. The effects of recipient genetic background on the expression of phenotypes by donor segments must be kept in mind (CHEN et al. 1989). Thus, general application of the results of limited reassortment tests must be approached with caution. However, if the researcher is aware of the potential pitfalls, reassortment studies can continue to be a fruitful avenue of research.

8.2 Rescue of Cloned and Genetically Manipulated Genes

Representatives of all 11 rotavirus genome segments have been cloned into cDNA copies (ESTES and COHEN 1989) and studies of the in vitro regulation, targeting, transport, and expression of the gene products have begun (STIRZAKAR et al. 1987; STIRZAKER and BOTH 1989; MACKOW et al. 1989). These studies promise to tell us a great deal about the molecular biology of the rotavirus genes and gene products, but they are much less likely to reveal how the molecular biology affects pathogenesis and host immune response (e.g., molecular pathogensis). Once reassortment experiments have been used to identify genes important in different stages of pathogenesis, it will be desirable to apply site-directed mutagenesis and other molecular techniques to map the fine structure of the pathogenesis determinants in vivo. This is not currently possible.

One of the most intriguing problems facing the rotavirologist is to devise methods for the rescue of copies of cloned and genetically manipulated genes into infectious virus particles. Until this is accomplished it will be impossible to evaluate the effects of mutations, engineered by site-directed mutagenesis or the other powerful techniques of molecular biology, on infection with the virus. Essentially the problem is to determine how to get exogenous RNA to enter into the pool of RNA available for reassortment into infectious progeny virus. Clearly, the problem is a complex one, as many laboratories have attempted rescue without success. A great deal more work is likely to be expended on this problem, and further work on the process of reassortment may provide the critical clue necessary for successful rescue.

Recently, the successful rescue of a functional RNA copy of a foreign gene into the segmented genome of influenza virus (LUYTJES et al. 1989) and the rescue of RNA copies of cDNAs of influenza virus genome segments (P. Palese, personal communication) has been reported, suggesting that rescue is also an attainable goals for rotaviruses.

8.3 Reassortment and Evolution of the Rotaviruses

It is clear that reassortment can move the genes of the rotaviruses into new combinations that have bot been previously detected. This movement of genes may be of particular importance for those genes that encode biologically relevant antigens (neutralization antigens, determinants of cell-mediated immunity, virulence determinants). In many cases, the phenotype of a reassortant is what would be expected from the presence of an antigen from a parent with a given phenotype. In the case of neutralizing antigens, for example, this simple direct expression of the parental antigenicity in a reassortant would have the potential to create new combinations of previously recognized antigens, but would be unlikely to create a virus recognized as a new serotype. However, it has now become clear that the recipient genetic background can affect the expression of antigenicity, presumably through the interactions of donor and recipient proteins (CHEN et al. 1992). In a case in which new combinations of antigens resulted in altered expression of antigenic epitopes and the possible masking or revealing of epitopes, the new combination of previously recognized antigens could be presented in a manner so as to be functionally recognized as a distinct serotype. Thus, reassortment not only has the capability or shuffling previously recognized genes, but also of affecting the expression of those genes in subtle but consequential ways. The potential for reassortment to generate viruses representing new serotypes remains to be demonstrated, but we must be aware that the potential exists.

8.4 Does Intramolecular Recombination Occur in Rotaviruses?

As noted above, genetic evidence does not currently support the notion of intramolecular recombination between markers in cognate genome segments. However, rearranged genome segments have recently been described that must have arisen through a process similar to copy-choice recombination. Copy-choice recombination has been demonstrated in a number of RNA viruses for which conventional wisdom held recombination did not occur. The observation of intramolecular recombination in other RNA viruses raises the question of whether it could also occur in rotaviruses or other dsRNA genome viruses.

Intramolecular recombination in unichromosomal RNA viruses has been the subject of interest, and the copy-choice mechanism proposed for these recombination events (KIRKEGAARD and BALTIMORE 1986) is very similar to the template-slipping or template-jumping mechanism proposed for the generation of rearranged genomes in a number of RNA viruses (LAZZARINI et al. 1981). Thus, the possibility of intramolecular recombination cannot be ruled out for any of the *Reoviridae*. Indeed, rare intramolecular recombination events could prove to be significant for virus evolution. However, within the context of single-cycle growth experiments such as generally carried out in the laboratory, the likelihood that intramolecular recombination events would occur at detectable levels is vanishingly small.

Given that a copy-choice mechanism has been proposed in all RNA virus systems in which intramolecular recombination has been observed, predictions can be made as to the nature of the recombinants if such a mechanism was ultimately demonstrated in rotaviruses: (1) Because copy-choice recombination mechanisms involve template-switching by an active polymerase, recombination would be expected to occur in particles in the rotavirus-infected cell. It could occur either early in infection during transcription of (+) strand RNA from input or progeny "transcriptase" particles, or it could occur late in infection during (−) strand synthesis in subviral "replicase" particles. (2) The template strand-switch during copy-choice recombination has been suggested to be directed by homology between the nascent RNA strand and the new template (FIELDS and WINTER 1982). The demonstration of concatemeric rearranged RNAs in rotaviruses (ALLEN and DESSELBERGER 1985) suggests that appropriate repeats of homology may exist in rotavirus genome segments and lead to the formation of rearranged segments. However, if similar homologies existed between non-cognate segments, mosaic segments (containing sequences from two or more RNA segments) could result. It is possible that such mosaic segments have not been in rotaviruses because they interrupt coding sequences and are nonviable. In contrast, mosaic segments have been identified in defective interfering particles of influenza virus (FIELDS and WINTER 1982). (3) Copy-choice recombinants between the cognate segments of two parental viruses would not be expected to be generated in mixed-infected cells. This follows from the fact that transcriptase particles and replicase particles are thought to contain only one copy of each of the 11 genome segments. Thus, within the confines of the particle in which copy-choice recombination must occur, the substrates for copy-choice recombination between cognate segments of two different viruses do not exist. (4) As a rare event, subviral replicase particles with the requisite substrates for copy-choice recombination may exist. These particles would not be expected to mature to infectious virus, since the presence of two copies of one cognate segment would probably cause the exclusion of some other segment due to packaging restrictions. However, if there particles were able to mature to the stage of actively transcribing subviral particles, they could synthesize recombinant transcripts that could enter into subsequent rounds of reassortment and maturation that would lead to the production of infectious virus. Thus, the generation of copy-choice recombinants between cognate RNAs during mixed infection would require a sequence of events that would demand the process be inefficient.

9 Concluding Remarks

Segregation analysis has been an invaluable technique in the study of rotaviruses. The ease and widespread applicability of this technique have resulted in the identification of proteins responsible for numerous viral properties and phenotypes and have provided insight into rotavirus replication and evolution. Reassortment between rotavirus mutants has furnished the means of genetically grouping several rotavirus ts mutant collections for which classical complementation test have failed. Six ts mutant groups have been mapped to specific genome segments and, together with the known coding assignments of these segments, has yielded information on the potential functions of several viral proteins. Clearly, genetics has played a major role in rotavirology during the past decade and has provided a firm foundation for future studies. This foundation will clearly facilitate further genetic and molecular studies of rotavirus protein structure and function.

References

Allen AM, Desselberger U (1985) Reassortment of human rotaviruses carrying rearranged genomes with bovine rotavirus. J Gen Virol 66: 2703–2714
Bass DM, Macknow ER, Greenberg HB (1989) The rotavirus gene 8 product, NS35, adheres to the suface of intestinal epithelial and MA 104 cells and is a structural viral protein, Gastroenterology 96 (5, 2): A30

Bican P, Cohen J, Charpilienne A, Scherrer R (1982) Purification and characterization of bovine rotavirus cores. J Virol 43: 1113–1117

Bishop RF, Davidson GP, Holmes IH, Ruck BJ (1973) Virus particles in the epithelial cells of duodenal mucosa from children with acute non-bacterial gastroenteritis. Lancet 2: 1281–1283

Both GW, Bellamy AR, Siegman LJ (1982) A general strategy for cloning double-stranded RNA: nucleotide sequence for the simian-11 rotavirus gene 8. Nucleic Acids Res 10: 7075–7088

Boyle JF, Holmes KV (1986) RNA-binding proteins of bovine rotavirus. J Virol 58: 561–568

Brown DWG, Mathan MM, Bartram J, Crookshanks-Newman F, Chanock RM, Kapikian AZ (1988a) Rotavirus epidemiology in Vellore, South India: group, subgroup, serotype, and electropherotype. J Clin Microbiol 26: 2410–2414

Brown SE, Gonzalez HA, Bodkin DK, Tesh RB, Knudson DL (1988b) Intra- and inter-serogroup genetic relatedness of orbiviruses. II Blot hybridization and reassortment in vitro of epizootic haemorrhagic disease serogroup, bluetongue type 10 and pata viruses. J Gen Virol 69: 135–147

Burns JW, Welch SKW, Nakata S, Estes MK (1988) Functional and topographical analyses of epitopes on the hemagglutinin (VP4) of the simian rotavirus SA11. J Virol 62: 2164–2172

Burns JW, Chen D, Estes MK, Ramig RF (1989a) Biological and immunological characterization of a simian rotavirus SA11 variant with an altered genome segment 4. Virology 169: 427–435

Burns JW, Welch SKW, Nakata S, Estes MK (1989b) Characterization of monoclonal antibodies to human group B rotavirus and their use in an antigen-detection enzyme-linked immunosorbent assay (ELISA). J Clin Microbiol 27: 245–250

Chakraborty PR, Ahmed R, Fields BN (1979) Genetics of reovirus: the relationship of interference to complementation and reassortment of temperature-sensitive mutants at nonpermissive temperature. Virology 94: 119–127

Chan W-K, Penaranda ME, Crawford SE, Estes MK (1986) Two glycoproteins are produced from the rotavirus neutralization gene. Virology 151: 243–252

Chen D, Burns JW, Estes MK, Ramig RF (1989) The Phenotypes of rotavirus reassortments depend upon the recipient background. Proc Natl Acad Sci USA 86: 3743–3747

Chen D, Estes MK, Ramig RF (1992) Specific interactions between rotavirus outercapsid proteins VP4 and VP7 determine expression of a cross-reactive, neutralizing VP4-specific epitope. J Virol 66: 432–439

Chen D, Gombold JL, Ramig RF (1990) Intracellular RNA synthesis directed by temperature-sensitive mutants of simian rotavirus SA11. Virology 178: 143–151

Clark B, Dellelberger U (1988) Myristylation of rotavirus proteins. J Gen Virol 69: 2681–2686

Cohen J, Laporte J, Charpilienne A, Scherrer R (1979) Activation of rotavirus RNA polymerase by calcium chelation Arch. Virol 60: 177–186

Coulson BS, Fowler KJ, Bishop RF, Cotton RGH (1985) Neutralizing monoclonal antibodies to human rotavirus and indications of antigenic drift among strains from neonates. J Virol 54: 14–20

Cross RK, Fields BN (1976) Use of an aberrant polypeptide as a marker in three-factor crossess: further evidence for independent reassortment as the mechanism of recombination between temperature-sensitive mutants of reovirus type 3. Virology 74: 345–362

Dahlberg JE, Simon EH (1969) Physical and genetic studies of Newcastle disease virus: Evidence for multiploid particles. Virology 38: 666–678

Dyall-Smith ML, Lazdins I, Tregear GW, Holmes IH (1986) Location of the major antigenic sites involved in rotavirus serotype-specific neturalization. Prov Natl Acad Sci USA 83: 3465–3468

El-Hussein AM, Ramig RF, Holbrook FR, Beaty BJ (1989) Asynchronous mixed infection of Culiocoides variipennis with bluetongue virus serotypes 10 and 17. J Gen Virol 70: 3355–3362

Estes MK, Cohen J (1989) Rotavirus gene structure and function. Microbiol Rev 53: 410–449

Estes MK, Graham DY, Mason BB (1981) Proteolytic enhancement of rotavirus infectivity: molecular mechanisms. J Virol 39: 879–888

Estes MK, Palmer EL, Obijeski JF (1983) Rotavirus: a review. In: Compans RW, Cooper M, voprowsky JF (eds) Current topics in microbiology and immunology, vol 105. Springer, Berlin Heidelberg New York, pp 123–184

Faulkner-Valle GP, Clayton AV, McCrae MA (1982) Molecular biology of Rotaviruses. III. Isolation and characterization of temperature-sensitive mutants of bovine rotavirus. J Virol 42: 669–677

Faulkner-Valle GP, Lewis J, Pedley S, McCrae MA (1983) Isolation and characterization of ts
    mutants of bovine rotavirus. In: Compans RW, Bishop DHL (eds) Double-stranded RNA
    viruses. Elsevier, New York, pp 303–312
Fields BN (1971) Temperature-sensitive mutants of reovirus type 3: Features of genetic recom-
    bination. Virology 46: 142–148
Fields BN, Joklik WK (1969) Isolation and preliminary genetic and biochemical characterization
    of temperature-sensitive mutants of reovirus. Virology 37: 335–342
Fields S, Winter G (1982) Nucleotide sequence of influenza virus segments 1 and 3 reveal
    mosaic structure of a small viral RNA segment. Cell 28: 303–313
Flewett TH, Bryden AS, Davies H (1973) Virus Particles in gastroenteritis. Lancet 2: 1497
Flores J, Greenberg HB, Myslinski J, Kalica AR, Wyatt RG, Kapikian AZ, Chanock RM
    (1982) Use of transcription probes for genotyping rotavirus reassortants. Virology 121:
    288–295
Flores J, Midthun K, Hoshino Y, Green K, Gorziglia M, Kapikian AZ, Chanock RM (1986)
    Conservation of the fourth gene among rotaviruses recovered from asymptomatic newborn
    infants and its possible role in attenuation. J Virol 60: 972–979
Fukuhara N, Yoshie O, Kitaoka S, Konno T (1988) Role of VP3 in human rotavirus internalization
    after target cell attachment via VP7. J Virol 62: 2209–2218
Gallegos CO, Patton JT (1989) Characterization of rotavirus replication inermediates: A model
    for the assembly of single-shelled particles. Virology 172: 616–627
Garbarg-Chenon A, Bricout F, Nicolas J-C (1984) Study of genetic reassortment between two
    human rotaviruses. Virology 139: 358–365
Garbarg-Chenon A, Bricout F, Nicolas J-C (1986) Serological characterization of human
    reassortant rotaviruses. J Virol 59: 510–513
Gombold JL, Ramig RF (1986) Analysis of reassortment of genome segments in mice mixedly
    infected with rotaviruses SA11 and RRV. J Virol 57: 110–116
Gombold JL, Ramig RF (1987) Assignment of simian rotavirus SA11 temperature-sensitive
    mutant groups A, C, F, and G, to genome segments. Virology 161: 463–473
Gombold JL, Estes MK, Ramig RF (1985) Assignment of simian rotavirus SA11 temperature-
    sensitive mutant groups B and E to genome segments. Virology 143: 309–320
Gonzalez SA (1987) The genetic relatedness of orbiviruses: RNA-RNA blot hybridization and
    in vitro gene reassortment. PhD thesis, Yale University
Gonzalez SA, Mattion NM, Bellinzoni R, Burrone OR (1989) Structure of rearranged genome
    segment 11 in two different rotavirus strains generated by a similar mechanism. J Gen Virol
    70: 1329–1336
Graham A, Kudesia G, Allen AM, Desselberger U (1987) Reassortment of human rotavirus
    possessing genome rearrangements with bovine rotavirus: Non-randomness and evidence
    for host cell selection. J Gen Virol 68: 115–122
Greenberg HB, Kalica AR, Wyatt RW, Jones RW, Kapikian AZ, Chanock RM (1981) Rescue of
    noncultivatable human rotavirus by gene reassortment during mixed infection with ts
    mutants of a cultivatable bovine rotavirus. Proc Natl Acad Sci USA 78: 420–424
Greenberg HB, Wyatt RG, Kapikian AZ, Kalica AR, Flores J, Jones R (1982) Rescue and
    serotypic characterization of noncultivable human rotavirus by gene reassortment. Infect
    Immun 37: 104–109
Greenberg HB, Flores J, Kalica AR, Wyatt RG, Jones R (1983) Gene coding assignments for
    growth restriction, reutralization and subgroup specificities of the W and DS-1 strains of
    human rotavirus. J Gen Virol 64: 313–320
Helmberger-Jones M, Patton JT (1986) Characterization of subviral particles in cells infected
    with simian rotavirus SA11. Virology 155: 655–665
Holmes IH (1983) Rotaviruses. In: Joklik WK (ed) The Reoviridae. Plenum, New York,
    pp 359–423
Hoshino Y, Wyatt RG, Greenberg HB, Flores J, Kapikain AZ (1984) Serotypic similarity and
    diversity of rotaviruses of mammalian and avian origin as studied by plaque reduction
    neutralization. J Infect Dis 149: 694–702.
Hoshino Y, Sereno MM, Midthun K, Flores J, Kapikian AZ, Chanock RM (1985) Independent
    segregation of two antigenic specificities (VP3 and VP7) involved in neutralization of
    rotavirus infectivity. Proc Natl Acad Sci USA 82: 8701–8704
Hoshino Y, Gorziglia M, Valdesuso J, Askaa J, Glass RI, Kapikian AZ (1987a) An equine
    rotavirus (FI-14 strain) which bears both subgroup I and subgroup II specificities on its VP6.
    Virology 157: 488–496

Hoshino Y, Sereno MM, Midthun K, Flores J, Chanock RM, Kapikian AZ (1987b) Analysis by plaque reduction neutralization assay of intertypic rotaviruses suggests that gene reassortment occurs in vivo. J Clin Microbiol 25: 290–294

Hoshino Y, Saif LJ, Sereno MM, Chanock RM, Kapikian AZ (1988) Infection immunity of piglets to either VP3 or VP7 outer capsid protein confers resistance to challenge with a virulent rotavirus bearing the corresponding antigen. J Virol 62: 744–748

Hrdy DB (1982) Investigation of genetic interaction between simian rotavirus SA11 and reovirus. Microbiologica 5: 207–213

Hudziak RM, Laski FA, Raj Bhandary UL, Sharp PA, Capecchi MR (1982) Establishment of mammalian cell lines containing multiple nonsense mutations and functional suppressor tRNA genes. Cell 31: 137–146

Hundley F, Biryahwaho B, Gow M, Desselberger U (1985) Genome rearrangements of bovine rotavirus after serial passage at high multiplicity of infection. Virology 143: 88–103

Kalica AR, Garon CF, Wyatt RG, Mebus CA, Van Kirk DH, Chanock RM, Kapikian AZ (1976) Differentiation of human and calf reovirus-like agents associated with diarrhea using polyacrylamide gel electrophoresis of RNA. Virology 74: 86–92

Kalica AR, Sereno MM, Wyatt RG, Mebus CA, Chanock RM, Kapikian AZ (1987a) Comparison of human and animal rotavirus strains by gel electrophoresis of RNA. Virology 87: 247–255

Kalica AR, Wyatt RG, Kapikian AZ (1987b) Detection of differences among human and animal rotaviruses using analysis of viral RNA. J Am Vet Med Assoc 173: 531–537

Kalica AR, Greenberg HB, Wyatt RG, Flores J, Sereno MM, Kapikian AZ, Chanock RM (1981a) Genes of human (strain Wa) and bovine (strain UK) rotaviruses that code for neutralization and subgroup antigens. Virology 112: 385–390

Kalica AR, Greenberg HB, Espejo RT, Flores J, Wyatt RG, Kapikian AZ, Chanock RM (1981b) Distinctive ribonucleic acid patterns of human rotavirus subgroups 1 and 2. Infect Immun 33: 958–961

Kalica AR, Flores J, Greenberg HB (1983) Identification of the rotaviral gene that codes for hemagglutination and protease-enhanced plaque formation. Virology 125: 194–205

Kaljot KT, Shaw RD, Rubin DH, Greenberg HB (1988) Infectious rotavirus enters cells by direct cell membrane penetration, not by endocytosis. J Virol 62: 1136–1144

Kapikian AZ, Cline WL, Greenberg HB, Wyatt RG, Kalica AR, Banks CE, James HD, Flores J, Chanock RM (1981) Antigenic characterization of human and animal rotaviruses by immune adherence hemagglutination assay (IAHA): evidence for distinctness of IAHA and neutralization antigens. Infect Immun 33: 415–425

Kapikian AZ, Flores J, Hoshino Y, Glass RI, Midthun K, Gorziglia M, Chanock RM (1986) Rotavirus: the major etiologic agent of severe infantile diarrhea may be controllable by a "Jennerian" approach to vaccination. J Infect Dis 153: 815–822

Kingsbury DW, Granoff A (1970) Studies on mixed infection with Newcastle disease virus. IV. On the structure of heterozygotes. Virology 42: 262–265

Kirkegaard K, Baltimore D (1986) The mechanism of RNA recombination in poliovirus. Cell 47: 433–441

Kotaoka S, Fukuhara N, Tazawa F, Suzuki H, Sato T, Konno T, Ebino T, Ishida N (1986) Characterization of monoclonal antibodies against human rotavirus hemagglutinin. J Med Virol 19: 313–323

Lazzarini RA, Keene JD, Schubert M (1981) The origins of defective interfering particles of the negative strand RNA viruses. Cell 26: 145–154

Liu M, Estes MK (1989) Nucleotide sequence of the simian rotavirus SA11 genome segment 3. Nucleic Acids Res 17: 7991

Liu M, Offit PA, Estes MK (1988) Identification of the simian rotavirus SA11 genome segment 3 product. J Virol 163: 26–32

Lourenco MH, Nicolas JC, Cohen J, Scherrer R, Bricout F (1981) Study of human rotavirus genome by electrophoresis: attempt of classification among strains isolated in France. Ann Virol (Inst. Pasteur) 132E: 161–173

Luytjes W, Krystal M, Enami M, Parvin JD, Palese P (1989) Amplification, expression and packaging of a foreign gene by influenza virus. Cell 59: 1107–1113

MacKenzie JS (1970) Isolation of temperature-sensitive mutants and the construction of a preliminary map for influenza virus. J Gen Virol 6: 63–75

Mackow ER, Shaw RD, Matsui SM, Vo PT, Benfield DA, Greenberg HB (1988a) Characterization of homotypic and heterotypic VP7 neutralization sites of rhesus rotavirus. Virology 165: 511–517

Mackow ER, Shaw RD, Matsui SM, Vo PT, Dang MN, Greenberg HB (1988b) Characterization of the rhesus rotavirus VP3 gene: location of amino acids involved in homologous and heterologous rotavirus neutralization and identification of a putative fusion region. Proc Natl Acad Sci USA 85: 645–649

Mackow ER, Barnett JW, Chan H, Greenberg HB (1989) The rhesus rotavirus outer capsid protein VP4 functions as a hemagglutinin and is antigenically conserved when expressed by a baculovirus recombinant. J Virol 63: 1661–1668

Mason BB, Graham DY, Estes MK (1983) Biochemical mapping of the simian rotavirus SA11 genome. J Virol 46: 413–423

Matsui SM, Offit PA, Vo PT, Mackow ER, Benfield DA, Shaw RD, Padilla-Noriega L, Greenberg HB (1989) Passive protection against rotavirus-induced diarrhea by monoclonal antibodies to the heterotypic neutralization domain of VP7 and the VP8 fragment of VP4. J Clin Microbial 27: 780–782

Matsuno S, Hasegawa A, Kalica AR, Kono R (1980) Isolation of a recombinant between simian and bovine rotaviruses. J Gen Virol 48: 253–256

Matthews REF (1979) The classification and nomenclature of viruses. Summary of results of mettings of the International Committee on Taxonomy of Viruses. The Hague, September 1978. Intervirology 11: 133–135

Mattion N, Gonzalez SA, Burrone O, Bellinzoni R, La Torre JL, Scodeller EA (1989) Rearrangement of genomic segment 11 in two swine rotavirus strains. J Gen Virol 69: 695–698

McCrae MA, McCorquodale JG (1982) The molecular biology of rotaviruses. II. Identification of the protein-coding assignments of calf rotavirus genome RNA species. Virology 117: 435–443

McPhillips TH, Ramig RF (1984) Extragenic suppression of temperature-sensitive phenotype in reovirus: mapping suppressor mutations. Virology 135: 428–439

Much DH, Zajac I (1972) Purification and characterization of epizootic diarrhea of infant mice virus. Infect Immun 6: 1019–1024

Nakagomi O, Nakagomi T, Hoshino Y, Flores J, Kapikian AZ (1987) Genetic analysis of a human rotavirus that belongs to subgroup I but has an RNA pattern typical of subgroup II human rotaviruses. J Clin Microbiol 25: 1159–1164

Nakata S, Estes MK, Graham DY, Loosle R, Hung T, Wang S, Saif LJ, Melnick JL (1986) Antigenic characterization and ELISA detection of adult diarrhea rotaviruses. J Infect Dis 154: 448–455

Newmann JFE, Brown F, Bridger JC, Woode GN (1975) Characterization of a rotavirus. Nature 258: 631–633

Nicolas JC, Pothier P, Cohen J, Lourenco MH, Thompson R, Guimbaud P, Chenon A, Dauvergne M, Bricout F (1984) Survey of human rotavirus propagation as studied by electrophoresis of genomic RNA. J Infect Dis 49: 688–693

Novo E, Esparza J (1981) Composition and topography of structural polypeptides of bovine rotavirus. J Gen Virol 56: 325–335

Nuttall SD, Hum CP, Holmes IH, Dyall-Smith ML (1989) Sequences of VP9 genes from short and supershort rotavirus strains. Virology 171: 453–457

Offit PA, Blavat G (1986) Identification of the two rotavirus genes determining neutralization specificities. J Virol 57: 376–378

Offit PA, Clark HF (1985a) Protection against rotavirus induced gastroenteritis in a murine model by passively acquired gastrointestinal but not circulating antibodies. J Virol 54: 58–64

Offit PA, Clark HF (1985b) Maternal antibody-mediated protection against gastroenteritis due to rotavirus in newborn mice is dependent on both serotype and titer of antibody. J Infect Dis 152: 1152–1158

Offit PA, Blavat G, Greenberg HB, Clark HF (1986a) Molecular basis of rotavirus virulence: role of gene segment 4. J Virol 57: 46–49

Offit PA, Clark HF, Blavat G, Greenberg HB (1986b) Reassortment rotaviruses containing structural proteins VP3 and VP7 from different parents induce antibodies protective against each parental serotype. J Virol 60: 491–496

Offit PA, Shaw RD, Greenberg HB (1986c) Passive protection against rotavirus-induced diarrhea by monoclonal antibodies to surface proteins VP3 and VP7. J Virol 58: 700–703

Palese P (1984) Reassortment continuum. In: Notkins AL, Oldstone MBA (eds) Concepts in viral pathogenesis. Springer, Berlin Heidelberg, New York, pp 144–151

Pedley S, Hundley F, Chrystie I, McCrae MA, Desselberger U (1984) The genomes of rotaviruses isolated from chronically infected immunodeficient children. J Gen Virol 65: 1141–1150

Periera HG, Azeredo RA, Fialho AM, Vidal MNP (1984) Genomic heterogeneity of simian rotavirus SA11. J Gen Virol 65: 815–818

Poruchynsky MS, Tyndall C, Both GW, Sato F, Bellamy AR, Atkinson PH (1985) Deletion into an NH_2-terminal hydrophobic domain results in secretion of rotavirus VP7, a resident endoplasmic reticulum membrane glycoprotein. J Cell Biol 101: 2119–2209

Ramig RF (1982) Isolation and genetic characterization of temperature-sensitive mutants of simian rotavirus SA11. Virology 120: 93–105

Ramig RF (1983a) Isolation and genetic characterization of temperature-sensitive mutants that define five additional recombination groups in simian rotavirus SA11. Virology 130: 464–473

Ramig RF (1983b) Genetic studies with simian rotavirus SA11. In: Compans RW, Bishop DHL (eds) Double-stranded RNA viruses. Elsevier, New York, pp 321–327

Ramig RF (1983c) Factors that affect genetic interaction during mixed infections with temperature-sensitive mutants of simian rotavirus SA11. Virology 127: 91–99

Ramig RF (1990) Superinfecting rotaviruses are not excluded from genetic interactions during asynchronous mixed infections in vitro. Virology 176: 308–310

Ramig RF, Fields BN (1979) Revertants of temperature-sensitive mutants of reovirus: Evidence for frequent extragenic suppression. Virology 92: 155–167

Ramig RF, Fields BN (1983) Genetics of reoviruses. In: Joklik WK (ed) The reoviridae, Plenum, New York, pp 197–228

Ramig RF, Galle KL (1990) Rotavirus genome segment 4 determines viral replication phenotype in cultured liver cells (HepG2). J Virol 64: 1044–1049

Ramig RF, Petrie BL (1984) Characterization of temperature-sensitive mutants of simian rotavirus SA11: protein synthesis and morphogenesis. J Virol 49: 665–673

Ramig RF, White RM, Fields BN (1977) Suppression of the temperature-sensitive phenotype of a mutant of reovirus type 3. Science 195: 406–407

Ramig RF, Garrison C, Chen D-Y, Bell-Robinson D (1989) Analysis of reassortment and superinfection during mixed infection of vero cells with bluetongue virus serotypes 10 and 17. J Gen Virol 70: 2595–2603

Rodger SM, Schnagl RD, Holmes IH (1975) Biochemical and biophysical characterization of diarrhea viruses of human and calf origin. J Virol 16: 1229–1235

Rubin DH, Fields BN (1980) Molecular basis of reovirus virulence: role of the M2 gene. J Exp Med 152: 853–868

Sabara M, Gilchrist JE, Hudson GR, Babiuk LA (1985) Preliminary characterization of an epitope involved in neutralization and cell attachment that is located on the major bovine rotavirus glycoprotein. J Virol 53: 58–66

Sabara M, Ready KFM, Frenchick PJ, Babiuk L (1987) Biochemical evidence for the oligomeric arrangement of bovine rotavirus nucleocapsid protein and its possible significance in the immunogenicity of this protein. J Gen Virol 68: 123–133

Saif LJ, Terrett LA, Miller KL, Cross RF (1988) Serial propagation of porcine group C rotavirus (pararotavirus) in a continuous cell line and characterization of the passaged virus. J Clin Microbiol 26: 1277–1282

Samal SK, El Hussein A, Holbrook FR, Beaty BJ, Ramig RF (1987a) Mixed infection of culicoides variipennis with bluetongue virus serotypes 10 and 17: Evidence for high frequency reassortment in the vector. J Gen Virol 68: 2319–2329

Samal SK, Livingston CW, McConnell S, Ramig RF (1987b) Analysis of mixed infection of sheep with bluetonge virus serotypes 10 and 17: evidence for reassortment in the vertebrate host. J Virol 61: 1086–1091

Sandino AM, Jashes M, Faundez G, Spencer E (1986) Role of the inner capsid on in vitro human rotavirus transcription. J Virol 60: 797–802

Schwarz KB, Moore TJ, Willoughby RE, Yolken RH (1989) Growth of rotaviruses in a human liver cell line. Pediatr Res 25: 184A

Sedivy JM, Capone JP, RajBhandary UL, Sharp PA (1987) An inducible mammalian amber suppressor: propagation of a poliovirus mutant. Cell 50: 379–389

Sethi SK, Olive DM, Strannegard OO, Al-Nakib W (1988) Molecular epidemiology of human rotavirus infections based on genome segment variations in viral strains. J Med Virol 26: 249–259

Sharpe AH, Ramig RF, Mustoe TA, Fields BN (1978) A genetic map of reovirus. I. Correlation of genome segments between serotypes 1, 2 and 3. Virology 84: 63–74

Shaw RD, Vo PT, Offit PA, Coulson BS, Greenberg HB (1986) Antigenic mapping of the surface proteins of rhesus rotavirus. Virology 155: 434–451

Shipham SO, de la Rey M (1976) The isolation and preliminary genetic classification of temperature-sensitive mutants of bluetongue virus. Ond erstepoort J Vet Res 43: 189–192

Simpson RW, Hirst GK (1968) Temperature-sensitive mutants of influenza A virus: isolation of mutants and preliminary observations on genetic recombination and complementation. Virology 35: 41–49

Spencer EG, Avendano LF, Garcia BI (1983) Analysis of huamn rotavirus mixed electrophero-types. Infect Immun 39: 569–574

Steele AD, Alexander JJ (1988) The relative frequency of subgroupa I and II rotaviruses in black infants in South Africa. J Med Virol 24: 321–327

Stirzaker SC, Both GW (1989) The signal peptide of the rotavirus glycoprotein is essential for its retention in the ET as an integral membrane protein. Cell 56: 741–747

Stirzaker SC, Whitfeld Pl, Christie DL, Bellamy AR, Both GW (1987) Processing of rotavirus glycoprotein VP7: implications for the retention of the protein in the endoplasmic reticulum. J Cell Biol 105: 2897–2903

Stott JL, Oberst RD, Channel MB, Osburn BI (1987) Genome reassortment between two serotypes of bluetongue virus in a natural host. J Virol 61: 2670–2674

Suzuki H, Kitaoka S, Konno T, Sato T, Ishida N (1985) Two modes of human rotavirus entry into MA 104 cells. Arch Virol 85: 25–34

Svensson L, Grahnquist L, Pettersson C-A, Grandien M, Stintzing G, Greenberg HB (1988) Detection of human rotaviruses which do not react with subgroup I- and II-specific monoclonal antibodies. J Clin Microbiol 26: 1238–1240

Taniguchi K, Urasawa S, Urasawa T (1985) Preparation and characterization of neutralizing monoclonal antibodies with different reactivity patterns to human rotaviruses. J Gen Virol 66: 1045–1053

Taniguchi K, Morita Y, Urasawa T, Urasawa S (1987) Cross-reactive neutralization epitopes on VP3 of human rotavirus: analysis with monoclonal antibodies and antigenic variants. J Virol 61: 1726–1730

Taniguchi K, Hoshino Y, Nishikawa K, Green KY, Maloy WL, Morita Y, Urasawa S, Kapikian AZ, Chanock RM, Gorziglia M (1988a) Cross-reactive and serotype-specific neutralization epitopes on VP7 of human rotavirus: nucleotide sequence analysis of antigenic mutants selected with monoclonal antibodies. J Virol 62: 1870–1874

Taniguchi K, Maloy WL, Nishikawa K, Green KY, Hoshino Y, Urasawa S, Kapikian AZ, Chanock RM, Gorziglia M (1988b) Identification of cross-reactive and serotype specific neutralization epitopes on VP3 of human rotavirus. J Virol 62: 2421–2426

Uhnoo I, Riepenhoff-Talty M, Chegas P, Fisher J, Greenberg HB, Ogra PL (1989) Development of hepatitis and extraintestinal spread with rotaviruses in an heterologous host: Implications in human immunodeficient states. Pediatr Res. 25: 192A

Urasawa S, Urasawa T, Taniguchi K (1986) Genetic reassortment between two human rota-viruses having different serotype and subgroup specificities. J Gen Virol 67: 1551–1559

Wallis C, Melnick JL (1967) Virus aggregation as the cause of the nonneutralizable persistent fraction. J Virol 1: 478–488

Ward RL, Knowlton DR (1989) Genotypic selection following coinfection of cultured cells with subgroup 1 and subgroup 2 human rotaviruses. J Gen Virol 70: 1691–1699

Ward RL, Knowlton DR, Schiff GM, Hoshino Y, Greenberg HB (1988a) Relative concentrations of serum neutralizing antibody to VP3 and VP7 proteins in adults infected with a human rotavirus. J Virol 62: 1543–1549

Ward RL, Knowlton DR, Greenberg HB (1988b) Phenotype mixing during coinfection of cells with two strains of human rotavirus. J Virol 62: 4358–4361

Ward RL, Knowlton DR, Hurst P-FL (1988c) Reassortant formation and selection following coinfection of cultured cells with subgroup 2 human rotaviruses. J Gen Virol 69: 149–162

Ward RL, Knowlton DR, Greenberg HB, Schiff GM, Bernstein DI (1990) Serum neutralizing antibody to VP4 and VP7 proteins in infants following vaccination with WC3 bovine rotavirus. J Virol 64: 2687–2691

Welch AB (1971) Purification, morphology and partial characterization of a reovirus-like agent associated with neonatal calf diarrhea. Can J Comp Med 35: 195–202

Welch AB, Thompson TL (1973) Physicochemical characterization of neonatal calf diarrhea virus. Can J Comp Med 37: 295–301

Wenske EA, Chanock SJ, Krata L, Fields BN (1985) Genetic reassortment of mammalian reoviruses in mice. J Virol 56: 613–616

White BT, McGeoch DJ (1987) Isolation and characterization of conditional lethal amber nonsense mutants of vesicular stomatitis virus. J Gen Virol 68: 3033–3044

Yolken R, Arango-Jaramillo S, Eiden J, Vonderfecht S (1988a) Lack of genomic reassortment following infection of infant rats with group A and group B rotaviruses. J Infect Dis 158: 1120–1123

Yolken RH, Wee SB, Eiden J, Kinney J, Vonderfecht S (1988b) Identification of a group-reactive epitope of group B rotaviruses recognized by monoclonal antibody and application to the development of a sensitive immunoassay for viral characterization. J Clin Microbiol 26: 1853–1858

Zheng S, Woode GN, Malendy DR, Raming RF (1989) Comparative studies of the antigenic polypeptide species VP3, VP6 and VP7, of three strains of bovine rotavirus. J Clin Microbial 27: 1939–1945

Rotavirus Antigens

Y. HOSHINO and A.Z. KAPIKIAN

1 Introduction

Rotaviruses are the major known etiologic agents of diarrhea in infants and young children worldwide (BARNETT 1983; ESTES et al. 1983; CUKOR and BLACKLOW 1984; KAPIKIAN and CHANOCK 1990). Thus, a global initiative to develop and evaluate candidate rotavirus vaccines is underway. In order to design a vaccine strategy and attempt to understand successes and failures of candidate vaccines, it has been important to elucidate the antigenic specificities of rotaviruses from an epidemiologic and a molecular biological perspective (HOLMES 1983; MATSUI et al. 1989a; ESTES and COHEN 1989; BELLAMY and BOTH 1990). The development of a rotavirus vaccine has had to take into account the existence of at least four epidemiologically important group A rotavirus serotypes among the nine human rotavirus VP7 serotypes described thus far. In this chapter, we will review major developments in the antigenic diversity of the group A rotaviruses. The nongroup A rotaviruses, which are not described in this chapter, are classified into groups B, C, D, E, F and G, because they each have a group antigen that is not only distinct from the group A rotaviruses but also distinct from each other (BRIDGER et al. 1987).

Epidemiology Section, Laboratory of Infectious Diseases, National Institute of Allergy and Infectious Diseases, National Institutes of Health, Building 7, Room 103, 9000 Rockville Pike, Bethesda, MD 20892, USA

Although all of the group A rotavirus proteins, both structural and nonstructural, encoded by the 11 RNA segments or genes act as antigens and are recognized by the immune system, special emphasis will be placed on the major inner capsid protein VP6 (gene 6 product) and the two outer capsid proteins VP7 (gene 7, 8 or 9 product) and VP4 (gene 4 product) because they are of importance biologically and clinically, can be detected rapidly by various assays and have been well characterized. The location of these proteins on the virus particle is shown schematically in Fig. 1.

2 Detection of Rotavirus

Before the discovery, in 1973, of human rotaviruses by electron microscopic examination of thin sections of duodenal biopsies obtained from infants and young children with gastroenteritis (BISHOP et al. 1973), morphologically similar agents had been described and shown to be causative agents of diarrhea in mice (ADAMS and KRAFT 1963) and in calves (MEBUS et al. 1969). Electron microscopy was established from the earliest studies as a mainstay for detection of rotaviruses because these agents that could not be propagated efficiently in cell culture had a distinctive morphologic appearance and were shed in large quantity in feces. Later, numerous techniques to detect rotavirus antigens directly from fecal specimens were developed including enzyme-linked immunosorbent assay (ELISA), latex agglutination, radioimmunoassay, counter immunoelectroosmophoresis, reverse passive hemagglutination, immune adherence hemagglutination assay and polyacrylamide gel electrophoresis of viral RNA (for reviews see KAPIKIAN and YOLKEN 1985; KAPIKIAN and CHANOCK 1990). The method of choice, because of its efficiency and practicality, is the confirmatory ELISA, in which a rotavirus positive or negative serum is incorporated into the procedure (for reviews see KAPIKIAN et al. 1979; KAPIKIAN and CHANOCK 1990).

Detection of rotavirus in stools can also be carried out by molecular biological techniques including hybridization by the use of RNA (FLORES et al. 1983; EIDEN et al. 1987; BELLINZONI et al. 1989), cloned cDNA (PEDLEY and McCRAE 1984; DIMITROV et al. 1985; LIN et al. 1985; ZHENG et al. 1989a; OLIVE and SETHI 1989), or synthetic oligodeoxyribonucleotide probes (SETHABUTR et al. 1990; CHAN et al. 1990) and more recently, by primer-directed nucleic acid amplification, commonly known as the polymerase chain reaction (XU et al. 1990; WILDE et al. 1990, 1992).

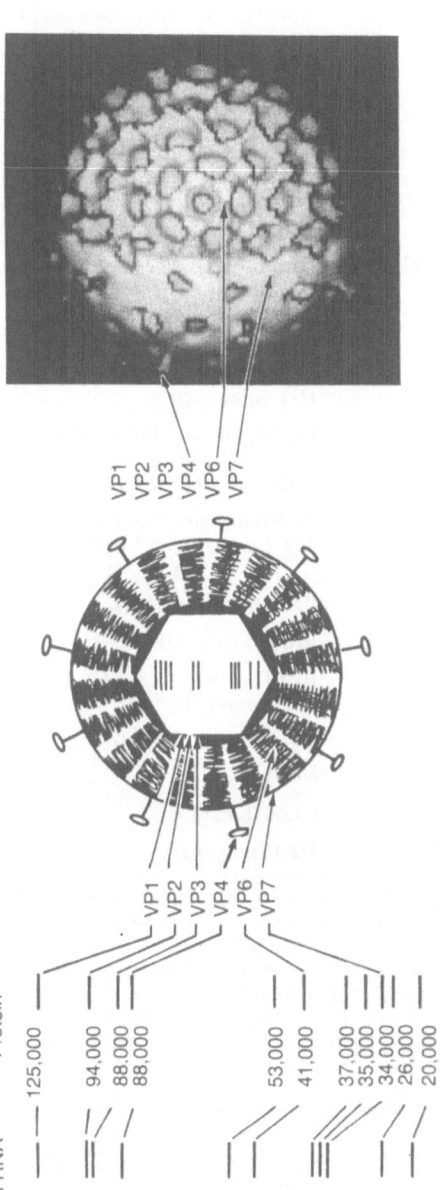

Fig. 1. Rotavirus gene products. *Left,* rotavirus double-shelled particle. *Right,* (from PRASAD et al. 1988) surface representations of the three-dimensional structures of a double-shelled particle (on the *left half*) and a particle (on the *right half*) in which most, if not all, of the outer shell and a small portion of the inner shell mass have been removed. (From KAPIKIAN and CHANOCK 1990)

Table 1. Rotavirus serotypes based on VP7 specificity, subgroup specificities, and representative strains[a]

Serotype	Human rotavirus strains (subgroup)	Animal rotavirus strains (subgroup)
1	Wa, D, KU, M37, RV-4, K8, WI79 (II)	C60, C86 (pig), T449 (cow) (I)
2	DS-1, S2, KUN, RV-5, Hu7, 1076 (I) 378/37 (II)	C134 (pig) (I)
3	P, MO, YO, Ito, RV-3, Nemoto, McN, WI78 (II), AU-1, AU228, Ro1845, HCR3 (I), 0264 (I and II)	SA11 (vervet monkey) (I) MMU18006 (rhesus monkey) (I) CU-1, RS15, K9 (dog) (I) TAKA, Cat 2, Cat 97, FRV-1 (Cat) (I) H-2, (horse) (not I or II) FI-14 (horse) (I and II) CRW-8, C176, ISU-65 (pig) (I) R-2 (II), Ala, C11 (rabbit) (I) EB (mouse) (I) EW (EDIM) (mouse) (not I or II)[b]
4	ST3, ST4, Hosokawa, Hochi, VA70, 57M (II)	Gottfried, SB-1A (pig) (II) BEN-144 (pig) (?) SB-2 (pig) (I)
5	None	OSU, EE, A580, TFR-41, A46 (pig) (I) H-1 (horse) (I)
6	PA151, PA169 (I)	NCDV, UK, RF, WC3, Q17, IND, ID, OK (cow) (I) B641, C486 (cow) (?)
7	None	Ch2 (chicken) (not I or II)[c] Ty1 (turkey) (not I or II)[c], PO-13 (pigeon) (I)
8	69M, B37, HAL1271 (I), PA171 (II)	678, J2538 (cow) (I)
9	WI61, F45 (II), 116E, Mc323 (I)	ISU-64 (pig) (I)
10	A64 (II), Mc35, I321 (I)	B223, V1005, KK3, 61A, B-11, Cr (cow) (I), Lp14 (sheep)
11	None	YM, A253.1 (pig) (I)
12	L26 (I)	None
13	None	L338 (horse) (I)
14	None	FI23 (horse) (I)

[a] Data from ABOUD et al. (1988); ALBERT et al. (1987a, b); ARIAS et al. (1989); BEARDS et al. (1980, 1992); BELLINZONI et al. (1990); BIRCH et al. (1985); BLACKHALL et al. (1992); BOHL et al. (1984); BROWNING et al. (1991a, b); BRUSSOW et al. (1990); CLARK et al. (1987); CORNAGLIA et al. (1992); DAS et al. (1993); DOLAN et al. (1985); GAUL et al. (1982); GENTSCH et al. (1993); GERNA et al. (1984b, 1990, 1993); GREENBERG et al. (1986); HOSHINO et al. (1984, 1985b, 1987a); HUANG et al. (1992); KITAOKA et al. (1987); KUTSUZAWA et al. (1982); LI et al. (1993a); LIPRANDI et al. (1991); MASCARENHAS et al. (1989); MATSUNO et al. (1985, 1988); MIDTHUN et al. (1989); MINAMOTO et al. (1988); MOCHIZUKI and YAMAZAKI (1987); MURAKAMI et al. (1983); NAGESHA and HOLMES (1988); NAKAGOMI et al. (1989); PARWANI et al. (1992a); PAUL et al. (1988); PONGSWANNA et al. (1990); RUIZ et al. (1988); SATO et al. (1982a, b); SHEN et al. (1993); SNODGRASS et al. (1984, 1990); TANAKA et al. (1988); TANIGUCHI et al. (1991); THOULESS et al. (1986); URASAWA et al. (1982, 1990a, b, 1992, 1993); WOODE et al. (1983); WYATT et al. (1982, 1983); PAUL et al. (unpublished studies, 1991).
[b] Human rotaviruses with neither subgroup I or II specificity have also been described but without serotype designation (TUFVESSON 1983; SVENSSON et al. 1988; GHOSH and NAIK 1989b).

3 Detection and Characterization of VP6

The inner capsid protein VP6 (gene 6 product), which forms a trimeric structure on the surface of a single-shelled particle (GORZIGLIA et al. 1985; SABARA et al. 1987), is the most abundant rotavirus protein and thus plays an important role as a major antigen in the detection of rotaviruses by the various serological assays mentioned above (Fig. 1). Group A rotaviruses of humans, animals and birds share common antigens which reside on the VP6. The common antigens can be detected by numerous serological techniques such as complement fixation, ELISA, immune-adherence hemagglutination (IAHA), immunofluorescence, immune electron microscopy, gel diffusion and radioimmunoassay (for review see KAPIKIAN and CHANOCK 1990). Broadly reactive monoclonal antibodies (mAbs) which recognize most group A rotaviruses are now available (GARY et al. 1982; GREENBERG et al. 1982a; ROSETO et al. 1983; TANIGUCHI et al. 1984; CUKOR et al. 1984; POTHIER et al. 1987; GERNA et al. 1989; KANG and SAIF 1991; MINAMOTO et al. 1993).

In addition to the group antigen, subgroup antigens are also located on VP6 (Fig.1). Subgroup specificities were identified initially by IAHA and ELISA using convalescent sera from gnotobiotic calves infected with different rotavirus strains (KAPIKIAN et al. 1981). Three subgroup specificities were found: subgroup I, which included certain human, bovine, porcine and simian rotaviruses; subgroup II, which included certain human rotaviruses; and a third group composed of a murine strain that failed to react with either a subgroup I or subgroup II reagent. Subgrouping thus allowed a unified system for antigenic classification of both human and animal rotaviruses that shared the common group (later termed "A") antigen (BACHMANN et al. 1984). Subsequently, subgroup-specific mAbs were generated (GREENBERG et al. 1982a; SINGH et al. 1983; TANIGUCHI et al. 1984; NAKAGOMI et al. 1984; GERNA et al. 1989; LIPRANDI et al. 1990) and used extensively in epidemiologic studies of both human and animal rotaviruses (WHITE et al. 1984; NAGUIB et al. 1984; NAKAGOMI et al. 1988a; STEELE and ALEXANDER 1988; SVENSSON et al. 1988; THEIL and MCCLOSKEY 1989a,b; GHOSH and NAIK 1989a,b). Currently, as shown in Table 1, four subgroup specificities are recognized: subgroup I (most animal and certain human rotavirus strains); subgroup II (most human and a few porcine and lapine rotavirus strains); subgroup I and II (rare human or animal rotavirus strains); and neither

---

[c] Although one-way neutralization relationship was demonstrated between chicken rotavirus Ch2 strain and turkey rotavirus Ty1 strain initially (HOSHINO et al. 1984), this relationship may be due to shared VP4 specificity of the two viruses. Recent nucleotide and deduced amino acid sequence analyses of the VP7-encoding gene of Ch2 (NISHIKAWA et al. 1991) and Ty1 (KOOL and HOLMES 1993) viruses showed that the two avian rotaviruses shared 70% amino acid homology. In addition, the two viruses differed significantly in amino acid homology in antigenic regions A, B, and C (DYALL-SMITH et al. 1986), which suggested that they may belong to different VP7 serotypes.

subgroup I nor II (rare human or animal strains and most avian rotavirus strains).

Using a panel of eight VP6-specific mAbs raised against bovine subgroup I rotavirus RF strain (three of the eight mAbs), human subgroup I rotavirus 46 strain (one of the eight mAbs) or human subgroup II rotavirus 308 strain (four of the eight mAbs), POTHIER et al. (1987) studied by a competitive binding assay the antigenic topology of VP6. They showed the presence of at least five epitopes organized into four nonoverlapping antigenic domains. Among the latter, three were shared by all rotavirus strains tested (five prototype strains and 54 rotavirus-positive stool suspensions). By immunizing mice with simian SA11, HORNG et al. (1989) produced four VP6-specific mAbs. All four mAbs cross-reacted in ELISA with human rotavirus Wa and S2. Competitive binding studies suggested that three of the four mAbs recognized three distinct epitopes on VP6, while the fourth mAb appeared to react with an epitope at or near the same epitope recognized by one of the three.[1]

In an attempt to delineate both group and subgroup antigenic determinants on VP6, three peptides (amino acids 5–21, 133–147 and 373–385), corresponding to conserved sequences among group A rotaviruses and two peptides (amino acids 93–106 and 103–115) encompassing a region of VP6 where a number of amino acid differences distinguish subgroup I (bovine RF and simian SA11) and subgroup II (human Wa) rotaviruses, were synthesized (BORRAS-CUESTA et al. 1987). Rabbit immune sera raised against peptides 92–106, 103–115, 133–147 or 373–385 conjugated to bovine serum albumin reacted with rotavirus by ELISA. None of the sera demonstrated neutralizing activity. The antisera to peptides 103–115 or 133–147 reacted with human Wa, simian SA11, murine EDIM and turkey IP6 rotaviruses. Thus, these two peptides appear to encompass the immunogenic linear regions on VP6 shared by certain human, animal and avian rotaviruses.[2]

Although certain early studies reported that monospecific hyperimmune antisera to VP6 had low levels of neutralizing activity (BASTARDO et al. 1981; KILLEN and DIMMOCK 1982; SABARA et al. 1985a), with one exception (SABARA et al. 1987) none of the mAbs raised against VP6 showed such

---

[1] By using a competitive binding assay and immunofluorescent assay, MINAMOTA et al. (1993) analyzed 15 mAbs raised against VP6 of pigeon rotavirus PO-13. They showed that the avian rotavirus VP6 possessed at least seven epitopes within four spatially distinct antigenic sites (I–IV). Epitopes common to all group A rotaviruses examined were localized in sites II and III. Mabs directed at one epitope in site IV were found to react with available avian and subgroup I rotavirus strains. It is of interest that KANG and SAIF (1991) generated mAbs by using turkey rotavirus O strain reactive only with avian rotaviruses.
[2] By using chemical cleavages of VP6 of bovine rotavirus RF strain, synthetic peptides, and VP6-specific mAbs, KOHLI et al. (1992) identified a region (amino acids 48–75) that was antigenic and immunogenic. Furthermore, by examining reactivity of VP6-specific mAbs with 391 consecutive heptapeptides derived from VP6 of bovine RF strain, the authors (1993) identified four major antigenic regions on the VP6 (amino acids 32–64, 155–167, 208–272, and 380–397).

activity. By evaluating the antigenic and molecular properties of a baculovirus-expressed VP6 protein of the simian rotavirus SA11 strains, ESTES et al. (1987) demonstrated that the VP6 was expressed in its native conformation. Hyperimmune antiserum to this protein did not have any neutralizing activity. SA11 VP6 was expressed also by a recombinant vaccinia virus in MA104 cells (PONCET et al. 1990). The expressed VP6 was shown to retain the conformation of the viral protein as judged by its reactivity with monoclonal and polyclonal antibodies. Mice inoculated via various routes with this recombinant vaccinia virus produced high levels of anti-VP6 antibodies lacking neutralizing activity.[3]

The complete nucleotide sequence of gene 6 of various rotaviruses belonging to the four different subgroup specificities has been determined (BOTH et al. 1984; ESTES et al. 1984; COHEN et al. 1984; HOFER et al. 1987; GORZIGLIA et al. 1988b; LOPEZ and ARIAS 1993a; ROHWEDDER et al. 1993). A comparison of the deduced amino acid sequence of the sixth gene product of these viruses indicated that subgroup II strains were most closely related to each other and, in addition, that the subgroup I strains, although not as closely related as the subgroup II strains, also formed a closely related family. A strain that did not belong to either subgroup was more closely related to subgroup I, whereas a strain belonging to both subgroup I and II was more closely related to subgroup II.[4] Five regions that may contribute to subgroup epitopes were identified (GORZIGLIA et al. 1988b). However, the finding that VP6 monoclonal antibodies with subgroup I or subgroup II specificity reacted with the trimeric and not with the monomeric form of VP6, suggests that subgroup specificity may be determined by conformational epitopes produced by the folding of VP6 or the interaction between VP6 monomers (GORZIGLIA et al. 1988b). In contrast, immunoblot analyses of both monomeric and trimeric forms of VP6 with "commonly reactive" mAbs suggest that common epitopes on VP6 have continuous determinants (GORZIGLIA et al. 1988b).[5]

---

[3] REDDY et al. (1992) reported the expression of a chimeric VP6 (VP6sc) gene of SA11 virus that incorporated the hemagglutinin signal sequence from influenza virus and an anchor sequence from the mouse gamma immunoglobulin gene using a recombinant vaccinia virus. VP6sc, which was glycosylated, transported, and anchored in the plasma membrane as a trimer, retained its subgroup I epitope. The recombinant vaccinia virus that expressed VP6sc induced serum antibody titers in mice that were approximately ten times higher than those generated by recombinant virus that expressed the wild-type gene.

[4] In addition, an avian-like group A rotavirus isolated from a diarrheic calf was found to be substantially "distant" in amino acid sequence from mammalian rotaviruses (ROHWEDDER et al. 1993).

[5] Epitope mapping studies performed by KOHLI et al. (1992, 1993) suggested also that epitopes recognized by group-specific mAbs are sequential and subgroup-specific epitopes are conformational. By using chimerical genes constructed between the VP6 genes of human rotavirus Wa (subgroup II) and porcine rotavirus YM (subgroup I) and subgroup I- and II-specific mAbs, LOPEZ et al. (1993b) recently mapped the subgroup II epitope of the Wa rotavirus in the region between amino acids 167 and 347.

Initially, it appeared that there was a linkage between the subgroup specificity of human rotaviruses and the electropherotype pattern of viral RNA : a "long" RNA pattern correlated with subgroup II and a "short" (i.e., slower moving) RNA pattern with subgroup I specificity (KALICA et al. 1981a; KUTSUZAWA et al. 1982a). Infrequent exceptions to this association have now been described as follows: subgroup I strains with the long RNA pattern (NAKAGOMI et al. 1987; STEELE and ALEXANDER 1988; BROWN et al. 1988; MASCARENHAS et al. 1989; GHOSH and NAIK 1989a; KOBAYASHI et al. 1989; ARISTA et al. 1990) and a subgroup II strain with the short RNA pattern (STEELE and ALEXANDER 1988; GATHERU et al. 1993).

An antigenic specificity on the inner capsid protein VP2 (gene 2 product) of many strains may cosegregate with the subgroup specificities on VP6 (TANIGUCHI et al. 1986). However, SVENSSON et al. (1990) showed that the VP2 and VP6 subgroup specificities are not linked and can segregate independently in nature; it was thus suggested that the term "subgroup" be reserved exclusively for antigens on VP6.

The antigenic composition on the VP6 of 49 rotavirus strains derived from 11 different animal and avian species was analyzed with six VP6-specific mAbs with varying reactivity and the evolution of group A rotaviruses was examined (HOSHINO et al. 1987a). Two ancestral lineages were suggested: one (subgroup II) with pig-human lineage and the other (subgroup I) with bovine-simian-human lineage. A possible subgroup II human-pig ancestral lineage for gene 2 was suggested following analysis of 33 rotavirus strains derived from 13 different animal and avian species with a single VP2-specific mAb (SVENSSON et al. 1990).

4 Detection and Characterization of VP7

Neutralizing antibodies are considered to play a major role in protection against many diseases, including rotavirus-induced diarrhea. Thus, knowledge of the serotypic similarity and diversity of rotaviruses is essential for vaccine development. Rotaviruses have been characterized according to serotype by various assays: reduction of plaques, fluorescent foci or virus yield in cultures and neutralization of cytopathic effect (for review see KAPIKIAN and CHANOCK 1990); viral interference (SUZUKI et al. 1984); immune electron microscopy (GERNA et al. 1984a; KAPIKIAN and CHANOCK 1990); ELISA using type-specific mAbs (SHAW et al. 1985; HEATH et al. 1986; AKATANI and IKEGAMI 1987; TANIGUCHI et al. 1987a; COULSON et al. 1987a; BEARDS 1987; BIRCH et al. 1988); nucleic acid probes (LIN et al. 1987; MIDTHUN et al. 1987; ZHENG et al. 1989a; FLORES et al. 1989, 1990; ROSEN et al. 1990; JOHNSON et al. 1990; BINGNAN et al. 1991; SHAHID et al. 1991); and oligonucleotide probes (SETHABUTR et al. 1990, 1992; CHAN et al. 1990; BERN et al. 1992).

In addition, with serotype-specific primers of VP7 genes, the polymerase chain reaction has been employed successfully to "serotype" rotaviruses in stool specimens (GOUVEA et al. 1990a, b; ROHWEDDER and WERCHAU 1990; NAKAGOMI et al. 1991; USHIJIMA et al. 1992; PARWANI et al. 1992a).

Early studies demonstrated that a glycoprotein VP7 (gene 7, 8 or 9 product depending on the strain) is a major neutralization protein (KALICA et al. 1981b) (Fig. 1). Many "non-cultivatable" field strains of human rotavirus were "rescued" by genetic reassortment with a cultivatable bovine rotavirus strain and then serotyped (GREENBERG et al. 1981, 1982b). Following the description of techniques for in vitro cultivation of fastidious human and animal rotaviruses (SATO et al. 1981; URASAWA et al. 1981), many strains became available for systematic serotypic characterization by the use of type-specific polyclonal hyperimmune antisera. Fourteen distinct serotypes are now recognized according to VP7 specificity (Table 1). The criterion for establishment of a new serotype is the demonstration of a 20-fold or greater difference in reciprocal neutralizing antibody titers between a reference serotype strain and a putative new serotype (WYATT et al. 1982). As shown in Table 1, various animal rotaviruses share serotype specificity via VP7 with human rotaviruses: (1) serotype 3 includes viruses derived from humans and many different animal species including vervet monkey, rhesus monkey, dog, cat, horse, pig, rabbit and mouse; (2) eight of nine human rotavirus serotypes share serotype specificity with animal species; (3) five of nine human rotavirus serotypes share serotype specificity with a porcine rotavirus.[6] In addition, bovine rotaviruses antigenically related to human rotavirus serotypes 2 and 3 were reported recently (BLACKHALL et al. 1990; HUSSEIN et al. 1993); they are not included in Table 1 because the bovine strain designations are not yet available. Since there are two neutralization specificities, it has recently been suggested that for facilitating the description of strains that the term "G" (for glycoprotein) be adopted for designation of VP7 serotype specificity and the term "P" (for protease) for VP4 serotype specificity (NAGESHA and HOLMES 1991b).

Neutralizing or nonneutralizing mAbs directed against the VP7 of various rotavirus strains have been generated and characterized (Tables 2 and 3). These mAbs show strain or serotype or cross-reactive specificity. Type-specific mAbs to all but two (types 13 and 14) of 14 serotypes are now available (Table 2). It should be noted that the reactivity of each VP7 mAb has not been evaluated with each of the 14 serotypes. It is of interest that many of the cross-reactive VP7 mAbs react with serotype 3 virus. Such cross-reactive antibodies, singly or in combination with other antibodies, may play a role in heterotypic immune responses and heterotypic cross-protection between distinct serotypes.

---

[6] In addition, (4) four of nine human rotavirus serotypes share serotypic specificity with a bovine rotavirus (Table 1).

Table 2. Homotypic VP7 monoclonal antibodies

Serotype	mAb	Immunizing virus (species of origin)	Neutralizing activity	Reference
1	AH49	Wa (human)	+	Akatani and Ikegami (1987)
	KU-4	KU (human)	+	Taniguchi et al. (1987a)
	KU-3C7	KU (human)	+	Taniguchi et al. (1985)
	KU-6A11	KU (human)	+	Taniguchi et al. (1987a)
	KU-5H1	KU (human)	+	Morita et al. (1988)
	RV-4:2	RV-4 (human)	+	Coulson et al. (1987a)
	2C9	Wa (human)	+[a]	Shaw et al. (1985)
	5C8	FH4232 (human)	?	Heath et al. (1986)
	2G10	FH6228 (human)	?	Birch et al. (1988)
	5E8	D × RRV (human × rhesus monkey)	+	Padilla-Noriega et al. (1990)
	K4-6BG	KU (human)	+	Taniguchi et al. (1987a)
	RV4:1	RV4 (human)	+	Coulson et al. (1987a)
	RV4:3	RV4 (human)	+	Coulson et al. (1987a)
	K8-6D6	K8 (human)	+	Kobayashi et al. (1991a)
	K8-3C10	K8 (human)	+	Kobayashi et al. (1991a)
	K8-1D15	K8 (human)	+	Kobayashi et al. (1991a)
	K8-4C4	K8 (human)	+	Kobayashi et al. (1991a)
2	AG12	F43 (human)	−	Akatani and Ikegami (1987)
	S2-2G10	S2 (human)	+	Taniguchi et al. (1985)
	RV-5:3	RV-5 (human)	+	Coulson et al. (1987a)
	1C3	Hu7 (human)	+[a]	Heath et al. (1986)
	5D2	FH1064 (human)	?	Birch et al. (1988)
	1C10	DS-1 × RRV (human × rhesus monkey)	+	Padilla-Noriega et al. (1990)
	2F1	DS-1 × RRV (human × rhesus monkey)	+	Shaw et al. (1987)
3	AC5	F32 (human)	+	Akatani and Ikegami (1987)
	YO-1E2	YO (human)	+	Taniguchi et al. (1987a)
	RV-3:1	RV-3 (human)	+	Coulson et al. (1985)
	RV-3:2	RV-3 (human)	+[a]	Coulson et al. (1985)
	2D7	FH6139 (human)	+[a]	Heath et al. (1986)

Group	Clone	Strain		Reference
	5F4	FH6139 (human)	?	Birch et al. (1988)
	McN-1	McN (human)	+	Kobayashi et al. (1991b)
	McN-10	McN (human)	+	Kobayashi et al. (1991b)
	96	MMU18006 (rhesus monkey)	+	Greenberg et al. (1983)
	3	MMU18006 (rhesus monkey)	+	Greenberg et al. (1983)
	5H3	MMU18006 (rhesus monkey)	+	Shaw et al. (1986)
	4F8	MMU18006 (rhesus monkey)	+	Shaw et al. (1986)
	4C3	MMU18006 (rhesus monkey)	+	Shaw et al. (1986)
	4F5	MMU18006 (rhesus monkey)	+	Shaw et al. (1986)
	B8/X1	SA11 (vervet monkey)	+	Sonza et al. (1983)
	A10/N3	SA11 (vervet monkey)	+	Sonza et al. (1983)
	A11/M9	SA11 (vervet monkey)	+	Sonza et al. (1983)
	2D7	SA11 (vervet monkey)	+	Heath et al. (1986)
	2A2	SA11 (vervet monkey)	+	Gerna et al. (1988a)
	3A6	SA11 (vervet monkey)	+	Gerna et al. (1988a)
	C1/1	CRW-8 (pig)	+	Nagesha et al. (1989)
4	AE18	F13 (human)	−	Akatani and Ikegami (1987)
	ST-2G7	ST3 (human)	+	Taniguchi et al. (1987a)
	ST-3:1	ST3 (human)	+	Coulson et al. (1987a)
	3A3	VA70 (human)	+	Gerna et al. (1988b)
	4G7	FH4154 (human)	?	Birch et al. (1988)
	B2/4	BEN-144 (pig)	+	Nagesha et al. (1989)
	RG38A3	Gottfried (pig)	+	Kang et al. (1993)
	RG39C12	Gottfried (pig)	+	Kang et al. (1993)
	RG42A6	Gottfried (pig)	+	Kang et al. (1993)
5	T1/1	TFR-41 (pig)	+	Nagesha et al. (1989)
	5B8	OSU (pig)	+	Snodgrass et al. (1990)
6	4B5-5	C486 (cow)	+	Sabara et al. (1985b)
	10D2-2	C486 (cow)	−	Sabara et al. (1985b)
	UK/7	UK (cow)	+	Snodgrass et al. (1990)
	B641-N2b	B641 (cow)	+[a]	Zheng et al. (1989)
	AB4	NCDV (cow)	−	Akatani and Ikegami (1987)

(Contd.)

Table 2. (*Continued*)

Serotype	mAb	Immunizing virus (species of origin)	Neutralizing activity	Reference
7	RQ115	Q17 (cow)	–	Cornaglia et al. (1993)
	2D11	Q (turkey)	+	Kang and Saif (1991)
	6E8	Q (turkey)	+	Kang and Saif (1991)
8	B37:1	B37 (human)	+	Tursi et al. (1987)
	69M-2D	69M (human)	+	Ahmed et al. (1991b)
9	AJ26	F45 (human)	+	Akatani and Ikegami (1987)
	WI61-6A1	WI-61 (human)	+	Midthun et al. (1989)
	F45:5	F45 (human)	+	Kirkwood et al. (1993)
	4A5F11	WI61 (human)	+	Genra et al. (Manuscript in preparation)
10	B223-N7	B223 (cow)	+	Zheng et al. (1989c)
	B223/3	B223 (cow)	+	Snodgrass et al. (1990)
11	8E8	A253 (pig)	+	Ciarlet et al. (in press)
	8D10	A253 (pig)	+	Ciarlet et al. (in press)
12	No designation	L26 (human)	+	Urasawa et al. (1990a)

[a] Negative by immuniprecipitation.

Table 3. Heterotypic VP7 monoclonal antibodies

Monoclonal antibody	Immunizing virus (species of origin) (VP7 serotype)	Neutralizing activity	Serotype recognizable by indicated monoclonal antibody	Reference
129	Wa (human) (1)	−	1, 2, 3, 4, 5, 6, 8, 9	Shaw et al. (1986)
60	Wa (human) (1)	−	1, 2, 3, 4	Shaw et al. (1986)
RV-4:3	RV-4 (human) (1)	+	1, 3	Coulson et al. (1986)
YO-4C2	YO (human) (3)	+	1, 3, 4	Taniguchi et al. (1988a)
S3-3E	S3 (human) (3)	+	3, 8	Kobayashi et al. (1991b)
McN-3	McN (human) (3)	+	3, 9	Kobayashi et al. (1991b)
McN-7	McN (human) (3)	+	3, 6, 9	Kobayashi et al. (1991b)
C4/B3	SA11 (vervet monkey) (3)	−	1, 3, 6	Sonza et al. (1983)
1P20C3	SA11 (vervet monkey) (3)	−	1, 3	Grunert et al. (1987)
159	MMU18006 (rhesus monkey) (3)	+	3, 4[a]	Greenberg et al. (1983)
C3/1	CRW-8 (pig) (3)	+	3, 5	Nagesha et al. (1989)
57-8	Gottfried (pig) (4)	+	3, 4, 6, 9, 10	Benfield et al. (1987)
RG36H9	Gottfried (pig) (4)	+	3, 4	Kang et al. (1993)
Mab24	Fecal rota (human) (4)	−	1, 3, 4, 6	Coulson et al. (1987b)
11D12-6	C486 (cow) (6)	+	6, 10	Sabara et al. (1985b)
B641-E3	B641 (cow) (6)	−	3, 6, 10	Zheng et al. (1989c)
F45:1	F45 (human) (9)	+	4, 9	Kirkwood et al. (1993)
F45:2	F45 (human) (9)	+	3, 4, 9	Kirkwood et al. (1993)
F45:6	F45 (human) (9)	+	5, 9	Kirkwood et al. (1993)
WI61:1	WI61 (human) (9)	+	5, 8, 9	Kirkwood et al. (1993)
B223/4	B223 (cow) (10)	+	3, 10	Browning et al. (1992)

[a] Does not react with human serotype 4 strain (reacts with porcine Gottfried and SB1A strains).

Simple, specific and sensitive ELISA procedures for serotyping rotaviruses in stools and in cell culture have been developed using serotype-specific mAbs (SHAW et al. 1985; HEATH et al. 1986; AKATANI and IKEGAMI 1987; TANIGUCHI et al. 1987b; COULSON et al. 1987a; BEARDS 1987; BIRCH et al. 1988; NAGESHA and HOLMES 1991a). Such immunoassays have been successfully employed to study the natural history and the geographical and temporal distribution of rotavirus serotypes (URASAWA et al. 1988, 1989; FLORES et al. 1988; BROWN et al. 1988; LINHARES et al. 1988; GEORGES-COURBOT et al. 1988; PONGSWANNE et al. 1989; UNICOMB and BISHOP 1989; UNICOMB et al. 1989; AHMED et al. 1989; BEARDS et al. 1989; BISHOP et al. 1989; BELLINZONI et al. 1990; SNODGRASS et al. 1990; MATSON et al. 1990; GERNA et al. 1990a, b; GOMEZ et al. 1990; PADILLA-NORIEGA et al. 1990; ARISTA et al. 1990; KAWAMOTO et al. 1990; KIM et al. 1990; AHMED et al. 1991a; BEGUE et al. 1992; BISHOP et al. 1991; BROWNING et al. 1992; FITZGERALD and BROWNING 1991; GATHERU et al. 1993; NAGESHA and HOLMES 1991a; NOEL et al. 1991; PIPITTAJAN et al. 1991; RASOOL et al. 1993; STEEL et al. 1992; URASAWA et al. 1992; VALEZQUEZ et al. 1993; WARD et al. 1991; WHITE et al. 1991; WOODS et al. 1992). They have also been used to evaluate the efficacy of candidate rotavirus vaccines (CHRISTY et al. 1988; KAPIKIAN et al. 1988a, b; GOTHEFORS et al. 1989; PEREZ-SCHAEL et al. 1990; VESIKARI et al. 1990; BERNSTEIN et al. 1990; SANTOSHAM et al. 1991; KAPIKIAN et al. 1992). Caution must be exercised in the selection of mAbs for serotyping because there is increasing evidence for the existence of antigenic variation or antigenic drift within a given serotype. Thus, a previously characterized serotype-specific mAb may fail to recognize certain strains belonging to that serotype (GREENBERG et al. 1983; COULSON et al. 1985; GREEN et al. 1988; UNICOMB and BISHOP 1989; UNICOMB et al. 1989; SNODGRASS et al. 1990). These intratypic antigenic variants have been referred to as "monotypes" (COULSON 1987). Recently, NISHIKAWA et al. (1989a) examined the relatedness of the VP7 of 27 serotype 3 strains derived from eight different animal species by both comparative sequence analysis and reactivity with three anti-VP7 serotype 3 neutralizing mAbs. These studies indicated that: (1) species-specific sequences are present in the VP7 protein; (2) the sequence divergence among serotype 3 strains derived from different animal species is found in regions which have been described as divergent among different serotypes; and (3) serotype 3 specificity is determined by two or more epitopes. Thus, pooled type-specific mAbs that recognize different epitopes on the VP7 protein may be necessary to serotype field strains (BIRCH et al. 1988; GREEN et al. 1990b; WARD et al. 1991; RAJ et al. 1992).[7]

---

[7] The term "subtype" was introduced by GERNA et al. (1988b) to distinguish two populations of human VP7 serotype 4 rotaviruses (subtypes A and B) by using cross-absorbed polyclonal hyperimmune antisera as well as VP7-specific mAb. Later, by analyzing nucleotide and deduced amino acid sequences of the VP7-encoding gene of six G serotype 4 viruses (2 subtype A and 4

The role of complement components or anti-immunoglobulin antibodies in the neutralization of rotavirus by mAbs, hyperimmune antisera, or infection sera has not been explored. Recently, NISHIKAWA et al. (1989a) showed that a neutralizing mAb, YO-1E2 (isotype IgG), raised against human serotype 3 YO strain failed to neutralize the homotypic human P strain. However, in the presence of anti-mouse IgG, this mAb was able to neutralize this strain. The VP7 encoding gene of antigenic mutants of the P strain that was resistant to neutralization by this mAb was sequenced and the amino acid mutation was mapped to amino acid 221. Thus, care must be used in the characterization of each mAb and the mapping of the neutralization topography on VP7 or VP4.

By serological analysis of mAbs by neutralization, hemagglutination-inhibition or competitive binding assay, or of mutants resistant to neutralizing mAbs, a single large conformational domain consisting of two or more regions on the VP7 protein has been identified (GREENBERG et al. 1983; SONZA et al. 1984; LAZDINS et al. 1985; TANIGUCHI et al. 1985; SHAW et al. 1986; MORITA et al. 1988; GERNA et al. 1988a; KOBAYASHI et al. 1991b). By sequencing the VP7 encoding gene of such antigenic mutants of simian rotavirus SA11, DYALL-SMITH et al. (1986) demonstrated three distinct regions which were involved in serotype-specific neutralization. These regions were designated A (amino acids 87–99), B (amino acids 145–150) and C (amino acids 211–223). Nine regions (VP1–VR9) which are divergent in amino acid homology were observed when the deduced amino acid sequences of VP7 of eight serotypically distinct rotaviruses were compared (GREEN et al. 1989) (Fig. 2). Regions A, B and C correspond to VR5, VR7 and VR8, respectively. Regions A and C appear to lie close together in the native glycoprotein because a single mAb is capable of selecting a mutation in either A or C region (DYALL-SMITH et al. 1986). Similar sequence analyses were performed on neutralization-resistant mutants of rhesus monkey rotavirus (MACKOW et al. 1988a) and human rotaviruses (TANIGUCHI et al. 1988a; NISHIKAWA et al. 1989a; GREEN et al. 1990a; KOBAYASHI et al. 1991a; COULSON and KIRKWOOD 1991; DUNN et al. 1993). Single amino acid substitution mutations in such mutants selected with either homotypic or heterotypic neutralizing mAbs were mapped to region A (VR5) or C (VR8)

---

subtype B, respectively), GREEN et al. (1992a) showed that several amino acid residues were conserved within a subtype and different between subtypes. They speculate that VP7 subtype differences may result from critical amino acid substitutions within an immunodominant serotype 4-specific antigenic site. BROWNING et al. (1992b) examined six equine G serotype 3 rotaviruses by cross neutralization employing polyclonal hyperimmune antisera, reactivity patterns with a panel of three VP7-specific mAbs, or liquid hybridization. They showed that there were at least two subtypes (G3A and G3B) of equine G serotype 3 rotaviruses. The existence of two subtypes among Japanese equine rotaviruses was also reported recently (IMAGAWA et al. 1993).

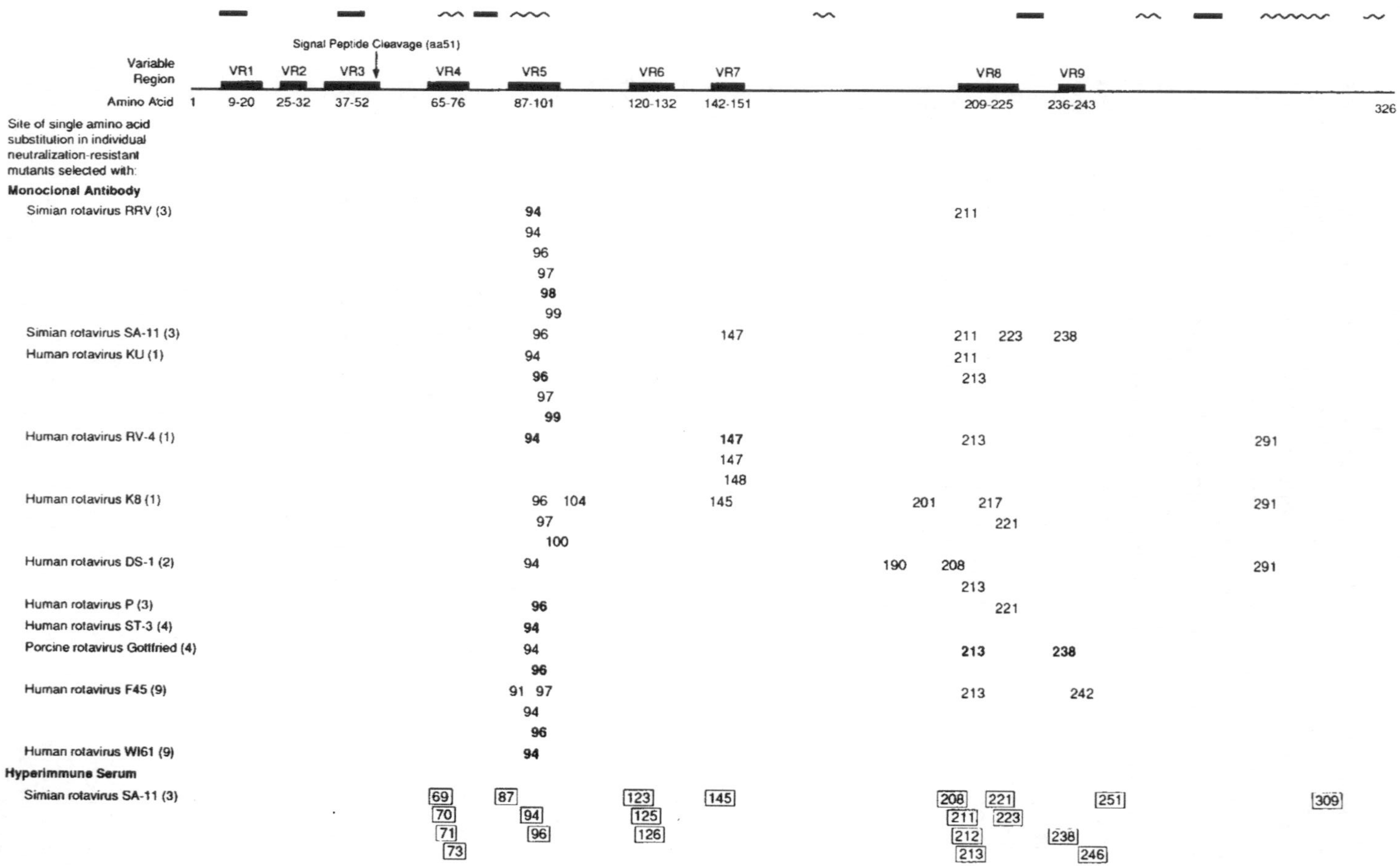
Signal Peptide Cleavage (aa51)
Variable Region
VR1
VR2
VR3
VR4
VR5
VR6
VR7
VR8
VR9
Amino Acid
1
9-20
25-32
37-52
65-76
87-101
120-132
142-151
209-225
236-243
326
Site of single amino acid substitution in individual neutralization-resistant mutants selected with:
Monoclonal Antibody
Simian rotavirus RRV (3)
94 94 96 97 98 99 211
Simian rotavirus SA-11 (3)
96 147 211 223 238
Human rotavirus KU (1)
94 96 97 99 211 213
Human rotavirus RV-4 (1)
94 147 147 148 213 291
Human rotavirus K8 (1)
96 104 97 100 145 201 217 221 291
Human rotavirus DS-1 (2)
94 190 208 213 291
Human rotavirus P (3)
96 221
Human rotavirus ST-3 (4)
94
Porcine rotavirus Gottfried (4)
94 96 213 238
Human rotavirus F45 (9)
91 97 94 96 213 242
Human rotavirus WI61 (9)
94
Hyperimmune Serum
Simian rotavirus SA-11 (3)
69 70 71 73 87 94 96 123 125 126 145 208 211 212 213 221 223 238 246 251 309

Fig. 2. Linear structure and antigenic organization of rotavirus outer capsid VP7; VP7 serotype. ***Boldface numbers*** indicate amino acid substitutions in mutants selected with monoclonal antibody exhibiting cross-reactive (i.e., heterotypic) neutralizing activity; *numbers not in boldface and not in boxes* indicate amino acid substitutions selected with monoclonal antibody with homotypic neutralizing activity. *Numbers in boxes* indicate amino acid substitutions in mutants selected with hyperimmune serum to SA-11 virus. At the *top* of the figure, hydrophobic regions are denoted by *closed box* whereas hydrophilic regions are shown by *wavy line* (Adapted from Kapikian and Chanock 1990, with addition. Data from Dyall-Smith et al. 1986; Caust et al. 1987; Mackow et al. 1988a; Taniguchi et al. 1988a; Green et al. 1990; Gorziglia et al. 1990c; Nishikawa et al. 1989a; Kobayashi et al. 1991a, b; Coulson and Kirkwood 1991; Kirkwood et al. 1993; Kang et al. 1993; Dunn et al. 1993; Hoshino et al. 1993)

(Fig. 2).[8] Later, a correlation between the sequences in regions A and C, which are divergent among different serotypes while highly conserved within a serotype (GREEN et al. 1987), and the virus serotype defined by neutralization was shown by direct sequence analysis of a total of 25 field isolates and ten laboratory strains. Furthermore, GREEN et al. (1988) demonstrated that a field isolate could be serotyped by comparison of its nucleotide sequences in the two regions mentioned above with those of a reference virus from each serotype.

Recently, NAGESHA et al. (1990) reported the existence of a porcine rotavirus which has "chimeric" amino acid sequences on VP7. The MDR-13 strain of porcine rotavirus was shown to have a two-way VP7 antigenic relationship with both VP7 serotype 3 (G3) and VP7 serotype 5 (G5) viruses and was thus designated as G 3/5. By sequence analysis of the three antigenic regions (A, B and C) of the MDR-13 VP7 gene, they detected the mosaic amino acid sequences in these regions. They speculate that these sequences resulted from sequential genetic mutations which led to amino acid substitutions rather than from genetic recombination between the two distinct VP7 genes.[9]

GORZIGLIA et al. (1990c) sequenced the VP7 and VP4 encoding genes of seven antigenic mutants of simian rotavirus SA11 4fM (VP7:3; VP4:NCDV type; NISHIKAWA et al. 1988) that were selected after 39 passages in the presence of polyclonal hyperimmune antiserum to SA11 (KNOWLTON and WARD 1985). They showed that: (1) twice as many amino acid substitutions occurred in the VP7 protein as in VP4, which has a molecular weight twice that of VP7; (2) most amino acid substitutions clustered in six variable regions of VP7; (3) most amino acid substitutions in VP7 of these mutants were also observed in antigenic mutants selected with neutralizing mAbs but that some of the amino acid substitutions were not selected for by neutralizing mAbs; and (4) some of the neutralizing epitopes appeared to be interrelated, while others appeared to be independent (Fig. 2). It is noteworthy that, in addition to the epitopes involved in neutralization, serotype-specific epitopes which are not involved in neutralization also exist on the VP7 (Table 2). Using mAbs directed to such epitopes, AKATANI and IKEGAMI (1987) and others (NAKAGOMI et al. 1988b, 1990) have successfully serotyped rotavirus field isolates. The genetic mapping of such mAbs and

---

[8] More recently, all three regions (A, B, and C) have been shown to contain monotype-specific, serotype-specific, or serotype-cross-reactive neutralization sites (COULSON and KIRKWOOD 1991; KOBAYASHI et al. 1991c; DUNN et al. 1993b; HOSHINO et al. in press). Furthermore, not only variable regions mentioned above but also constant regions have been shown to be involved in the formation of neutralization epitopes (COULSON and KIRKWOOD 1991; KOBAYASHI et al. 1991a, GORZIGLIA et al. 1990c; DUNN et al. 1993b). In this context, it is noteworthy that by analyzing endonucleic restriction patterns of VP7-encoding genes from 110 field isolates and 40 cell-grown strains of human and animal rotavirus, GOUVEA et al. (1993) reported that (i) there were strong associations between restriction patterns and G type or animal species of origin, and (ii) G type-specific or species-specific nucleic acid sequences were located inside as well as outside hypervariable regions.

[9] Antigenic drift has been reported in human as well as in additional porcine rotaviruses (COULSON et al. 1985; NAGESHA et al. 1992).

their role, if any, in protection against rotavirus disease needs further investigation.

In order to identify immunogenic regions capable of inducing neutralizing antibodies, GUNN et al. (1985) synthesized six peptides (amino acids 66–76, 90–103, 174–183, 208–225, 247–259 and 275–295) which span the hydrophilic regions of SA11 VP7, coupled them to keyhole limpet hemocyanin, and hyperimmunized rabbits with each of them. Although antisera to each of these peptides recognized the respective homologous peptides in a solid-phase radioimmunoassay and bound to denatured VP7 in a Western immunoblot, none of the antisera recognized the virus, suggesting that conformation (as maintained by disulfide bonds) is important in functional neutralization epitopes.

STRECKERT et al. (1986) raised antibodies in rabbits using as immunogen a synthetic peptide corresponding to the eight COOH-terminal amino acids of SA11 VP7 protein (VP7 serotype 3). They demonstrated that the antipeptide sera recognized the intact SA11 VP7 molecules by ELISA and immunoprecipitation. The specificity of the sera was also confirmed in a competition experiment between the octapeptide and the viral protein. The antipeptide sera showed cross-reactivity by ELISA against human Wa (VP7 serotype 1) and Hochi (VP7 serotype 4) viruses.[10]

An important feature that has emerged from the neutralization epitope-mapping studies is the recognition of the importance of glycosylation in determining the antigenicity of the VP7 protein. CAUST et al. (1987) studied SA11 mutants selected with certain neutralizing mAbs and found that a new glycosylation site in region C (VR8) led to marked antigenic alterations. They also showed that removal of the attached carbohydrate on the mutant VP7 reversed the antigenic changes. The addition of a carbohydrate side-chain resulting from an amino acid substitution in region A (amino acids 87–99) of a neutralization escape mutant of rhesus monkey rotavirus, however, did not affect the reactivity of the mutant with selected mAbs or homologous hyperimmune antiserum (MACKOW et al. 1988a). Additional studies are needed to define the importance of glycosylation in the modification of antigenicity and immunogenicity of the VP7 and to clarify the "immunodominance" of the neutralization epitopes on the glycoprotein (DYALL-SMITH et al. 1986).

With various homotypic anti-VP7 neutralizing mAbs to serotypes 1, 2, 3 and 4, serotype-specific antibody responses following infection with a monovalent vaccine in infants and adults were examined by epitope blocking assay (EBA) (SHAW et al. 1987; GREEN et al. 1990). Vaccinees of different age groups exhibited divergent immune responses to two serotype 1 VP7 epitopes defined by mAbs 2C9 (mapped to amino acid 94 in VR5)

---

[10] They (1992) also raised mAb to a synthetic peptide (amino acids 319–326), which was shown to be reactive with purified SA11 as well as determinants of SA11-infected MA104 cells. This anti-VP7 peptide mAb had no neutralizing activity.

and KU-4 (mapped to amino acid 213 in VR8). Only one of 43 vaccinees <6 months of age had a response to KU-4 epitope whereas many of the adult vaccinees responded to both epitopes. Blocking antibodies in serum samples defined by the EBA may not always reflect precise epitope-specific responses because many epitopes on VP7 are conformational and overlapping. However, since numerous VP7-specific mAbs, both homotypic and heterotypic with or without neutralizing activity, are now available (Tables 2 and 3), the EBA may prove to be a powerful tool not only to assess epitope-specific antibody responses (SNODGRASS et al. 1991; TANIGUCHI et al. 1991; MATSON et al. 1992) and the relative importance of each epitope against rotavirus disease (GREEN and KAPIKIAN 1992) but also to dissect the antigenic topology on VP7.

Attempts have been made to express in bacteria the VP7 of various rotavirus strains including SA11 (VP7 serotype 3; ARIAS et al. 1986), UK (VP7 serotype 6; McCRAE and McCORQUODALE 1987), and NCDV (VP7 serotype 6 FRANCAVILLA et al. 1987); however, induction of neutralizing antibodies with the expressed proteins was either nil (NCDV VP7) or poor (SA11 VP7 and UK VP7). In early studies in rabbits, wild-type vaccinia virus-expressed intracellular VP7 proteins of SA11 induced serotype-specific neutralizing antibodies but of low titer (ANDREW et al. 1987). Recently, ANDREW et al. (1990) using a recombinant vaccinia vector with the coding sequences of SA11 VP7 and the signal peptide and transmembrane anchor sequences from the influenza hemagglutinin were able to express a chimeric protein. The COOH-terminal anchored form of the hybrid protein was expressed at the cell surface in tissue culture. When expressed in mice and rabbits, high levels of neutralizing antibodies were induced. T cells were also more effectively stimulated by this expressed protein than by intracellularly expressed VP7 by the wild-type vaccinia virus recombinant.[11]

5 Detection and Characterization of VP4

PRASAD et al. (1988, 1990) and YEAGER et al. (1990) recently demonstrated by cryoelectron microscopy that VP4 protrudes from the outer capsid of the rotavirus particle in the form of 60 short spikes (Fig. 1). These spikes were approximately 10 to 12 nm in length with a knob at the distal end.[12]

---

[11] By using the same strategy, they expressed the SA11 VP7sc in recombinant adenovirus type 5. Mouse pups born to the dam vaccinated intranasally with the recombinant were completely protected against diarrhea upon challenge with homologous virus (BOTH et al. 1993). The SA11 VP7 was expressed also by recombinant herpes simplex virus-1; however, the protein lacked most neutralizing epitopes on it (DORMITZER et al. 1992a, b).
[12] In the presence of trypsin, the VP4 protein is cleaved into two fragments designated VP5 and VP8 (Fig. 3).

In early studies, VP4 separated from the other gene products by polyacrylamide gel electrophoresis induced neutralizing antibodies in rabbits (BASTARDO et al. 1981; SABARA et al. 1985a, b); however, the neutralizing specificity of these antisera was not extensively examined.

Genetic studies of reassortants generated from two serotypically distinct parental rotavirus strains showed that not only VP7 but also VP4 reacted with neutralizing antibodies (GREENBERG et al. 1983). Genetic analyses of naturally occurring intertypic rotaviruses (HOSHINO et al. 1985a, 1987b) or reassortant rotaviruses generated in vitro (OFFIT and BLAVAT 1986) provided strong evidence that VP7 and VP4 neutralization specificities segregate independently and that VP4 is as potent as VP7 in inducing neutralizing antibodies following parenteral or oral immunization. In in vivo studies, mouse dams hyperimmunized orally with a reassortant rotavirus containing a VP4 of one parental strain and the VP7 of the other, protect their pups passively against diarrhea induced by either parental rotavirus (OFFIT et al. 1986a). Furthermore, in piglets, a reassortant rotavirus with the VP4 of one serotype and the VP7 of another, elicited sufficient neutralizing antibodies to both serotypes following enteric infection to prevent diarrhea following challenge with either parental virus (HOSHINO et al. 1988). In addition, in a passive-transfer model, mice can be protected against rotavirus challenge by homotypic or heterotypic neutralizing mAbs directed against VP4 or VP7 (OFFIT et al. 1986b; MATSUI et al. 1989b). With the role of VP4 in neutralization established, it soon became apparent that a binary system of rotavirus classification would be needed to designate the neutralization specificity of both VP4 and VP7 outer capsid proteins (HOSHINO et al. 1985a); later the G and P designations were suggested as noted earlier.[13]

Although the serotypic specificity of VP7 has been well characterized, analysis of the specificity of the VP4 protein has not been carried out in a systematic manner. For example, the diversity, if any, of serotypic specificities on the VP4 of rotavirus strains belonging to the various VP7 serotype 3 strains derived from different animal species including vervet monkey, rhesus monkey, horse, pig, dog, cat, rabbit and mouse is not known. The VP4 specificity of a few strains has been elucidated primarily by analysis of reassortant rotaviruses. In this way, two VP7 serotype 6 strains, NCDV and UK, were shown by neutralization to have serotypically distinct VP4s (HOSHINO et al. 1985a). Human rotavirus P and rhesus monkey rotavirus MMU18006, both VP7 serotype 3 strains, have distinct VP4 neutralization specificities (Hoshino et al., unpublished observation; MACKOW et al. 1990a).

---

[13] The mechanisms of neutralization by antibodies directed against VP4 or VP7 were studied by RUGGERI and GREENBERG (1991) by using neutralizing mAbs raised against trypsin cleavage subunits VP8 or VP5, or VP7 of rhesus monkey rotavirus (RRV). They showed that (i) all mAbs tested were able to neutralize RRV after its binding to the cell and (ii) VP8-specific mAbs blocked binding of radiolabeled virions to cell monolayers, and this inhibition was responsible for their neutralization activity.

The porcine rotavirus Gottfried strain (VP7 serotype 4) and OSU strain (VP7 serotype 5) also have distinct VP4 specificities but the porcine rotavirus SB1A strain shares VP7 specificity with the Gottfried strain and VP4 specificity with the OSU strain (HOSHINO et al. 1987a, 1988).[14]

Numerous mouse mAbs, both neutralizing and nonneutralizing, directed against the VP4 protein of rotaviruses belonging to different VP7 serotypes have been generated (Table 4). Although in general, mABs to VP4 appear to be more broadly cross-reactive than those made to VP7, some are strain-specific and others recognize only strains belonging to a single VP7 serotype. In this regard, it is noteworthy that VP4 mAbs have been generated that recognize only strains belonging to a single VP7 serotype (Table 4). It should be noted, however, that the reactivity pattern of each mAb in Table 4 is not complete, since none of the listed mAbs have been tested against all of the known rotavirus VP7 serotypes.[15] By analyzing the neutralization and ELISA reactivity of 13 mAbs (nine neutralizing and four nonneutralizing) directed at the VP4 protein of porcine Gottfried virus, with various human and animal rotaviruses, KANG et al. (1989) found that the VP4 neutralization epitopes of Gottfried and various human rotavirus strains are highly conserved. On this basis, along with the sharing of VP7 specificity with human rotavirus VP7 serotype 4, they suggested that the Gottfried virus may represent a naturally occurring pig-human rotavirus reassortant or a rotavirus which is human in origin but pathogenic for swine. In addition, each of six VP4-specific neutralizing mAbs raised against various human rotavirus strains neutralize the Gottfried virus (GORZIGLIA et al. 1990a).[16]

Approximately 45% of the SA11 VP4 encoding gene, including the cleavage region, was expressed by ARIAS et al. (1987) in Escherichia coli as a fusion protein. The expressed VP4 protein induced both homotypic neutralizing and hemagglutination-inhibiting antibodies in mice. Heterotypic reactivity of the antisera was not examined. Using a baculovirus vector, MACKOW et al. (1989) expressed the complete VP4 of rhesus monkey rotavirus (RRV) in insect cells. The expressed protein was conformationally conserved and antigenically indistinguishable from the native viral VP4 as

---

[14] The existence of at least two VP4 types (OSU-like and Gottfried-like) was also shown among Australian porcine rotaviruses (NAGESHA and HOLMES 1991b). More recently, a third porcine rotavirus VP4 type was reported from Australia (HUANG et al. 1993). By analysis of hyperimmune antisera raised against various bovine rotavirus reassortants, at least three VP4 serotypes were established among bovine rotaviruses (MATSUDA et al. 1990; SNODGRASS et al. 1992).

[15] COULSON (1993) developed an enzyme-linked immunosorbent assay (ELISA) for typing of human rotavirus VP4 by using VP4-specific mAbs. Ninety-eight percent of 118 rotavirus-positive stools were successfully P-typed. They noticed that P-type specificity in ELISA was highest when the G-type specificity of the capture antiserum was matched to the G type of the rotavirus in the test sample.

[16] LIPRANDI et al. (1991) analyzed the binding reactivity of eight VP4-specific neutralizing mAbs raised against porcine OSU and demonstrated the existence of at least five VP4 monotypes among the 11 porcine rotaviruses belonging to four VP7 serotypes that shared the same VP4 serotype.

Table 4. Representative VP4 monoclonal antibodies

Monoclonal antibody	Immunizing virus (species of origin) (VP7 serotype)	Neutralizing activity	VP7 serotype recognizable with indicated Monoclonal antibody	References
KU-4D7	KU (human) (1)	+	1, 2, 3, 4	Taniguchi et al. (1987b)
KU-6B11	KU (human) (1)	+	1, 3, 4	Taniguchi et al. (1987b)
KU-10C	KU (human) (1)	+	1, 3, 4, 9	Kobayashi et al. (1990)
KU-12H	KU (human) (1)	+	1, 3, 4, 9	Kobayashi et al. (1990)
KU-2A	KU (human) (1)	+	1, 2, 3, 4, 9	Kobayashi et al. (1990)
1A10	Wa (human) (1)	+	1, 2, 3, 9	Padilla-Noriega et al. (1993)
1C6	Wa (human) (1)	+	1, 2, 3, 4, 9	Padilla-Noriega et al. (1993)
1E4	Wa (human) (1)	+	1, 3, 4	Padilla-Noriega et al. (1993)
2A3	Wa (human) (1)	+	1, 3, 4	Padilla-Noriega et al. (1993)
2C11	Wa (human) (1)	+	1, 3, 4	Padilla-Noriega et al. (1993)
2G1	Wa (human) (1)	+	1, 2, 3	Padilla-Noriega et al. (1993)
3D6	Wa (human) (1)	+	1,2,3	Padilla-Noriega et al. (1993)
K8-2C12	K8 (human) (1)	+	1	Kobayashi et al. 1991d
S2-2F2	S2 (human) (2)	+	2	Taniguchi et al. (1987b)
RV-5:2	RV-5 (human) (2)	+	2	Coulson et al. (1985)
K-1532	KUN (human) (2)	+	2	Kitaoka et al. (1986)
YO-2C2	YO (human) (3)	+	1, 3, 4	Taniguchi et al. (1987b)
YO-1S3	YO (human) (3)	+	1, 3, 4	Taniguchi et al. (1987b)
YO-1E6	YO (human) (3)	+	1, 3, 4	Taniguchi et al. (1987b)
S3-5E	S3 (human) (3)	+	1, 2, 3, 9	Kobayashi et al. (1990)
M11	MMU18006 (rhesus monkey) (3)	+	3	Mackow et al. (1988b)
1A9	MMU18006 (rhesus monkey) (3)	+	3	Mackow et al. (1988b)
5D9	MMU18006 (rhesus monkey) (3)	+	3	Mackow et al. (1988b)
5C4	MMU18006 (rhesus monkey) (3)	+	3	Shaw et al. (1988)
7A12	MMU18006 (rhesus monkey) (3)	+	3	Shaw et al. (1988)
M2	MMU18006 (rhesus monkey) (3)	+	1, 3, 4, 6	Shaw et al. (1986)
M7	MMU18006 (rhesus monkey) (3)	+	3, 6	Mackow et al. (1988b)
A1	MMU18006 (rhesus monkey) (3)	+	3	Shaw et al. (1986)
A15	MMU18006 (rhesus monkey) (3)	+	3	Mackow et al. (1988b)
M14	MMU18006 (rhesus monkey) (3)	+	3, 6	Shaw et al. (1986)

3F7	MMU18006 (rhesus monkey) (3)	−	3	Shaw et al. (1986)
23	MMU18006 (rhesus monkey) (3)	+	3	Greenberg et al. (1983)
17	MMU18006 (rhesus monkey) (3)	+	3	Greenberg et al. (1983)
3C8	SA11 (vervet monkey) (3)	−	3, 8	Burns et al. (1988)
8D4	SA11 (vervet monkey) (3)	−	3, 4, 8	Burns et al. (1988)
3D8	SA11 (vervet monkey) (3)	−	3, 8	Burns et al. (1988)
7G6	SA11 (vervet monkey) (3)	+	3	Burns et al. (1988)
1P14E2	SA11 (vervet monkey) (3)	+	3	Grunert et al. (1987)
1P8E2	SA11 (vervet monkey) (3)	−	3	Grunert et al. (1987)
C6/3	CRW-8 (pig) (3)	+	3, 5	Nagesha and Holmes (1991b)
ST-1F2	ST3 (human) (4)	+	1, 3, 4	Taniguchi et al. (1987b)
ST3:3	ST3 (human) (4)	+	1, 2, 3, 4	Coulson (1993)
HS3	ST3 (human) (4)	+	1, 3, 4	Padilla-Noriega et al. (1993)
HS6	ST3 (human) (4)	+	1, 3, 4	Padilla-Noriega et al. (1993)
HS8	ST3 (human) (4)	+	1, 3, 4	Padilla-Noriega et al. (1993)
B1/3	BEN-144 (pig) (4)	+	4	Nagesha et al. (1989)
RG-23G10	Gottfried (pig) (4)	+	1, 2, 3, 4, 9	Kang et al. (1989)
RF29B3	Gottfried (pig) (4)	+	1, 2, 3, 4, 9	Kang et al. (1989)
RF29F3	Gottfried (pig) (4)	+	1, 2, 3, 4, 9	Kang et al. (1989)
RG16D2	Gottfried (pig) (4)	+	2, 4[a]	Kang et al. (1989)
RG21A1	Gottfried (pig) (4)	+	4[b]	Kang et al. (1989)
RG22D12	Gottfried (pig) (4)	−	1, 2, 4[a]	Kang et al. (1989)
RG16G10	Gottfried (pig) (4)	−	4[b]	Kang et al. (1989)
2G4	OSU (pig) (5)	+	3, 5, 6	Shaw et al. (1986)
2D5	OSU (pig) (5)	+	5, 11	Liprandi et al. (1991)
1C11	OSU (pig) (5)	+	5, 11	Liprandi et al. (1991)
5D9	OSU (pig) (5)	+	5, 11	Liprandi et al. (1991)
2C9	OSU (pig) (5)	+	5, 11	Liprandi et al. (1991)
3E7	OSU (pig) (5)	+	5, 11	Liprandi et al. (1991)
11D10-1	C486 (cow) (6)	+	6	Sabara et al. (1985b)
3641-N1	B641 (cow) (6)	+	3, 6	Zheng et al. (1989c)
88E22	Thiverral (cow) (6?)	−	1, 6?	Roseto et al. (1983)
RQ4	Q17 (cow) (6)	+	6	Cornaglia et al. (1993)
RQ31	Q17 (cow) (6)	+	6	Cornaglia et al. (1993)
RQ104	Q17 (cow) (6)	−	6	Cornaglia et al. (1993)

(Contd.)

Table 4. (*Continued*)

Monoclonal antibody	Immunizing virus (species of origin) (VP7 serotype)	Neutralizing activity	VP7 serotype recognizable with indicated Monoclonal antibody	References
1G1	Q (turkey) (7)	+	7	Kang and Saif (1991)
5B8	Q (turkey) (7)	+	7	Kang and Saif (1991)
4E2	Q (turkey) (7)	+	7	Kang and Saif (1991)
3G1	Q (turkey) (7)	+	7	Kang and Saif (1991)
2E3	Q (turkey) (7)	+	7	Kang and Saif (1991)
4E9	Q (turkey) (7)	−	7	Kang and Saif (1991)
F45:4	F45 (human) (9)	+	1, 3, 4, 9	Coulson (1993)
3223-N1	B223 (cow) (10)	+	10	Zheng et al. (1989c)
3223-N6	B223 (cow) (10)	+	3, 10	Zheng et al. (1989c)

[a] Porcine rotavirus type 4.
[b] Not human rotavirus type 4.

judged by its reactivity with a library of neutralizing VP4-specific mAbs. Mouse dams hyperimmunized parenterally with this protein developed high levels of serum neutralizing and hemagglutination-inhibiting antibodies to the homologous RRV (MACKOW et al. 1990a, b). Immunization with the RRV VP4 also generated low levels of cross-reactive neutralizing antibodies to various human and animal rotaviruses including Wa, ST3, S2, SA11, NCDV, and EB. Neutralizing activity was not detected against human P, DS-1 or porcine Gottfried viruses. VP4-immunized dams passively protected the newborn mice against a virulent dose of the RRV or murine EB viruses. NISHIKAWA et al. (1989b) expressed the VP4 protein of porcine rotavirus OSU strain in insect cells using a baculovirus vector also. Reactivity with polyclonal and neutralizing mAbs suggested that neutralization epitopes were functionally unaltered on the expressed proteins. The expressed VP4 protein induced hemagglutination-inhibiting and neutralizing antibodies to OSU in guinea pigs. The OSU VP4 appeared to be distinct serotypically from the VP4 of human (Wa, DS-1, VA70, ST3 and M37), simian (RRV), and porcine (Gottfried) strains.[17]

Hemagglutination is another property associated with the VP4 (KALICA et al. 1983) of certain human, animal and avian rotavirus strains (Fig. 1). Human rotaviruses demonstrate hemagglutination activity only with fixed 1-day-old chicken erythrocytes; trypsin activation of virions abolishes this activity (KITAOKA et al. 1984). In contrast, certain animal rotaviruses agglutinate both fixed and fresh erythrocytes from various species. Trypsin activation of animal rotaviruses destroys hemagglutination activity against fixed erythrocytes but not against fresh erythrocytes (FUKUDOME et al. 1989). Hemagglutination of fresh erythrocytes by animal rotaviruses appears to be dependent on sialic acids and to be mediated by glycophorin, the major erythrocyte sialoglycoprotein, since neuraminidase treatment of erythrocytes or glycophorin completely inhibits their hemagglutination activity against fresh erythrocytes (BASTARDO and HOLMES 1980; FUKUDOME et al. 1989). Furthermore, hemagglutination activity of the baculovirus-expressed RRV VP4 protein is shown to be inhibited by glycophorin (MACKOW et al. 1989). Hemagglutination of fixed chicken erythrocytes by nontrypsin-treated human and animal rotaviruses, however, appears to be independent of sialic acid, since neuraminidase treatment of fixed erythrocytes does not inhibit their hemagglutination activities (FUKUDOME et al. 1989).[18]

---

[17] Recently, GORZIGLIA and KAPIKIAN (1992) generated an adenovirus type 5-porcine rotavirus OSU VP4 gene recombinant, which expressed apparently authentic VP4 protein. Upon intranasal infection of the recombinant, cotton rats' developed neutralizing antibodies to OSU virus.

[18] Recently, NAKAGOMI et al. (1992) reported that strain Ro1845 (G serotype 3) isolated from a three-week-old infant with diarrhea in Israel hemagglutinated guinea pig, sheep, chicken, and human (group O) erythrocytes, whereas, glycophorin inhibited hemagglutination of the virus. They speculated that this virus may be of animal origin. In addition, a similar G serotype 3 rotavirus was recently isolated from a clinically normal infant in the USA (Li et al.et al. 1993a). This virus (HCR3 strain) grew extremely well in cell cultures and demonstrated hemagglutination activity.

Both VP5-specific and VP8-specific neutralizing mAbs have been shown to inhibit hemagglutination of the baculovirus-expressed homologous VP4 of RRV (MACKOW et al. 1989). By neutralization or hemagglutination-inhibition assay with selected mAbs, three antigenic domains (hemagglutination and neutralization, hemagglutination alone or neither hemagglutination nor neutralization) have been found on the VP4 of SA11 virus (BURNS et al. 1988).[19] Whether a fourth domain which is involved in neutralization alone exists is uncertain. Although certain VP7 mAbs exhibit hemagglutination-inhibition activity, studies with reassortant rotaviruses indicate that this is due to steric interference rather than to direct binding to a hemagglutinin site on the VP4 (GREENBERG et al. 1983).[20]

RNA-RNA hybridization (FLORES et al. 1986; LARRALDE and FLORES 1990) and sequence analysis (GORZIGLIA et al. 1988a; TANIGUCHI et al. 1989; QIAN and GREEN 1991; ISEGAWA et al. 1992) indicate that at least five different VP4 gene groups (alleles) exist among the human rotaviruses: (1) a Wa-like VP4 gene group consisting of VP7 serotypes 1, 3, 4 and 9 (from symptomatic infections); (2) a DS-1-like VP4 gene group consisting of VP7 serotype 2 (from symptomatic infections); (3) a M37-like VP4 gene group consisting of VP7 serotype 1, 2, 3 or 4 neonatal strains (from asymptomatic infections); (4) a K8-like gene group consisting of VP7 serotype 1 or 3; and (5) a 69M-like VP4 gene group (consisting of VP7 serotype 4 or 8).[21] Recently, by analyzing polyclonal hyperimmune antisera raised against recombinant VP4 proteins of strains KU (VP7 serotype 1), DS-1 (VP7 serotype 2) or 1076 (VP7 serotype 2, neonatal strain) in neutralization assays, GORZIGLIA et al. (1990b) and LI et al. (1993b) classified the human VP4s into three serotypes and one subtype: serotype 1A (strain with VP7 serotype 1, 3, 4 or 9 specificity); serotype 1B (strains with VP7 serotype 2 specificity); serotype 2 (neonatal strains with VP7 serotype 1, 2, 3 or 4 specificity); and serotype 3 (the K8 strain with VP7 serotype 1

---

[19] Similarly, KANG and SAIF (1991) later demonstrated the existence of the three domains by analyzing 8 mAbs directed to the VP4 of turkey rotavirus O strain.

[20] Recently, the VP8 subunit of VP4 protein of SA11 virus was expressed in recombinant baculovirus (FIORE et al. 1991) or E. coli (LIZANO et al. 1991) and was shown to have HA activity.

[21] More recently, the existence of a sixth VP4 gene group was reported by LI et al. (1993a). The deduced amino acid sequence of the VP4 of this human rotavirus (HCR3 strain, G serotype 3) was shown to share 89.4% homology with that of rhesus monkey rotavirus (MMU18006 strain, G serotype 3) (LI et al. 1993a).

[22] Since VP4-specific mAbs for serotyping rotavirus VP4 are still not widely available, gene group (allele) typing has been developed as a proxy method for P typing with the use of PCR (GENTSCH et al. 1992) and dot or Northern-blot hybridization (LARRALDE and FLORES 1990; PARWANI et al. 1992b; ROSEN et al. 1992). By using PCR probes capable of distinguishing the three to five human VP4 gene alleles, BERN et al. (1992), STEELE et al. (1993), and GUNASENA et al. (1993) recently demonstrated that the majority of rotavirus field isolates from Bangladesh, Europe, Latin America, and Japan belong to either the Wa or DS-1 gene 4 allele.

specificity). A direct correlation between VP4 amino acid homology and the newly assigned VP4 serotype was shown.[23]

Among human rotaviruses, only two neutralizing VP4 mAbs have been mapped to the VP8 fragment (KITAOKA et al. 1986; PADILLA-NORIEGA et al. 1993) (Table 4). BURNS et al. (1988) generated VP4-specific neutralizing mAbs which neutralize VP7 serotype 3 rotaviruses (human, simian and canine strains) but do not neutralize human rotaviruses belonging to VP7 serotype 1, 2, 4 or 8. This is the first report to demonstrate a region on the VP4 protein which is specific for certain strains belonging to VP7 serotype 3. This finding implies that serotype-specific regions may exist on the VP4 of human rotavirus VP7 serotypes 1, 4, 8 and 9. Sequence analysis of the VP4 encoding gene of these VP7 serotype 3 virus mutants resistant to neutralization by these mAbs should identify VP7 serotype 3-specific sites on the VP4 gene.

Functional and antigenic topography on the VP4 protein have been studied by analyzing serologic specificities of VP4-specific mAbs and their ability to recognize antigenic mutants grown in the presence of neutralizing

---

[23] In order to circumvent the lack of appropriate and readily-available reagents for serotyping and classification of rotavirus VP4, a tentative typing scheme was proposed by ESTES and COHEN (1989) based on nucleic acid hybridization and sequence analysis of the VP4-encoding gene, in which nine inferred VP4(P) types (gene 4 alleles) of human and other mammalian rotaviruses were defined. Serological results obtained from reciprocal or one-way cross-neutralization assays which incorporated hyperimmune antisera raised against various laboratory-generated reassortant rotaviruses as well as recombinant VP4s have largely validated their VP4 typing proposal. To date, 15 such genotypes (or inferred VP4 serotypes) have been identified (SERENO and GORZIGLIA 1993). However, a genotype established by nonserological methods is not always identical to a serotype defined by serological methods. For example, human rotavirus strain DS-1 VP4 (G serotype 2) was defined as a distinct inferred VP4 type (genotype 4) (ESTES and COHEN 1989), however, it was shown later by cross-neutralization to belong to a subtype of VP4 serotype 1 (GORZIGLIA et al. 1990b). Hence, it has become obvious that a rotavirus VP4(P) serotype scheme has to be established based on the data obtained by serological methods. When the validity of a genotyping is once confirmed by serological methods, the typing of gene 4 alleles could be used as a proxy method for VP4(P) serotyping (GENTSCH et al. 1992) in common with the now widely-used typing of VP7-encoding gene alleles for VP7(G) serotyping. Recently, SERENO and GORZIGLIA (1993) have proposed a tentative VP4(P) serotype classification scheme and attempted to unify it with the genotype classification scheme of gene 4 alleles (Table 5), in which they established ten VP4(P) serotypes based on the data obtained by reciprocal or one-way cross-neutralization performed by various investigators. Thus, the standardization of classification and nomenclature of rotavirus VP4(P) serotype is urgently needed to unify the numbering system and avoid further confusion among veterinary and medical researchers. Ideally both the VP4(P) serotype and the VP4 genotype should follow the same numbering system as is the case for VP7(G) serotype and VP7 genotype.

By analyzing neutralizing hyperimmune antisera to the *E. coli*-expressed VP5 or VP8 subunit of strains KU, DS-1, and 1076, LARRALDE et al. (1991) showed that the VP8 subunit contained the major antigenic site(s) responsible for serotype-specific neutralization, whereas the VP5 subunit was responsible for cross-reactivity among strains belonging to different VP4 serotypes. In order to define further regions in the VP8 subunit that are responsible for neutralizing activity, they expressed three peptides that were used in an immunoblot assay. They showed that peptides A (amino acid 1 to 102) and C (amino acids 150 to 246 plus 247 to 251 from VP5) cross-reacted with heterotypic human rotavirus VP4 antisera, whereas peptide B (amino acids 84 to 180) appeared to be involved in the VP4 serotype and subtype specificities (LARRALDE and GORZIGLIA 1992).

Table 5. Serotypic and genotypic classification of group A rotavirus VP4 (adapted from SERENO and GORZIGLIA (1993) with additions)

VP4(P)	Serotype[a]	VP4 Genotype[b]	Human Rotavirus Strains (G serotype)	Animal Rotavirus Strains (G serotype [species]
1	A	8	KU, Wa (1); P, YO, MO (3); VA70, Hochi, Hosokawa (4); WI61, F45 (9)	none
	1B	4	DS-1, S2, RV-5[c] (2); L26[c] (12)	none
2	A	6	M37 (1); 1076 (2) WcN RV-3[c] (3); ST3 (4)	none
	2B	6	none	Gottfried (4) [pig]
3	A	9	K8 (1); AU-1[c] (3); PA151[c] (6)	FRV-1[c] (3) [cat]
	3B	9	Mc35 (10)	none
	4	10	57M[c] (4); 69M (8)	none
	5	3	HCR3[c] (3)	MMU18006 (3) [rhesus monkey]
	6	1	none	NCDV, C486, J2538 (6); A5[c] (8) (cow); SA11-4fM (3) [vervet monkey]
	7	5	none	UK, B641, IND[c], OK[c], (6); 61A[c] (10) [cow]
	8	11	116E[c] (9); 1321[c] (10)	B223, B-11, A44[c], KK3[c], Cr[c] (10) [cow]
	9	7	none	CRW-8, Ben-307 (3); BMI-1, SB-1A (4); OSU, TFR-41 (5); YM, A253.1 (11) [pig]
	10	15	none	Eb (3) [mouse]
	?	2	none	SA11 (3) [vervet monkey]
	?	12	none	H-2 (3) [horse]
	?	13	none	MDR-13 (3/5) [pig]
	?	14	none	Lp14 (10) [sheep]

[a] VP4(P) serotype as determined by reciprocal or one-way cross-neutralization.
[b] VP4 genotype as determined by comparative amino acid sequence analysis and/or nucleic acid hybridization.
[c] Not tested by neutralization; relationship suggested by amino acid sequence and/or nucleic acid hybridization analyses.

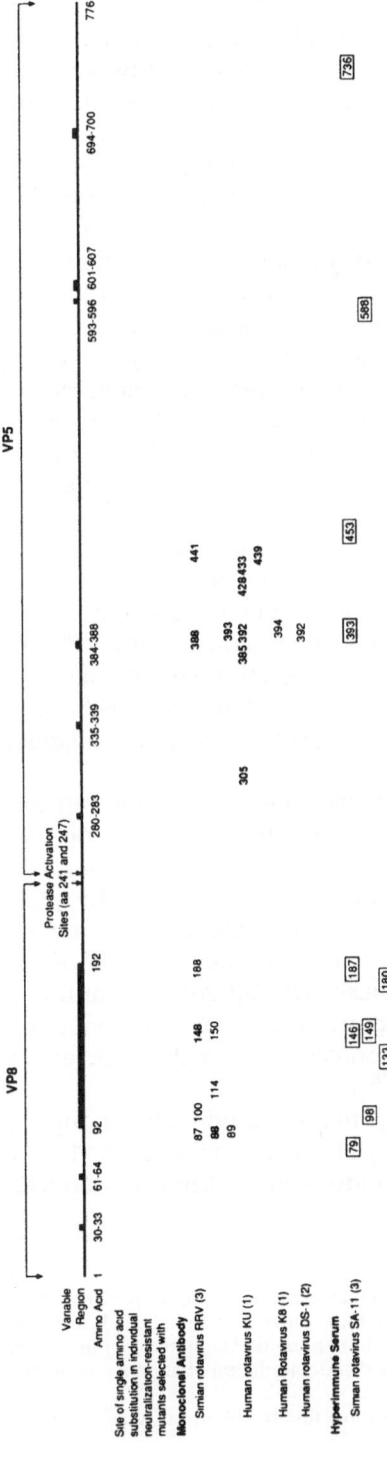

Fig. 3. Linear structure and antigenic organization of rotavirus outer capsid VP4; VP7 serotype; *Boldface numbers* indicate amino acid substitutions in mutants selected with monoclonal antibody exhibiting cross-reactive (i.e., heterotypic) neutralizing activity; *numbers not in boldface and not in boxes* indicate amino acid substitutions in mutants selected with monoclonal antibody with homotypic neutralizing activity. *Numbers in boxes indicate amino acid substitutions in mutants selected with hyperimmune serum to SA-11 virus.* (Adapted from KAPIKIAN and CHANOCK 1990, with additions. Data from MACKOW et al. 1988b; TANIGUCHI et al. 1988b; GORZIGLIA et al. 1990c; KOBAYASHI et al. 1991d)

mAbs (GREENBERG et al. 1983; TANIGUCHI et al. 1985, 1987b; SHAW et al. 1986; BURNS et al. 1988; KOBAYASHI et al. 1990). Serotype-specific and serotype-cross-reactive epitopes on VP4 have been identified by analysis of nucleotide and deduced amino acid sequences of such neutralization escape mutants (MACKOW et al. 1988b; TANIGUCHI et al. 1988b; KOBAYASHI et al. 1991d) (Fig. 3). In this way, five regions on VP8 and three on VP5 have been identified with some appearing to be conformational and others to be linear or sequential. By sequence analysis of the VP4 gene of seven antigenic mutants of simian rotavirus SA11 4fM that were selected after 39 passages in the presence of hyperimmune antiserum to SA11, GORZIGLIA et al. (1990c) showed that: (1) most amino acid substitutions clustered in two variable regions of VP4; (2) although most amino acid changes of these mutants were also observed in mutants selected in the presence of neutralizing mAbs, some amino acid changes occurred that were not selected for by neutralizing mAbs; (3) amino acid changes were not found in six of the nine previously described VP4 variable regions; (4) amino acid changes were not present at residues 305, 433 and 441, which have been shown to be involved in the binding of anti-VP4 neutralizing mAbs; and (5) amino acid substitutions were not found within or close to the cleavage sites. The latter have been suggested to represent antigenic sites by STRECKER et al. 1988; ARIAS et al. 1987. The mAbs that map to the regions on the VP8 are almost exclusively serotype- or strain-specific whereas those mapping to the VP5 regions are mostly serotype-cross-reactive. Consistent with this finding is the observation that the VP8 region of the VP4 protein shows the greatest sequence variability.[24]

TANIGUCHI et al. (1988b) demonstrated that one of the three regions on VP5 of human rotavirus KU strain is sequential or linear by showing that a synthetic peptide (amino acids 296–313) which included the sequence of the first region reacted with its corresponding neutralizing mAb (YO-2C2) by ELISA. Two other synthetic peptides that included the sequence of either the second or third region failed to react with their corresponding neutralizing mAbs, which suggested that these two regions are conformational. They also showed that these two regions, each of which contained three functional groups, overlapped operationally whereas the first region was independent (KOBAYASHI et al. 1990).[25]

Hyperimmunization of rabbits with a synthetic peptide corresponding to the cleavage region (amino acid 228–241) of simian rotavirus SA11 VP4 was shown by STRECKERT et al. (1988) to induce neutralizing antibodies.

---

[24] Recently, KOBAYASHI et al. (1993) examined the reactivity of five neutralizing mAbs raised against the VP5 subunit of the VP4 protein of various human rotavirus strains with 20 rotavirus strains derived from 8 animal species as well as various human rotavirus strains. They found that the composition of neutralization epitopes on the VP5 subunit protein of human rotavirus strains (except for three unique strains K8, 57M, and 69M) was distinct from that of animal rotaviruses except for the porcine rotavirus Gottfried strain.

[25] Recently, a fourth functional region distinct from the three functional groups was identified (PADILLA-NORIEGA et al. 1993).

The anti-peptide serum was specific for SA11 in neutralization and immune-electron microscopy studies. Neutralizing activity was not demonstrated against human rotavirus Wa, S2, YO, Hochi, bovine rotavirus UK or NCDV.[26] ARIAS et al. (1987) showed that, although a synthetic peptide comprised of amino acids 220-233 of the SA11 VP4 did not induce neutralizing antibodies in mice, the peptide was able to prime mice for a neutralizing antibody response to SA11 or ST3 following secondary inoculation with either simian SA11 or human ST3 virus. Since peptides spanning amino acids 231–241 or 231–247 of the SA11 VP4 were unable to induce neutralizing antibodies or to prime the immune response of mice, the authors speculated that the overlapping region between the two peptides (amino acids 228–233) might be responsible for the observed priming effect.

With the use of DNA amplification-restricted transcription-translation, MACKOW et al. (1990b) analyzed the coding region of gene segment 4 of RRV to identify the minimum polypeptide required for antibody recognition. They demonstrated that: (1) a minimum polypeptide of 228 amino acids (amino acids 247–474) of VP5 is required for binding of heterotypic neutralizing mAbs M2, M7 or 2G4; and (2) the minimum essential binding domain for eight largely serotype-specific anti-VP8 neutralization mAbs consists of amino acids 55–222 of VP4. PRASAD et al. (1990) analyzed localization of VP4 neutralization sites in rotavirus by three-dimensional cryoelectron microscopy. They demonstrated that two Fab fragments of heterotypic VP4-specific neutralizing mAb 2G4, which has been mapped to the VP5 region of VP4, bind to the distal ends of each spike of simian SA11 VP4 and suggested that the distal end of the spike contains the VP5 region.[27]

An interesting and recurring unexplained observation is that the VP4 protein of a reassortant rotavirus is characteristically more immunogenic than that of its parental strain (HOSHINO et al. 1985a; 1988; OFFIT and BLAVAT 1986; OFFIT et al. 1986a; URASAWA et al. 1986). It has been suggested that a conformational change in VP4 and VP7 produces an alteration in the interaction of these outer capsid proteins that might influence the antigenicity and immunogenicity and the antibody binding capacity of one or both

---

[26] Recently, HANSEN et al. (1992) raised mAb to the synthetic peptide (amino acid residues 228–241). This mAb had a moderate level of neutralizing activity against SA11 virus and reacted by immunofluorescence with determinants of SA11-infected cells.

[27] By using a neutralizing mAb (RQ31, mAb1) raised against VP4 of G serotype 6 bovine rotavirus Q17 strain, CORNAGLIA et al. (1992) generated an anti-idiotypic mAb (mAb2) in mice. The mAb2 inhibited the binding of mAb1 to the virus as well as virus neutralization mediated by mAb1 in a dose-dependent fashion. Hyperimmune antiserum (anti-anti-idiotypic antibodies, Ab3) from guinea pigs immunized with mAb2 was shown to be mAb1-like by ELISA reactivity pattern against various rotavirus strains, by virus neutralization as well as by immunoprecipitation. Since mAb2 can be produced in unlimited amounts, they speculate that the mAb2 could be used for vaccination of cows to develop a transferable passive immunity or for oral vaccination of calves to develop an active local immunity.

proteins.[28] Although reassortant of rotaviruses appears to occur with frequency in nature, as determined by examination of electropherotypes in the clinical setting, it seems that certain genes such as those encoding VP7 and VP4 do not reassort at random (for review see RAMIG and WWARD 1991). This observation deserves further investigation (for example study of the role of glycosylation which appears to be an important modifier of the VP7 protein) not only from the point of view of characterization of two surface proteins and of rotavirus vaccine development but also from the aspect of rotavirus evolution in nature.

6 Conclusions

A variety of mAbs directed to both structural and nonstructural proteins including VP2, VP4, VP6, NS34, NS35, VP7 and NS28 of various rotaviruses belonging to different serotypes has been generated. Such mAbs, in combination with various serological and molecular biological techniques, have been utilized successfully to dissect the structure and function of the proteins mediating three important rotavirus specificities: group, subgroup, and serotype. In addition, all the structural and nonstructural rotaviral proteins have been expressed in various expression vector systems and characterized. Thus, the topography of various immunological specificities such as neutralization of these proteins has been elucidated. With increasing knowledge of the antigenic contribution of each of the constituent proteins comprising the virus particle, advances in the understanding of the structure and function of the virion itself should be forthcoming. As noted previously, substitution of one outer capsid protein of a virus for the corresponding protein of another virus through genetic reassortment influences the antigenicity and immunogenicity of one or both proteins of a resulting reassortant. It is not known, however, how the VP7 and VP4 interact and influence each other in the virion in terms of the qualitative nature and quantitative distribution of antigenic determinants comprising the serological specificities of both proteins. Knowledge of the mechanisms underlying such phenomena will give insight into the observations made recently by: (1) WARD et al. (1988, 1990, 1993) that host immune systems responded almost exclusively

---

[28] The interaction of VP4–VP7 outer capsid proteins of rotavirus affecting the expression of phenotypes was reported by CHEN et al. (1989). They examined various reassortants to analyze protease-associated phenotypes of simian rotavirus SA11 (SA11–4fM) and showed that the phenotypes segregated with two genes (VP4 and VP7) or a single gene (VP4), depending on the recipient genetic background. Furthermore, they examined the reactivity of VP4-specific cross-reactive neutralizing mAb 2G4 with various ressortants generated between simian (SA11) and bovine (B223) rotaviruses, and demonstrated that the presentation of an epitope on SA11 VP4 recognizable by mAbs 2G4 was affected in subtle ways by specific physical interactions of VP4 and VP7 of heterologous rotaviruses (CHEN et al. 1992). A similar phenomenon was observed by KIRKWOOD et al. (1993) who described a cross-reactive VP7-specific neutralizing mAb F45:2 which did not neutralize human rotavirus ST3 (G serotype 4) but neutralized reassortant ST3-SA which had 10 genes from SA11 and only VP7-encoding gene from ST3.

to either VP4 or VP7; and (2) ZHENG et al. (1988, 1989b), GERNA et al. (1990a) and PADILLA-NORIEGA et al. (1990), that primary infection of neonates with different human rotavirus serotypes resulted in a spectrum of varying heterotypic neutralizing responses. Such information as well as current knowledge should provide logical and practical strategies to controlling rotavirus disease.

References

Aboudy Y, Shif I, Ziberstein I, Gotlieb-Stematsky T (1988) Use of polyclonal and monoclonal antibodies and analysis of viral RNA in the detection of unusual group A human rotaviruses. J Med Virol 25: 351–359

Adams WR, Kraft LM (1963) Epizoolic diarrhea of infant mice: identification of the etiologic agent. Science 141: 359–360

Ahmed MU, Taniguchi K, Kobayashi N, Urasawa T, Wakasugi F, Islam M, Shaikh H, Urasawa S (1989) Characterization by enzyme-linked immunosorbent assay using subgroup- and serotype-specific monoclonal antibodies of human rotavirus obtained from diarrheic patients in Bangladesh. J Clin Microbiol 27: 1678–1681

Ahmed MU, Urasawa S, Taniguchi K, Urasawa T, Kobayashi N, Wakasugi F, Islam AIMM, Sahikh HA (1991a) Analysis of human rotavirus strains prevailing in Bangladesh in relation to nationwide floods brought by the 1988 monsoon. J Clin Microbiol 29: 2273–2279

Ahmed MU, Taniguchi K, Kobayashi N, Urasawa T, Urasawa S (1991b) Preparation of a neutralizing monoclonal antibody specific to serotype 8 rotavirus strains with a super-short RNA pattern. Southeast Asian J Trop Med Public Health 22: 41–45

Akatani K, Ikegami N (1987) Typing of fecal rotavirus specimens by an enzyme linked immunosorbent assay using monoclonal antibodies (in Japanese). Clin Virol 15: 61–68

Albert MJ, Unicomb LE, Bishop RF (1987a) Cultivation and characterization of human rotaviruses with "super short" RNA patterns. J Clin Microbiol 25: 183–185

Albert MJ, Unicomb LE, Tzipori SR, Bishop RF (1987b) Isolation and serotyping of animal rotaviruses and antigenic comparison with human rotaviruses. Arch Virol 93: 123–130

Andrew ME, Boyle DB, Coupar BEH, Whitfeld PL, Both GW, Ballamy AR (1987) Vaccinia virus recombinants expressing the SA11 rotavirus VP7 glycoprotein gene induce serotype-specific neutralizing antibodies. J Virol 61: 1054–1060

Andrew ME, Boyle DB, Whitfeld PL, Lockett LJ, Anthony ID, Bellamy AR, Both GW (1990) The immunogenicity of VP7, a rotavirus antigen resident in the endoplasmic reticulum, is enhanced by cell surface expression. J Virol 64: 4776–4783

Arias CF, Ballado T, Plebanski M (1986) Synthesis of the outer-capsid glycoprotein of the simian rotavirus SA11 in Escherichia coli. Gene 47: 211–219

Arias CF, Lizano M, López S (1987) Synthesis in Escherichia coli and immunological characterization of a polypeptide containing the cleavage sites associated with trypsin enhancement of rotavirus SA11 infectivity. J Gen Virol 68: 633–642

Arias CF, Ruiz AM, López S (1989) Further antigenic characterization of porcine rotavirus YM. J Clin Microbiol 27: 2871–2873

Arista S, Giovannelli L, Pistoia D, Cascio A, Parea M, Gerna G (1990) Electropherotypes, subgroups and serotypes of human rotavirus strains causing gastroenteritis in infants and young children in Palermo, Italy, from 1985 to 1989. Res Virol 141: 435–448

Bachmann PA, Bishop RF, Flewett TH, Kapikian AZ, Mathan MM, Zissis G (1984) Nomenclature of human rotaviruses: designation of subgroups and serotypes. Bull WHO 62: 501–503

Barnett B (1983) Viral gastroenteritis. Med Clin North Am 67: 1031–1058

Bastardo JW, Holmes IH (1980) Attachment of SA-11 rotavirus to erythrocyte receptors. Infect Immun 29: 1134–1140

Bastardo JW, McKimm-Breschkin JL, Snoza S, Mercer LD, Holmes IH (1981) Preparation and characterization of antisera to electrophoretically purified SA11 virus polypeptides. Infect Immun 34: 641–647

Beards GM (1987) Serotyping of rotavirus by NADP-enhanced enzyme immunoassay. J Virol Methods 18: 77–85

Beards GM, Thouless ME, Flewett TH (1980) Rotavirus serotypes by serum neutralization. J Med Virol 5: 231–237

Beards GM, Desselberger U, Flewett TH (1989) Temporal and geographical distributions of human rotavirus serotypes, 1983 to 1988. J Clin Microbiol 27: 2827–2833

Beards G, Xu L, Ballard A, Desselberger U, McCrae MA (1992) A serotype 10 human rotavirus. J Gen Virol 30: 1432–1435

Begue RE, Dennehy PH, Huang J, Martin P (1992) Serotype variation of group A rotaviruses over nine winter epidemics in southeastern New England. J Clin Microbiol 30: 1592–1594

Bellamy AR, Both GW (1990) Molecular biology of rotaviruses. Adv Virus Res 38: 1–43

Bellinzoni R, Jiang X, Tanaka T, Scodellar E, Estes MK (1989) Rotavirus gene detection with biotinylated single-stranded RNA probes. Mol Cell Probes 3: 233–244

Bellinzoni RB, Mattion NM, Matson DO, Blackhall J, LaTorre JL, Scodeller EA, Urasawa S, Taniguchi K, Estes MK (1990) Porcine rotaviruses antigenically related to human rotavirus serotypes 1 and 2. J Clin Microbiol 28: 633–636

Benfield DA, Nelson EA, Hoshino Y (1987) A monoclonal antibody to the Gottfried strain of porcine rotavirus which neutralizes rotavirus serotypes 3, 4 and 6. Abstracts of the 7th international congrees on virology, Berlin, p 111

Bern C, Unicomb L, Gentsch JR, Banul N, Yunus M, Sack RB, Glass RI (1992) Rotavirus diarrhea in Bangladeshi children: correlation of disease severity with serotypes. J Clin Microbiol 30: 3234–3238

Bernstein DI, Smith VE, Sander DS, Pax KA, Schiff GM, Ward RL (1990) Evaluation of WC3 rotavirus vaccine and correlates of protection in healthy infants. J Infect Dis 162: 1055–1062

Bingnan F, Unicomb L, Rahim Z, Banu N, Podder GP, Clemens J, VanLoon FPI, Rao MR, Malek A, Tzipori S (1991) Rotavirus-associated diarrhea in rural Bangladesh: two-year study of incidence and serotype distribution. J Clin Microbiol 29: 1359–1363

Birch CJ, Health RL, Marshall JA, Liu S, Gust ID (1985) Isolation of feline rotaviruses and their relationship of human and simian isolates by electropherotype and serotype. J Gen Virol 60: 2731–2735

Birch CJ, Heath RJ, Gust ID (1988) Use of serotype-specific monoclonal antibodies to study epidemiology of rotavirus infection. J Med Virol 24: 45–53

Bishop RE, Davidson GP, Holmes IH, Ruck BJ (1973) Virus particles in epithelial cells of duodenal mucosa from children with viral gastroenteritis. Lancet 2: 1281–1283

Bishop RF, Unicomb LE, Soenarto Y, Suwardji H, Ristanto, Barnes GL (1989) Rotavirus serotypes causing acute diarrhoea in hospitalized children in Yogyakarta, Indonesia during 1978–1979. Arch Virol 107: 207–213

Bishop RF, Unicomb LE, Barnes GL (1991) Epidemiology of rotavirus serotypes in Melbourne, Australia, from 1979 to 1989. J Clin Microbiol 29: 862–868

Blackhall JO, Estes MK, Matson D, Mattion NM, LaTorre JL, Bellinzoni RC (1990) Bovine rotaviruses antigenically related to human rotavirus serotype 1, 2 and 3. Abstracts of the 8th international congress on virology, Berlin, p 25–022

Blackhall J, Bellinzoni R, Mattion N, Estes MK, LaTorre J, Magnusson G (1992) A bovine rotavirus serotype 1: serologic characterization of the virus and nucleotide sequence determination of the structural glycoprotein VP7 gene. Virology 189: 833–837

Bohl EH, Thiel KW, Saif LJ (1984) Isolation and serotyping of porcine rotavirus and antigenic comparison with other rotaviruses. J Clin Microbiol 19: 105–111

Borras-Cuesta F, Petit A, Pery P, Fedon Y, Garnier J, Cohen J (1987) Immunogenicity of synthetic peptides corresponding to regions of the major inner capsid protein of bovine rotavirus (BRV). Ann Inst Pasteur/Virol 138: 437–450

Both GW, Siegman LJ, Bellamy AR, Ikegami N, Shatlom AJ, Furuichi Y (1984) Comparative sequence analysis of rotavirus genomic segment 6 – the gene specifying viral subgroups 1 and 2. J Virol 51: 97–101

Both GW, Lockett LJ, Janardhana V, Edwards SJ, Bellamy AR, Graham FL, Prevec L, Andrew ME (1993) Protective immunity to rotavirus-induced diarrhea is passively transferred to newborn mice from naive dams vaccinated with a single dose of a recombinant adenovirus expressing rotavirus VP7sc. Virology 193: 940–949

Bridger JC (1981) Novel rotaviruses in animals and man. Ciba Found Symp 128: pp 5–23

Brown DWG, Mathan MM, Mathew M, Martin R, Beards GM, Mathan VI (1988) Rotavirus epidemiology in Vellore, South India: group, subgroup, serotype and electropherotype. J Clin Microbiol 26: 2410–2414

Browning GF, Chalmers RM, Fitzgerald TA, Snodgrass DR (1991a) Serological and genomic characterization of L338, a novel equine group A rotavirus G serotype. J Gen Virol 72: 1059–1064

Browning GF, Fitzgerald TA, Chalmers RM, Snodgrass DR (1991b) A novel group A rotavirus G serotype: serological and genomic characterization of equine isolate FI23. J Clin Microbiol 29: 2043–2046

Browning GF, Chalmers RM, Fitzgerald TA, Corley KTT, Campbell I, Snodgrass DR (1992a) Rotavirus serotype G3 predominates in horses. J Clin Microbiol 30: 59–62

Browning GF, Chalmers RM, Fitzgerald TA, Snodgrass DR (1992b) Evidence for two serotype G3 subtypes among equine rotaviruses. J Clin Microbiol 30: 485–491

Brussow H, Snodgrass D, Fitzgerald T, Eichhorn W, Gerhards R, Bruttin A (1990) Antigenic and biochemical characterization of bovine rotavirus V1005, a new member of rotavirus serotype 10. J Gen Virol 71: 2625–2630

Burns JW, Greenberg HB, Shaw RD, Estes MK (1988) Functional and topographical analysis of epitopes on the hemagglutinin (VP4) of the simian rotavirus SA11. J Virol 62: 2164–2172

Caust J, Dyall-Smith ML, Lazdins I, Holmes IH (1987) Glycosylation, an important modifier of rotavirus antigenicity. Arch Virol 96: 123–134

Chan CK, Tam IS, French GL (1990) Serotyping of human rotaviruses by oligonucleotide probes hybridization. Abstracts VIIIth Internatl Congress Virol, Berlin, No. P71-030

Chen D, Burns JW, Estes MK, Ramig RF (1989) Phenotypes of rotavirus reassortants depend upon the recipient genetic background. Proc Natl Acad Sci USA 86: 3743–3747

Chen D, Estes MK, Ramig RF (1992) Specific interactions between rotavirus outer capsid proteins VP4 and VP7 determine expression of a cross-reactive, neutralizing VP4-specific epitope. J Virol 66: 432–439

Christy C, Madore HP, Pichichero ME, Gala C, Pincus P, Vosefski D, Hoshino Y, Kapikian A, Dolin R, Elmwood and Panorama Pediatrics Groups (1988) Field trial of rhesus rotavirus vaccine in infants. Pediatr Infect Dis J 7: 645–650

Ciarlet M, Hidalgo M, Gorziglia M, Liprandi F (1993) Characterization of neutralization epitopes on VP7 of porcine rotavirus serotype G11. Virology (in press)

Clark HF, Hoshino Y, Bell LM, Groff J, Hess G, Bachman P, Offit PA (1987) Rotavirus isolate WI61 representing a presumptive new human serotype. J Clin Microbiol 25: 1757–1762

Cohen J, Lefevre F, Estes MK, Bremont M (1984) Cloning of bovine rotavirus (RF strain): nucleotide sequence of the gene coding for the major capsid protein. Virology 138: 178–182

Cornaglia EM, Elazhary YMAS, Brodeur BR, Talbot BG (1992) Monoclonal anti-idiotype induces antibodies against bovine Q17 rotavirus. J Virol 66: 5763–5769

Cornaglia E, Elazhary Y, Talbot B (1993) Bovine rotavirus type detection by neutralizing monoclonal antibodies. Arch Virol 129: 243–250

Coulson BS (1987) Variation in neutralization epitopes of human rotavirus in relation to genomic RNA polymorphism. Virology 159: 209–216

Coulson BS (1993) Typing of human rotavirus VP4 by an enzyme immunoassay using monoclonal antibodies. J Clin Microbiol 31: 1–8

Coulson BS, Fowler KJ, Bishop RF, Cotton RGH (1985) Neutralizing monoclonal antibodies to human rotavirus and indications of antigenic drift among strains from neonates. J Virol 54: 14–20

Coulson BS, Tursi JM, McAdam WJ, Bishop RF (1986) Derivation of neutralizing monoclonal antibodies to human rotaviruses and evidence that an immunodominant neutralization site is shared between serotypes 1 and 3. Virology 154: 302–312

Coulson BS, Unicomb LE, Pitson GA, Bishop RF (1987a) Simple and specific enzyme immunoassay using monoclonal antibodies for serotyping human rotaviruses. J Clin Microbiol 25: 509–515

Coulson BS, Fowler KJ, White JR, Cotton RGH (1987b) Non-neutralizing monoclonal antibodies to a trypsin-sensitive site on the major glucoprotein of rotavirus which discriminate between virus serotypes. Arch Virol 93: 199–211

Coulson BS, Kirkwood C (1991) Relation of VP7 amino acid sequence to monoclonal antibody neutralization of rotavirus and rotavirus monotype. J Virol 65: 5968–5974

Cukor G, Blacklow NR (1984) Human viral gastroenteritis. Microbiol Rev 48: 157–179

Cukor G, Peron DM, Hudson R, Blacklow NR (1984) Detection of rotavirus in human stools by using monoclonal antibody. J Clin Microbiol 19: 888–892

Das BK, Gentsch JR, Hoshino Y, Ishida S, Nakagomi O, Bhan MK, Kumar R, Glass RI (1993) Characterization of the G serotype and genogroup of New Delhi newborn rotavirus strain (116E. Virology 197: 99–102

Das M, Dunn SJ, Woode GN, Greenberg HB, Rao CD (1993) Both surface proteins (VP4 and VP7) of an asymptomatic neonatal rotavirus strain (I321) have high levels of sequence identity with the homologous proteins of a serotype 10 bovine rotavirus. Virology 194: 374–379

Dimitrov DH, Graham DV, Estes MK (1985) Detection of rotaviruses by nucleic acid hybridization with cloned DNA of simian rotavirus SA11 genes. J Infects Dis 152: 293–300

Dolan KT, Twist ME, Horton-Slight F, Forrer C, Bell LM, Plotkin SA, Clark HF (1985) Epidemiology of rotavirus electropherotypes determined by a simplified diagnostic technique with RNA analysis. J Clin Microbiol 21: 753–758

Dormitzer PR, Ho DY, Mackow ER, Mocarski ES, Greenberg HB (1992a) Neutralizing epitopes on herpes simplex-expressed rotavirus VP7 are dependent on coexpression of other rotavirus proteins. Virology 187: 18–32

Dormitzer PR, Greenberg HB (1992b) Calcium chelation induces a conformational change in recombinant herpes simplex virus-1-expressed rotavirus VP7. Virology 189: 828–832

Dunn SJ, Greenberg HB, Ward RL, Nakagomi O, Burns JW, Vo PT, Pax KA, Das M, Gowda K, Rao CD (1993a) Serotypic and genomic characterization of human serotype 10 rotaviruses from asymptomatic neonates. J Clin Microbiol 31: 165–169

Dunn SJ, Ward RL, McNeal MM, Cross TL, Greenberg HB (1993b). Identification of new neutralization epitope on VP7 of human serotype 2 rotavirus and evidence for electropherotype differences caused by single nucleotide substitutions. Virology 197: 397–404

Dyall-Smith ML, Lazdins I, Tregear GW, Holmes IH (1986) Location of the major antigenic sites involved in rotavirus serotype-specific neutralization. Proc Natl Acad Sci USA 83: 3465–3468

Eiden J, Sato S, Yolken R (1987) Specificity of dot hybridization assay in the presence of rRNA for detection of rotaviruses in clinical specimens. J Clin Microbiol 25: 1809–1811

Estes MK, Cohen J (1989) Rotavirus gene structure and function. Microbiol Rev 53: 410–449

Estes ME, Palmer EL, Obijeski JF (1983) Rotaviruses: a review. In: Compans RW, Cooper M, Koprowski H et al. (eds) Current topics in microbiology and immunology, vol. 105. Springer, Berlin Heidelberg New York, pp 123–184

Estes MK, Mason BB, Crawford S, Cohen J (1984) Cloning and nucleotide sequence of the simian rotavirus gene 6 that codes for the major inner capsid protein. Nucleic Acids Res 12: 1875–1887

Estes MK, Crawford SE, Penaranda ME, Petric BL, Burns JW, Chan W-K, Ericson B, Smith GE, Summers MD (1987) Synthesis and immunogenicity of the rotavirus major capsid antigen using a baculovirus expression system. J Virol 61: 1488–1494

Fiore L, Greenberg HB, Mackow ER (1991) The VP8 fragment of VP4 is the rhesus rotavirus hemagglutinin. Virology 181: 553–563

Fitzgerald TA, Browning GF (1991) Increased sensitivity of a rotavirus serotyping enzyme-linked immunosorbent assay by the incorporation of $CaCl_2$. J Virol Methods 33: 299–304

Flores J, Boeggeman E, Purcell RH, Sereno M, Perez I, White L, Wyatt RG, Chanock RM, Kapikian AZ (1983) A dot hybridization assay for detection of rotaviruses. Lancet i: 555–559

Flores J, Midthun K, Hoshino Y, Green K, Gorziglia M, Kapikian AZ, Chanock RM (1986) Conservation of the fourth gene among rotaviruses recovered from asymptomatic newborn infants and its possible role in attenuation. J Virol 60: 972–979

Flores J, Taniguchi K, Green K, Pérez-Schael I, Garcia D, Sears J, Urasawa S, Kapikian AZ (1988) Relative frequencies of rotavirus serotypes 1, 2, 3 and 4 in Velezuelan infants with gastroenteritis. J Clin Microbiol 26: 2092–2095

Flores J, Green KY, Garcia D, Sears J, Pérez-Schael I, Avendano LF, Rodriquez WB, Taniguchi K, Urasawa S, Kapikian AZ (1989) Dot hybridization assay for distinction of rotavirus serotypes. J Clin Microbiol 27: 29–34

Flores J, Sears J, Pérez-Schael I, White L, Garcia D, Lanata C, Kapikian AZ (1990) Identification of human rotavirus serotype by hybridization to polymerase chain reaction-generated probes derived from a hyperdivergent region of the gene encoding outer capsid protein VP7. J Virol 64: 4021–4024

Francavilla M, Miranda P, DiMatteo A, Sarasini A, Gerna G, Milanesi G (1987) Expression of bovine rotavirus neutralization antigen in Escherichia coli. J Gen Virol 68: 2975–2980

Fukudome K, Yoshie O, Konno T (1989) Comparison of human, simian and bovine rotaviruses for requirement of sialic acid in hemagglutinin and cell adsorption. Virology 172: 196–205

Gary GW Jr, Black DR, Palmer E (1982) Monoclonal IgG to the inner capsid of human rotavirus. Arch Virol 72: 223–227

Gatheru Z, Kobayashi N, Adachi N, Chiba S, Mali J, Ogaja P, Nyangao J, Kiplagat E, Tukei PM (1993) Characterization of human rotavirus strains causing gastroenteritis in Kenya. Epidemiol Infect 110: 419–423

Gaul SK, Simpson TF, Woode GN, Fulton RW (1982) Antigenic relationships among some animal rotaviruses: virus neutralization in vitro and cross-protection in piglets. J Clin Microbiol 16: 495–503

Gentsch JR, Glass RI, Woods P, Gouvea V, Gorziglia M, Flores J, Das BK, Bhan MK (1992) Identification of group A rotavirus gene 4 types by polymerase chain reaction. J Clin Microbiol 30: 1365–1373

Gentsch JR, Das BK, Jiang B, Bhan MK, Glass RI (1993) Similarity of the VP4 protein of human rotavirus 116E to that of the bovine B223 strain. Virology 194: 424–430

Georges-Courbot MC, Berand AM, Beards GM, Compbell AD, Gonzales JP, George AJ, Flewett TH (1988) Subgroups, serotypes and electropherotypes of rotavirus isolated from children in Bangui, Central African Republic. J Clin Microbiol 26: 668–671

Gerna G, Battaglia M, Milenesi G, Passarani N, Percivalle E, Cattaneo E (1984a) Serotyping of cell culture-adapted subgroup 2 human rotavirus strains by neutralization. Infect Immun 43: 722–729

Gerna G, Passarani N, Battaglia M, Percivalle E (1984b) Rapid serotyping of human rotavirus strains by solid-phase immune electron microscopy. J Clin Milansi 19: 273–278

Gerna G, Sarasini A, di Matteo A, Passarani N, Gagliardi V, Milanesi G, Astaldi Ricotti GCB, Battaglia M (1988a) The outer capsid glycoprotein VP7 of simian rotavirus SA11 contains two distinct neutralization epitopes. J Gen Virol 69: 937–944

Gerna G, Sarasini A, di Matteo A, Parea M, Orsolini P, Battaglia M (1988b) Identification of two subtypes of serotype 4 human rotavirus by using VP7-specific neutralizing monoclonal antibodies. J Clin Microbiol 26: 1388–1392

Gerna G, Sarasini A, Torsellini M, di Matteo A, Baldanti F, Parea M, Battaglia M (1989) Characterization of rotavirus subgroup-specific monoclonal antibodies and use in single-sandwich ELISA systems for rapid subgrouping of human strains. Arch Virol 107: 315–322

Gerna G, Sarasini A, Torsellini M, Torre D, Parea M, Battaglia M (1990a) Group- and type-specific serologic response in infants and children with primary rotavirus infections and gastroenteritis caused by a strain of known serotype. J Infect Dis 161: 1105–1111

Gerna G, Sarasini A, Arista S, di Matteo A, Giovanelli L, Parea M, Halonen P (1990b) Prevalence of human rotavirus serotypes in some European countries 1981–1988. Scand J Infect Dis 22: 5–10

Gerna G, Sarasini A, Zentilin L, DiMatteo A, Miranda P, Parea M, Battaglia M, Milanesi G (1990c) Isolation in Europe of 69M-like (serotype 8) human rotavirus strains with either subgroup I or II specificity and a long RNA electropherotype. Arch Virol 112: 27–40

Gerna G, Sarasini A, Parea M, Arista S, Miranda P, Brissow H, Hoshino Y, Flores J (1992) Isolation and characterization of two distinct human rotavirus strains with G6 specificity. J Clin Microbiol 30: 9–16

Ghosh SK, Naik TN (1989a) Detection of a large number of subgroup I human rotaviruses with a "long" RNA electropherotype. Arch Virol 105: 119–127

Ghosh SK, Naik TN (1989b) Evidence for a new rotavirus subgroup in India. Epidemiol Infect 102: 523–530

Gomez J, Estes MK, Matson DO, Bellinzoni R, Alvarez A, Gristein S (1990) Serotyping of human rotaviruses in Argentina by ELISA with monoclonal antibodies. Arch Virol 112: 249–259

Gorziglia M, Green K, Nishikawa K, Taniguchi K, Jones R, Kapikian AZ, Chanock RM (1988a) Sequence of the fourth gene of human rotaviruses recovered from asymptomatic and symptomatic infections. J Virol 62: 2978–2984

Gorziglia M, Hoshino Y, Nishikawa K, Maloy WL, Jones RW, Kapikian AZ, Chanock RM (1988b). Comparative sequence analysis of the genomic segment 6 of four rotaviruses each with a different subgroup specificity. J Gen Virol 69: 1659–1669

Gorziglia M, Lanen C, Liprandi F, Esparza J (1985) Biochemical evidence for the oligomeric (possibly trimeric) structure of the major inner capsid polypeptide (45K) of rotavirus. J Gen Virol 66: 1889–1900

Gorziglia M, Nishikawa K, Hoshino Y, Taniguchi K (1990a) Similarity of the outer capsid protein VP4 of the Gottfried strains of porcine rotavirus to that of asymptomatic human rotavirus strains. J Virol 64: 414–418

Gorziglia M, Larralde G, Kapikian AZ, Chanock RM (1990b) Antigenic relationships among human rotaviruses as determined by outer capsid VP4. Proc Natl Acad Sci USA 87: 7155–7159

Gorziglia M, Larralde G, Ward RL (1990c) Neutralization epitopes on rotavirus SA11 4fM outer capsid proteins. J Virol 64: 4534–4539

Gorziglia M, Kapikian AZ (1992) Expression of the OSU rotavirus outer capsid protein VP4 by an adenovirus recombinant. J Virol 66: 4407–4412

Gothefors L, Wadell G, Juto P, Taniguchi K, Kapikian AZ, Glass RI (1989) Prolonged efficacy of rhesus rotavirus vaccine in Swedish children. J Infect Dis 159: 753–757

Gouvea V, Glass RI, Woods P, Taniguchi K, Clark HF, Forrester B, Fang Z-Y (1990a) Polymerase chain reaction amplification and typing of rotavirus nucleic acid from stool specimens. J Clin Microbiol 28: 276–282

Gouvea V, Ho M-H, Glass RI, Woods P, Forester B, Robinson C, Ashley R, Riepenhoff-Talty M, Clark HF, Taniguchi K, Meddix E, McKellar B, Pickering L (1990b) Serotypes and electropherotypes of human rotavirus in the USA: 1987–1989. J Infect Dis 162: 362–367

Gouvea V, Ramirez C, Li B, Santos N, Saif L, Clark HF, Hoshino Y (1993) Restriction endonuclease analysis of the VP7 genes of human and animal rotaviruses. J Clin Microbiol 31: 917–923

Green KY, Midthun K, Gorziglia M, Hoshino Y, Kapikian AZ, Chanock RM, Flores J (1987) Comparison of the amino acid sequences of the major neutralization protein of four human rotavirus serotype. Virology 168: 153–159

Green KY, Sears J, Taniguchi K, Midthun K, Hoshino Y, Gorziglia M, Nishikawa K, Urasawa S, Kapitian AZ, Chanock RM, Flores J (1988) Prediction of human rotavirus serotype by nucleotide sequence analysis of the VP7 protein gene. J Virol 62: 1819–1823

Green KY, Hoshino Y, Ikegami N (1989) Sequence analysis of the gene encoding the serotype-specific glycoprotein (VP7) of two new human rotavirus serotypes. Virology 168: 429–433

Green KY, Taniguchi K, Mackow ER, Kapikian AZ (1990a) Homotypic and heterotypic epitope-specific antibody responses in adult and infant rotavirus vaccines: implications for vaccine development. J Infect Dis 161: 667–679

Green KY, James HDJr, Kapikian AZ (1990b) Evaluation of three panels of monoclonal antibodies for the identification of human rotavirus VP7 serotype by ELISA. Bull WHO 68: 601–610

Green KY, Sarasini A, Qian Y, Gerna G (1992). Genetic variation in rotavirus serotype 4 subtypes. Virology 188: 362–368

Green KY, Kapikian AZ (1992) Identification of VP7 epitopes associated with protection against human rotavirus illness or shedding in volunteers. J Virol 66: 548–553

Greenberg HB, Kalica AR, Wyatt RG, Jones RW, Kapikian AZ, Chanock RM (1981) Rescue of non-cultivatable human rotavirus by gene reassortment during mixed infection with ts mutants of a cultivatable bovine rotavirus. Proc Natl Acad Sci USA 78: 420–424

Greenberg HB, McAuliff V, Valdesuso J, Wyatt R, Flores J, Kalica A, Hoshino Y, Singh N (1982a) Serologic analysis of the subgroup protein of rotavirus using monoclonal antibody. Infect Immun 39: 91–99

Greenberg HB, Wyatt RG, Kapikian AZ, Kalica AR, Flores J, Jones R (1982b) Rescue and serotype characterization of noncultivable human rotavirus by gene reassortment. Infect Immun 37: 104–109

Greenberg HB, Valdesuso J, van Wyke K, Midthun K, Walsh M, McAuliffe V, Wyatt RG, Kalica AR, Flores J, Hoshino Y (1983) Production and preliminary characterization of monoclonal antibodies directed at two surface proteins of rhesus rotavirus. J Virol 47: 267–275

Greenberg HB, Vo PT, Jones R (1986) Cultivation and characterization of three strains of murine rotavirus. J Virol 57: 585–590

Grunert B, Streckert H-J, Liedtke W, Houly C, Mietens C, Werchan H (1987) Development of a monoclonal antibody specific for serotype 3 rotavirus strains. Eur J Clin Microbiol 6: 136–141

Gunasena S, Nakagomi O, Isegawa Y, Kaga E, Nakagomi T, Steele AD, Flores J, Ueda S (1993) Relative frequency of VP4 gene alleles among human rotaviruses recovered over a 10-year period (1982–1991) from Japanese children with diarrhea. J Clin Microbiol 31: 2195–2197

Gunn PR, Sato F, Powell KFH, Bellamy AR, Napier JR, Harding DRK, Hancock WS, Siegman LJ, Both GW (1985) Rotavirus neutralizing protein VP7: Antigenic determinants investigated by sequence analysis and peptide synthesis. J Virol 54: 791–797

Hansen G, Mehnert F, Streckert H-J, Werchau H (1992) Monoclonal antipeptide antibodies recognize epitopes upon VP4 and VP7 of simian rotavirus SA11 in infected MA104 cells. Arch Virol 122: 281–291

Heath R, Birch C, Gust I (1986) Antigenic analysis of rotavirus isolates using monoclonal antibodies specific for human serotypes 1, 2, 3, 4 and SA11. J Gen Virol 67: 2455–2466

Hofer JM, Sato F, Street JE, Bellamy AR (1987) Nucleotide sequence for gene 6 of rotavirus strain S2. Nucleic Acid Res 15: 175

Holmes IH (1983) Rotaviruses. In: Joklik WK (ed) The reoviridae. Plenum, New York, pp 359–423

Horng WJ, Spiezia KS, Mushahwar IK (1989) Preparation and characterization of monoclonal antibodies to rotavirus. J Virol Methods 23: 241–252

Hoshino Y, Wyatt RG, Greenberg HB, Flores J, Kapikian AZ (1984) Serotypic similarity and diversity of rotaviruses of mammalian and avian origin as studied by plaque reduction neutralization. J Infect Dis 149: 694–702

Hoshino Y, Sereno MM, Midthun K, Flores J, Kapikian AZ, Chanock RM (1985a) Independet segregation of two antigenic specificities (VP3 and VP7) involved in neutralization of rotavirus infectivity. Proc Natl Acad Sci USA 82: 8701–8704

Hoshino Y, Wyatt RG, Flores J, Midthun K, Kapikian AZ (1985b) Serotypic characterization of rotaviruses derived from asymptomatic human neonatal infections. J Clin Microbiol 21: 425–430

Hoshino Y, Gorziglia M, Valdesuso J, Askaa J, Glass RI, Kapikian AZ (1987a) An equine rotavirus (FI–14 strain) which bears both subgroup I and subgroup II specificities on its VP6. Virology 157: 488–496

Hoshino Y, Sereno MM, Midthun K, Flores J, Chanock RM, Kapikian AZ (1987b) Analysis by plaque reduction neutralization assay of intertypic rotaviruses suggests that gene reassortment occurs in vivo. J Clin Microbiol 25: 290–294

Hoshino Y, Saif LJ, Sereno MM, Chanock RM, Kapikian AZ (1988) Infection immunity of piglets to either VP3 or VP7 outer capsid protein confers resistance to challenge with a virulent rotavirus bearing the corresponding antigen. J Virol 62: 744–748

Hoshino Y, Nishikawa K, Benfield DA, Gorziglia M (1994) Mapping of antigenic sites involved in serotype-cross-reactive neutralization on group A rotavirus outer capsid glycoprotein VP7. Virology (in press)

Huang J-A, Nagesha HS, Snodgrass DR, Holmes IH (1992) Molecular and serological analyses of two bovine rotaviruses (B-11 and B-60) causing calf scours in Australia. J Clin Microbiol 30: 85–92

Huang J-A, Nagesha HS, Holmes IH (1993) Comparative sequence analysis of VP4s from five Australian porcine rotaviruses: implication of an apparent new P type. Arch Virol 196: 319–327

Hussein HA, Parwani AV, Rosen BI, Luechelli A, Saif LJ (1993) Detection of rotavirus serotypes G1, G2, G3, and G11 in feces of diarrheic calves by using polymerase chain reaction-derived cDNA probes. J Clin Microbiol 31: 2491–2496

Imagawa H, Tanaka T, Sekiguchi K, Fukunaga Y, Anzai T, Minamoto N, Kamada M (1993) Electropherotypes, serotypes, and subgroups of equine rotaviruses isolated in Japan. Arch Virol 131: 169–176

Isegawa Y, Nakagomi O, Nakagomi T, Ueda S (1992) A VP4 sequence highly conserved in human rotavirus strain AU-1 and feline rotavirus strain FRV-1. J Gen Virol 73: 1939–1946

Johnson ME, Paul PS, Gorziglia M, Rosenbusch R (1990) Development of specific nucleic acid probes for the differentiation of porcine rotavirus serotypes. Vet Microbiol 24: 307–326

Kalica AR, Greenberg HB, Espejo RT, Flores J, Wyatt RG, Kapikian AZ, Chanock RM (1981a) Distinctive ribonucleic acid patterns of human rotavirus subgroups 1 and 2. Infect Immun 33: 958–961

Kalica AR, Greenberg HB, Wyatt RG, Flores J, Sereno MM, Kapikian AZ, Chanock RM (1981b) Genes of human (strain Wa) and bovine (strain UK) rotavirus that code for neutralization and subgroup antigens. Virology 112: 385–390

Kalica AR, Flores J, Greenberg HB (1983) Identification of the rotaviral gene that codes for hemagglutination and protease-enhanced plaque formation. Virology 125: 194–205

Kang S-Y, Saif LJ, Miller KL (1989) Reactivity of VP4-specific monoclonal antibodies to a serotype 4 porcine rotavirus with distinct serotypes of human (symptomatic and asymptomatic) and animal rotaviruses. J Clin Microbiol 27: 2744–2750

218   Y. Hoshino and A.Z. Kapikian

Kang S-Y, Saif LJ (1991) Production and characterization of monoclonal antibodies against an avian group A rotavirus. Avian Dis 35: 563–571

Kang SY, Benfield DA, Gorziglia M, Saif LJ (1993) Characterization of the neutralizing epitopes of VP7 of the Gottfried strain of porcine rotavirus. J Clin Microbiol 31: 2291–2297

Kapikian AZ, Chanock RM (1989) Viral gastroenteritis. In: Evans AS (ed) Viral infections of humans, 3rd ed. Plenum, New York, pp–293–340

Kapikian AZ, Yolken RH (1985) Rotavirus. In: Mandell GL, Douglas RG Jr, Bennett JE (eds) Principles and practice of infectious diseases, 2nd ed. Wiley, New York, pp 933–944

Kapikian AZ, Chanock RM (1990) Rotaviruses, In: Fields BN, Knipe M, Chanock RM, Hirsch MS, Melnick JL, Monath TP, Roizman B (eds) Virology, 2nd edn. Raven, New York, pp. 1353–1404

Kapikian AZ, Cline WL, Greenberg HB, Wyatt RG, Kalica AR, Banks CE, James HDJr, Flores J, Chanock RM (1981) Antigenic characterization of human and animal rotaviruses by immune adherence hemagglutination assay (IAHA): evidence for distinctness of IAHA and neutralization antigens. Infect Immun 33: 415–425

Kapikian AZ, Yolken RH, Greenberg HB, Wyatt RG, Kalica AR, Chanock RM, Kim HW (1979) Gastroenteritis viruses. In: Lennet H, Schmidt NJ (eds) Diagnostic procedures for viral, rickettsial and chlamydial infections, 5th ed. American Public Health Association, Washington DC, pp 927–995

Kapikian AZ, Flores J, Green KY, Hoshino Y, Gorziglia M, Nishikawa K, Chanock RM, Pérez-Schael I (1988a) Prospects for development of a rotavirus vaccine against rotavirus diarrhea by a Jennerian and a modified Jennerian strategy. In: Norrby SR, Mills J, Norrby E, Whitton LJ (eds) Frontiers of infectious diseases: new antiviral strategies. Churchill Livingstone, Edinburgh, pp 217–232

Kapikian AZ, Flores J, Midthun K, Hoshino Y, Green KY, Gorziglia M, Taniguchi K, Nishikawa K, Chanock RM, Potash L, Pérez-Schael I, Dolin R, Christy C, Santoshan M, Halsey N, Clements M, Sears J, Black R, Levine M, Losonsky G, Rennels M, Gothesfors L, Wadell G, Glass R, Vesikari T, Anderson E, Belshe R, Wright P, Urasawa S (1988b) Development of a rotavirus vaccine by a "Jennerian" and a modified "Jennerian" approach. In: Chanock RM, Lerner RA, Brown F, Ginsberg H (eds) Vaccines 88: modern approaches to new vaccines including prevention of AIDS. Cold Spring Harbor Laboratory, New York, pp 151–159

Kapikian AZ, Vesikari T, Ruuska T, Madore HP, Christy C, Dolin R, Flores J, Green KY, Davidson BL, Gorziglia M, Hoshino Y, Chanock RM, Midthun K, Perez-Schael I (1992). An update on the "Jennerian" and modified "Jennerian" approach to vaccination of infants and young children against rotavirus diarrhea. Adv Exp Med Biol. In: Ciardi JE et al. (eds.) Genetically Engineered Vaccines. Plenum Press: New York 327: 59–69

Kawamoto H, Tanaka H, Urasawa S, Urasawa T, Taniguchi K (1990) Serotype analysis of group A rotavirus related to acute gastroenteritis in winter in Gifu City. Microbiol Immunol 34: 675–681

Killen HM, Dimmock NJ (1982) Identification of a neutralization-specific antigen of a calf rotavirus. J Gen Virol 62: 297–311

Kim K-H, Yang J-M, Joo S-I, Cho Y-G, Glass RI, Cho Y-J (1990) Importance of rotavirus and adenovirus 40 and 41 in acute gastroenteritis in Korean children. J Clin Microbiol 28: 2279–2284

Kirkwood C, Masendycz PJ, Coulson BS (1993). Characteristics and location of cross-reactive and serotype-specific neutralization sites on VP7 of human G type 9 rotaviruses. Virology 196: 79–88

Kitaoka S, Suzuki H, Numazaki T, Sato T, Konno T, Ebina T, Ishida N, Nakagomi O, Nakagomi T (1984) Hemagglutination by human rotavirus strains. J Med Virol 13: 215–222

Kitaoka S, Fukuhara N, Tazawa F, Suzuki H, Saro T, Konno T, Ebina T, Ishida N (1986) Characterization of monoclonal antibodies against human rotavirus hemagglutinin. J Med Virol 19: 313–323

Kitaoka S, Nakagomi T, Fukuhara N, Hoshino Y, Suzuki H, Nakagomi O, Kapikian AZ, Ebina T, Konno T, Ishida N (1987) Serologic characteristics of a human rotavirus isolate, AU-1, which has a long RNA pattern and subgroup I specificity. J Med Virol 23: 351–357

Knowlton DK, Ward RL (1985) Effect of mutation in immunodominant neutralization epitopes on the antigenicity of rotavirus SA11. J Gen Virol 66: 2375–2381

Kobayashi N, Lintag IC, Urasawa T, Taniguchi K, Saniel MC, Urasawa S (1989) Unusual human rotavirus strains having subgroup I specificity and "long" RNA electropherotype. Arch Virol 109: 11–23

Kobayashi M, Taniguchi K, Urasawa S (1990) Identification of operationally overlapping and independent cross-reactive neutralization regions on human rotavirus VP4. J Gen Virol 71: 2615–2623

Kobayashi N, Taniguchi K, Urasawa S (1991a) Analysis of the newly identified neutralization epitopes on VP7 of human rotavirus serotype 1. J Gen Virol 72: 117–124

Kobayashi N, Taniguchi K, Urasawa S (1991b) Operational overlapping of cross-reactive and serotype-specific neutralization epitopes on VP7 of human rotavirus serotype 3. Arch Virol 117: 73–84

Kobayashi N, Taniguchi K, Urasawa T, Urasawa S (1991c) Analysis of the neutralization epitopes on human rotavirus VP7 recognized by monotype-specific monoclonal antibodies. J Gen Virol 72: 1855–1861

Kobayashi N, Taniguchi K, Urasawa T, Urasawa S (1991d) Preparation and characterization of a neutralizing monoclonal antibody directed to VP4 of rotavirus strain K8 which has unique VP4 neutralization epitopes. Arch Virol 121: 153–162

Kobayashi N, Taniguchi K, Urasawa T, Urasawa S (1993) Reactivity of anti-human rotavirus VP4 neutralizing monoclonal antibodies with animal rotaviruses and with unusual human rotaviruses having different P and G serotypes. Res Virol 144: 201–207

Kohli E, Maurice L, Vautherot JF, Bourgeois C, Bour J, Pothier P (1992) Localization of group-specific epitopes on the major capsid protein of group A rotavirus. J Gen Virol 73: 907–914

Kohli E, Maurice L, Bourgeois C, Bour JB, Pothier P (1993). Epitope mapping of the major inner capsid protein of group A rotavirus using peptide synthesis. Virology 194: 110–116

Kool DA, Holmes IH (1993) The avian rotavirus Ty-1 VP7 nucleotide and deduced amino acid sequences differ significantly from those of Ch-2 rotavirus. Arch Virol 129: 227–234

Kutsuzawa T, Konno T, Suzuki H, Ebina T, Ishida N (1982a) Two distinct electrophoretic migration patterns of RNA segments of human rotaviruses prevalent in Japan in relation to their serotypes. Microbiol Immun 26: 271–273

Kutsuzawa T, Konno T, Susuki H, Kapikian AZ, Ebina I, Ishida N (1982b) Isolation of human rotavirus subgroups 1 and 2 in cell culture. J Clin Microbiol 16: 727–730

Lambert J-P, Marbehant P, Marissens D, Zissis G (1984) Monoclonal antibodies directed against different antigenic determinants of rotavirus. J Virol 51: 47–51

Larralde G, Flores J (1990) Identification of gene 4 alleles among human rotaviruses by polymerase chain reaction-derived probes. Virology 179: 469–473

Larralde G, Li B, Kapikian AZ, Gorziglia M (1991) Serotype-specific epitope(s) present on the VP8 subunit of rotavirus VP4 protein. J Virol 65: 3213–3218

Larralde G, Gorziglia M (1992) Distribution of conserved and serotype-specific epitopes on the VP8 subunit of rotavirus VP4. J Virol 66: 7438–7443

Lazdins I, Sonza S, Dyall-Smith ML, Coulson BS, Holmes IH (1985) Demonstration of an immunodominant neutralization site by analysis of antigenic variants of SA11 rotavirus. J Virol 56: 317–319

Li B, Clark HF, Gouvea V (1993a) Nucleotide sequence of the VP4-encoding gene of an unusual human rotavirus (HCR3). Virology 196: 825–830

Li B, Larralde G, Gorziglia M (1993b) Human rotavirus K8 strain represents a new VP4 serotype. J Virol 67: 617–620

Lin M, Imai M, Bellamy AR, Ikegami N, Furuichi Y, Simmons D, Nuss DL, Diebel R (1985) Diagnosis of rotavirus infection with cloned cDNA copies of viral genome segments. J Virol 55: 509–512

Lin M, Imai M, Bellamy AR, Ikegami N, Furuichi Y, Simmons D, Nuss DL, Diebel R (1985) Diagnosis of rotavirus infection with cloned cDNA copies of viral genome segments. J Virol 55: 509–512

Lin M, Imai M, Ikegami N, Bellamy AR, Summers D, Nuss DL, Deibel R, Furuichi Y (1987) cDNA probes of individual genes of human rotavirus distinguish viral subgroups and serotypes. J Virol Methods 15: 285–289

Linhares AC, Gabbay YB, Mascarenhas JDP, Freitas RB, Flewett TH, Beards GM (1988) Epidemiology of rotavirus subgroups and serotypes in Belem, Brazil: a three-year study. Ann Inst Pasteur/Virol 139: 89–99

Liprandi F, Lopez G, Rodriquez I, Hidalgo M, Lundert JE, Mattion N (1990) Monoclonal antibodies to the VP6 of porcine subgroup I rotaviruses reactive with subgroup I and non-subgroup I non-subgroup II strains. J Gen Virol 71: 1395–1398

Liprandi F, Rodriquez I, Pina C, Larralde G, Gorziglia M (1991) VP4 monotype specificities among porcine rotavirus strains of the same VP4 serotype. J Virol 65: 1658–1661

Lizano M, Lopez S, Arias C (1991) The amino-terminal half of rotavirus SA114fm VP4 protein contains a hemagglutination domain and primes for neutralizing antibodies to the virus. J Virol 65: 1383–1391

Lopez S, Arias CF (1993a) Sequence analysis of rotavirus YM VP6 and NS28 proteins. J Gen Virol 74: 1223–1226

Lopez S, Greenberg HB, Arias CF (1993b) Mapping the subgroup epitopes of rotaviruses. In Abstracts of IXth International Congress of Virology, Glasgow, Scotland, p 87

Mackow ER, Shaw RD, Matsui SM, Vo PT, Benfield DA, Greenberg HB (1988a) Characterization of homotypic and heterotypic VP7 neutralization sites of rhesus rotavirus. Virology 165: 511–517

Mackow ER, Shaw RD, Matsui SM, Vo PT, Dang M-N, Greenberg HB (1988b) The rhesus rotavirus gene encoding protein VP3: location of amino acids involved in homologous and heterologous rotavirus neutralization and identification of a putative fusion region. Proc Natl Acad Sci USA 85: 645–649

Mackow ER, Barnett JW, Chan H, Greenberg HB (1989) The rhesus rotavirus outer capsid protein VP4 functions as a hemagglutinin and is antigenically conserved when expressed by a baculovirus recombinant. J Virol 63: 1661–1668

Mackow ER, Vo PT, Broome R, Bass D, Greenberg HB (1990a) Immunization with baculovirus-expressed VP4 protein passively protects against simian and murine rotavirus challenge. J Virol 64: 1698–1703

Mackow ER, Yamanaka MY, Dang MN, Greenberg HB (1990b) DNA amplification restricted transcription-translation: rapid analysis of rhesus rotavirus neutralization sites. Proc Natl Acad Sci USA 87: 518–522

Mascarenhas JDP, Linhares AC, Gabbay YB, de Freitas RB, Mendes E, Lopes S, Arias CF (1989) Naturally occurring serotype 2/subgroup II rotavirus reassortants in Northern Brazil. Virus Res 14: 235–240

Matson DO, Estes MK, Burnes JW, Greenberg HB, Taniguchi K, Urasawa S (1990) Serotype variation of human group A rotaviruses in two regions of the USA. J Infect Dis 162: 605–614

Matson DO, O'Ryan ML, Pickering LK, Chiba S, Nakata S, Raj P, Estes MK (1992) Characterization of serum antibody responses to natural rotavirus infections in children by VP7-specific epitope-blocking assay. J Clin Microbiol 30: 1056–1061

Matsuda Y, Nakagomi O, Offit PA (1990) Presence of three P types (VP4 serotypes) and two G types (VP7 serotypes) among bovine rotavirus strains. Arch Virol 115: 199–207

Matsui SM, Mackow ER, Greenberg HB (1989a) Molecular determinant of rotavirus neutralization and protection. Adv Virus Res 36: 181–214

Matsui SM, Offit PA, Vo PT, Mackow ER, Benfield DA, Shaw RD, Padilla-Noriega L, Greenberg HB (1989b) Passive protection against rotavirus-induced diarrhea by monoclonal antibodies to the heterotypic neutralization domain of VP7 and the VP8 fragment of VP4. J Clin Microbial 27: 780–782

Matsuno S, Hasegawa A, Mukoyama A, Inouye S (1985) A candidate for a new serotype of human rotavirus. J Virol 54: 623–624

Matsuno S, Mukoyama A, Hasegawa A, Taniguchi K, Inouye S (1988) Characterization of a human rotavirus strain which is possibly a naturally-occurring reassortant virus. Virus Res 10: 167–175

McCrae MA, McCorquodale (1987) Expression of a major bovine rotavirus neutralization antigen (VP7c) in Escherichia coli. Gene 55: 9–18

Mebus CA, Underdahl NR, Rhodes MB, Twiehaus MJ (1969) Calf diarrhea (scours): reproduced with a virus from a field outbreak. Univ Nebraska Res Bull 233: 1–16

Midthun K, Flores J, Taniguchi K, Urasawa S, Kapikian AZ, Chanock RM (1987) Genetic relatedness among human rotavirus genes coding for VP7, a major neutralization protein, and its application to serotype identification. J Clin Microbiol 25: 1269–1274

Midthun K, Valdesuso J, Kapikian AZ, Hoshino Y, Green KY (1989) Identification of serotype 9 human rotavirus by enzyme-linked immunosorbent assay with monoclonal antibodies. J Clin Microbiol 27: 2112–2114

Minamoto N, Oki K, Tomita M, Kinjo T, Suzuki Y (1988). Isolation and characterization of rotavirus from fecal pigeon in mammalian cell cultures. Epidemiol Infect 100: 481–492

Minamoto N, Sugimoto O, Yokota M, Tomita M, Goto H, Sugiyama M, Kinjo T (1993) Antigenic analysis of avian rotavirus VP6 using monoclonal antibodies. Arch Virol 131: 293–305

Mochizuki M, Sameshima R, Ata M, Minami K, Okabayashi K, Harasawa R (1985) Characterization of canine rotavirus RS15 strain and comparison with other rotaviruses. Jpn J Vet Sci 47: 531–538

Mochizuki M, Yamakawa M (1987) Detection of rotavirus in cat feces. Jpn J Vet Sci 49: 159–160

Morita Y, Taniguchi K, Urasawa T, Urasawa S (1988) Analysis of serotype-specific neutralization epitopes on VP7 of human rotavirus by the use of neutralizing monoclonal antibodies and antigenic variants. J Gen Virol 69: 451–458

Murakami Y, Nishioka N, Hashiguchi Y, Kuniyasu C (1983) Serotype of bovine rotaviruses distinguished by serum neutralization. Infect Immun 40: 851–855

Nagesha HS, Holmes IH (1988) New porcine rotavirus serotype antigenically related to human rotavirus serotype 3. J Clin Microbiol 26: 171–174

Nagesha HS, Brown LE, Holmes IH (1989) Neutralizing monoclonal antibodies against three serotypes of porcine rotavirus. J Virol 63: 3545–3549

Nagesha HS, Huang J, Hum CP, Holmes IH (1990) A porcine rotavirus strain with dual VP7 serotype specificity. Virology 175: 319–322

Nagesha HS, Holmes IH (1991a) Direct serotyping of porcine rotaviruses using VP7-specific monoclonal antibodies by an enzyme immunoassay. J Med Virol 35: 206–211

Nagesha HS, Holmes IH (1991b) VP4 relationships between porcine and other rotavirus serotypes. Arch Virol 116: 107–108

Negesha HS, Huang J, Holmes IH (1992) A variant serotype G3 rotavirus isolated from an unusually severe outbreak of diarrhea in piglets. J Med Virol 38: 79–85

Naguib T, Wyatt RG, Mohieldin MS, Taki AM, Imam IT, Dupont HL (1984) Cultivation and subgroup determination of human rotaviruses from Egyptian infants and young children. J Clin Microbiol 19: 210–212

Nakagomi O, Kutsuzawa-Nakagomi T, Oyamada H, Suto T, Ochi A (1984) Enzymeimmunoassay for subgrouping human rotaviruses using monoclonal antibodies. Tohoku J Exp Med 144: 105–106

Nakagomi O, Nakagomi T, Oyamada H, Suto T (1985) Relative frequency of human rotavirus subgroups 1 and 2 in Japanese children with acute gastroenteritis. J Med Virol 17: 29–34

Nakagomi O, Nakagomi T, Hoshino Y, Flores J, Kapikian AZ (1987) Genetic analysis of a human rotavirus that belongs to subgroup I but has an RNA pattern typical of subgroup II human rotaviruses. J Clin Microbiol 25: 1159–1164

Nakagomi T, Katsushima N, Nakagomi O (1988a) Relative frequency of human rotavirus subgroups I and II in relation to "short" and "long" electropherotypes of viral RNA. Ann Inst Pasteur/Virol 139: 295–300

Nakagomi T, Akatani K, Ikegami N, Katsushima N, Nakagami O (1988b) Occurrence of changes in human rotavirus serotypes with concurrent changes in genomic RNA electropherotypes. J Clin Microbiol 26: 2586–2592

Nakagomi T, Matsuda Y, Oshima A, Mochizuki M, Nakagomi O (1989) Characterization of a canine rotavirus strain by neutralization and molecular hybridization assays. Arch Virol 106: 145–150

Nakagomi O, Nakagomi T, Akatani K, Ikegami N, Katsushima N (1990) Relative frequency of rotavirus serotypes in Yamagata, Japan, over four consecutive rotavirus seasons. Res Virol 141: 459–463

Nakagomi O, Oyamada H, Nakagomi T (1991) Experience with serotyping rotavirus strains by reverse transcription and two-step polymerase chain reaction with generic and type-specific primers. Mol Cell Probes 5: 285–289

Nakagomi O, Mochizuki M, Aboudy Y, Shif I, Silberstein I, Nakagomi T (1992) Hemagglutination by a human rotavirus isolate as evidence for transmission of animal rotaviruses to humans. J Clin Microbiol 30: 1011–1013

Nishikawa K, Taniguchi K, Torres Y, Hoshino Y, Green K, Kapikian AZ, Chanock RM, Gorziglia M (1988) Comparative analysis of the VP3 gene of divergent strains of rotaviruses simian SA11 and bovine Nebraska calf diarrhea virus. J Virol 62: 4022–4026

Nishikawa K, Hoshino Y, Taniguchi K, Green KY, Greenberg HB, Kapikian AZ, Chanock RM, Gorziglia M (1989a) Rotavirus VP7 neutralization epitopes of serotype 3 strains. Virology 171: 503–515

Nishikawa K, Fukuhara N, Liprandi F, Green K, Kapikian AZ, Chanock RM, Gorziglia M (1989b) VP4 protein of porcine rotavirus strain OSU expressed by a baculovirus recombinant induces neutralizing antibodies. Virology 173: 631–637

Nishikawa K, Hoshino Y, Gorziglia M (1991) Sequence of the VP7 gene of chicken rotavirus Ch2 strain of serotype 7 rotavirus. Virology 185: 853–856

Noel JS, Beards GM, Cubitt WD (1991) Epidemiological survey of human rotavirus serotypes and electropherotypes in young children admitted to two children's hospitals in northeast London from 1984 to 1990. J Clin Microbiol 29: 2213–2219

Offit PA, Blavat G (1986) Identification of the two rotavirus genes determining neutralization specificities. J Virol 57: 376–378

Offit PA, Clark HF, Blavat G, Greenberg HB (1986a) Reassortant rotaviruses containing structural proteins VP3 and VP7 from different parents induce antibodies protective against each parental serotype. J Virol 60: 491–496

Offit PA, Shaw RD, Greenberg HB (1986b) Passive protection against rotavirus-induced diarrhea by monoclonal antibodies to surface proteins VP3 and VP7. J Virol 58: 700–703

Olive DM, Sethi SK (1989) Detection of human rotavirus by using an alkaline phosphatase-conjugated DNA probe in comparison with enzyme-linked immunoassay and polyacrylamide gel analysis. J Clin Microbiol 27: 53–57

Padilla-Noriega L, Arias C, Lopez S, Puerto F, Snodgrass DR, Taniguchi K, Greenberg HB (1990) Diversity of rotavirus serotypes in Mexican infants with gastroenteritis. J Clin Microbiol 28: 1114–1119

Padilla-Noriega LP, Werner-Eckert R, Mackow ER, Gorziglia M, Larralde G, Taniguchi K, Greenberg HB (1993) Serologic analysis of human rotavirus serotypes P1A and P2 using monoclonal antibodies. J Clin Microbiol 31: 622–628

Parwani AV, Rosen BI, Flores J, McCrae MA, Gorziglia M, Saif LJ (1992a) Detection and differentiation of bovine group A rotavirus serotypes using polymerase chain reaction-generated probes to the VP7 gene. J Vet Diagn Invest 4: 148–158

Parwani AV, Rosen BI, McCrae MA, Saif LJ (1992b) Development of cDNA probes for typing group A rotaviruses on the basis of VP4 specificity. J Clin Microbiol 30: 2717–2721

Paul PS, Lyoo YS, Andrews JJ, Hill HT (1988) Isolation of two new serotypes of porcine rotavirus from pigs with diarrhea. Arch Virol 100: 139–143

Pedley S, McCrae MA (1984) A rapid screening assay for detecting individual RNA species in field isolates of rotaviruses. J Virol Methods 9: 173–181

Perez-Schael I, Garcia D, Gonzalez M, Gonzalez R, Daoud N, Perez M, Cunto W, Kapikian AZ, Flores J (1990) Prospective study of diarrheal diseases in Venezuelan children to evaluate the efficacy of rhesus rotavirus vaccine. J Med Virol 30: 212–229

Pipittajan P, Kassempimolporn S, Ikegami N, Akatani K, Wasi C, Sinarachatanant P (1991) Molecular epidemiology of rotavirus associated with pediatric diarrhea in Bangkok, Thailand. J Clin Microbiol 29: 617–624

Poncet D, Corthier G, Charpilienne A, Cohen J (1990) A recombinant vaccinia virus expressing the major capsid protein of simian rotavirus induced anti-rotavirus antibodies. Virus Res 15: 267–274

Pongsuwanna Y, Taniguchi K, Choonthanom M, Chiwakul M, Susansook T, Saguanwongse S, Jayavasu C, Urasawa S (1989) Subgroup and serotype distributions of human, bovine, and porcine rotavirus in Thailand. J Clin Microbiol 27: 1956–1960

Pongsuwanna Y, Taniguchi K, Choonthanom M, Chiwakul M, Jayavasu C, Snodgrass DR, Urasawa S (1990) Serological and genetic characterization of bovine rotaviruses in Thailand by ELISA and RNA–RNA hybridization: detection of numerous non-serotype 6 strains. South Asian J Trop Med Pub Health 21: 607–613

Pothier P, Kohli E, Drouet E, Ghims S (1987) Analysis of the antigenic sites on the major inner capsid protein (VP6) of rotavirus with monoclonal antibodies. Ann Inst Pasteur/Virol 138: 285–295

Prasad BV, Wang GJ, Clerx JPM, Chiu W (1988) Three-dimensional structure of rotavirus. J Mol Biol 199: 269–275

Prasad BV, Burns JW, Marietta E, Estes MK, Chiu W (1990) Localization of VP4 neutralization sites in rotavirus by three-dimentional cryo-electron microscopy. Nature 343: 476–479

Qian Y, Green KY (1991) Human rotavirus strain 69M has a unique VP4 as determined by amino acid sequence analysis. Virology 182: 407–412

Raj P, Matson DO, Coulson BS, Bishop RF, Taniguchi K, Urasawa S, Greenberg HB, Estes MK (1992) Comparison of rotavirus VP7-typing monoclonal antibodies by competition binding assay. J Clin Microbiol 30: 704–711

Ramig RF, Ward RL (1991) Genomic segment reassortment in rotaviruses and other reoviridae. Adv Virus Res 39: 164–199

Rasool NBG, Green KY, Kapikian AZ (1993) Serotype analysis of rotaviruses from different locations in Malaysia. J Clin Microbiol 31: 1815–1819

Reddy DA, Bergmann CC, Meyer JC, Berriman J, Both JW, Coupar BH, Boyle DB, Andrew ME, Bellamy AR (1992) Rotavirus VP6 modified for expression on the plasma membrane forms arrays and exhibits enhanced immunogenicity. Virology 189: 423–434

Rohwedder A, Werchau H (1990) Serotyping of rotaviruses with PCR amplified cDNA probes derived from gene segment 9. Abstracts of the 8th international congress on virology, Berlin, p 25-006

Rohwedder A, Irmak H, Werchau H, Brussow H (1993) Nucleotide sequence of gene 6 of avian-like group A rotavirus 993/83. Virology 195: 820–825

Rosen BI, Saif LJ, Jackwood DJ, Gorziglia M (1990) Serotypic differentiation of group A rotaviruses with porcine rotavirus gene 9 probes. J Clin Microbiol 28: 2526–2533

Rosen BI, Parwani AV, Gorziglia M, Larralde G, Saif LJ (1992) Characterization of full-length and polymerase chain reaction-derived partial-length Gottfried and OSU gene 4 probes for serotypic differentiation of porcine rotaviruses. J Clin Microbiol 30: 2644–2652

Roseto A, Scherrer R, Cohen J, Guillemin MC, Charpilienne A, Feynerol C, Peries J (1983) Isolation and characterization of anti-rotavirus immunoglobulins secreted by cloned hybridoma cell lines. J Gen Virol 64: 237–240

Ruggeri FM, Greenberg HB (1991) Antibodies to trypsin cleavage peptide VP8 neutralize rotavirus by inhibiting binding of virions to target cells in culture. J Virol 65: 2211–2219

Ruiz AM, López IV, López S, Espejo RT, Arias CF (1988) Molecular and antigenic characterization of porcine rotavirus YM, a possible new rotavirus serotype. J Virol 62: 4331–4336

Sabara M, Gilchrist JE, Hudson GR, Babink LA (1985a) Preliminary characterization of an epitope involved in neutralization and cell attachment that is located on the major bovine rotavirus glycoprotein. J Virol 53: 58–66

Sabara M, Barrington A, Babink LA (1985b) Immunogenicity of a bovine rotavirus glycoprotein fragment. J Virol 56: 1037–1040

Sabara M, Ready KFM, Frenchick PJ, Babiuk LA (1987) Biochemical evidence for the oligomeric aggangement of bovine rotavirus nucleocapsid protein and its possible significance in the immunogenicity of this protein. J Gen Virol 68: 123–133

Santosham M, Letson GW, Wolff M, Reid R, Gahagan S, Adams R, Callahan C, Sack RB, Kapikian AZ. A field study of the safety and efficacy of two candidate rotavirus vaccines in a native American population. J Infect Dis (in press)

Sato K, Inaba Y, Shinozaki T, Fujii R, Matumoto M (1981) Isolation of human rotavirus in cell cultures. Arch Virol 69: 155–160

Sato K, Inaba Y, Miura Y, Tokuhisa S, Matumoto M (1982a) Antigenic relationships between rotaviruses from different species as studied by neutralization and immunofluorescence. Arch Virol 73: 45–50

Sato K, Inaba Y, Miura Y, Tokuhisa S, Matumoto M (1982b) Isolation of lapine rotavirus in cell cultures. Arch Virol 71: 267–271

Sereno MM, Gorziglia MI (1993) The outer capsid protein VP4 of murine rotavirus strain Eb represents a tentative new P type. Virology (in press)

Sethabutr O, Unicomb LE, Holmes IH, Taylor DN, Bishop RF, Echeverria P (1990) Serotyping of human group A rotavirus with oligonucleotide probes. J Infect Dis 162: 368–372

Sethabutr O, Hanchalay S, Lexonboon U, Bishop RF, Holmes IH, Echeverria P (1992) Typing of human group A rotavirus with alkaline phosphatase-labeled oligonucleotide probes. J Med Virol 37: 192–196

Shahid NS, Banu NN, Bingnan F, Tzipori SR, Unicomb LE (1991) Rotavirus infection detected in enonates from hospitals in urban Bangladesh. Arch Virol 119: 135–140

Shaw RD, Stoner-Ma DL, Estes MK, Greenberg HB (1985) Specific enzyme-linked immunoassay for rotavirus serotypes 1 and 3. J Clin Microbiol 22: 286–291

Shaw RD, Vo PT, Offit PA, Coulson BS, Greenberg HB (1986) Antigenic mapping of the surface proteins of rhesus rotavirus. Virology 155: 434–451

Shaw RD, Fong KJ, Losonsky GA, Levine MM, Maldonado Y, Yolken R, Flores J, Kapikian AZ, Vo PT, Greenberg HB (1987) Epitope-specific immune responses to rotavirus vaccination. Gastroenterol 93: 941–950

Shaw RD, Mackow ER, Dyall-Smith ML, Lazdins I, Holmes IH, Greenberg HB (1988) Serotypic analysis of VP3 and VP7 neutralization escape mutants of rhesus rotavirus. J Virol 62: 3509–3512

Shen S, Burke B, Desselberger U (1993) Nucleotide sequences of the VP4 and VP7 genes of a Chinese lamb rotavirus: evidence for a new P type in a G10 type virus. Virology 197: 497–500

Singh N, Sereno MM, Flores J, Kapikian AZ (1983) Monoclonal antibodies to subgroup I rotavirus. Infect Immun 42: 835–837

Snodgrass DR, Ojeh CK, Campbell I, Herring AJ (1984) Bovine rotavirus serotypes and their significance for immunization. J Clin Microbiol 20: 342–346

Snodgrass DR, Fitzgerald T, Campbell I, Scott FMM, Browning GF, Miller DL, Herring AJ, Greenberg HB (1990) Rotavirus serotypes 6 and 10 predominate in cattle. J Clin Microbiol 28: 504–507

Snodgrass DR, Fitzgerald TA, Campbell I, Browning GF, Scott FMM, Hoshino Y, Davies RC (1991) Homotypic and heterotypic serological responses to rotavirus neutralization epitopes in immunologically naive and experienced animals. J Clin Microbiol 29: 2668–2672

Snodgrass DR, Hoshino Y, Fitzgerald TA, Smith M, Browning GF, Gorziglia M (1992) Identification of four VP4 serological types (P serotypes) of bovine rotavirus using viral reassortants. J Gen Virol 73: 2319–2325

Sonza S, Breschkin AM, Holmes IH (1983) Derivation of neutralizing monoclonal antibodies against rotavirus. J Virol 45: 1143–1146

Sonza S, Breschkin AM, Holmes IH (1984) The major surface glycoprotein of simian rotavirus (SA11) contains distinct epitopes. Virology 134: 318–327

Steel HM, Garnham S, Beards GM, Brown DWG (1992) Investigation of an outbreak of rotavirus infection in geriatric patients by serotyping and polyacrylamide gel electrophoresis (PAGE). J Med Virol 37: 132–136

Steele AD, Alexander JJ (1988) The relative frequency of subgroups I and II rotaviruses in black infants in South Africa. J Med Virol 24: 321–327

Steele AD, Garcia D, Sears J, Gerna G, Nakagomi O, Flores J (1993) Distribution of VP4 gene alleles in human rotaviruses by using probes to the hyperdivergent region of the VP4 gene. J Clin Microbiol 31: 1735–1740

Streckert H-J, Grunert B, Werchau H (1986) Antibodies specific for the carboxy-terminal region of the major surface glycoprotein of simian rotavirus (SA11) and human rotavirus (Wa). J Cell Biochem 30: 41–49

Streckert H-J, Brussow H, Werchau H (1988) A synthetic peptide corresponding to the cleavage region of VP3 from rotavirus SA11 induces neutralizing antibodies. J Virol 62: 4265–4269

Streckert H-J, Werchau H (1992) Monoclonal antipeptide antibodies recognizes epitopes upon VP4 and VP7 of simian rotavirus SA11 in infected MA104 cells. Arch Virol 122: 281–291

Suzuki H, Konno T, Numazaki Y, Kitaoka S, Sato T, Imai A, Tazawa F, Nakagomi T, Nakagomi O, Ishida N (1984) Three different serotypes of human rotavirus determined using an interference test with coxsackievirus B. J Med Virol 13: 41–44

Svensson L, Grahnquist L, Pattersson C-A, Grandien M, Stintzing G, Greenberg HB (1988) Detection of human rotaviruses which do not react with subgroup I- and II-specific monoclonal antibodies. J Clin Microbiol 26: 1238–1240

Svensson L, Padilla-Noriega L, Taniguchi K, Greenberg HB (1990) Lack of cosegregation of the subgroup II antigens on genes 2 and 6 in porcine rotaviruses. J Virol 64: 411–413

Tanaka TN, Conner ME, Graham DY, Estes MK (1988) Molecular characterization of three rabit rotavirus strains. Arch Virol 98: 253–265

Taniguchi K, Urasawa S, Yasuhara T (1984) Production of subgroup specific-monoclonal antibodies against human rotaviruses and their application to an enzyme-linked immunosorbent assay for subgroup determination. J Med Virol 14: 115–125

Taniguchi K, Urasawa S, Urasawa T (1985) Preparation and characterization of neutralizing monoclonal antibodies with different reactivity patterns to human rotaviruses. J Gen Virol 66: 1045–1053

Taniguchi K, Urasawa T, Urasawa S (1986) Reactivity patterns to human rotavirus strains of a monoclonal antibody against VP2, a component of the inner capsid of rotavirus. Arch Virol 87: 135–141

Taniguchi K, Urasawa T, Morita Y, Greenberg HB, Urasawa S (1987a) Direct serotyping of human rotavirus in stools by enzyme-linked immunosorbent assay using serotype 1-, 2-, 3- and 4-specific monoclonal antibodies to VP7. J Infect Dis 155: 1159–1166

Taniguchi K, Morita Y, Urasawa T, Urasawa S (1987b) Cross-reactive neutralization epitopes on VP3 of human rotavirus. Analysis with monoclonal antibodies and antigenic variants. J Virol 61: 1726–1730

Taniguchi K, Hoshino Y, Nishikawa K, Green KY, Maloy WL, Morita Y, Urasawa S, Kapikian AZ, Chanock RM, Gorziglia M (1988a) Cross-reactive and serotype-specific neutralization epitopes on VP7 of human rotaviruses: nucleotide sequence analysis of antigenic mutants selected with monoclonal antibodies. J Virol 62: 1870–1874

Taniguchi K, Maloy WL, Nishikawa K, Green KY, Hoshino Y, Urasawa S, Kapikian AZ, Chanock RM, Gorziglia M (1988b) Identification of cross-reactive and serotype 2-specific neutralization epitopes on VP3 of human rotaviruses. J Virol 62: 2421–2426

Taniguchi K, Nishikawa K, Urasawa T, Urasawa S, Midthun K, Kapikian AZ, Gorziglia M (1989) Complete nucleotide sequence of the gene encoding VP4 of a human rotavirus (strain K8) which has unique VP4 neutralization epitopes. J Virol 63: 4101–4106

Taniguchi K, Urasawa T, Pongsuwanna Y, Choonthanoum M, Jayavasu C, Urasawa S (1991a) Molecular and antigenic analyses of serotype 8 and 10 of bovine rotaviruses in Thailand. J Gen Virol 72: 2929–2937

Taniguchi K, Urasawa T, Kobayashi N, Ahmed M, Adachi N, Chiba S, Urasawa S (1991b) Antibody response to serotype-specific and cross-reactive neutralization epitopes on VP4 and VP7 after rotavirus infection or vaccination. J Clin Microbiol 29: 483–487

Theil KW, McCloskey CM (1989a) Molecular epidemiology and subgroup determination of bovine group A rotaviruses associated with diarrhea in dairy and beef calves. J Clin Microbiol 27: 126–131

Theil KW, McCloskey CM (1989b) Nonreactivity of American avian group A rotaviruses with subgroup-specific monoclonal antibodies. J Clin Microbiol 27: 2846–2848

Thouless ME, DiGiacomo RF, Neuman DS (1986) Isolation of two lapine rotaviruses: characterization of their subgroup, serotype and RNA electropherotypes. Arch Virol 89: 161–170

Tufvesson B (1983) Detection of a human rotavirus strain different from types 1 and 2 – a new subgroup? Epidemiology of subgroups in a Swedish and an Ethiopian community. J Med Virol 12: 111–117

Tursi JM, Albert MJ, Bishop RF (1987) Production and characterization of neutralizing monoclonal antibody to a human rotavirus strain with a "super short" RNA pattern. J Clin Microbiol 25: 2426–2427

Unicomb LE, Bishop RF (1989) Epidemiology of rotavirus strains infecting children throughout Australia during 1986–1987. A study of serotype and RNA electropherotype. Arch Virol 106: 23–34

Unicomb LE, Coulson BS, Bishop RF (1989) Experience with an enzyme immunoassay for serotyping human group A rotaviruses. J Clin Microbiol 27: 586–588

Urasawa S, Urasawa T, Taniguchi K (1982) Three human serotypes demonstrated by plaque neutralization of isolated strains. Infect Immun 38: 781–784

Urasawa S, Urasawa T, Taniguchi K (1986) Genetic reassortment between two human rotaviruses having different serotype and subgroup specificities. J Gen Virol 67: 1551–1559

Urasawa S, Urasawa T, Taniguchi K, Morita Y, Sakurada N, Saeki Y, Morita O, Hasegawa S (1988) Validity of an enzyme-linked immunosorbent assay with serotype-specific monoclonal antibodies for serotyping human rotavirus in stool specimens. Microbiol Immunol 32: 699–708

Urasawa S, Urasawa T, Taniguchi K, Wakasugi F, Kabayashi N, Chiba S, Sakurada N, Morita M, Moriya O, Tokieda M, Kawamoto H, Minekawa Y, Ohseto M (1989) Survey of human rotavirus serotypes in different locales in Japan by enzyme-linked immunosorbent assay with monoclonal antibodies. J Infect Dis 160: 44–51

Urasawa S, Urasawa T, Wakasugi F, Kobayashi N, Taniguchi K, Lintag IC, Saniel MC, Goto H (1990a) Presumptive seventh serotype of human rotavirus. Arch Virol 113: 279–282

Urasawa S, Hasegawa A, Urasawa T, Taniguchi K, Wakazugi F, Suzuki H, Inouye S, Pongprot B, Supawadee J, Suprasert S, Rangsiyanond P, Ponusin S, Yamazi Y (1992) Antigenic and genetic analyses of human rotaviruses in Chiang Mai, Thailand: evidence for a close relationship between human and animal rotaviruses. J Infect Dis 166: 227–234

Urasawa T, Urasawa S, Taniguchi K (1981) Sequential passages of human rotavirus in MA104 cells. Microbiol Immunol 25: 1025–1035

Urasawa T, Taniguchi K, Kobayashi N, Wakasugi F, Oishi I, Minekawa Y, Oseto M, Ahmed MU, Urasawa S (1990b) Antigenic and genetic analysis of human rotavirus with dual subgroup specificity. J Clin Microbiol 28: 2837–2841

Urasawa T, Taniguchi K, Kobayashi N, Mise K, Hasegawa A, Yamazi Y, Urasawa S (1993) Nucleotide sequence of VP4 and VP7 genes of a unique human rotavirus strain Mc35 with subgroup I and serotype 10 specificity. Virology 195: 766–771

Ushijima H, Koike H, Mukoyama A, Hasegawa A, Nishimura S, Gentsch J (1992) Detection and serotyping of rotaviruses in stool specimens by using reverse transcription and polymerase chain reaction amplification. J Med Virol 38: 292–297

Velazquez FR, Calva JJ, Guerrero ML, Mass D, Glass RI, Pickering LK, Ruiz-Palacios GM (1993) Cohort study of rotavirus serotype patterns in symptomatic and asymptomatic infections in Mexican children. Pediatr Infect Dis J 12: 54–61

Vesikari T, Rautanen T, Varis T, Beards GM, Kapikian AZ (1990) Clinical trial of rhesus rotavirus candidate vaccine (strain MMU18006) in children vaccinated between 2 and 5 months of age. Am J Dis Child 144: 285–289

Ward RL, Knowlton DR, Schiff GM, Hoshino Y, Greenberg HB (1988) Relative concentrations of serum neutralizing antibody to VP3 and VP7 proteins in adults infected with a human rotavirus. J Virol 62: 1543–1549

Ward RL, Knowlton DR, Greenberg HB, Schiff GM, Bernstein DI (1990) Serum-neutralizing antibody to VP4 and VP7 proteins in infants following vaccination with WC3 bovine rotavirus. J Virol 64: 2687–2691

Ward RL, McNeal MM, Clemens JD, Sack DA, Rao M, Huda N, Green KY, Kapikian AZ, Coulson BS, Bishop RF, Greenberg HB, Gerna G, Schiff GM (1991). Reactivities of serotyping monoclonal antibodies with culture-adapted human rotaviruses. J Clin Microbiol 29: 449–456.

Ward RL, McNeal MM, Sander DS, Greenberg HB, Bernstein DI (1993). Immunodominance of the VP4 neutralization protein of rotavirus in protective natural infection of young children. J Virol 67: 464–468

White L, Perez I, Perez M, Urbina B, Greenberg H, Kapikian A, Flores J (1984) Relative frequency of rotavirus subgroups 1 and 2 in Venezuelan children with gastroenteritis as assayed with monoclonal antibodies. J Clin Microbiol 19: 516–520

White L, Garcia D, Boher Y, Blanco M, Perez M, Romer H, Flores J, Perez-Schael I (1991) Temporal distribution of human rotavirus serotypes 1, 2, 3, and 4 in Venezuelan children with gastroenteritis during 1979–1989. J Med Virol 34: 79–84

Wilde JA, Yolken RH, Eiden JJ (1990) Rotavirus detection in fecal specimens by means of polymerase chain reaction (PCR). Abstracts of the viith-international congress on virology, Berlin, p 25-013

Wilde J, Van R, Pickerng L, Eiden J, Yolken R (1992) Detection of rotaviruses in the day care environment by reverse transcriptase polymerase chain reaction. J Infect Dis 166: 507–511

Woode GN, Bridger JC, Jones JM, Flewett TH, Bryden AS, Davies HA, White GBB (1976) Morphological and antigenic relationships between viruses (rotaviruses) from acute enteritis of children, calves, piglets, mice and foals. Infect Immun 14: 804–810

Woode GN, Kelso NE, Simpson TF, Gaul SK, Evans LE, Babiuk L (1983) Antigenic relationships among some bovine rotavirus: serum neutralization and cross-protection in gnotobiotic calves. J Clin Microbiol 18: 358–364

Woods PA, Gentsch J, Gouvea V, Mata L, Simhon A, Santosham M, Bai Z-S, Urasawa S, Glass RI (1992) Distribution of serotypes of human rotavirus in different populations. J Clin Microbiol 30: 781–785

Wyatt RG, Greenberg HB, James HDJr, Pittman AL, Kalica AR, Flores J, Chanock RM, Kapikian AZ (1982) Definition of human rotavirus serotypes by plaque reduction assay. Infect Immun 37: 110–115

Wyatt RG, James HDJr, Pittman AL, Hoshino Y, Greenberg HB, Kalica AR, Flores, J, Kapikian AZ (1983) Direct isolation in cell culture of human rotaviruses and their characterization into four serotypes. J Clin Microbiol 18: 310–317

Xu L, Harbour D, McCrae MA (1990) The application of polymerase chain reaction to the detection of rotaviruses in faeces. J Virol Methods 27: 29–38

Yeager M, Dryden KA, Olson NH, Greenberg HB, Baker TS (1990) Three-dimensional structure of rhesus rotavirus by cryoelectron microscopy and image reconstruction. J Cell Biol 110: 2133–2144

Zheng BJ, Han SX, Yan YK, Liang XR, Ma GZ, Yang Y, Ng MH (1988) Development of neutralizing antibodies and group common antibodies against natural infections with human rotavirus. J Clin Microbiol 26: 1506–1512

Zheng BJ, Lam WP, Yan YK, Lo SKF, Lung ML, Ng MH (1989a) Direct identification of serotypes of natural human rotavirus isolates by hybridization using cDNA probes derived from segment 9 of the rotavirus genome. J Clin Microbiol 27: 552–557

Zheng BJ, Lo SKF, Tam JSL, Lo M, Yeung CY, Ng MH (1989b) Prospective study of community-acquired rotavirus infections. J Clin Microbiol 27: 2083–2090

Zheng S, Woode GN, Melendy DR, Ramig RF (1989c) Comparative studies of the antigenic polypeptide species VP4, VP6 and VP7 of three strains of bovine rotavirus. J Clin Microbiol 27: 1939–1945

Immunologic Determinants of Protection Against Rotavirus Disease

P.A. OFFIT

Division of Infectious Diseases, The Children's Hospital of Philadelphia and The University of
Pennsylvania School of Medicine, 34th St. and Civic Center Blvd., Philadelphia, PA 19104, USA

1 Introduction

Infants and young children previously infected with rotaviruses are protected against severe disease after reinfection (BISHOP et al. 1983). However, despite two decades of study, the immunologic mechanisms by which these infants are protected against disease remain unclear. Two important questions remain unanswered. First, which immunologic effector functions best predict protection against disease? Second, what is the importance of rotavirus serotype in inducing a protective immune response? To address these questions, this chapter will review data collected from human infants and experimental animals infected with rotavirus.

How does the host respond to rotavirus infection? A likely scenario would be the following: Rotaviruses replicate in mature villus epithelial cells of the small intestine. Infectious virus or viral proteins probably cross the basement membrane and enter the lamina propria or attach to membranous epithelial cells (M cells) and enter Peyer's patches. In either of these two sites, rotaviruses would first come in contact with the immune system. Rotaviruses are most likely processed by antigen presenting cells (probably B cells) and presented to helper T (T_h) lymphocytes, cytotoxic T lymphocytes (CTLs), and B lymphocytes within Peyer's patches and lamina propria. After virus-specific T_h cells within the intestine release B cell growth and differentiation factors, rotavirus-specific IgM and IgA appear at the intestinal mucosal surface. Coincident with the appearance in the circulation of rotavirus-specific plasma cells, B cells, and T_h cells, rotavirus-specific IgM and IgA appear in the serum 4–6 days after infection. Within 1 month of infection, virus-specific B and T cell precursors are probably distributed throughout intestinal and nonintestinal lymphocyte populations.

Is it possible to sort through this complex interplay of effector functions and determine those best associated with protection against rotavirus disease? Although investigators often refer to the relative importance of humoral as compared to cellular immunity, these two effector functions are inseparable. Virtually all viruses evoke a B cell response which is T cell-dependent. T_h cells and their cytokines are necessary for the production of virus-specific antibodies. In this review we will discuss the relative importance of different aspects of the humoral immune response (e.g., circulating vs intestinal fluid antibodies, serotype-specific vs group-specific antibodies, and rotavirus-specific IgA vs IgG) and the possible importance of CTLs in protection against disease.

There are several strategies used by investigators to determine immunologic predictors of amelioration of acute infection and protection against reinfection. Each strategy permits only a partial view of events which occur during natural infection. These strategies and their limitations are listed below:

1. Study the active immune response of infants naturally infected and naturally reinfected with rotavirus. Unfortunately, obtaining adequate specimens to study the immune response in infants is difficult. First, it is difficult to obtain permission to collect duodenal or jejunal fluid from young infants; virus-specific antibodies detected in the feces may not accurately reflect the timing or quantity of small intestinal virus-specific antibodies. In addition, analysis of serum antibodies may not accurately reflect events occurring at the intestinal mucosal surface. Second, it is difficult to obtain adequate numbers of T lymphocytes from peripheral blood to analyze virus-specific activity and cytokine production.

2. Study the active immune response of adults experimentally infected and experimentally reinfected with pathogenic rotavirus strains. Studies of jejunal fluid antibodies are possible and collection of adequate numbers of T cells from peripheral blood may offer a window to events occurring in the intestine. However, owing to the relative maturity of the adult immune system, probable previous exposure of adults to numerous rotavirus strains, and morphologic and physiologic differences between the adult and infant gastrointestinal tract, studies of the adult response to rotavirus challenge may not accurately predict determinants of protection against disease in infants.

3. Study the active immune response in animals infected with homologous host strains. Similar to findings in infants infected with human rotaviruses, homologous host rotavirus strains in animals undergo multiple cycles of replication and are shed in feces for several days. Intestinal B and T lymphocytes can be isolated and studied; inbred animals allow for easy study of virus-specific CTLs; and the relationship between antigen distribution and B and T cell activation and memory can be accurately measured. Unfortunately, several animal models used to study rotavirus pathogenesis and immunogenesis have unique anatomic or immunologic characteristics which limit their ability to mimic rotavirus infection and disease in humans. For example, unlike the human intestine, the intestinal surface of suckling mice and calves allows for the ready transport of antibodies from maternal colostrum and milk into the newborn circulation (BRAMBELL 1970). In addition, owing possibly to maturation of a virus-bending receptor on villus epithelial cells, mice are only susceptible to rotavirus disease between 2 and 12 days of age (RIEPENHOFF-TALTY et al. 1982b); adult mice (whether immunocompetent or immunocompromised) simply do not develop disease (EIDEN et al. 1986; RIEPENHOFF-TALTY et al. 1987). This pattern of disease in mice is distinct from that found in humans. Although children between 6 and 24 months of age are most susceptible to rotavirus disease (RODRIGUEZ et al. 1977). rotavirus-induced gastroenteritis has been reported among adults (ECHEVERRIA et al. 1983; KIM et al. 1977; WENMAN et al. 1979).

4. Study the active immune response in animals infected with heterologous (including human) rotavirus strains. Heterologous host rotaviruses

(including human rotaviruses) can induce diarrhea in animals. However, heterologous host strains are less virulent than homologous strains (i.e., larger quantities of virus are necessary to induce disease) and heterologous strains do not undergo multiple cycles of replication or viral amplification. Therefore, studies of heterologous host strains may not accurately predict determinants of protection against homologous host challenge.

5. Study animals passively inoculated with rotavirus-specific antibodies or rotavirus-specific T cells. The advantage of this approach is that individual effector arms of the immune system can be studied. However, because an active immune response enlists effector arms which are numerous and mutually dependent, this approach can never accurately reflect events which occur during an active immune response. For example, there are a number of studies which investigated the relative importance of antibodies directed against outer capsid proteins VP4 or VP7 in protection against challenge (MATSUI et al. 1989; OFFIT et al. 1986b). Monoclonal antibodies secreted by hybridomas were amplified either in ascites fluid or tissue culture, and suckling mice were orally inoculated with these preparations prior to rotavirus challenge. Although passive admininstration of monoclonal antibodies clearly protected suckling mice against disease, an intestine confronted with massive quantities of antibodies secreted by one clone of virus-specific B cells prior to rotavirus infection is at best and an opaque window on the complex interplay of effector functions evoked during an active immune response.

2 Assays Used to Detect the Immunologic Response to Rotavirus Infection

A number of assays have been developed to determine the humoral immune response to rotavirus infection. Rotavirus-specific antibodies have been detected by enzyme-linked immunoadsorbant assay (ELISA) (YOLKEN et al. 1978a), radioimmunoassay (RIA) (BABIUK et al. 1977), complement-fixation (CF) assay (GUST et al. 1977), hemagglutination-inhibition assay (HIA) (MARTIN et al. 1979), immune-adherence hemagglutination (IAHA) assay (MATSUNO et al. 1982), radioimmunoprecipitation (RIP) assay (OFFIT et al. 1983), plaque-reduction neutralization (PRN) assay (MATSUNO et al. 1977), and fluorescent-focus neutralization (FFN) assay (COULSON et al. 1985). Rotavirus-specific antibodies detected by these assays recognize different rotavirus proteins. Whereas ELISA, RIA, IAHA, and CF assays primarily detect antibodies directed against rotavirus inner capsid proteins, PRN, HIA, and FFN assays detect antibodies directed against outer capsid proteins. RIP assay may detect antibodies directed against all rotavirus-specific structural and nonstructural proteins.

Assays have also been developed to detect the cellular immune response to rotavirus infection. Rotavirus-specific T_h cells have been detected by lymphoproliferation assay (TOTTERDELL et al. 1988a), and virus-specific CTLs by ^{51}Cr-release assay (OFFIT and DUDZIK 1988). Rotavirus proteins recognized on the surface of virus-infected cells by human effector cells have not been determined.

3 Immune Response of Animals to Experimental Rotavirus Infection and Immunization

3.1 Animal Models Used To Study Rotavirus Immunogenesis

Several animal models have been used to study rotavirus pathogenesis and immunogenesis; none represents a perfect system to investigate immunologic factors associated with protection against infection and disease.

Most of the original studies determining both the relative importance of circulating as compared to intestinal antibodies and the importance of virus serotype in vaccine-induced protection were done in large animals, i.e., calves (SNODGRASS et al. 1980; SAIF et al. 1983), lambs (SNODGRASS and WELLS 1976), and pigs (BRIDGER and BROWN 1981). Because these animals were susceptible to rotavirus-induced disease for several weeks after birth, it was possible to study determinants of the active immune response associated with protection against disease. In addition, the adaptation of bovine, ovine, and porcine strains to growth in tissue culture facilitated studies of rotavirus immunogenesis. However, the expense and limited availability make studies of large animals difficult. In addition, the existence of other pathogens which induce diarrhea in these animals may confound studies of rotavirus.

Mice have been used extensively to study rotavirus pathogenesis and immunogenesis. Initially, murine rotavirus strains not adapted to growth in tissue culture were used (SHERIDAN et al. 1983). This model provided an excellent system in which to study both kinetics and site of virus replication and the humoral immune response (as determined by ELISA). However, because murine strains were not well adapted to plaquing in tissue culture, studies of rotavirus-neutralizing antibodies were difficult. In 1984 a murine model for oral infection with the tissue culture adapted simian rotavirus (strain SA11) was developed (OFFIT et al. 1984). Recently, murine rotavirus strains have been adapted to growth in tissue culture (GREENBERG et al. 1986). Similar to studies of large animals, the murine model has allowed for studies of the relative importance of circulating and intestinal fluid antibodies and the importance of immunoglobulin isotype in protection against disease. The short gestation period and modest price of these animals has allowed for studies of large numbers of animals. Also the relative ease of isolation of small as compared to large animals has allowed studies of adult animals not previously exposed to rotaviruses; previous exposure of large adult animals and human infants to rotavirus has confounded studies of the importance of rotavirus serotype in protection against disease. In addition, the availability of inbred animals has allowed for easy study of the CTL response. Unfortunately, mice are only susceptible to rotavirus-induced enteritis between 2 and 12 days of age. The immunologic immaturity of suckling mice and narrow window of susceptibility have not allowed for studies of determinants of the active immune response associated with protection against disease. To circumvent the narrow window of susceptibility, RIEPENHOFF-TALTY and coworkers (1987) found that mice with severe combined immunodeficiency (SCID) shed virus for several months after infection. Although this system clearly extends the length of time one can study protection against shedding in these animals, such mice can be used only to study determinants of protection after passive immunization.

Rabbits have recently been used to study rotavirus immunogenesis (CONNER et al. 1988; THOULESS et al. 1988). Rabbits (up to 4 months after birth) shed rotaviruses for several weeks after inoculation with either homologous or heterologous host rotavirus strains. Therefore, similar to large animals, studies of the active immune response are possible. In addition, rabbits are relatively inexpensive and easy to maintain. However, due probably to the large absorptive capacity of the lapine cecum, rabbits do not develop diarrhea after either homologous or heterologous host rotavirus infection. Rabbits are therefore an excellent small animal model to study determinants of protection against virus shedding after an active immune response, but are not a model to study determinants of protection against disease.

3.2 Relationship Between Site of Rotavirus Replication and Isotype of Virus-Specific Antibodies Elicited

T and B lymphocytes are abundant within the small intestine. Intestinal lymphocytes are located either at the intestinal surface among intestinal epithelial cells (intraepithelial lymphocytes) among lymphatic capillaries and connective tissue close to the epithelial surface but below the basement membrane (lamina propria lymphocytes), or within lymphatic nodules at the base of villus crypt epithelial cells (Peyer's patch lymphocytes). About 50%–60% of intra-epithelial lymphocytes are T cells, most of which have surface markers consistent with the functions of cytotoxicity or suppression (CERF-BENSUSSAN et al. 1985; PETIT et al. 1985). Less than 10% of IELs are

B cells (PARROTT et al. 1983). In lamina propria and Peyer's patch, by contrast, B cells outnumber T cells by ratios of 2:1 and 3:1, respectively (MCWILLIAMS et al. 1974; PARROTT et al. 1983). T_h cells are numerous throughout Peyer's patch and the lamina propria (PARROTT et al. 1983).

The function, memory, and location of virus-specific B and T cells after intestinal infection are probably related to presentation and distribution of viral antigens within intestinal and nonintestinal lymphocyte populations. The distribution of rotavirus antigens among intestinal and nonintestinal tissues after infection of humans is not known. However, DHARAKUL and coworkers (1988) examined the distribution of rotavirus antigens in lymphoid tissues of suckling mice orally inoculated with murine rotavirus. Rotavirus antigens were detected by immunofluorescence in villus epithelial cells 24 h after oral inoculation. Some 3–7 days after inoculation, rotavirus antigens were detected in M cells overlying Peyer's patches. Intracellular viral antigens were detected in subepithelial and interfollicular areas of Peyer's patches and in mesenteric lymph nodes between 3 and 20 days after infection. In addition, rotavirus antigens were detected in spleen beginning 5 days after infection. The mechanism by which rotavirus processing and presentation to the immune system determines virus-specific T and B lymphocyte stimulation, activation, differentiation, distribution, and memory remains to be determined.

The distribution of rotavirus antigens within the small intestine determines the kinetics, site, and isotype of the virus-specific humoral immune response; rotavirus-specific IgA is induced at the intestinal mucosal surface after oral inoculation. In mice orally inoculated with murine rotaviruses, viral replication occurs initially in the duodenum followed by the jejunum and ileum (SHERIDAN et al. 1983). Consistent with that observation, rotavirus-specific IgA-secreting plasma cells were first detected in duodenal lamina propria from suckling mice orally inoculated with murine rotavirus (DHARAKUL et al. 1988). IgA-secreting plasma cells were detected as early as 10 days after infection, and the number of virus-specific cells was consistently higher than in the jejunum or ileum for 30 days after infection. Similarly, in mice orally inoculated with a simian rotavirus strain (RRV), in the lamina propria there were 10- to 100-fold more plasma cells which secreted RRV-specific IgA than RRV-specific IgG (SHAW personal communication). Both of these studies are consistent with observations made previously in mice orally or parenterally inoculated with cholera toxin or cholera toxoid (FUHRMAN and CEBRA 1981). The number of lamina propria lymphocytes containing cholera-specific immunoglobulins on their surface did not differ after oral or parenteral inoculation. However, cholera-specific IgA-secreting plasma cells in the lamina propria were more numerous after oral inoculation and cholera-specific IgG-secreting plasma cells more numerous after parenteral inoculation. The consistent finding of pathogen-specific IgA-secreting plasma cells after oral as compared to parenteral inoculation is most likely explained by the presence of T_h cells located within intestinal lymphocyte populations which secrete cytokines (possibly interleukin-5) specific for proliferation and differentiation of IgA-secreting plasma cells (HARRIMAN and STROBER 1987).

3.3 Immune Response of Animals Inoculated with Homologous or Heterologous Host Rotaviruses, Inactivated Rotaviruses, or Purified Rotavirus Proteins

The humoral immune response in animals orally inoculated with homologous host rotaviruses is typical of that response found after other naturally acquired viral infections; virus-specific IgM is followed by IgA and IgG. Rotavirus-specific IgM followed by virus-specific IgA appeared in the feces of 5-day-old gnotobiotic calves beginning 5 days after oral inoculation with bovine rotavirus (SAIF 1987). Virus-specific IgM appeared in the serum 10 days after infection followed by IgG1, IgA, and IgG2. These findings were similar to those observed in intestinal fluids and sera of suckling mice orally inoculated with murine rotavirus (SHERIDAN et al. 1983).

Rotavirus-specific T_h cells and CTLs are detected after oral inoculation of mice with homologous or heterologous host strains and are distributed throughout the lymphoid system soon after infection. Virus-specific T_h cells were detected in the spleens of suckling mice orally inoculated with murine rotavirus beginning 2 days and peaking 10 days after inoculation (RIEPENHOFF-TALTY et al. 1982a). In addition, virus-specific CTLs were detected 6 days after oral inoculation among intraepithelial lymphocytes, Peyer's patches, mesenteric lymph nodes and spleens of adult mice orally inoculated with RRV (OFFIT and DUDZIK 1989a). Detection of virus-specific Th cells and CTLs in the spleen soon after rotavirus infection is explained either by (1) distribution of infectious virus, rotavirus proteins, or protein fragments to the spleen, (2) stimulation of T cells in the Peyer's patches, lamina propria or mesenteric lymph nodes with trafficking of lymphocytes to the spleen, or (3) infection of antigen-presenting cells in the lamina propria, Peyer's patches, or mesenteric lymph nodes with trafficking of antigen-presenting cells to the spleen and presentation of viral antigen to splenic lymphocytes.

Viral replication at the intestinal mucosal surface is probably not necessary to induce rotavirus-specific T and B cells among intestinal lymphocytes. There are several observations which support this hypothesis. First, adult mice orally inoculated with homologous or heterologous host rotaviruses develop virus-specific T and B cells among intestinal lymphocytes despite the fact that these animals do not develop diarrhea, viral replication is not detected, and few villus epithelial cells contain rotavirus antigens (SHERIDAN et al. 1983; OFFIT and DUDZIK 1989a). Rotavirus antigens may initially be presented to the intestinal surface (and ultimately to Peyer's patches or lamina propria) at quantities sufficient to induce a virus-specific immune response. Second, rotavirus-specific neutralizing antibodies are detected in the colostrum of mice parenterally inoculated with rotaviruses; rotavirus-specific IgG and not IgA is induced after parenteral inoculation (OFFIT and CLARK 1985a). Immunoglobulin secretion in colostrum is a reflection of events occurring at the intestinal mucosal surface (via the enteromammary axis) (GOLDBLUM et al. 1975). Third, rotavirus-specific neutralizing antibodies are induced in colostrum of mice parenterally inoculated with VP4 (purified from cells infected with a recombinant baculovirus containing gene segment 4 of RRV) (MACKOW et al. 1990). Lastly, adult mice parenterally inoculated with noninfectious bovine or simian rotaviruses develop both rotavirus-specific neutralizing antibodies in the serum and colostrum and rotavirus-specific CTLs among intestinal and nonintestinal lymphocytes (IJAZ et al. 1987; OFFIT and DUDZIK 1989b). The degree to which rotavirus replication at the intestinal mucosal surface determines rotavirus protein specificities of virus-specific T_h and CTLs and lymphocyte function, memory, distribution, and trafficking remains to be determined.

3.4 Immunologic Determinants of Protection Against Challenge

Attempts to collate data on the immunologic determinants of protection against challenge after active or passive immunization of animals have been frustrated by a number of problems. First, an inexpensive animal model to study determinants of protection against challenge after active immunization with tissue culture-adapted homologous host strains does not exist (see above). Second, studies differ in virulence of rotavirus strains used to challenge animals. For example, both homologous and heterologous host strains induce diarrhea in suckling mice. However, the dose required to induce disease with murine strains is approximately 100 000-fold less than that required to induce disease with nonmurine strains (GREENBERG et al. 1986). In addition, murine strains undergo multiple cycles of replication and viral amplification in the murine intestine; this does not occur after inoculation of mice with nonmurine strains. Therefore, immunologic determinants of protection against challenge with homologous host viruses may differ from those after heterologous host virus challenge. Third, studies differ in the methods used for quantitation of rotavirus strains used to immunize or challenge animals. Virus preparations used are often crude extracts from feces of animals or humans with diarrhea or are quantitated variously as plaque-forming units, fluorescent focus-forming units, tissue culture infective doses, or dose required to induce disease in 50% of inoculated animals. Fourth, measurement of the humoral immune response in different studies by either ELISA, PRN, FFR, HAI, CF, RIA, or RIP makes comparisons among studies difficult. Fifth, investigators differ in their definition of protection against challenge. Protection may mean protection against virus shedding or protection against relatively severe disease.

3.4.1 Determinants of Protection After Active Immunization

It remains unclear whether rotavirus-specific antibodies are important in protection against rotavirus challenge. In several studies investigators found that protection against diarrhea correlated with appearance of rotavirus-specific neutralizing antibodies in serum or feces or both. Gnotobiotic piglets were protected against challenge with a virulent wild-type porcine rotavirus strain 10–14 days after immunization with a human rotavirus strain (BISHOP et al. 1986). Hyperimmune antisera to human rotavirus neutralized porcine rotavirus by FFN assay, but the converse was not true. Development of porcine rotavirus-specific neutralizing antibodies in serum of human rotavirus-immunized animals predicted protection against severe diarrhea and dehydration. Similarly, serum neutralizing antibodies predicted protection against challenge in gnotobiotic piglets and gnotobiotic calves challenged with virulent homologous host rotaviruses (GAUL et al. 1982; WOODE et al. 1983).

Either VP4 or VP7 have been shown to evoke antibodies which protected against virulent rotavirus challenge. Gnotobiotic piglets were orally immunized at 4–5 days of age with reassortant rotaviruses containing either gene segments 4 or 9 from porcine strains OSU or Gottfried and challenged with either OSU or Gottfried rotaviruses 3 weeks after immunization (HOSHINO et al. 1988). Virus-neutralizing antibodies directed against either VP4 or VP7 of OSU or Gottfried protected piglets against diarrhea.

However, protection against rotavirus disease can occur in the absence of virus-specific neutralizing antibodies in the serum and feces. Gnotobiotic calves orally inoculated with an avirulent bovine rotavirus strain were protected against challenge with a virulent bovine strain of different serotype (BRIDGER and OLDHAM 1987). Protection occurred in the absence of rotavirus-specific neutralizing antibodies directed against the challenge strain in the serum and feces.

3.4.2 Determinants of Protection After Passive Immunization

Rotavirus-specific neutralizing antibodies passively protect against rotavirus diarrhea. Antibodies must be active at the intestinal mucosal surface; circulating antibodies do not alone protect against disease. Oral adminstration of milk from dams orally immunized with simian rotavirus (strain SA11) protected suckling mice against challenge (OFFIT and CLARK 1985a). In newborn mice foster-nursed by seronegative dams, circulating rotavirus-specific neutralizing antibodies in high titer (transferred transplacentally from parenterally immunized dams) did not protect against SA11 virus challenge. Similarly, lambs fed either colostrum or serum containing rotavirus-specific neutralizing antibodies were protected against challenge with homologous host rotavirus (SNODGRASS and WELLS 1976).

Parenteral inoculation with either infectious or noninfectious rotavirus induces antibodies in colostrum and milk (and presumably at the intestinal mucosal surface) capable of protection against challenge. Oral inoculation of mice with SA11 evokes a greater quantity of virus-specific IgA than IgG in milk; conversely, parenteral inoculation evokes a greater quantity of virus-specific IgG than IgA (OFFIT and CLARK 1985a). The quantity of IgG required to protect 50% of suckling mice against SA11-induced diarrhea was 0.021 ug compared with a quantity of only 0.002 ug required for IgA. Therefore, milk IgA (presumably secretory IgA) was approximately tenfold more potent in vivo than was IgG. Oral administration of milk from parenterally immunized dams also protected suckling mice against challenge; in this case, protective activity was detected in the IgG fraction. Although IgA is more potent in vivo than IgG, both appear to protect against disease. Because immunoglobulin secretion in colostrum and milk is a reflection of events occurring at the intestinal mucosal surface (via the enteromammary axis) (GOLDBLUM et al. 1975), oral immunization may not be necessary to induce virus-specific neutralizing antibodies at the intestinal surface. In addition, newborns from dams parenterally immunized with noninfectious simian rotavirus (strain RRV) were protected against RRV challenge (OFFIT and DUDZIK 1989b).

Similar to studies of active immunization, rotavirus-neutralizing antibodies directed against either VP4 or VP7 passively protect against challenge. Newborns from dams orally inoculated with reassortant rotaviruses containing either gene segments 4 or 9 from simian strain SA11 or bovine strain NCDV (two distinct serotypes) were protected against challenge with either SA11 or NCDV (OFFIT et al. 1986a). Therefore, VP4 and VP7 were independently capable of inducing an immune response which protected against challenge. In addition, monoclonal antibodies directed against either VP4 or VP7 passively protected suckling mice against challenge (MATSUI et al. 1989; OFFIT et al. 1986b).

Rotavirus-specific CTLs may protect suckling mice against homologous host rotavirus challenge. We found that suckling mice were protected against murine rotavirus challenge after intraperitoneal inoculation with splenic lymphocytes from adult mice inoculated with murine rotavirus (OFFIT and DUDZIK 1990). Protection was MHC-restricted, eliminated by treatment with anti-Thy 1 plus complement or anti-CD8 plus complement, and not associated with the presence of rotavirus-specific neutralizing antibodies in the serum of protected animals. In addition, Dharakul and coworkers (DHARAKUL et al. 1990) found that rotavirus shedding was ablated in SCID mice chronically infected with murine rotavirus after passive transfer of lymphocytes obtained from MHC-compatible adult mice previously inoculated with murine rotavirus Thy 1$^+$ and CDB$^+$ cells were responsible for ablation of rotavirus shedding. The function, trafficking, and memory of intestinal and nonintestinal CTLs after rotavirus infection remain to be determined.

3.5 Importance of Rotavirus Serotype in Protection Against Challenge

A number of animal models have been used to determine whether immunization with one rotavirus serotype induces protection against challenge with a different serotype (heterotypic protection). Several problems (in addition to those described for immunologic factors associated with protection against challenge described above) have confounded these studies. First, rotaviruses are similar to influenza viruses in that two different rotavirus surface proteins (i.e., VP4 and VP7) determine serotype (HOSHINO et al. 1985; OFFIT and BLAVAT 1986). However, unlike influenza viruses, nomenclature of rotavirus serotypes does not always include an individual description of these two proteins. For example, although simian strain RRV and human strain P are both considered to be serotype 3 strains, their gene segment 4 products are serotypically distinct (Flores, personal communication). Unfortunately, most studies which determine protection against challenge do not include a complete description of antigenic relationships between both VP4 and VP7 of homotypic and heterotypic strains. Second, the relative capacity of an epitope to evoke neutralizing antibodies may differ after natural infection as compared to parenteral immunization. For example, suckling mice from dams parenterally immunized with human, bovine, or simian rotavirus strains were protected against either homotypic or heterotypic challenge, whereas oral inoculation of dams afforded protection only against homotypic challenge (OFFIT and CLARK 1985b).

3.5.1 Serotype Specificity of Immune Response

Serotype specificity of the immune response is dependent upon previous exposure to rotavirus strains. Animals not previously exposed to rotaviruses will develop type-specific neutralizing antibodies to the immunizing strain after inoculation. Although immunization with one serotype will boost neutralizing antibodies against different serotypes to which the animal has previously been exposed, the neutralizing antibody response will not broaden to include serotypes to which the animal has not been exposed. Adult mice not previously exposed to rotaviruses developed type-specific neutralizing antibodies after either oral or parenteral inoculation with simian, bovine, or human strains (OFFIT and CLARK 1985b). Similarly, rabbits responded to sequential parenteral inoculation with bovine and human rotavirus serotypes with neutralizing antibodies specific to each serotype; the response did not broaden to include serotypes to which the animal had not been exposed

(SNODGRASS et al. 1984). Also adult cows with neutralizing antibodies to two bovine and three human rotavirus serotypes produced a significant increase in neutralizing antibody titers to these five serotypes but not to two other serotypes (one bovine and one human) after immunization with a bovine rotavirus serotype to which these cows had not previously been exposed (SNODGRASS et al. 1984).

3.5.2 Importance of Serotype in Protection Against Disease as Determined by Studies of Animals After Active Immunization

Studies which support the hypothesis that active immunization with one rotavirus serotype protects against challenge with a different rotavirus serotype are balanced by those which do not. Gnotobiotic calves, 5–15 days old and orally immunized with an avirulent bovine rotavirus, were protected against challenge with a virulent rotavirus of different serotype; protection occurred in the absence of neutralizing antibodies directed against the challenge virus in either the serum or feces (BRIDGER and OLDHAM 1987). Similarly, gnotobiotic calves orally immunized with a virulent bovine strain were protected against challenge with a virulent strain of different serotype (WOODE et al. 1987). Again, rotavirus-neutralizing antibodies in the serum and feces did not predict protection against disease suggesting that there may be an as yet undefined effector arm of the immune system associated with protection against challenge.

By contrast, piglets orally inoculated at 4–5 days of age with porcine strain OSU (serotype 5) were not protected against challenge with porcine strain Gottfried (serotype 4) and vice versa (HOSHINO et al. 1988). Similarly, studies in gnotobiotic calves did not show cross-protection among different bovine serotypes (WOODE et al. 1983).

3.5.3 Importance of Serotype in Protection Against Disease as Determined by Studies of Animals After Passive Immunization

Similar to studies of active immunization, studies which support the concept of heterotypic protection using a passive immunization scheme are counterbalanced by those which do not. Suckling mice from dams parenterally immunized with human, bovine, or simian rotavirus strains were protected against either homotypic or heterotypic challenge; protection was closely correlated with in vitro neutralizing activity of maternal sera against the challenge virus (OFFIT and CLARK 1985b). However, oral immunization of mice afforded protection only against rotavirus strains homotypic to the challenge virus (OFFIT and CLARK 1985b).

Several studies have examined the molecular basis of heterotypic protection. Dams parenterally immunized with baculovirus-expressed VP4 derived from simian strain RRV developed high titers of neutralizing anti-bodies against RRV and low titers of neutralizing antibodies against different rotavirus serotypes (MACKOW et al. 1990). Newborns suckled by these dams were protected against challenge with both RRV and a murine rotavirus strain of different serotype. In addition, monoclonal antibody preparations directed against either VP4 or VP7 passively protected suckling mice against diarrhea induced by different rotavirus serotypes (MATSUI et al. 1989; OFFIT et al. 1986b).

4 Immune Respone of Humans to Natural Rotavirus Infection and Immunization

4.1 Immunologic Response to Natural Infection

Similar to studies in animals the humoral immune response to rotavirus infection in humans is characterized by the appearance in serum and jejunal fluid of rotavirus-specific IgM followed by IgG and IgA. Rotavirus-specific IgM appeared in serum and duodenal fluid 6 days after onset of acute gastroenteritis in infants and young children admitted to the hospital; virus-specific IgG and IgA appeared in serum and duodenal fluids 1 and 4 months after infection but not during the acute phase of illness (DAVIDSON et al. 1983). Peak titers of rotavirus-specific IgA and IgG in the feces were detected approximately 2 weeks after infection. Sera obtained during the convalescent phase of illness contained antibodies directed against inner capsid proteins VP1, VP2, VP3, and VP6, outer capsid proteins VP4 and VP7, and nonstructural proteins NS_1, NS_2 and NS_3 (SVENSSON et al. 1987).

There are few data on the virus-specific cellular immune response to rotavirus infection in humans. Elderly patients mounted detectable rotavirus-specific proliferative T cells (probably a measure of both T_h and CTLs) in blood during an outbreak of rotavirus gastroenteritis in a geriatric ward (TOTTERDELL et al. 1988b). Similarly, infants and young children mount a rotavirus-specific T_h-cell response in blood lymphocytes (OFFIT et al. 1993).

4.2 Immunologic Determinants of Protection Against Challenge, Including Importance of Rotavirus Serotype, After Natural Infection

4.2.1 Immunologic Determinants of Protection After Natural Infection

Infants and young children previously infected with rotaviruses are protected against relatively severe disease upon reinfection. Bishop and coworkers studied both newborns known to be excreting rotaviruses in the first 2 weeks of life and newborns who did not excrete rotavirus (BISHOP et al. 1983). Over the next 3 years many of these infants and young children developed rotavirus-induced gastroenteritis. Symptoms associated with rotavirus infections were significantly less frequent and less severe in newborns known to have been excreting rotavirus in the first 2 weeks of life than in those not infected as newborns. The development of rotavirus-specific antibodies or rotavirus-specific T cells after neonatal infection as a predictor of protection against reinfection was not studied.

Subsequent studies found that the presence of rotavirus-specific antibodies in the serum or feces predicted protection against reinfection. In a study over a 7 months period of Cuna Indians living on two isolated islands off Panama's Carribean coast, 8% of persons with rotavirus-specific IgG detected by CF developed rotavirus infection (determined by a four-fold increase in rotavirus-specific antibodies) as compared to 46% of persons without detectable virus-specific antibodies (RYDER et al. 1985). In a study of over 200 childrens in Denmark during a rotavirus season, the presence of neither rotavirus-specific IgA nor IgG predicted protection against disease (HJELT et al. 1987). However, the presence of rotavirus-specific IgA correlated with less severe symptoms. Lastly, WRIGHT and coworkers (1987) found that infants without rotavirus-specific antibodies in their serum, as detected by IAHA, CF, and tube neutralization, shed greater quantities of rotavirus in their feces after challenge with candidate vaccine strain RRV than those with detectable antibodies.

The importance of the immune response in both amelioration of acute rotavirus infection and protection against reinfection is indirectly supported by the increased frequency and severity of rotavirus infections in immuno-deficient patients (SAULSBURY et al. 1980). Immnodeficient children with X-linked agammaglobulinemia or SCID may shed rotavirus in the feces for longer than 6 weeks. Alternatively, these findings may be explained by unique physiologic or morphologic characteristics of the gastrointestinal tract associated with altered intestinal microbial flora or concomitant bacterial infections.

4.2.2 Serotype Specificity of the Immune Response

It is unclear whether children naturally infected with rotaviruses develop antibodies which neutralize the infecting serotype only or, in addition, also neutralize infectively of different rotavirus serotypes. As in studies of animals, the serotype specificity of the immune response is probably dependent upon previous exposure to rotavirus strains. Unfortunately, unlike studies of animals, it is impossible to assure that humans have not been previously exposed to rotaviruses. Lack of detectable circulating rotavirus-specific antibodies does not preclude the presence of rotavirus-specific memory T or B cells generated after previous exposure.

Studies which support the hypothesis that rotavirus infection of infants induces antibodies specific for the infecting serotype are balanced by those which do not. ZHENG and coworkers (1988) studied acute and convalescent sera and RNA electrophoretic patterns of rotaviruses excreted in the feces of 38 infants admitted to the hospital with rotavirus-induced gastroenteritis. Of 19 infants excreting rotaviruses with 'short' RNA electropherotypes (consistent with serotype 2 rotavirus strains), 18 showed a fourfold increase in neutralizing antibodies to serotype 2 by FFN, but not to serotypes 1, 3, or 4. Conversely, of ten infants excreting rotaviruses with a "long" electrophoretic pattern (consistent with serotypes 1, 3, or 4) nine showed a fourfold neutralizing response to serotype 3 and one patient a response to serotype 1 rotavirus. These authors concluded that within 2 weeks of an acute rotavirus-induced gastroenteritis the humoral immune response is specific for the infecting virus. However, studies by CLARK and coworkers (1985) contradicted these results. Eight infants studied during a 32 day period in Philadelphia presented with rotavirus-induced gastroenteritis as determined by either excretion of rotaviruses in feces or detection of a rotavirus-specific antibodies in convalescent sera as determined by PRN, ELISA, RIP, and HAI. Three originally seropositive infants showed an increase in neutralizing antibody titer to two or more serotypes by PRN assay. However, two of five originally seronegative infants also developed neutralizing antibodies to two or more serotypes. Therefore, no consistent pattern of neutralizing antibody response was detected after acute infection. This finding has been confirmed by others (CHIBA et al. 1986; PUERTO et al. 1987; CLARK et al. 1988).

In humans, similar to findings in animals, it appears that both VP4 and VP7 evoke neutralizing antibodies after oral inoculation of reassortant rotaviruses. A predominance (approximately 95%) of the neutralizing antibody response was directed against VP4 in sera obtained from adults orally inoculated with a reassortant rotavirus strain containing gene segments from two different human rotavirus serotypes (WARD et al. 1988). These results were similar to those found by other investigators (CLARK et al. 1990; FLORES et al. 1989). However, subsequent studies with reassortants made between bovine and either simian or murine strains showed that VP7 was the dominant immunogen after intestinal infection of infants orally inoculated with bovine rotavirus (WARD et al. 1990a).

4.2.3 Importance of Serotype in Protection Against Disease After Natural Infection

Infection with one rotavirus serotype may not induce protection against challenge with a different rotavirus serotype. There are several studies which support this hypothesis. RODRIGUEZ et al. (1978) reported the case of an infant with two separate episodes of rotavirus-induced gastroenteritis at 3 months and 14 months of age caused by serotypes 1 and 2, respectively. FRIEDMAN and coworkers (1988) studied two sequential outbreaks over a 1 year period of rotavirus-induced gastroenteritis in Israel apparently caused by serotypes 3 and 1, respectively. Six of 12 childrens who developed symptomatic rotavirus infection during the first outbreak also developed symptomatic infection during the second outbreak. Finally, CHIBA et al. (1986) studied institutionalized children less than 2 years of age exposed to consecutive outbreaks of serotype 3 rotavirus gastroenteritis over a 20 month period in Japan. Protection against rotavirus-induced diarrhea was dependent upon the serotype specificities and levels of neutralizing antibodies detected in the serum. Children with high titers of serotype 3 rotavirus-specific neutralizing antibodies were protected against serotype 3 rotavirus challenge. However, children with high titers of neutralizing antibodies against serotypes 1 or 4 or both but low neutralizing titers against serotype 3 were not protected against serotype 3 rotavirus challenge.

4.2.4 Importance of Breast-Feeding in Protection Against Disease

Breast-feeding is important in reducing the morbidity and mortality of intestinal bacterial infections. The importance of breast-feeding is best documented in protection against enteropathogenic *E. Coli*, *Salmonella*, *Shigella*, and cholera-induced enteritis (WELSH and MAY 1979). The beneficial effects of breast-feeding are associated with production and maintenance of a lactobacillary flora and low pH in feces secondary to the high lactose, low protein, low phosphate and poor buffering capacity of breast milk (BULLEN and WILLIS 1971). In addition, immunologic factors such as pathogen-specific IgG, secretory IgA (sIgA), and T cells probably play a role in protection against disease (WELSH and MAY 1979).

A large percentage of lactating women in both developed and developing countries have rotavirus-specific antibodies in their milk and colostrum. Several investigators have documented the presence of rotavirus-specific IgG, IgM, and sIgA by ELISA in colostrum and breast milk (BELL et al. 1988; COOK et al. 1978; CUKOR et al. 1979; RINGENBERGS et al. 1988; YOLKEN et al. 1978b). Rotavirus-specific antibodies are detected up to 2 years postpartum, neutralize infectivity of different rotavirus serotypes in vitro, and agglutinate rotavirus particles when assayed by immune electron microscopy. In addition, 30% of healthy lactating women have rotavirus-specific lymphoproliferative cells in breast milk (TOTTERDELL et al. 1988a).

Despite the presence of rotavirus-specific immunoglobulins and lymphoproliferative cells in breast milk and colostrum, breast-feeding probably does not afford protection against either rotavirus infection or severe rotavirus disease. This statement is based on the following studies: (1) GLASS and workers (1986) studied approximately 2000 children less than 4 years of age admitted to a hospital with acute gastroenteritis in rural Bangladesh. Breast-feeding was more common among children with rotavirus-induced diarrhea than among children with non-rotavirus diarrhea. However, breast-feeding was less common among children infected with cholera or *Shigella*. Some 20% of breast milks consumed by infants less than 1 year of age had serotype 1 rotavirus-specific neutralizing antibodies detected at high titers by PRN. However, virus-specific neutralizing antibody titers in the serum of infants with rotavirus diarrhea did not significantly differ from those of infants with diarrhea of other causes. Finally, despite the prolonged breast-feeding common in Bangladesh, the mean age of hospitalization with rotavirus disease is approximately the same as in countries where the duration of breast-feeding is far shorter. (2) DUFFY et al. (1986) followed approximately 200 infants through a single rotavirus season in Buffalo, New York. There were no apparent differences in the incidence of rotavirus infections in breast-fed as compared to bottle-fed infants. (3) TOTTERDELL and coworkers (1980) studied infants during an outbreak of serotype 2 rotavirus infection in a neonatal nursery in London. There was no difference in the titers of rotavirus-specific IgG or IgA detected by ELISA in the breast milks of mothers of babies excreting rotavirus as compared to those of infants not excreting rotavirus. In addition, among breast-fed infants with serotype 2 rotavirus in their feces, there was no correlation between rotavirus serotype 2-specific IgA in expressed breast milk and quantity of serotype 2 rotavirus detected in the feces. (4) Lastly, WEINBERG and coworkers (1984) studied infants less than 1 year of age with rotavirus gastroenteritis seen as outpatients in Rochester, New York. The incidence of breast-feeding in patients infected with rotavirus was similar to that in an aged-matched control population. There were no significant differences between groups in the average age of children infected with rotavirus, mean duration of diarrhea, frequency of bowel movements per day, or in frequency of fever or irritability.

4.3 Immunologic Determinants of Protection Against Challenge, Including Importance of Rotavirus Serotype, After Active Immunization

Investigators in both developed and developing countries found that active immunization of infants with heterologous host (i.e., nonhuman) rotaviruses mimics the protection against disease found after natural infection (BISHOP et al. 1983); infants are protected against severe rotavirus disease but are not protected against rotavirus reinfection. The immunologic basis of protection against challenge after active immunization has not been clearly defined.

4.3.1 Immunization with Live, Attenuated Bovine Rotaviruses (Strains RIT 4237 or WC3)

Immunization of infants with WC3 or RIT 4237 was found to protect against disease caused by serotype 1 rotavirus in trials of protective efficacy in infants (CLARK et al. 1988; VESIKARI et al. 1985). Protection occurs despite the fact that WC3 and RIT 4237 are serotypically distinct from any known human rotavirus strains (heterotypic protection). The immunologic basis of heterotypic protection is not clearly explained by the humoral immune response.

Oral immunization with bovine rotavirus strain RIT 4237 protected infants against relatively severe rotavirus-induced diarrhea in Finland (Vesikari et al. 1985). Approximately 300, 6–12 month old infants were orally inoculated with two doses of RIT 4237 or placebo several months prior to a predominantly serotype 1 rotavirus epidemic. Some 3% of vaccine recipients and 16% of placebo recipients developed severe diarrhea. Many infants inoculated with bovine rotavirus were protected against rotavirus challenge despite the absence of virus-specific antibodies in the serum measured by ELISA. Protection in this study was most likely afforded by rotavirus-specific intestinal immunoglobulins (IgG or IgA), virus-specific cellular immunity (e.g., CTLs), or both. Consistent with this hypothesis, mice orally inoculated with WC3 develop CTLs which lyse target cells infected with serotype 1 rotavirus (OFFIT and SVOBODA 1989c). Alternatively, because virus-specific antibodies in this study were detected by ELISA using RIT 4237 as the detecting antigen, protection against challenge may have been afforded by serotype 1 rotavirus-specific neutralizing antibodies which were present in the serum but not detected by ELISA.

The importance of rotavirus-specific neutralizing antibodies in protection against serotype 1 rotavirus challenge after immunization with bovine rotaviruses was addressed in the investigations of CLARK and coworkers (1988). Using a double-blind study design, approximately 100 infants ages 3–12 months were inoculated with either WC3 or placebo prior to a predominantly serotype 1 rotavirus epidemic in Philadelphia. Some 20% of placebo-inoculated and 0% of WC3-inoculated infants developed moderate to severe diarrhea. Using strains Wa (serotype 1), SA11 (serotype 3), and WC3 in PRN assays, 71% of vaccines developed neutralizing antibodies to WC3, 45% to serotype 3 and only 8% to serotype 1. There are two possible hypotheses to explain these findings. First, infants are protected against serotype 1 rotavirus challenge without developing serotype 1 rotavirus-specific neutralizing antibodies. As was stated above, protection may be mediated by an as yet undefined effector arm of the immune system. Alternatively, intertypic variation between Wa and epidemic strains may not have allowed for an accurate determination of neutralizing antibodies directed against challenge isolates.

4.3.2 Immunization with Live, Attenuated Simian Rotavirus (Strain RRV)

Immunization of infants with RRV was found to protect against serotype 3 but not serotype 1 rotavirus infection in efficacy trials in infants (FLORES et al. 1987). RRV is serotypically similar to human serotype 3 strains, and protection correlated with development of serotype 3-but not serotype 1-specific neutralizing antibodies in the serum of immunized infants (homotypic protection).

FLORES and coworkers (1987) studied the efficacy of RRV in a double-blind longitudinal field trial in Caracas, Venezuela during a predominantly serotype 3 epidemic. Approximately 250 1–10 month old infants were inoculated with either 1×10^4 pfu of RRV or placebo. Some 4% of vaccinees developed rotavirus diarrhea as compared to 17% of infants receiving placebo. In 1–5 month-old infants rotavirus diarrhea occurred in 1.5% of vaccinees as compared to 23% of controls. For the entire group the vaccine efficacy against severe diarrhea was 100%.

4.3.3 Immunization with Live, Human Rotavirus (Strain CJN)

Perhaps the most thorough examination of immunologic determinants of protection against challenge was performed in adult volunteers in Cincinnati experimentally challenged with a pathogenic human rotavirus strain (WARD et al. 1989). There were 38 (18–45 year old) volunteers who were orally inoculated with 9×10^4 focus-forming units of human rotavirus strain CJN. Strain CJN was obtained from the feces of an 8 month old infant with gastroenteritis and passaged twice in African green monkey kidney cells. Strain CJN appears to be a serotype 1 rotavirus. Rotavirus-specific serum and jejunal neutralizing antibodies, serum IgG and IgA, jejunal IgA, and fecal IgA were measured for their capacity to predict protection against rotavirus infection and illness. A volunteer was considered to be infected with rotavirus if rotavirus was detected in the feces, or if there was a greater than four fold rise in rotavirus-specific antibodies, or both. Some 75% of subjects became infected and 55% of those developed gastroenteritis. The level of preinoculation neutralizing antibodies in jejunal fluid pre-dicted protection against illness as determined by a stepwise logistic regres-sion analysis. In addition, the titer of rotavirus-specific IgG detected in the serum predicted protection against infection. Inoculation of adults with CJN induced a broad, cross-reactive, neutralizing antibody re-sponse detected in the serum directed against rotavirus serotypes 1, 2, 3 and 4.

To determine whether inoculation with strain CJN protected adults against rechallenge with the same strain, 20 previously inoculated subjects were reinoculated with strain CJN 9-12 months after the original inoculation (WARD et al. 1990b). In addition, four previously uninoculated adults were included as controls. One of eight subjects not infected after the initial inoculation and two of 12 subjects infected after the initial inoculation were infected upon rechallenge. None of the three subjects infected after reinoculation developed gastroenteritis. However, three of four controls were infected and one of these developed diarrhea after inoculation with strain CJN. Because mean titers of serum, jejunal, and fecal rotavirus-specific antibodies in these 20 subjects were higher at the time of reinoculation than prior to the initial inoculation, rotavirus-specific antibodies predicted protection against infection and illness upon reinoculation. However, the extent of protection against challenge was greater than would have been predicted based on the relationship between serum rotavirus-specific IgG and the probability of infection determined after initial inoculation. Based on the serum rotavirus-specific IgG titers found prior to reinoculation with strain CJN, one would have predicted that 13 of 20 subjects would have been infected on rechallenge, but, only three were. There are several hypotheses to explain these observations. First, lack of detection of rotavirus-specific antibodies in the serum or jejunal fluid does not preclude the presence of rotavirus-specific B cell precursors. Second, there may be an as yet undetected effector arm of the immune system associated with protection against challenge.

5 Summary

5.1 Which Immunologic Effector Functions Best Predict Protection Against Disease?

Since an active immune response elicits a variety of effector functions which are mutaully dependent, it is difficult to sort out their relative importance. However, based on studies in experimental animals and human vaccine recipients we draw the following conclusions: (1) The presence of rotavirus-specific neutralizing antibodies at the intestinal mucosal surface protects against disease. Rotavirus-specific antibodies in jejunal fluid more accurately reflect events occurring at the intestinal surface than serum antibodies, and are therefore a better predictor of protection against challenge. (2) Protective neutralizing antibodies are directed against epitopes located on either VP4 or VP7. (3) Rotavirus-specific neutralizing antibodies can be induced at the intestinal surface after either oral or parenteral immunization. Rotavirus-specific sIgA is evoked primarily after oral immunization and is probably more potent in vivo than is IgG. However, intestinal rotavirus-specific IgG, evoked primarily after parenteral immunization, also protects against challenge. (4) It does not appear that, under natural conditions of infection, breast milk and colostrum contain quantities of rotavirus-specific neutralizing antibodies necessary to protect against disease. (5) Although it is intriguing that rotavirus-specific CTLs are present at the intestinal surface 6 days after virus infection and may, in part, explain the immunologic basis of heterotypic protection, their importance in protection against rotavirus disease remains undetermined.

5.2 What is the Importance of Rotavirus Serotype in Inducing a Protective Immune Response?

Serotype specificity of the humoral immune response is probably dependent upon previous exposure to rotavirus strains. Animals not previously exposed to rotaviruses develop type-specific neutralizing antibodies to the immunizing strain. Although immunization with one rotavirus serotype will boost neutralizing antibodies against different serotypes to which the animal has previously been exposed, the neutralizing response will not broaden to include serotypes to which the animal has not been exposed.

Because it is difficult to prove that infants have not been previously exposed to rotaviruses, studies examining serotype specificity of the humoral immune response in humans have been contradictory. The immunologic basis of protection against heterotypic challenge found in studies of human infants inoculated with WC3 remains undetermined. It remains unclear whether infants need to be immunized with all serotypes to which they are likely to be exposed.

References

Babiuk LA, Acres SD, Rouse BT (1977) Solid-phase radioimmunoassay for detecting bovine (neonatal calf diarrhea) rotavirus antibody. J Clin Microbiol 6: 10–15
Bell LM, Clark HF, Offit PA, Horton-Slight P, Arbeter AM, Plotkin SA (1988) Rotavirus serotype-specific neutralizing activity in human milk. Am J Dis Child 142: 275–278
Bishop RF, Barnes GL, Cipriani E, Lund JS (1983) Clinical immunity after neonatal rotavirus infection: a prospective longitudinal study in children. N Engl J Med 309: 72–76
Bishop RF, Tzipori SR, Coulson BS, Unicomb LE, Albert MJ, Barnes GL (1986) Heterologous protection against rotavirus-induced diesease in gnotobiotic piglets. J Clin Microbiol 24: 1023–1028
Brambell FWR (1970) The transmission of passive immunity from mother to young. North Holland, Amsterdam (Frontiers of biology, vol 18)
Bridger JC, Brown JF (1981) Development of immunity to porcine rotavirus in piglets protected from disease by bovine colostrum. Infect Immun 31: 906–910
Bridger JC, Oldham G (1987) Avirulent rotavirus infections protect calves from disease with and without inducing high levels of neutralizing antibody. J Gen Virol 68: 2311–2317
Bullen CL, Willis AT (1971) Resistance of the breast-fed infant to gastroenteritis. Br Med J 3: 338–343
Cerf-Bensussan N, Guy-Grand D, Griscelli C (1985) Intraepithelial lymphocytes of human gut: isolation, characterization and study of natural killer activity. Gut 26: 81–88
Chiba S, Nakata S, Urasawa T, Urasawa S, Yokoyama T, Morita Y, Taniguchi K, Nakao T (1986) Protective effect of naturally acquired homotypic and heterotypic rotavirus antibodies. Lancet i: 417–421
Clark HF, Dolan KT, Horton-Slight P, Palmer J, Plotkin SA (1985) Diverse serologic response to rotavirus infection of infants in a single epidemic. J Pediatr Infect Dis 4: 626–631
Clark HF, Borian FE, Bell LM, Modesto K, Gouvea V, Plotkin SA (1988) Protective effect of WC3 vaccine against rotavirus diarrhea in infants during a predominantly serotype 1 rotavirus season. J Infect Dis 158: 570–587
Clark HF et al. (1990) Serotype 1 reassortant of bovine rotavirus WC3, strain WI 79-9, induces a polytypic antibody response in infants. Vaccine (in press).

Conner ME, Estes MK, Graham, DY (1988) Rabbit model of rotavirus infection. J Virol 62: 1625–1633

Cook CA, Zbitnew A, Dempster G, Gerrard JW (1978) Detection of antibody to rotavirus by counterimmunoelectrophoresis in human serum, colostrum, and milk. J Pediatr 93: 967–970

Coulson BS, Fowler KJ, Bishop RF, Cotton RGH (1985) Neutralizing monoclonal antibodies to human rotavirus and indications of antigenic drift among strains from neonates. J Virol 54: 14–20

Cukor G, Blacklow NR, Capozza FE, Panjvani ZFK, Bednarek F (1979) Persistence of antibodies to rotavirus in human milk. J Clin Microbiol 9: 93–96

Davidson GP, Hogg RJ, Kirubakaran CP (1983) Serum and intestinal immune response to rotavirus enteritis in children. Infect Immun 40: 447–452

Dharakul T, Riepenhoff-Talty M, Albini B, Ogra PL (1988) Distribution of rotavirus antigen in intestinal lymphoid tissues: potential role in development of the mucosal immune response to rotavirus. Clin Exp Immunol 74: 14–19

Dharakul T, Rott L, Greenberg H (1990) Recovery from chronic rotavirus infection in mice with severe combined immunodeficiency: virus clearance mediated by adaptive transfer of CD8[+] T lymphocytes. J Virol 64: 4375–4382

Duffy LC, Riepenhoff-Talty M, Byers TE, LaScolea LJ, Zielezny MA, Dryja DM, Ogra PL (1986) Modulation of rotavirus enteritis during breast feeding. Am J Dis Child 140: 1164–1168

Echeverria P, Blacklow NR, Cukor CG, Vibulbandhitkit S, Changchawalit S, Boonthai P (1983) Rotavirus as a cause of severe gastroenteritis in adults. J Clin Micorbiol 18: 663–667

Eiden J, Lederman HM, Vonderfecht S, Yolken R (1986) T-cell-deficient mice display normal recovery from experimental rotavirus infection. J Virol 57: 706–708

Flores J, Gonzalez M, Perez M, Cunto W, Perez-Schael I, Garcia D, Daoud N, Chanock RM, Kapikian AZ (1987) Protection against severe rotavirus diarrhoea by rhesus rotavirus vaccine in Venezuelan infants. Lancet i: 882–884

Flores J, Perez-Schael I, Blanco M, Vilar M, Garcia D, Perez M, Daoud N, Midthun K, Kapikian AZ (1989) Reactions to and antigenicity of two human-rhesus rotavirus reassortant vaccine candidates of serotypes 1 and 2 in Venezuelan infants. J Clin Microbiol 27: 512–518

Friedman MG, Galil A, Sarov B, Margalith M, Katzir G, Midthun K, Taniguchi K, Urasawa S, Kapikian AZ, Edelman R, Sarov I (1988) Two sequential outbreaks of rotavirus gastroenteritis; evidence for symptomatic and asymptomatic reinfections. J Infect Dis 158: 814–822

Fuhrman JA, Cebra JJ (1981) Special features of the priming process for a secretory IgA response: B cell priming with cholera toxin. J Exp Med 153: 534–544

Gaul SK, Simpson TF, Woode GN, Fulton RW (1982) Antigenic relationships among some animal rotaviruses: virus neutralization in vitro and cross-protection in piglets. J Clin Microbiol 16: 495–503

Glass RI, Stoll BJ, Wyatt RG, Hoshino Y, Banu H, Kapikian AZ (1986) Observations questioning a protective role for breast-feeding in severe rotavirus diarrhea. Acta Paediatr Scand 75: 713–718

Goldblum RM, Ahlstedt S, Carlsson B, Hanson LA, Jodal V, Lider-Janson G, Sohl-Akerlund A (1975) Antibody-forming cells in human colostrum after oral immunization. Nature 257: 797–799

Greenberg HB, Vo PT, Jones R (1986) Cultivation and characterization of three strains of murine rotavirus. J Virol 57: 585–590

Gust ID, Pringle RC, Barnes GL, Davidson GP, Bhshop RF (1977) Complement fixing antibody response to rotavirus infection. J Clin Microbiol 5: 125–130

Harriman GR, Strober WS (1987) Interleukin 5, a mucosal lymphokine? J Immunol 139: 3553–3555

Hoshino Y, Sereno MM, Midthun K, Flores J, Kapikian AZ, Chanock RM (1985) Independent segregation of two antigenic specificities (VP3 and VP7) involved in neutralization of rotavirus infectivity. Proc Natl Acad Sci USA 82: 8701–8704

Hoshino Y, Saif LJ, Sereno MM, Chanock RM, Kapikian AZ (1988) Infection immunity of piglets to either VP3 or VP7 outer capsid confers resistance to challenge with a virulent rotavirus bearing the corresponding antigen. J Virol 62: 744–748

Hjelt K, Grauballe PC, Paerregaard A, Nielson OH, Krasilnikoff PA (1987) Protective effect of preexisting rotavirus-specific immunoglobulin A against naturally acquired rotavirus infection in children. J Med Virol 21: 39–47

Ijaz MK, Sabara MI, Frenchik PJ, Babiuk LA (1987) Effect of different routes of immunization with bovine rotavirus on lactogenic antibody response in mice. Antiviral Res 8: 283–298

Kim HW, Brandt CD, Kapikian AZ, Wyatt RG, Arrobio JO, Rodriguez WJ, Chanock RM, Parrott RH (1977) Human reovirus-like agent infection: Occurrence in adult contacts of pediatric patients with gastroenteritis. JAMA 238: 404–407

Mackow ER, Vo PT, Broome R, Bass D, Greenberg HB (1990) Immunization with baculovirus expressed VP4 protein passively protects against homologous and heterologous rotavirus challenge. J Virol (in press)

Martin ML, Gary GW Jr., Palmer EL (1979) Comparison of hemagglutination-inhibition, complement-fixation and enzyme-linked immunosorbent assay for quantitation of human rotavirus antibodies. Arch Virol 62: 131–136

Matsui SM, Offit PA, Vo PT, Mackow ER, Benfield DA, Shaw RD, Padilla-Noriega L, Greenberg HB (1989) Passive protection against rotavirus-induced diarrhea by monoclonal antibodies to the heterotypic neutralization domain of VP7 and the VP8 fragment of VP4. J Clin Microbiol 27: 780–782

Matsuno S, Inouye S, Kono R (1977) Plaque assay of neonatal calf diarrhea virus and the neutralizing antibody in human sera. J Clin Microbiol 5: 1–4

Matsuno S, Inouye S, Hasegawa A, Kono R (1982) Assay of human rotavirus antibody by immune adherence hemagglutination with a cultivable human rotavirus as antigen. J Clin Microbiol 15: 163–165

McWilliams M, Lamm ME, Phillips-Quagliata JM (1974) Surface and intracellular markers of mouse mesenteric and peripheral lymph node and Peyer's patch cells. J Immunol 113: 1326–1333

Offit PA, Blavat G (1986) Identification of the two rotavirus genes determining neutralization specificities. J Virol 57: 376–378

Offit PA, Clark HF (1985a) Protection against rotavirus-induced gastroenteritis in a murine model by passively-acquired gastrointestinal but not circulating antibodies. J Virol 54: 58–64

Offit PA, Clark HF (1985b) Maternal antibody-mediated protection against gastroenteritis due to rotavirus in newborn mice is dependent on both serotype and titer of antibody. J Infect Dis 152: 1152–1158

Offit PA, Dudzik KI (1988) Rotavirus-specific cytotoxic T lymphocytes cross-react with target cells infected with different rotavirus serotypes. J Virol 62; 127–131

Offit PA, Dudzik KI (1989a) Rotavirus-specific cytotoxic T lymphocytes appear at the intestinal mucosal surface after rotavirus infection. J Virol 63: 3507–3512

Offit PA, Dudzik KI (1989b) Noninfectious rotavirus (strain RRV) induces an immune response in mice which protects against rotavirus challenge. J Clin Microbiol 27: 885–888

Offit PA, Svoboda YM (1989c) Rotavirus-specific cytotoxic T lymphocyte response of mice after oral inoculation with candidate rotavirus vaccine strains RRV or WC3. J Infect Dis 160: 783–788

Offit P, Dudzik K (1990) Rotavirus-specific cytotoxic T lymphocytes passively protect against gastroenteritis in suckling mice. J Virol 64: 6325–6328

Offit PA, Clark, HF, Plotkin, SA (1983) Response of mice to rotaviruses of bovine or primate origin assessed by radioimmunoassay, radioimmunoprecipitation, and plaque-reduction neutralization. Infect Immun 42: 293–300

Offit PA, Clark HF, Kornstein MJ, Plotkin SA (1984) A murine model for oral infection with a primate rotavirus (simian SA11). J Virol 51: 233–236

Offit PA, Clark HF, Blavat G, Greenberg HB (1986a) Reassortant rotaviruses containing structural proteins VP3 and VP7 from different rotavirus parents induce antibodies protective against each parental serotype. J Virol 60: 491–496

Offit PA, Shaw RD, Greenberg HB (1986b) Passive protection against rotavirus-induced diarrhea by monoclonal antibodies to surface proteins VP3 and VP7. J Virol 58: 700–703

Offit P, Hoffenberg E, Santos N, Gouvea VS (1993) Rotavirus-specific humoral and cellular immune response after primary symptomatic infection. J Infect Dis 167: 1436–1440

Parrott DMV, Tait C, McKenzie S, Mowat AM, Davies MDJ, Mickle HS (1983) Analysis of the effector functions of different populations of mucosal lymphocytes. Ann NY Acad Sci 409: 307–320

Petit A, Ernst PB, Befus AD, Clark DA, Rosenthal KL, Ishizaka T, Bienenstock J (1985) Murine intestinal intraepithelial lymphocytes. I. Relationship of a novel Thy 1⁻, Lyt 1⁻, Lyt 2⁺ granulated subpopulation to natural killer cells and mast cells. Eur J Immunol 15: 211–215

Puerto FI, Padilla-Noriega L, Zamora-Chavez A, Briceno A, Puerto M, Arias CF (1987) Prevalent patterns of serotype-specific seroconversion in Mexican children infected with rotavirus. J Clin Microbiol 25: 960–963

Riepenhoff-Talty M, Suzuki H, Ogra PL (1982a) Characteristics of the cell-mediated immune response to rotavirus in suckling mice. Dev Biol Stand 53; 263–268

Riepenhoff-Talty M, Lee P-C, Carmody PJ, Barrett HJ, Ogra PL (1982b) Age-dependent rotavirus-enterocyte interactions. Proc Soc Exp Biol Med 170: 146–154

Riepenhoff-Talty M, Dharakul T, Kowalski E, Michalak S, Ogra PL (1987) Persistent rotavirus infection in mice with severe combined immunodeficiency. J Virol 61: 3345–3348

Ringenbergs M, Albert MJ, Davidson GP, Goldsworthy W, Haslam R (1988) Serotype-specific antibodies to rotavirus in human colostrum and breast milk and in maternal and cord blood. J Infect Dis 158: 477–480

Rodriguez WJ, Kim HW, Arrobio JO, Brandt CD, Chanock RM, Kapikian AZ, Wyatt RG, Parrott RH (1977) Clinical features of acute gastroenteritis associated with human reovirus-like agent in infants and young children. J Pediatr 91: 188–193

Rodriquez WJ, Kim HW, Brandt CD, Yolken RH, Arrobio JO, Kapikian AZ, Chanock RM, Parrott RH (1978) Sequential enteric illnesses associated with different rotavirus serotypes. Lancet i: 37

Ryder RW, Singh N, Reeves WC, Kapikian AZ, Greenberg HB, Sack RB (1985) Evidence of immunity induced by naturally acquired rotavirus and Norwalk virus infection on two remote Panamanian islands. J Infect Dis 151: 99–105

Saif LJ (1987) Development of nasal, fecal, and serum isotype-specific antibodies in calves challenged with bovine coronavirus or rotavirus. Vet Immunol Immunopathol 17: 425–439

Saif LJ, Redman DR, Smith KL, Theil KW (1983) Passive immunity to bovine rotavirus in newborn calves fed colostrum supplements from immunized or nonimmunized cows. Infect Immun 41: 1118–1131

Saulsbury FT, Winkelstein JA, Yolken RH (1980) Chronic rotavirus infection in immuno-deficiency. J Pediatr 97: 61–65

Sheridan JF, Eydelloth RS, Vonderfecht SL, Aurelian L (1983) Virus-specific immunity in neonatal and adult mouse rotavirus infection. Infect Immun 39: 917–927

Snodgrass DR, Wells PW (1976) Rotavirus infection in lambs: studies on passive protection. Arch Virol 52: 201–205

Snodgrass DR, Fahey KJ, Wells PW, Campbell I, Whitelaw A (1980) Passive immunity in calf rotavirus infections: maternal vaccination increases and prolongs immunoglobulin G1 antibody secretion in milk. Infect Immun 28: 344–349

Snodgrass DR, Ojeh CK, Campbell I, Herring AJ (1984) Bovine rotavirus serotypes and their significance for immunization. J Clin Microbiol 20: 342–346

Svensson L, Sheshberadaren H, Vene S, Norrby E, Grandien M, Wadell G (1987) Serum antibody responses to individual viral polypeptides in human rotavirus infection. J Gen Virol 68: 643–651

Thouless ME, Digiacomo RF, Deeb BJ, Howard H (1988) Pathogenicity of rotavirus in rabbits. J Clin Microbiol 26: 943–947

Totterdell BM, Chrystie IL, Banatvala JE (1980) Cord blood and breast-milk antibodies in neonatal rotavirus infection. Br Med J ii: 828–830

Totterdell BM, Patel S, Banatvala JE, Chrystie IL (1988a) Development of a lymphocyte transformation assay for rotavirus in whole blood and breast milk. J Med Virol 25: 27–36

Totterdell BM, Banatvala JE, Chrystie IL, Ball G, Cubitt WD (1988b) Systemic lympho-proliferative responses to rotavirus. J Med Virol 25: 37–44

Vesikari T, Isolauri E, Delem A, d'Hondt E, Andre FE, Beards GM, Flewett TH (1985) Clinical efficacy of the RIT 4237 live attenuated bovine rotavirus vaccine in infants vaccinated before a rotavirus epidemic. J Pediatr 107: 189–194

Ward RL, Knowlton DR, Schiff GM, Hoshino Y, Greenberg HB (1988) Relative concentrations of serum neutralizing antibody to VP3 and VP7 proteins in adults infected with a human rotavirus. J Virol 62: 1543–1549

Ward RL, Bernstein DI, Shukla R, Young EC, Sherwood JR, McNeal MM, Walker MC, Schiff GM (1989) Effects of anitbody to rotavirus on protection of adults challenged with a human rotavirus. J Infect Dis 159: 79–88

Ward RL, Knowlton DR, Greenberg HB, Schiff GM, Bernstein DI (1990a) Serum neutralizing antibody to VP4 and VP7 proteins in infants following vaccination with WC3 bovine rotavirus. J Virol (in press)

Ward RL, Bernstein DI, Shukla R, McNeal MM, Sherwood JR, Young EC, Schiff GM (1990b) Protection of adults rechallenged with a human rotavirus. J Infect Dis (in press)

Weinberg RJ, Tipton G, Klish WJ, Brown MR (1984) Effect of breast-feeding on morbidity in rotavirus gastroenteritis. Pediatrics 74: 250–253

Welsh JK, May JT (1979) Anti-infective properties of breast milk. J Pediatr 94: 1–9

Wenman WM, Hinde D, Feltham S, Gurwith M (1979) Rotavirus infection in adults: results of a prospective family study. N Engl J Med 301: 303–306

Woode GN, Kelso NE, Simpson TF, Gaul SK, Evans LE, Babiuk L (1983) Antigenic relationships among some bovine rotaviruses: serum neutralization and cross-protection in gnotobiotic calves. J Clin Microbiol 18: 358–364

Woode GN, Zheng S, Rosen BI, Knight N, Kelso Gourley NE, Ramig RF (1987) Protection between different serotypes of bovine rotavirus in gnotobiotic calves: specificity of serum antibody and coproantibody responses. J Clin Microbiol 25: 1052–1058

Wright PF, Tajima T, Thompson J, Kokubun K, Kapikian A, Karzon DT (1987) Candidate rotavirus vaccine (rhesus rotavirus strain) in children: an evaluation. Pediatrics 80: 473–480

Yolken RH, Wyatt RG, Kim HW, Kapikian AZ, Chanock RM (1978a) Immunologic response to infection with human reovirus-like agent: measurement of anti-human reovirus-like agent immunoglobulin G and M levels by the method of enzyme-linked immunosorbent assay. Infect Immun 19: 540–546

Yolken RH, Wyatt RG, Mata L, Urrutia JJ, Garcia B, Chanock RM, Kapikian AZ (1978b) Secretory antibody directed against rotavirus in human milk-measurement by means of enzyme-linked immunosorbent assay. J Pediatr 93: 916–921

Zheng B-J, Han S-H, Yan Y-K, Liang X-R, Ma G-Z, Yang Y, Ng MH (1988) Development of neutralizing antibodies and group A common antibodies against natural infections with human rotavirus. J Clin Microbiol 26: 1506–1512

Rotavirus Pathology and Pathophysiology

H.B. Greenberg[1], H.F. Clark[2], and P.A. Offit[2]

Rotaviruses represent the single most important enteric viral pathogens of humans. They are also frequent causes of morbidity and mortality in many animal species. The virus has a rather limited tissue tropism, in most cases only infecting the villus epithelium of the small intestine. In addition, symptomatic rotavirus infection appears to be generally restricted to the

---

[1] Departments of Medicine and Microbiology and Immunology, Stanford University School of Medicine, Stanford, CA 94305, USA
[2] Division of Infectious Diseases, The Children's Hospital of Philadelphia and The University of Pennsylvania School of Medicine, 34th St and Civic Centre Blvd., Philadelphia, PA 19104, USA

Current Topics in Microbiology and Immunology, Vol. 185
© Springer-Verlag Berlin · Heidelberg 1994

young of a species. The virus manifests a substantial host range restriction as well, which forms the genetic basis for a variety of human vaccine strategies. In the following discussion, we will review: (1) histopathologic changes of the small intestine induced by rotavirus infection, (2) pathophysiologic mechanisms by which rotaviruses induce disease, (3) factors of the virus and host which determine why rotavirus is primarily a virulent pathogen of infants and young children, and (4) factors of the host which determine why rotavirus virulence is enhanced in developing as compared to developed countries.

1 Pathologic Lesions Produced by Rotavirus Infection

A large number of investigators have studied the histologic changes that take place in the bowel following rotavirus infection. In most cases these studies have involved naturally and experimentally infected animals as opposed to humans. By and large the studies in various animal species arrive at similar conclusions, although in some cases the severity, sequence, and exact localization of the infection vary. It must be remembered when comparing histologic studies of different species that variations exist in the inoculum used, the immune status of the host, the age of the host at the time of challenge, and the method of assessing rotavirus effect. These studies can most easily be discussed and compared by reviewing the findings germane to several different animal species.

1.1 Changes in the Bovine Intestine Following Rotavirus Infection

Much of the original information regarding histologic changes induced by rotavirus infection came from studies of experimentally and naturally infected calves (MEBUS 1976; MEBUS et al. 1974; PEARSON et al. 1978). In colostrum-deprived calves, rotavirus infection led to a change in the villus epithelium from columnar to cuboidal. Villi became stunted and shortened. In addition, the villus tip cells became denuded and the lamina propria had an increase in reticulum-like cells (Fig. 1–4). In the calf it would appear that infection starts proximally and rapidly progresses caudally in the small intestine. Changes are most pronounced in the proximal small intestine. Others have found the calf ileum to be the most affected segment of bowel (TORRES-MEDINA 1984a). Changes occur within 24 h of infection and, by the time clinic diarrhea begins, many of the infected villus tips cells have already been shed. CARPIO and colleagues (1981) compared the histologic changes

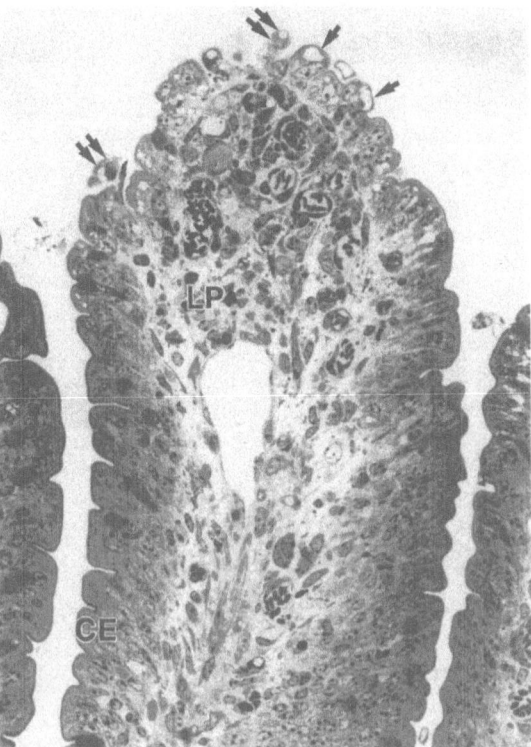

Fig. 1. Photomicrograph of a villus in the terminal ileum of a gnotobiotic calf experimentally inoculated with bovine rotavirus. Enterocytes at the villus tip are disarranged, vacuolated (*single arrows*) and some have exfoliated (*double arrow*). Apparently normal columnar enterocytes (*CE*) are present on the sides of the villus. The lamina propria (*LP*) appears to contain increased numbers of mononuclear cells, but this may be because it is condensed as a consequence of villus stunting. (Figure was kindly supplied by Dr. G.A. Hall, AFRC Institute for Animal Health, Compton, United Kingdom)

in calves exposed to bovine rotaviruses of differing virulence. They also noted that infection produced the most pronounced changes in the jejunum rather than the ileum. Changes consisted of villus atrophy, reduced villus/crypt ratios, and focal flattening of epithelial cells. Inflammation was virtually absent. Strains causing milder illness were characterized by milder histologic changes. Histologic changes are not, however, limited to calves made ill by their rotavirus infection. Asymptomatically infected calves (REYNOLDS et al. 1985) also demonstrated villus stunting and fusion and cuboidal enterocytes. DUBOURGUIER et al. (1981) observed similar findings in asymptomatic calves and also noted that rotavirus infection was associated with release of mucin from uninfected goblet cells.

The effect of rotavirus infection on the specialized follicle-associated epithelium or M cell has been studied in gnotobiotic calves (TORRES-MEDINA 1984b). Infections leads to a loss of M cell microvilli although it did not appear that the rotavirus replicates within these cells.

The histologic changes observed in the calf when rotavirus infection is accompanied by enteric bacterial infection are similar but more extensive than those observed with rotavirus alone (see Sect. 4.2 for further discussion).

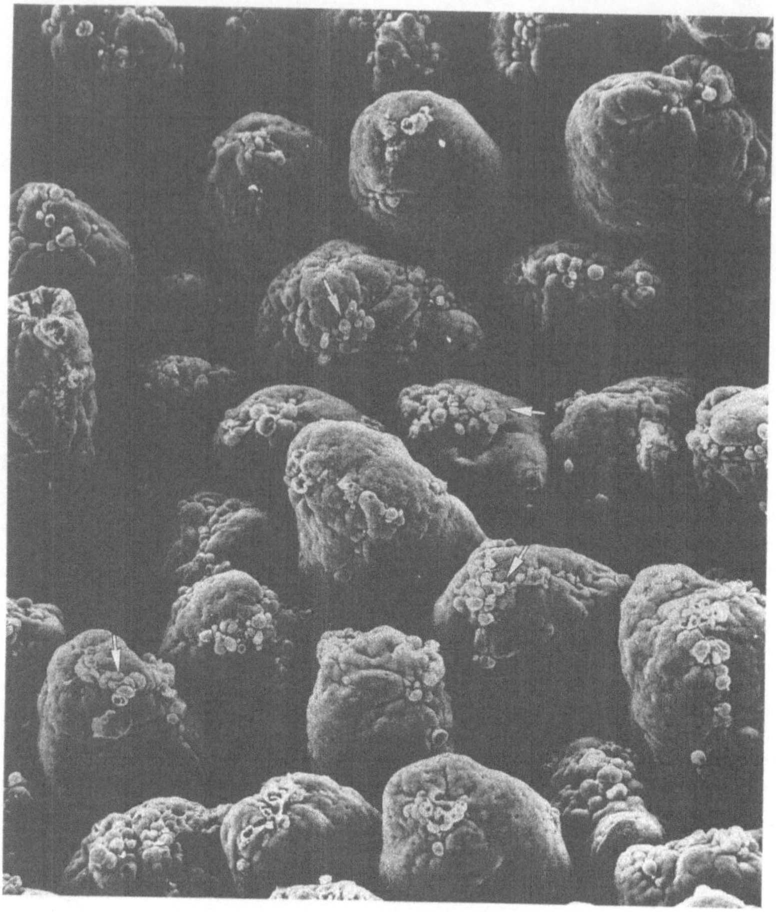

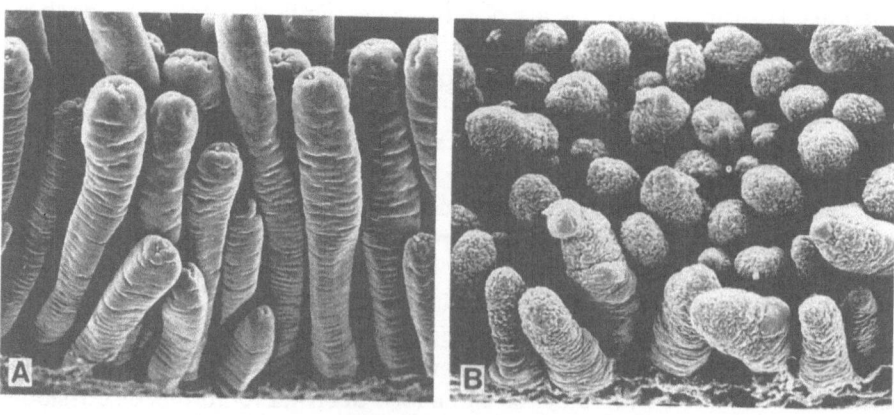

Fig. 2. Scanning electron micrograph of the terminal ileum of a gnotobiotic calf inoculated experimentally with bovine rotavirus. Numerous degenerate enterocytes (*arrows*), apparently about to exfoliate, are visible at the tips of the villi. (Figure was kindly supplied by Dr. G.A. Hall, AFRC Institute for Animal Health, Compton, United Kingdom)

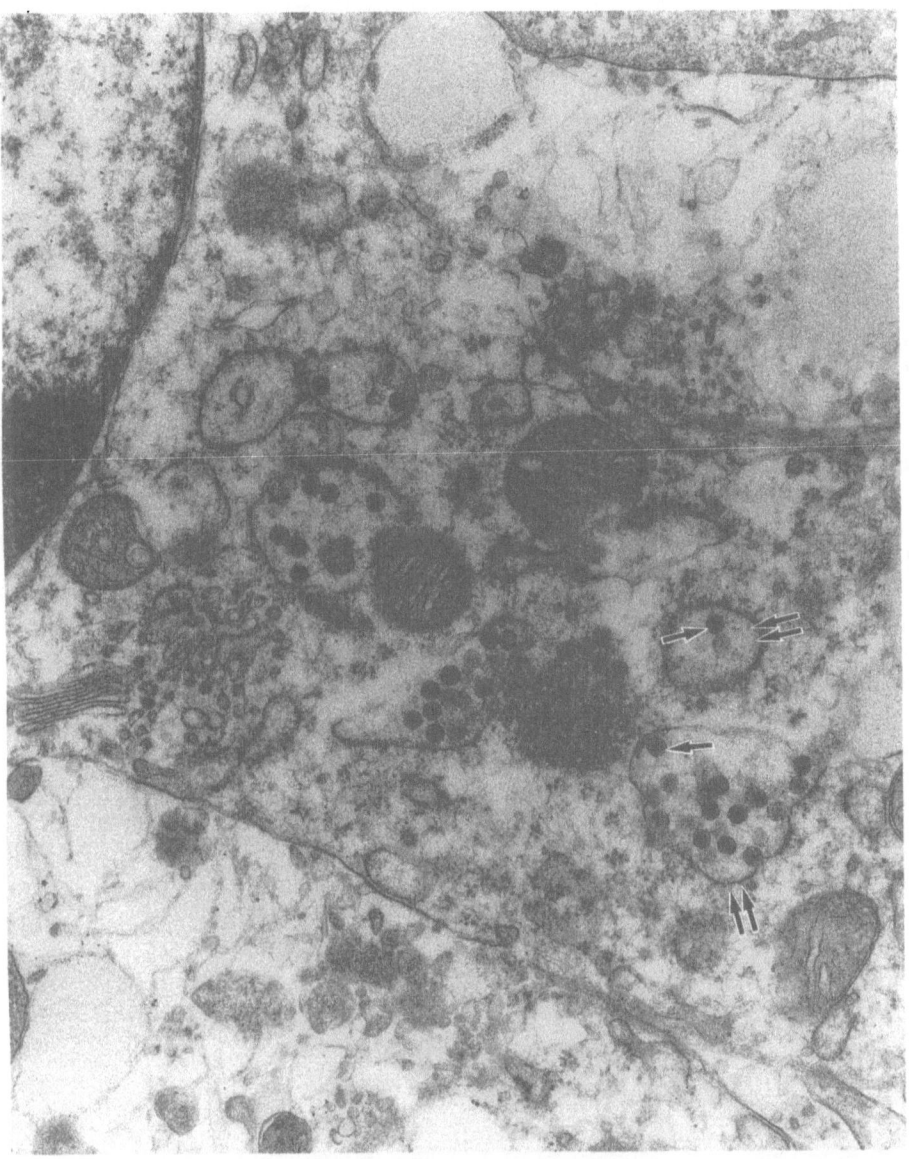

Fig. 4. Transmission electron micrograph of an enterocyte in the mid-small intestine of a gnotobiotic calf inoculated experimentally with bovine rotavirus. Rotavirus particles (*single arrow*) can be seen with dilated rough endoplasmic reticulum (*double arrows*) and the cytoplasm is abnormally electron-lucent. (Figure was kindly supplied by Dr. G.A. Hall, AFRC Institute for Animal Health, Compton, United Kingdom)

◄────────────────────────────────

Fig. 3A, B. Composite of two scanning electron micrographs. **A** mid-small intestine of a gnotobiotic calf inoculated experimentally with saline. **B** mid-small intestine of a gnotobiotic calf inoculated experimentally with bovine rotavirus. The normally long, finger-like villi have become stunted and misshapen and the enterocytes covering the upper half of the villi are disarranged and swollen. (Figure was kindly supplied by Dr. G.A. Hall, AFRC Institute for Animal Health, Compton, United Kingdom)

Whether the changes are simply additive or in fact synergistic remains to be clarified (RUNNELS et al. 1986; STIGLMAIR-HERB et al. 1986; TORRES-MEDINA 1984a).

In the calf, the site of viral replication has been localized by electron microscopy (EM) immunofluorescence, and tissue culture experiments. By and large investigators have detected virus in the same place that histologic abnormalities have been found, i.e., the tip of the villus epithelium. In an early study using a sensitive tissue culture detection assay, infectious virus at low titer was isolated from mesenteric lymph nodes and lung (MEBUS 1976).

1.2 Changes in the Porcine Intestine Following Rotavirus Infection

The pig intestine undergoes pronounced histologic changes following rotavirus infection. The intensity of histologic alternation correlates with the high level of susceptibility of the pig to rotavirus infection. The macroscopic changes characteristic of rotavirus infection in the pig include thinning of the intestinal wall and disappearance of chyle within mesenteric lymphatics. Microscopic changes (Figs. 5–7) are characterized by villus atrophy, which is most severe in the distal small bowel, and, as in the calf, villus blunting and conversion to a cuboidal epithelium (CROUCH and WOODE 1978; LEECE and KING 1978; LEECE et al. 1976; PEARSON and McNULTY 1977; THEIL et al.

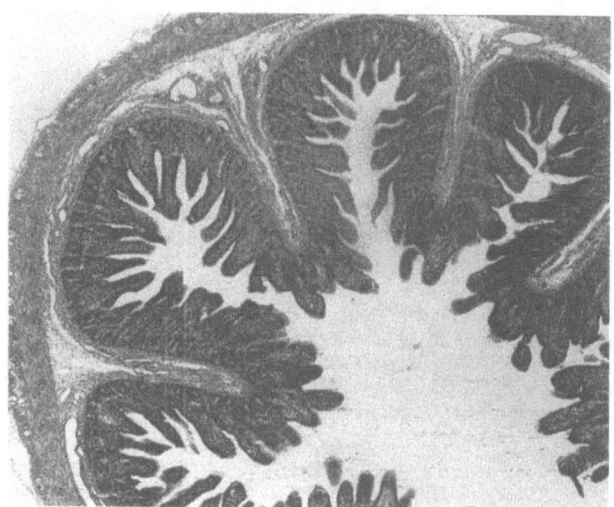

Fig. 5. Jejunum of a pig 24 h after experimental inoculation with porcine rotavirus. Villi have cuboidal epithelial cells. (Figure reproduced from THEIL et al. 1978 with permission, from the American Journal of Veterinary Research) × 200

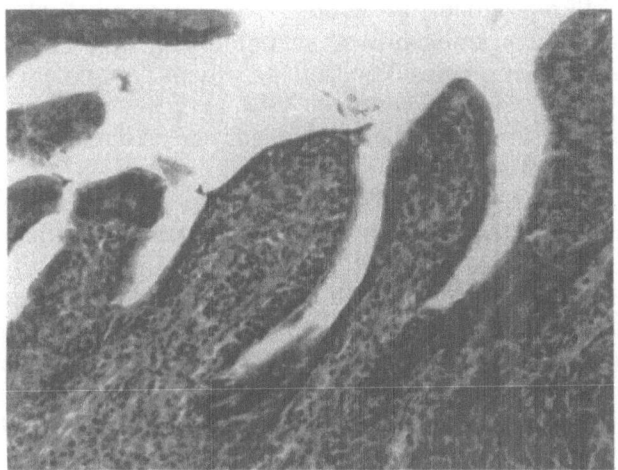

Fig. 6. Jejunum of a pig 24 h after experimental inoculation with porcine rotavirus. Villi are short and blunt. (Figure reproduced from THEIL et al. 1978 with permission, from the American Journal of Veterinary Research) × 30

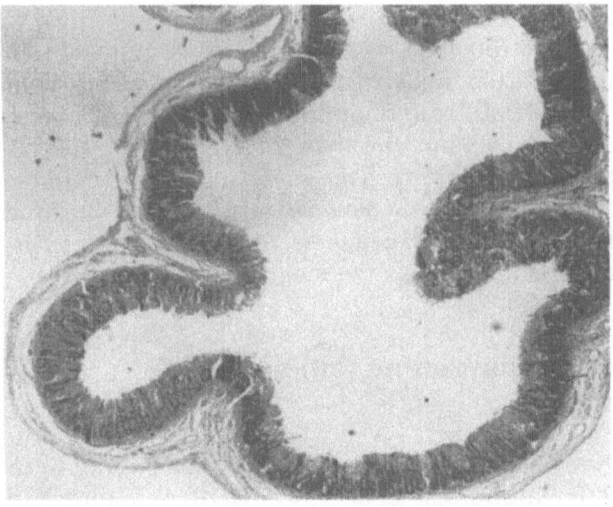

Fig. 7. Jejunum of a pig 48 h after experimental inoculation with porcine rotavirus. Villi are absent. (Figure reproduced from THEIL et al. 1978 with permission, from the American Journal of Veterinary Research) × 30

1978). Although immunofluorescence studies demonstrate that most rotavirus is localized to the porcine small bowel, scattered foci of infected
epithelium have been seen in the porcine colon, and mesenteric lymph nodes
have been shown to contain rotaviral antigens. Viral antigens were not
identified in porcine stomach, lung, liver, kidney, spleen, or pancreas after
experimental infection (THEIL et al. 1978).

The changes induced by rotavirus infection in the pig have also been
studied at the ultrastructural level (NARITA et al. 1982; PEARSON and McNULTY
1979; POSPISCHIL et al. 1982; TORRES-MEDINA and UNDERDAHL 1980). By
scanning EM the ileum would appear to be the most susceptible region of
the small bowel. The earliest changes in the enterocyte include swelling
and separation, with loss of microvilli occurring on the upper third of the
villi. These alterations are followed later in infection by cell rounding,
endoplasmic reticulum (ER) distension, villus fusion, loss of tip cells, and
a small amount of inflammatory infiltrate in the submucosa. Some investigators have found rotavirus within both goblet cells and epithelial cells in
the porcine intestine (NARITA et al. 1982), but this has not been a universal
observation.

Pigs have a well documented development resistance to rotavirus
infection similar to that seen in mice and cattle (see Sects. 1.4 and 3.1).
In pigs, newborns are much more susceptible to severe illness than pigs
over the age of 7 days. Using ligated intestinal loops, KIRSTEIN et al. (1985)
found that epithelial cells from older pigs were less susceptible to rotavirus
infection than cells from newborns. This change in susceptibility correlated
temporally with the loss of ability to transport macromolecules through
enterocytes. Of interest was the observation that the loss of enterocyte
susceptibility to rotavirus progressed caudally with time.

As is the case with bovine rotavirus, naturally occurring porcine rotaviruses have been shown to differ somwhat in their ability to cause histologic
abnormalities (COLLINS et al. 1989). Although the SDSU strain of porcine
rotavirus caused more severe villus atrophy than the OSU strain, the two
strains produced similar clinical symptoms. This study again points out that
there is not an exact one to one correlation between histologic or ultrastructural change and clinical course, although in general rotavirus strains causing
the most severe pathologic damage cause the most illness.

1.3 Changes in the Ovine Intestine Following
   Rotavirus Infection

Rotavirus infection in the lamb is less severe than in either the pig or the
calf (McNULTY et al. 1976; SNODGRASS et al. 1977, 1979). Viral infection
in the lamb, like the pig, appears to be centered in the caudal portion of
the small intestine with little or no evidence of infection in the duodenum.
Microscopic abnormalities are limited to villus blunting, tip denudation, and
villus infiltration with macrophages and a few polymorphonuclear leukocytes.

Viral Antiges are seen in the same location as histologic changes, except that some virally infected cells are found in the cecum and colon and some nonspecific histologic abnormalities, but no viral antigens, are found in the liver and lung. Morphometric studies of the lamb small intestine demonstrate both villus atrophy and crypt hypertrophy changes that are most pronounced in the ileum. Epithelial kinetic studies of lamb intestine during rotavirus infection demonstrate a sustained and substantial increase in crypt cell division which may be responsible, in part, for fluid loss (see Sect. 2).

1.4 Changes in the Murine Intestine Following Rotavirus Infection

Rotavirus infection in the mouse has been studied more extensively than in any other species. Murine rotaviruses were the first to be discovered and the mouse has proven to be an extremely useful animal model (ADAMS and KRAFT 1963, 1967; CHEEVER and MUELLER 1947). However, the restriction of illness to the first 15–17 days of age and the great difficulty of propagating murine rotavirus strains efficiently in vitro have reduced the utility of this animal model system for many studies. Recent studies indicating that adult mice are highly susceptible to rotavirus infection but not disease should increase the utility of the mouse model (WARD et al. 1990, 1992).

The pathology of murine rotavirus is generally similar to that observed in calves, pigs, and lambs; but it does, in fact, differ in certain respects (Figs. 8–10). Within 24 h after a mouse is infected with a murine rotavirus, small intestinal enterocytes appear swollen and vacuolated when studies by scanning or transmission EM or by light microscopy (COELHO et al. 1981; OSBORNE et al. 1988; RODRIGUEZ-TORO 1980; STARKEY et al. 1986). Infection and histologic change are concentrated in the upper small intestine and spread caudally with time as is seen in calves. Some investigators have found evidence for scattered viral infection in the colon (BANFIELD et al. 1968; LITTLE and SHADDUCK 1982, 1983; WILSNACK et al. 1969), but others have not (COELHO et al. 1981; STARKEY et al. 1986). With the exception of recent studies (UHNOO et al. 1990b) and early studies by ADAMS and KRAFI (1963), showing isolation of virus from nonintestinal tissues, most investigators have concluded that infection of mice with rotavirus is limited to the intestine and to intestinal lymphoid tissue (DHARAKUL et al. 1988). When rotavirus is administered by aerosol to infant mice, antigen (but not infection) can be demonstrated in the lung and spleen (PRINCE et al. 1986). Vacuolation of enterocytes is most prominent on the villus tips but occurs in enterocytes throughout the villus except for the crypts. The vacuolation is more extensive in mice than in other species. In mice, unlike other species, villus blunting is limited and very transient and crypt cell hyperplasia is not seen. Loss of villus tip cells is also more limited in mice than in other animals. The mononuclear infiltrate is modest and less extensive than in calves or pigs. This lack of extensive pathologic changes found in the mouse intestine,

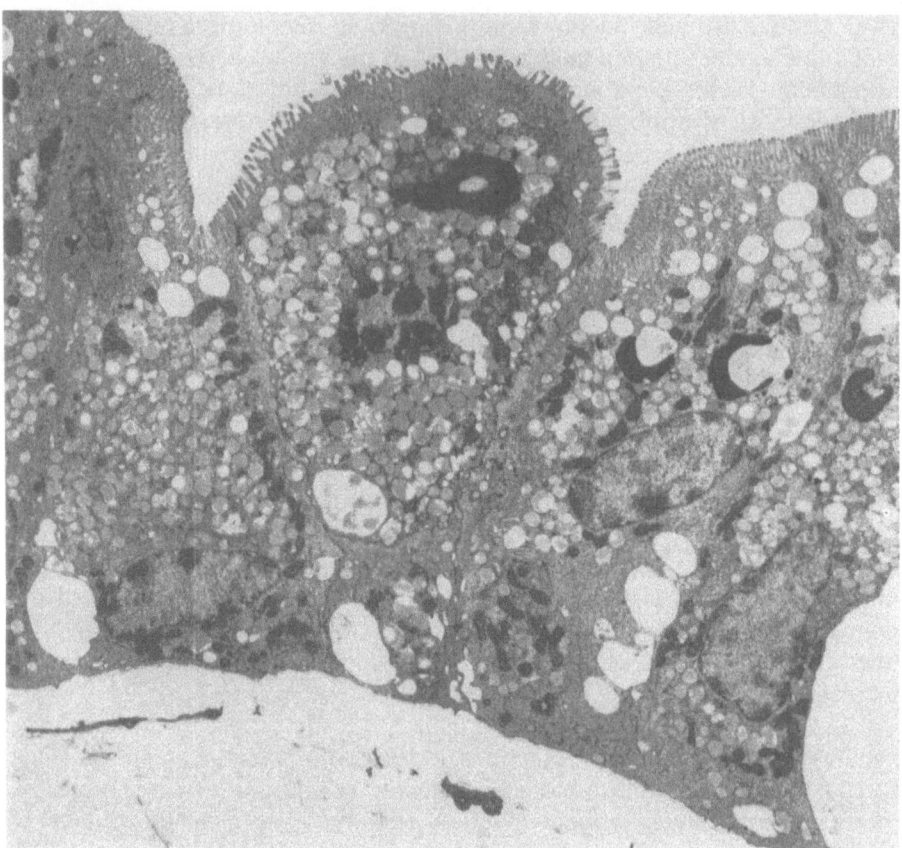

Fig. 8. Low power electron micrograph of a mouse small intestinal epithelial cell approximately 1 day after murine rotavirus infection. The mucosal brush border is damaged and the cell contains lipoid-like (*L*) material in vesicles and/or the rough endoplasmic reticulum. (The micrograph was kindly supplied by Dr. J. Wolf, Boston, MA)

coupled with the observation that many enterocytes appear vacuolated but do not show evidence of viral replication, has led OSBORNE et al. (1988) to postulate that, in the mouse, infection leads to local villus ischemia and that diarrhea is secondary to vascular damage, i.e., is of ischemic origin rather than due to direct destruction of enterocytes.

A variety of investigators have localized the site of viral replication in mice and studied the progression of infection using either immunofluorescent or EM techniques (EYDELLOTH et al. 1984; LITTLE and SHADDUCK 1982; SHERIDAN et al. 1983; STARKEY et al. 1986). Viral shedding frequently occurs in a biphasic pattern with peaks occuring approximately 2 and 6 days after inoculation. Infection is most pronounced in the proximal and mid-small bowel and can be localized to all areas of the villus with the exception of the crypt.

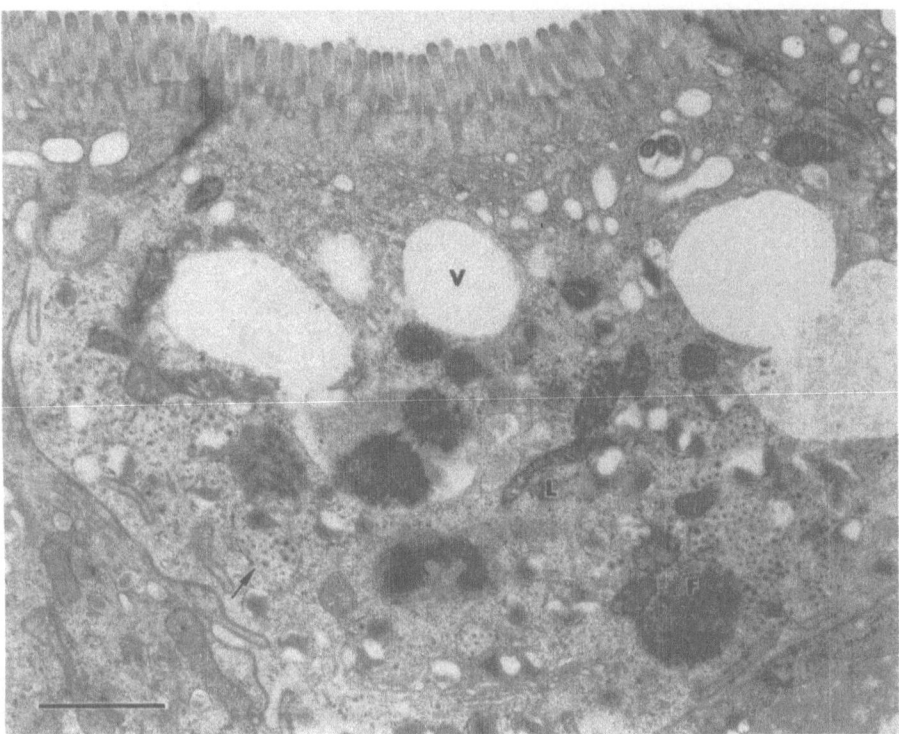

Fig. 9. Electron micrograph of a typical mouse ileal absorptive cell infected with epizootic diarrhea of infant mice (EDIM) virus. Features shown are: viral factories (*F*), endoplasmic reticulum containing virus particles (*arrow*), vacuolization of the cytoplasm (*V*), and vacuoles containing lipoid-like material (*L*). *Bar*, 1 um. (From WOLF et al. 1981).

The pathogenic potential of heterologous host rotaviruses has been studied most extensively in the mouse. Although human rotavirus can cause only mild illness in calves and pigs (WYATT et al. 1978) and simian and bovine rotaviruses would appear to be attenuated in humans (KAPIKIAN et al. 1989), quantitative comparisions between homologous host and heterologous host rotaviruses have not been carried out except in mice. The simian SA11 rotavirus is capable, when administered to infant mice in large amounts, of causing diarrhea and histologic changes in the intestine similar to murine strains (OFFIT et al. 1984). Selected human, bovine, and other simian can also cause illness in mice (BELL et al. 1987; GOUVEA et al. 1986; IJAZ et al. 1989; RAMIG 1988b). In all cases of heterologous rotavirus infection in mice, viral replication is quite restricted. In the mouse the murine rotavirus dose required to induce disease in 50% of inoculated animals (DD_{50}) appears to be 10^4- to 10^5-fold lower than the DD_{50} of heterologous host strains. In addition, heterologous host strains do not spread among infant mice while murine rotaviruses spread with great efficiency (GREENBERG

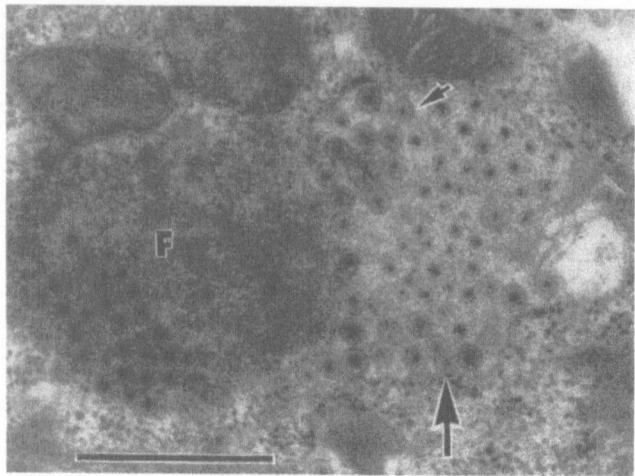

Fig. 10. Higher magnification of a portion of Fig. 8 showing viral factories (*F*) and a profile of endoplasmic reticulum (*large arrow*) containing viral particles 65-130 nm in diameter. Some, but not all, particles appear to have an outer membrane (*small arrow*). *Bar*, 0.5 μm. (From WOLF et al. 1981).

et al. 1986). Neither the molecular nor the genetic basis for the obvious host range restriction of rotavirus illness in the mouse and in many other species has yet been determined although recent studies have implicated several non-structural as well as structural proteins (BROOME et al. 1993; HOSHINO et al. 1993).

Recently, the dogma that rotavirus replication is highly restricted to the gastrointestinal tract has been called into question by the observation that some heterologous simian rotavirus strains can spread to the liver of severely immunodeficient (SCID) mice and cause a lethal hepatitis (UHNOO et al. 1990b). Rotavirus hepatotropism in the mouse appears to be limited to some heterologous host strains and to be greatly enhanced in, but not limited to, SCID mice. Since there has been some evidence of abnormal liver function during natural rotavirus infection of humans (GRIMWOOD et al. 1988; GRUNOW et al. 1985; KOVACS et al. 1986), this observation is potentially relevant to human vaccine studies. A recent in vitro study indicates that gene 4 encodes the ability of selected rotavirus strains to replicate efficiently in a continuous line of human hepatocytes (RAMIG and GALLE 1990). However, the basis for hepatotropism in vivo remains unclear.

1.5 Changes in the Avian and Mammalian Intestine Following Rotavirus Infection with Other Strains of Group A Rotavirus

Studies of the intestinal lesions seen following rotavirus infection in humans are limited (HOLMES et al. 1975; SUZUKI and KONNO 1975). Findings are similar to those reported in other animal species and include both loss of the microvilli and swelling of the ER. Obviously, most human specimens are obtained from the proximal part of the small bowel and distal small bowel pathology is not well characterized.

Rotavirus infection in dogs appears to spare the duodenum and be limited to the jejunal and ileal regions of the small intestine (JOHNSON et al. 1986). At the macroscopic and light microscopic level changes consist of thinning of the bowel wall and lowering of the villus/crypt ratio. Scanning and transmission EM studies show villus swelling and loss of tip cells followed by mild villus atrophy (JOHNSON et al. 1983).

Changes in the rabbit intestine after rotavirus infection are similar to those seen in other species with villus shortening and blunting, epithelial cell flattening, and slight increase in crypt depth (CONNER et al. 1988; HAMBRAEUS et al. 1989; SCHOEB et al. 1986). Although rabbits develop histologic lesions and illness during field epidemics, in the laboratory rotavirus infection of rabbit does not cause illness.

It is interesting to note that group A rotavirus infection of certain avian species such as turkeys is also associated with characteristic pathologic changes in the small intestine. In birds, these changes are most severe in older rather than younger animals (YASON et al. 1987). Macroscopically, after rotavirus infection the turkey intestine appears pallid and distended. Microscopic changes include basal vacuolation, villus atrophy, villus fusion and mild leukocyte infiltration into the lamina propria. The villus crypt ratio is decreased and villus tip cells are lost into the lumen of the small intestine.

1.6 Changes in the Avian or Mammalian Intestine Following Group B Rotavirus Infection

The nongroup A rotaviruses are covered more completely in Chapter of L.J. SAIF and B. JIANG (pp 339). It should be noted, however, that the histologic changes associated with infection with these agents differ to some degree from those seen in group A infection (ASKAA and BLOCH 1984; CHASEY and BANKS 1986; CHASEY et al. 1989; HUBER et al. 1989; THEIL et al. 1985; VONDERFECHT et al. 1984, 1986). Group B rotavirus infection is restricted to the small intestinal villus tip cells and appears to be most abundant in the distal small bowel. Changes are somewhat milder that those seen in group A infection. The single major distinction between group A and group B infection involves the presence of syncytial cell formation on the villus tip during group B rotavirus replication. Syncytia have been observed in rats, pigs, calves, and lambs and appear to be a fundamental property of group B infection. The structural basis for the giant cell formation is not currently understood.

2 Pathophysiology of Rotavirus infection

2.1 Mechanism of Diarrhea

Despite the importance of rotavirus illness in humans and animals, studies of the mechanism by which viral infection actually causes diarrhea have been infrequent. In an early investigation, DAVIDSON and colleagues (1977) used human rotavirus to infect conventional 8–10 day old piglets. Some 72 h after infection they sacrificed the animals and studied Na^+ and Cl^- fluxes in an Ussing chamber system using a short-circuited jejunal epithelium. They determined that net ion fluxes were not different from those in control uninfected animals but that glucose-mediated sodium absorption was diminished. In addition, they observed that rotavirus-infected epithelium did not have an altered Cl^- secretion in response to theophylline and that cyclic AMP levels were not changed by rotavirus infection. An increased thymidine kinase activity and decreased sucrase activity at the time of study was also observed. They interpreted these findings to indicate that retarded differentiation of uninfected enterocytes migrating at an accelerated rate up the villus is responsible for the absorptive abnormalities seen during rotavirus illness. These studies did not directly evaluate carbohydrate absorption.

In vivo perfusion of the jejunum in miniature swine was used to directly assess absorption during acute rotavirus infection (GRAHAM et al. 1984). Absorption of Na^+, water, and 3-0-methylglucose was decreased in infected pigs. Intestinal lactase and sucrase, and, to a lesser degree, Na^+–K^+ ATPase were also reduced during infection. Concurrent with histologic abnormalities in the intestine, an increased fecal osmotic gap and increased fecal lactose concentration in rotavirus infected piglets was found. These findings were felt to be most compatible with the conclusion that destruction of mature villus tip cells leads to carbohydrate malabsorption and hence osmotic diarrhea. This study did not, however, evaluate crypt cell secretion as a possible additional factor in rotavirus illness.

Most recently, studies of the mouse have been used to clarify the basis for diarrhea during rotavirus infection (COLLINS et al. 1988). During murine rotavirus infection intestinal alkaline phosphatase and lactase are clearly decreased, and as observed in the pig, thymidine kinase is increased. Unlike the pig, however, intestinal sucrase and maltase levels are found to rise rather than decrease. This "precocious" maturation of selected brush border hydrolases might be due to a steroid effect occurring during acute infection. As expected, there are no changes in adenylate or guanylate cyclase during acute murine rotavirus infection.

Although mice developed lactase deficiency during acute infection, they did not in fact demonstrate lactose malabsorption during acute infection. COLLINS and colleagues (1988) postulated that, in the mouse, carbohydrate malabsorption may not be the cause of diarrhea as it appears to be in the pig.

They also noted that lambs are not lactose intolerant during rotavirus infection (FERGUSON et al. 1981). It was thus suggested that in mice and perhaps lambs, crypt cell secretion may play a more important role as a cause of fluid loss. It is interesting to note that lactose malabsorption does occur during rotaviral infection of humans (HYAMS et al. 1981). In calves, lactose deficiency during rotavirus infection is exacerbated by concurrent toxigenic *E. coli* infection (STIGLMAIR-HERB et al. 1986).

2.2 Changes in Absorption of Macromolecules Across the Intestinal Mucosal Surface During Rotavirus Infection

Rotavirus infection appears to alter the permeability of the intestine to a variety of molecules. D-xylose absorption is decreased in mice and humans during rotavirus infection (IJAZ et al. 1987; MAVROVICHALIS et al. 1977). In order to further investigate small bowel permeability during rotavirus infection, nine acutely infected children were studied to determine their ability to absorb a series of different size polyethylene glycols (PEGs) (STINTZING et al. 1986). Rotavirus infection decreased the ability of the intestine to absorb all these molecules (molecular weight 282–1250) but the effect was greatest on the smaller PEGs.

The intestinal permeability to lactose, lactulose, and L-rhamnose was evaluated simultaneously in 18 rotavirus-infected children and 20 controls (NOONE et al. 1986). Administering the absorption test markers together and comparing the ratio of marker uptake improved the specificity of the assay and avoided problems of gastric emptying and intestinal transit. These studies demonstrated that intestinal permeability to lactose and lactulose was increased during infection while uptake of L-rhammose was reduced. In addition, these studies again demonstrated in humans that these was a relative reduction in lactose hydrolysis during rotavirus infection.

The transport and degradation of a macromolecule such as horseradish peroxidase (HRP) is also altered in both conventional and germ-free mice during rotavirus infection (HEYMAN et al. 1987). Absorption of intact HRP is increased during or shortly after rotavirus infection. This absorption is influenced by the presence or absence of bowel flora. It was postulated that the increased permeability of the gut to undegraded proteins during rotavirus infection might effect the subsequent immune status of the host to intraluminal contents.

It seems clear that there is no unifying hypothesis governing the effect of rotavirus infection on intestinal absorption and transport. Uptake of some molecules including HRP and L-rhamnose is increased while uptake of many others including PEG, lactulose, and D-xylose is lowered. It also seems likely that these diverse affects may vary somewhat from species to species and from one region to another in the gut.

3 Factors of the Virus and Host Which may Determine why Rotavirus is Primarily a Virulent Pathogen of Infants and Young Children

A number of studies investigating the age at which patients develop symptomatic or asymptomatic rotavirus infections allows one to draw the following conclusions: First, patients hospitalized with rotavirus-induced gastroenteritis are primarily between 6 and 24 months of age. Second, neonates may be more likely to shed rotavirus without symptoms than older infants and young children. Third, although adult contacts of children with symptomatic rotavirus infection often shed rotavirus in their feces, rotaviruses are an unusual cause of severe diarrhea in this age group.

A number of investigators found that most patients admitted to the hospital with diarrhea secondary to rotavirus-induced gastroenteritis were between 6 and 24 months of age. KAPIKIAN and coworkers (1976) studied infants and children up to 13 years of age admitted to the hospital with diarrhea and dehydration. Rotaviruses were associated with diarrhea in 21% of children less than 6 months of age, 68% of children 6–24 months of age, and only 8% of children 24 months to 13 years of age. These findings were similar to those of RODRIGUEZ and coworkers (1977).

Studies which found that neonates may be more likely than older infants and young children to develop asymptomatic as compared to symptomatic rotavirus infection are balanced by those which did not. PEREZ-SCHAEL and coworkers (1984) studied rotavirus shedding by neonates in nurseries of a large maternity hospital in Caracas, Venezuela over a 1 year period. Of 108 neonates studied, 62 (57%) shed rotavirus within the first few days of life. Only four (6%) of the neoates who shed rotavirus developed diarrhea; one child required oral rehydration therapy. Similarly, CHRYSTIE and others (1978) found that 343 (32.5%) of 1056 newborns in London excreted rotaviruses in their feces; 92% of neonates were asymptomatic and the remaining 8% of newborns developed only mild symptoms of gastroenteritis. However, the findings of PEREZ-SCHAEL et al. (1984) and CHRYSTIE et al. (1978) are balanced by studies which did not find differences in the ages of children who developed asymptomatic or symptomatic rotavirus infections. BARRON-ROMERO and coworkers (1985) detected rotaviruses and other enteropathogenic agents in 288 (42%) of 684 children in daycare centres in Mexico City. Children with asymptomatic and symptomatic rotavirus infection were found, respectively, in 26% and 31% of 2–11 month old infants, 26% and 23% of 12–23 month old children, 30% and 23% of 24–35 month old children, 34% and 29% of 36–47 month old children, and 33% and 40% of 48–60 month old children. Unfortunately, the 2–11 month old age group was not further stratified to include children less than or older than 5 months of age. Combining infants 0–5 months of age with those 6–11 months of age may have obscured differences between these two groups in percentages of infants who developed asymptomatic infection. MURPHY and others (1977) found that 304 (49%) of 628 newbrons in Australia were excreting rotaviruses within the first 7 days of life. In contrast to previously cited studies, 84 (28%) newborns developed diarrhea.

Although adult contacts of children with symptomatic rotavirus infection often shed rotavirus in their feces, rotaviruses are less commonly a cause of severe diarrhea in adults than in infants and young children. WENMAN and coworkers (1979) prospectively studied diarrhea in 98 families in central Canada for approximately 16 months. A total of 43 rotavirus infections were identified in 188 adults. Fourteen (32%) of these infections were associated with symptoms of mild gastroenteritis. Rotavirus infections occured in 36 (35%) of 102 adults whose children had rotavirus infection, as compared with four (5%) of 86 adults without infected children ($p < 0.001$). Similarly, KIM and others (1977) found that 26 (41%) of 64 adult contacts of children with rotavirus-induced gastroenteritis developed evidence of infection; 88% of adults with rotavirus infection were asymptomatic and 12% developed symptoms of mild gastroenteritis. KAPIKIAN and coworkers (1976) found that adult contacts of infants and young children with rotavirus were likely to develop an asymptomatic rotavirus infection. Of 40 adult contacts of children with rotavirus-induced gastroenteritis, 13 (32.5%) developed sero- logic evidence of rotavirus infection. Three of these 13 adults developed symptoms of mild diarrhea, the remainder were asymptomatic. Lastly, BARRON-ROMERO et al. (1985) found that 62 (20%) of 302 adults in daycare centers shed rotavirus in their feces; all of these infections were asymptomatic.

Several hypotheses which may in part explain these epidemiologic findings are summarized below.

3.1 Age-Related Differences in the Quantity of Rotavirus-Binding Receptors on Villus Epithelial Cells

The quantity of rotavirus-specific binding receptors on mature villus epithelial cells of the small Intestine may be influenced by age of the host. RIEPENHOFF- TALTY and coworkers (1982) used a murine model to examine differences in putative rotavirus-binding receptors on villus epithelial cells of suckling mice. Suckling mice from 2–75 days of age were orally inoculated with murine rotavirus. Similar to findings of other investigators, only mice less than 14 days of age developed diarrhea. Rotavirus antigens were detected in 2%–30% of enterocytes from suckling mice (less than 3 weeks of age) but not detected in enterocytes of adult mice. The peak age at which rotavirus antigens were detected in enterocytes corresponded to the age at which animals were most likely to develop diarrhea (6–11 days of age). To determine the quantity of putative rotavirus-binding receptors on villus epithelial cells, enterocytes from rotavirus-inoculated adult or suckling mice were incubated with murine rotavirus-coated sheep red blood cells (MRV-

SRBCs). Binding of three or more MRV-SRBCs to an enterocyte was considered to be an enterocyte-MRV-SRBC rosette. MRV-SRBCs bound to approximately 18%, 20%, 10%, 1%, and 0% of enterocytes from 2-5, 6-11, 12-19, 20-26, and greater than 75 day old mice, respectively. These findings were consistent with the hypothesis that the capacity of murine rotaviruses to induce diarrhea in suckling but not adult mice may be related to the quantity of rotavirus-binding receptors on the surface of villus epithelial cells. WOLF and others (1981) studied mice less than 40 days of age after intragastric inoculation with murine rotavirus. Similar to studies by RIEPENHOFF-TALTY et al. (1982), 23%, 95%, 41%, and 0% of 1-3, 3-11, 15-17, and greater than 34 day old mice developed diarrhea. Mice administered cortisone acetate developed premature maturation of intestinal epithelial cells manifested by increased activity of the intestinal brush border enzymes maltase, sucrase, and alkaline phosphatase. Associated with increased intestinal epithelial cell maturation, 8 day old mice receiving cortisone were less susceptible to murine rotavirus-induced diarrhea than were 8 day old mice not receiving cortisone acetate.

A problem with using mice in studies of rotavirus pathogenesis is that mice are only susceptible to rotavirus-induced gastroenteritis between 2 and 14 days of age (RIEPENHOFF-TALTY et al. 1982; WOLF et al. 1981). Unlike adult humans, adult mice simply do not develop diarrhea after inoculation with homologous host rotaviruses. Therefore, the degree to which a murine model is predictive of events occurring in the human gastrointestinal tract remains unclear.

3.2 Age-Related Differences in Exposure to Virulent Strains

As stated above, it remains unclear whether neonates are more likely than older infants and young children to develop asymptomatic or symptomatic rotavirus infections. With this caveat in mind, several investigators found that neonates shed rotaviruses (nursery strains) which were distinct from strains shed by older infants and young children. Investigators postulated that differences in likelihood of developing disease after rotavirus infection may be related to virulence of rotavirus strains to which the infant is exposed. ALBERT and coworkers (1987) examined 23 rotavirus isolates obtained from 71 newborns in Melbourne, Australia between 1975 and 1979. All strains collected were serotype 3 as determined by fluorescent-focus neutralization assay using monoclonal antibodies directed against the outer capsid protein VP7. RNA electrophoretic patterns of neonatal strains differed from other serotype 3 strains circulating in the community at that time. The serotypic similarity and electrophoretic dissimilarity of neonatal strains from community strains suggested that rotaviruses obtained from newborns may form a unique group of viruses. PEREZ-SCHAEL et al. (1984) examined 52 rotavirus isolates obtained from newborn children in Caracas, Venezuela. RNA gene segments from these strains showed a similar pattern of migration on SDS-polyacrylamide gels. In addition, cross-hybridization analysis revealed a strong degree of genetic homology among the neonatal strains tested which did not extend to laboratory strains originally isolated from symptomatic infants and older children.

Recent studies found that alteration of rotavirus virulence in the neonate may be related to differences in the viral gene segment 4 protein product. FLORES and coworkers (1986) performed RNA:RNA cross-hybridization analysis on three rotavirus strains obtained from asymptomatic newborns (serotypes 1, 2, and 3) and four laboratory strains originally obtained from symptomatic infants and young children (serotypes 1, 2, 3 and 4). When analyzed by northern blot hybridization at a low level of stringency, all genes from strains tested cross-hybridized. However, at higher levels of stringency, a difference in gene segment 4 sequence was detected between strains recovered from asymptomatic newborns and those obtained from symptomatic infants and older children. Gene segment 4 appeared to be highly conserved among strains obtained from asymptomatic infants even though these strains represented distinct rotavirus serotypes. GORZIGLIA et al. (1986) extended the observations of FLORES and coworkers (1986) by examining amino acid homologies between the VP5 and VP8 regions of gene segment 4. Alignment of amino acid sequences of VP8 protein, the downstream cleavage region, and the NH_2-terminal of VP5 of rotaviruses obtained from asymptomatic and symptomatic infants indicated a high degree of homology (96% or more) among 'asymptomatic' rotaviruses while homology between asymptomatic and symptomatic strains was less (68%–72%). In addition, a high degree of coservation of amino acid sequences (92%–97%) was observed among virulent strains. At 48 positions in the protein sequence of VP8, the cleavage region, and the NH_2-terminal region of VP5, an amino acid was conserved among asymptomatic rotaviruses, while a different amino acid was conserved among virulent rotaviruses.

Studies of FLORES et al. (1986) GORZIGLIA et al. (1986) suggested that the gene segment 4 protein product (VP4) may be related to virulence. Results of animal model studies suggest that VP4 may play a role in viral virulence. OFFIT and coworkers (1986) found that bovine rotavirus strain NCDV and simian rotavirus strain SA11 exhibited markedly different patterns of gastrointestinal tract virulence when orally inoculated into suckling mice. SAII rotavirus was more virulent than NCDV; the does of SA11 required to induce disease in 50% of inoculated animals was 50-fold less that required for NCDV. Using reassortment rotaviruses derived by coinfection of MA-104 cells with SA11 and NCDV, this difference in dose of SA11 and NCDV required to induce disease was determined by gene segment 4. However, heterologous host rotaviruses in mice do not appear to undergo multiple cycles of replication (GREENBERG et al. 1986). In addition, approximately 10^5-fold more rotavirus is required to induce disease after heterologous than after homologous host infection (GREENBERG et al. 1986). The murine model may be predictive of events occurring early in virus infection, prior to or after a single round of replication (e.g., attachment, entry, or uncoating). However, this model may not accurately predict factors associated with virulence when disease is induced after multiple cycles of virus replication.

To more accurately determine the genetics of rotavirus virulence, HOSHINO and coworkers (1988) examined pigs inoculated with homologous or heterologous host strains. They found that newborn gnotobiotic piglets were more susceptible to disease after oral inoculation with porcine (strain SB-1A) than with human rotavirus (strain DS-1). Oral inoculation of piglets with reassortant rotaviruses from these two parent strains did not reveal a single gene associated with viral virulence. Rotavirus virulence appeared to be associated with more than one gene and not easily sorted out using a reassortant virus approach. It appears that viral replication in vivo requires efficient function of several rotavirus gene products.

However, studies using reassortant rotaviruses must be interpreted with caution. RAMING and coworkers (1988a) found that the genetic background onto which donor genes were reassorted affected phenotypes conferred by the presence of donor gene segments. These investigators isolated a simian rotavirus (SA11 4F) which was slightly less virulent than SA11 but significantly more virulent than bovine strain B223 when inoculated orally into suckling mice. The dose required to induced disease in 50% of inoculated animals was approximately 50-fold greater for bovine B223 than that required for simian SA11 4F. To determine the gene or genes associated with virulence, reassortant rotaviruses were generated by coinfection of MA-104 cells with SA11 4F and B223. Virulence did not segregate with any single gene segment of SA11 4F/B223 reassortants. These data suggested that phenotype conferred by a single gene segment depends on the genetic background in which the segment is being tested. Therefore, virulence studies using only two parent strains to generate reassortant rotaviruses should be interpreted with caution.

3.3 Neonates are Protected Against Relatively Severe Disease by Passive Transfer of Maternal Antibodies Transplacentally or in Colostrum and Milk

Newborns acquire maternal antibodies transplacentally prior to birth. Because rotaviruses are ubiquitous, most newborns will have rotavirus-specific antibodies in serum at the time of birth. Consistent with the fact that IgG has a half-life in serum of approximately 25 days, investigators found that rotavirus-specific antibodies decline over the first 6 months of life (ELIAS 1977). It remains unclear whether rotavirus-specific antibodies acquired transplacentally are important in protection of the neonate against relatively severe disease. Consistent with the hypothesis that circulating, maternally

acquired, rotavirus-specific neutralizing antibodies protect against symptomatic infection, WRIGHT and coworkers (1987) found that infants with
rotavirus-specific neutralizing antibodies in their serum shed lesser quantities
of rotavirus in their feces after challenge with candidate vaccine strain RRV
than those without detectable antibodies. Unfortunately, in human infants
it is difficult to ascertain whether rotavirus-specific antibodies in serum are
generated in response to previous infection or are the result of passive
acquisition. Using cross-fostering techniques with a murine model for oral
infection with a heterologous host rotavirus strain, OFFIT and CLARK (1985)
found that circulating antibodies did not alone protect against disease. In
newborn mice foster-nursed by seronegative dams, circulating rotavirus-
specific neutralizing antibodies in high titer (transferred transplacentally from
parenterally immunized dams) did not protect against SA11 virus challenge.
Therefore, rotavirus-specific antibodies may need to be active at the intestinal
mucosal surface to protect against disease. Rotavirus-specific antibodies
acquired transplacentally may not reach the intestinal surface in adequate
quantities to protect against challenge.

Rotavirus-specific antibodies may be passively transferred to the newborn
in colostrum or breast milk in quantities sufficient to afford protection against
disease. Despite the presence of rotavirus-specific immunoglobulins and
lymphoproliferative cells in breast milk and colostrum (BELL et al. 1988;
COOK et al. 1988; CUKOR et al. 1979; RIGENBERGS et al. 1988; TOTTERDELL
et al. 1988; YOLKEN et al. 1978), breast feeding probably does not afford
protection against either rotavirus infection or severe rotavirus disease.
Studies which support this hypothesis are dicussed in Chap. 8 of this volume.

3.4 Newborns Lack Intestinal Proteases at Quantities Sufficient to Allow Efficient Entry of Rotaviruses into Small Intestinal Villus Epithelial Cells

Rotavirus entry into target cell cytoplasm is facilitated by proteolytic cleavage
of VP4 (FUKUHARA et al. 1988; KALJOT et al. 1988). Cleavage of VP4 to VP5
and VP8 occurs in the presence of trypsin, elastase, or pancreatin (ESTES et al.
1981). The presence of typsin in viral growth media is required for efficient
replication of rotaviruses in tissue culture (ALMEIDA et al. 1978; BABIUK et al.
1977; GRAHAM and ESTES 1980; THEIL et al. 1977). However, levels of
exopeptidases are decreased in intestinal fluid secretions of newborns as
compared to older infants and young children. Decreased entry of rotaviruses
into small intestinal epithelial cells in the absence of proteases may in part
explain decreased virulence in newborns. LEBENTHAL and LEE (1980) studied
both pancreatic enzyme activity in duodenal fluids and the response of the
exocrine pancreas to secretogogues (pancreozymin, PZM, and secretin, SEC)
of full-term infants at birth and at 30 days of age. Findings were compared
to identical studies of children 2 years of age and above. Trypsin levels in

duodenal fluids obtained at 1 day of age were approximately one-half those obtained at 1 month of age. In addition, trypsin levels in 1 day old infants obtained after administration of PZM or SEC were approximately one-half those of 1 month old infants. Levels of other exopeptidases (chymotrypsin and carboxypeptidase B) were decreased in both 1 day old and 1 month old infants as compared to levels obtained at 2 years of age.

4 Factors of the Host Which may Determine why Rotavirus Virulence is Enhanced in Developing as Compared to Developed Countries

Diarrahea due to rotavirus infection account for 65 000–70 000 hospitalizations and approximately 200 deaths per year in the United States (HO et al. 1988). In Asia, Africa, and Latin America, it is estimated that approximately 850 000 children die of rotavirus-induced gastroenteritis each year (BLACK et al. 1980; WALSH and WARREN 1979). Infants and young children in developing countries are more likely to have inadequate medical access, poor nutrition, and concomitant enteric bacterial infections than those in developed countries. The degree to which the latter two factors contribute to relatively severe rotavirus disease is discussed below.

4.1 Effect of Malnutrition on Severity of Rotavirus Infection

Several animal model studies support the hypothesis that poor nutrition is a risk factor for development of severe and occasionally fatal rotavirus infection. OFFOR and coworkers (1985) examined the effect of malnutrition on development of severe rotavirus disease in suckling mice orally inoculated with murine rotavirus. Infant mice were separated within 12–16 h of birth so that litters contained seven to nine mice (normal group) or 18–20 mice (malnourished group). In comparison with well nourished animals, infection of malnourished mice was characterized by a significant decrease in viral incubation period and in minimal infectious dose required to induce disease. Viral replication in dispersed enterocytes was observed 6–12h earlier and clinical disease was more severe in malnourished than in well nourished animals. The authors concluded that malnutrition of the host induced alternations in virulence of rotavirus infection.

Specific aspects of poor nutrition associated with increased rotavirus virulence were studied by NOBLE et al. (1983) and MORREY et al. (1984). These investigators found that murine rotavirus-infected newborns from dams fed diets low in protein, low in folic acid, or low in nutrient to calorie ratios were more likely to develop severe rotavirus disease than newborns from dams fed normal diets. Although newborns inoculated with murine rotavirus rarely die of dehydration secondary to diarrhea, these authors observed a number of deaths in malnourished animals.

UHNOO and coworkers (1990a) studied the effect of malnutrition on intestinal barrier function in suckling mice orally inoculated with murine rotavirus. All mice developed diarrhea within 48 of inoculation. Malnourished suckling mice developed more severe disease than well nourished mice. Severe disease was associated with an increased number of rotavirus-containing villus epithelial cells and more pronounced villus atrophy and vacuolization. In addition, infection of malnourished animals resulted in severe intestinal mucosal damage manifested as a disruption of microvillus borders. After a single oral dose of 100 ug of ovalbumin 3 days after infection, malnourished animals developed significantly higher serum levels of oval-bumin than did well nourished controls. The peak uptake of ovalbumin per gram of body weight was approximately 4.5 times greater in malnourished-infected than in nourished-infected animals and 2.5 times greater in mal-nourished than in well nourished controls. The authors concluded that increased virulence of enteric pathogens in malnutrition may be secondary to a loss of intestinal barrier function allowing for direct penetration of intestinal organisms and toxins into the systemic circulation.

4.2 Effect of Concomitant Bacterial Infection on Severity of Rotavirus Infection

Several investigators found enhanced virulence after dual infection of animals with rotavirus and *E. coli*. NEWSOME and CONEY (1985) orally inoculated 4 day old mice with either murine rotavirus or 10^6 colony-forming units (cfu) of *E. coli* B44. Suckling mice inoculated with rotavirus developed mild diarrhea; no animals died secondary to rotavirus-induced gastroenteritis. Mice inoculated with *E. coli* developed severe diarrhea; 45% of the animals died. Mice inoculated with both rotavirus and *E. coli* developed severe gastroenteritis and 84% of animals died. The authors concluded that enhance-ment of virulence observed after infection of infant mice with both *E. coli* and rotavirus was explained by the mechanisms by which these two agents induced diarrhea. *E. coli* B44 secretes a heat-stable enterotoxin A which induces increased secretion of water from ileal crypt epithelial cells. Rota-viruses, by contrast, damage absorptive cells located at the tips of small intestinal villi. In animals infected with both *E. coli* and rotavirus, a combi-nation of increased secretion and decreased resorption of water could result in a greater loss of total body water and death secondary to dehydration. Similary, TZIPORI and coworkers (1982) inoculated foals up to 50 days of age with either foal rotavirus, bovine enterotoxigenic *E. coli*, or both. Neither rotavirus nor enterotoxigenic *E. coli* induced diarrhea in foals when inoculated alone. However, up to 16 days of age, inoculation with both pathogens induced diarrhea in all animals tested. Lastly, TORRES-MEDINA et al. (1984b) orally inoculated 1 day old gnotobiotic calves with either virulent bovine rotavirus or nonenterotoxigenic *E. coli* or both. A significant increase in the quantity of Nonenterotoxigenic bacteria was detected in calves infected with both nonenterotoxigenic *E. coli* and virulent bovine rotavirus as compared to calves infected with the latter alone.

References

Adams WR, Kraft LM (1963) Epizootic diarrhea of infant mice: identification of the etiologic agent. Science 141: 359–360

Adams WR, Kraft LM (1967) Electron microscopic study of the intestinal epithelium of mice infected with the agent of epizootic diarrhea of infant mice (EDIM virus). Am J Pathol 51: 39–60

Albert MJ, Unicomb LE, Barnes GL, Bishop RF (1987) Cultivation and characterization of rotavirus strains infecting newborn babies in Melbourne, Australia, from 1975 to 1979. J Clin Microbiol 25: 1635–1640

Almeida JD, Hall T, Banatvala JE, Totterdell BM, Chrystie IL (1978) The effect of trypsin on the growth of rotavirus. J Gen Virol 40: 213–218

Askaa J, Bloch B (1984) Infection in piglets with a porcine rotavirus-like virus. Experimental inoculation and ultrastructural examination. Arch Virol 80: 291–303

Babiuk LA, Mohammad K, Spence L, Fauvel M, Petro R (1977) Rotavirus isolation and cultivation in the presence of trypsin. J Clin Microbiol 6: 610–617

Banfield WG, Kasnic G, Blackwell JH (1968) Further observations on the virus of epizootic diarrhea of infant mice: an electron microscopic study. Virology 36: 411–417

Barron-Romero BL, Barreda-Gonzalez J, Doval-Ugalde, Liz JZ, Huerta-Pena M (1985) Asymptomatic rotavirus infections in day care centers. J Clin Microbiol 22: 116–118

Bell LM, Clark HF, O'Brien EA, Kornstein MJ, Plotkin SA, Offit PA (1987) Gastroenteritis caused by human rotaviruses (serotype three) in a suckling mouse model. Proc Soc Exp Biol Med 184: 127–132

Bell LM, Clark HF, Offit PA, Horton-Slight P, Arbeter AM, Plotkin SA (1988) Rotavirus serotype-specific neutralizing activity in human milk. Am J Dis Child 142: 275–278

Black REM, Merson MH, Rahman ASSM, Yunis M, Alim ARMA, Huq I, Yolken RH, Curlin GT (1980) A two-year study of bacterial, viral and parasitic agents associated with diarrhea in rural Bangladesh. J Infect Dis 142: 660–664

Broome RL, Vo PT, Ward RL, Clark HF, Greenberg HB (1993) Murine rotavirus genes encoding VP4 and VP7 are not major determinants of host range and virulence. J Virol 67: 2448–2455

Carpio M, Bellamy JEC, Babiuk LA (1981) Comparative virulence of different bovine rotavirus isolates. Can J Comp Med 45: 38–42

Chasey D, Banks J (1986) Replication of atypical ovine rotavirus in small intestine and cell culture. J Gen Virol 67: 567–576

Chasey D, Higgens RJ, Jeffrey M, Banks J (1989) Atypical rotavirus and villous epithelial cell syncytia in piglets. J Comp Pathol 100: 217–222

Cheever FS, Mueller JH (1947) Epidemic diarrhoeal disease of suckling mice: I. Manifestations, epidemiology and attempts to transmit the disease. J Exp Med 85: 405–416

Chrystie IL, Totterdell BM, Banatvala JE (1978) Asymptomatic endemic rotavirus infections in the newborn. Lancet i: 1176–1178

Coelho KIR, Bryden AS, Hall C, Flewitt TH (1981) Pathology of rotavirus infection in suckling mice: a study by conventional histology, immunofluorescence, ultrathin sections, and scanning electron microscopy. Ultrastruct Pathol 2: 59–69

Collins J, Starkey WG, Wallis TS, Clarke GJ, Workton KJ, Spencer AJ, Haddon SJ, Osborne MP, Candy DCA, Stephen J (1988) Intestinal enzyme profiles in normal and rotavirus-infected mice. J Pediatr Gastroenterol Nutr 7: 264–272

Collins J, Benfield DA, Duimstra JR (1989) Comparative virulence of two porcine group-A rotavirus isolates in gnotobiotic pigs. Am J Vet Res 50: 827–835

Conner ME, Estes MK, Graham DY (1988) Rabbit model of rotavirus infection. J Virol 62: 1625–1633

Cook CA, Zbitnew A, Dempster G, Gerrard JW (1978) Detection of antibody to rotavirus by counterimmunoelectrophoresis in human serum, colostrum, and milk. J Pediatr 93: 967–970.

Crouch CF, Woode GN (1978) Serial studies of virus multiplication and intestinal damage in gnotobiotic piglets infected with rotavirus. J Med Microbiol 11: 325–334

Cukor G, Blacklow NR, Capozza FE, Panjvani ZFK, Bednarek F (1979) Persistence of antibodies to rotavirus in human milk. J Clin Microbiol 9: 93–96

Davidson GP, Gall DG, Petric M, Butler DG, Hamilton JR (1977) Human rotavirus enteritis induced in conventional piglets: intestinal structure and transport. J Clin Invest 60: 1402–1409

Dharakul T, Riepenhoff-Talty M, Albini B, Ogra PL (1988) Distribution of rotavirus antigen in intestinal lymphoid tissues: potential role in development of the mucosal immune response to rotavirus. Clin Exp Immunol 74: 14–19

Dubourguier HC, Mandard O, Contrepois M, Gouet P (1981) Light and electronic microscopic studies of the changes in the intestinal mucosa of gnotobiotic calves infected with a wild rotavirus. Ann Virol (Inst Pasteur) 132E: 217–227

Elias MM (1977) Distribution and titres of rotavirus antibodies in different age groups. J Hyg 79: 365–372

Estes MK, Graham DY, Mason BB (1981) Proteolytic enhancement of rotavirus infectivity: molecular mechanisms. J Virol 39: 879–888

Eydelloth RS, Vonderfecht SL, Sheridan JF, Enders LD, Yolken RH (1984) Kinetics of viral replication and local and systemic immune responses in experimental rotavirus infection. J Virol 50: 947–950

Ferguson A, Paul G, Snodgrass DR (1981) Lactose tolerance in lambs with rotavirus diarrhea. Gut 22: 114–119

Flores J, Midthun K, Hoshino Y, Green K, Gorziglia M, Kapikian AZ, Chanock RM (1986) Conservation of the fourth gene segment among rotaviruses recovered from asymptomatic newborn infants and its possible role in attenuation. J Virol 60: 972–979

Fukuhara N, Yoshie O, Kitaoka S, Konno T (1988) Role of vp3 in human rotavirus internalization after target cell attachment via VP7. J Virol 62: 2209–2218

Gorziglia M, Hoshino Y, Buckler-White A, Blumentals I, Glass R, Flores J, Kapikian AZ, Chanock RM (1986) Conservation of amino acid sequence of vp8 and cleavage region of 84-kDa outer capsid protein among rotaviruses recovered from asymptomatic neonatal infection. Proc Natl Acad Sci USA 83: 7039–7043

Gouvea VS, Alencar AA, Barth OM, De Castro L, Fialho AM, Araujo HP, Majerowicz S, Pereira HG (1986) Diarrhoea in mice infected with a human rotavirus. J Gen Virol 67: 577–581

Graham D, Estes MK (1980) Proteolytic enhancement of rotavirus infectivity: biological mechanisms. Virology 101: 432–439

Graham DY, Sackman JW, Estes MK (1984) Pathogenesis of rotavirus-induced diarrhea — Preliminary studies in miniature swine piglet. Dig Dis Sci 29: 1028–1035

Greenberg HB, Vo PT, James R (1986) Cultivation and characterization of three strains of murine rotavirus. J Virol 57: 585–590

Grimwood K, Coakley JC, Hudson IL, Bishop RF, Barnes GL (1988) Serum aspartate amino-transferase levels after rotavirus gastroenteritis. J Pediatr 112: 597–600

Grunow JE, Dunton SF, Waner JL (1985) Human rotavirus-like particles in a hepatic abscess. J Pediatr 106: 73–76

Hambraeus AM, Hambraeus LEJ, Wadell G (1989) Animal model of rotavirus infection in rabbits – protection obtained without shedding of viral antigen. Arch Virol 107: 237–251

Heyman M, Corthier G, Petit A, Meslin J-C, Moreau C, Desjeux J-F (1987) Intestinal absorption of macromolecules during viral enteritis: an experimental study on rotavirus-infected conventional and germ-free mice. Ped Res 22: 72–78

Ho M, Glass RI, Pinsky PF (1988) Diarrheal deaths in American children: are they preventable? JAMA 260: 3281–3285

Holmes IH, Ruck BJ, Bishop RF, Davidson GP (1975) Infantile enteritis viruses: morphogenesis and morphology. J Virol 16: 937–943

Hoshino Y (1988) Proceedings of the 3rd NIH Rotavirus Vaccine Workshop, Bethesda MD

Hoshino Y, Sereno M, Kapikian AZ, Saif LJ, Kang S (1993) Genetic of determinants of rotavirus virulence studied in gnotobiotic piglets vaccines 93. Cold Spring Harbor Press p 277–282

Huber AC, Yolken RH, Mader LC, Strandberg JD, Vonderfecht SL (1989) Pathology of infectious diarrhea of infant rats (IDIR) induced by an antigenically distinct rotavirus. Vet Pathol 26: 376–385

Hyams JS, Krause PJ, Gleason PA (1981) Lactose malabsorption following rotavirus infection in young children. J Pediatr 99: 916–918

Ijaz MK, Sabara MI, Frenchick PJ, Babiuk LA (1987) Assessment of intestinal damage in rotavirus infected neonatal mice by a D-xylose absorption test. J Virol Methods 18: 153–157

Ijaz MK, Dent D, Haines D, Babiuk LA (1989) Development of a murine model to study the pathogenesis of rotavirus infection. Exp Mol Pathol 51: 186–204

Johnson CA, Snider TG III, Fulton RW, Cho D (1983) Gross and light microscopic lesions in neonatal gnotobiotic dogs inoculated with a canine rotavirus. Am J Vet Res 44: 1687–1693

Johnson CA, Snider TG III, Henk WG, Fulton RW (1986) A scanning and transmission electron microscopic study of rotavirus-induced intestinal lesions in neonatal gnotobiotic dogs. Vet Pathol 23: 443–453

Kaljot KT, Shaw RD, Rubin DH, Greenberg HB (1988) Infectious rotavirus enter cells by direct cell membrane penetration, not by endocytosis. J Virol 62: 1136–1144

Kapikian AZ, Wha H, Wyatt RG, Cline WL, Arrobio JO, Brandt CD, Rodriguez WJ, Sack DA, Chanock RM, Parrott RH (1976) Human reovirus-like agent as the major pathogen associated with "winter" gastroenteritis in hospitalized infants and young children. N Engl J Med 294: 965–972

Kapikian AZ, Flores J, Hoshino Y, Midthun K, Gorziglia M, Green KY, Chanock RM, Potash L, Sears SD, Clements ML, Halsey NA, Black RE, Perez-Schael I (1989) Prospects for development of a rotavirus vaccine against rotavirus diarrhea in infants and young children. Rev Infect Dis 11: S539-S546

Kim HW, Brandt CD, Kapikian AZ, Wyatt RG, Arrobio JO, Rodriguez WJ, Chanock RM, Parrott RH (1977) Human reovirus-like agent infection: occurrence in adult contacts of pediatric patients with gastroenteritis. JAMA 238: 404–407

Kirstein CG, Clare DA, Lecce JG (1985) Development of resistance of enterocytes to rotavirus in neonatal, agammaglobulinemic piglets. J Virol 55: 567–573

Kovacs A, Chan L, Hotrakitya C, Overturf G, Portnoy B (1986) Serum transaminase elevations in infants with rotavirus gastroenteritis. J Pediatr Gastroenterol Nutr 5: 873–877

Lebenthal E, Lee PC (1980) Development of functional response in human exocrine pancreas. Pediatrics 66: 556–560

Leece JG, King MW (1978) Role of rotavirus (reo-like) in weanling diarrhea of pigs. J Clin Microbiol 8: 454–458

Leece JG, King MW, Mock R (1976) Reovirus-like agent associated with fatal diarrhea in neonatal pigs. Infect Immun 14: 816–825

Little LM, Shadduck JA (1982) Pathogenesis of rotavirus infection in mice. Infect Immun 38: 755–763

Little LM, Shadduck JA (1983) Pathogenesis of rotavirus infections. Prog Food Nutr Sci 7: 179–187

McNulty MS, Allan GM, Pearson GR, McFerran JB, Curran WL, McCracken RM (1976) Reovirus-like agent (rotavirus) from lambs. Infect Immun 14: 1332–1338

Mavrovichalis J, Evans N, McNeish AS, Bryden AS, Davies HA, Flewett TH (1977) Intestinal damage in rotavirus and adenovirus gastroenteritis assessed by D-xylose malabsorption. Arch Dis Child 52: 589–591

Mebus CA (1976) Reovirus-like calf enteritis. Dig Dis 21: 592–598

Mebus CA, Stair EL, Underdahl NR, Twiehaus MJ (1974) Pathology of neonatal calf diarrhea induced by a reo-like virus. Vet Pathol 8: 490–505

Morrey JD, Sidwell RW, Noble RL, Barnett BB, Mahoney AW (1984) Effects of folic acid malnutrition on rotaviral infection in mice. Proc Soc Exp Biol Med 176: 77–83

Murphy AM, Albrey MB, Crewe EB (1977) Rotavirus infections of neonates. Lancet ii: 1149–1150

Narita M, Fukusho A, Shimizu Y (1982) Electron microscopy of the intestine of gnotobiotic piglets infected with porcine rotavirus. J Comp Path 92: 589–597

Newsome PM, Coney KA (1985) Synergistic rotavirus and Escherichia coli diarrheal infection of mice. Infect Immun 47: 573–574

Noble RL, Sidwell RW, Mahoney AW, Barnett BB, Spendlove RS (1983) Influence of malnutrition and alterations in dietary protein on murine rotavirus disease. Proc Soc Exp Biol Med 173: 417–426

Noone C, Menzies IS, Banatvala JE, Scopes JW (1986) Intestinal permeability and lactose hydrolysis in human rotaviral gastroenteritis assessed simultaneously by non-invasive differential sugar permeation. Eur J Clin Invest 16: 217–225

Offit PA, Clark HF (1985) Protection against rotavirus-induced gastroenteritis in a murine model by passively-acquired gastrointestinal but not circulating antibodies. J Virol 54: 58–64

Offit PA, Clark HF Kornstein MJ, Plotkin SA (1984) A murine model for oral infection with a primate rotavirus (simian SA11). J Virol 51: 233–236

Offit PA, Blavat G, Greenberg HB, Clark HF (1986) Molecular basis of rotavirus virulence: role of gene segment 4. J Virol 57: 46–49

Offor E, Riepenhoff-Talty M, Ogra PL (1985) Effect of malnutrition on rotavirus infection in suckling mice: kinetics of early infection. Proc Soc Exp Biol Med 178: 85–90

Osborne MP, Haddon SJ, Spencer AJ, Collins J, Starkey WG, Wallis TS, Clarke GJ, Worton KJ, Candy DC, Stephen J (1988) An electron microscopic investigation of time-related changes in the intestine of neonatal mice infected with murine rotavirus. J Pediatr Gastroenterol Nutr 7: 236–248

Pearson GR, McNulty MS (1977) Pathological changes in the small intestine of neonatal pigs infected with a pig reovirus-like agent (rotavirus). J Comp Pathol 87: 363–375

Pearson GR, McNulty MS (1979) Ultrastructural changes in small intestinal epithelium of neonatal pigs infected with pig rotavirus. Arch Virol 59: 127–136

Pearson GR, McNulty MS, Logan EF (1978) Pathological changes in the small intestine of neonatal calves naturally infected with reo-like virus (rotavirus). Vet Rec 102: 454–458

Perez-Schael I, Daoud G, White L, Urbina G, Daoud N, Perez M, Flores J (1984) Rotavirus shedding by newborn children. J Med Virol 14: 127–136

Pospischil A, Hess RG, Bachmann PA (1982) Morphology of intestinal changes in pigs experimentally infected with porcine rota-virus and two porcine corona viruses. Scand J Gastroenterol [Suppl] 74: 167–169

Prince DS, Astry CA, Vonderfecht S, Jakab G, Shen F, Yolken RH (1986) Aerosol transmission of experimental rotavirus infection. Pediatr Infect Dis J 5: 218–222

Ramig RF (1988a) Proceedings of the 3rd NIH Rotavirus Vaccine Workshop, Bethesda MD

Ramig RF (1988b) The effects of host age, virus dose, and virus strain on heterologous rotavirus infection of suckling mice. Microb Pathog 4: 189–202

Ramig RF, Galle KL (1990) Rotavirus genome segment 4 determines viral replication phenotype in cultured liver cells (HepG2). J Virol 64: 1044–1049

Reynolds DJ, Hall GA, Debney TG, Parsons KR (1985) Pathology of natural rotavirus infection in clinically normal calves. Res Vet Sci 38: 264–269

Riepenhoff-Talty M, Lee PC, Carmody PJ, Barrett HJ, Ogra PL (1982) Age-dependent rotavirus-enterocyte interactions. Proc Soc Exp Biol Med 170: 146–154

Ringenbergs M, Albert MJ, Davidson GP, Goldsworthy W, Haslam R (1988) Serotype-specific antibodies to rotavirus in human colostrum and breast milk and in maternal and cord blood. J Infect Dis 158: 477–480

Rodriguez WJ, Kim HW, Arrobio JO, Brandt CD, Chanock RM, Kapikian AZ, Wyatt RG, Parrott RH (1977) Clinical features of acute gastroenteritis associated with human reovirus-like agent in infants and young children. J Pediatr 91: 188–193

Rodriguez-Toro G (1980) Natural epizootic diarrhea of infant mice (EDIM)-A light and electron microscope study. Exp Mol Pathol 32: 241–252

Runnels PL, Moon HW, Matthews PJ, Whipp SC, Woode GN (1986) Effects of microbial and host variables on the interaction of rotavirus and Escherichia coli infections in gnotobiotic calves. Am J Vet Res 47: 1542–1550

Schoeb TR, Casebolt DB, Walker VE, Potgieter LND, Thouless ME, DiGiacomo RF (1986) Rotavirus-associated diarrhea in a commercial rabbitry. Lab Anim Sci 36: 149–152

Sheridan JF, Eydelloth RS, Vonderfecht SL, Aurelian L (1983) Virus-specific immunity in neonatal and adult mouse rotavirus infection. Infect Immun 39: 917–927

Snodgrass DR, Angus KW, Gray EW (1977) Rotavirus in lambs: pathogenesis and pathology. Arch Virol 55: 263–271

Snodgrass DR, Ferguson A, Allan F, Angus KW, Mitchell B (1979) Small intestinal morphology and epithelial cell kinetics in lamb rotavirus infections. Gastroenterology 76: 477–481

Starkey WG, Collins J, Wallis TS, Clarke GJ. Spencer AJ, Haddon SJ, Osborne MP, Candy DC, Stephen J (1986) Kinetics, tissue specificity and pathological changes in murine rotavirus infection of mice. J Gen Virol 67: 2625–2634

Stiglmair-Herb MT, Pospischil A, Hess RG, Bachmann PA, Baljer G (1986) Enzyme histochemistry of the small intestinal mucosa in experimental infections of calves with rotavirus and enterotoxigenic Escherichia coli. Vet Pathol 23: 125–131

Stintzing G, Johansen K, Magnusson KE, Svensson L, Sundqvist T (1986) Intestinal permeability in small children during and after rotavirus diarrhoea assessed with different-size polyethyleneglycols (PEG 400 and PEG 1000). Acta Paediatr Scand 75: 1005–1009

Suzuki H, Konno T (1975) Reovirus-like particles in jejunal mucosa of a Japanese infant with acute infections non-bacterial gastroenteritis. Tohoku J Exp Med 115: 199–211

Theil KW, Bohl EH, Agnes AG (1977) Cell culture propagation of porcine rotavirus (reovirus-like agent). Am J Vet Res 38: 1765–1768

Theil KW, Bohl EH, Cross RF, Kohler EM, Agnes AG (1978) Pathogenesis of porcine rotaviral infection in experimentally inoculated gnotobiotic pigs. Am J Vet Res 39: 213–220

Theil KW, Saif LJ, Moorhead PD, Whitmoyer RE (1985) Porcine rotavirus-like virus (Group B rotavirus): characterization and pathogenicity for gnotobiotic pigs. J Clin Microbiol 21: 340–345

Torres-Medina A (1984a) Effect of combined rotavirus and Escherichia coli in neonatal gnotobiotic calves. Am J Vet Res 45: 643–651

Torres-Medina A (1984b) Effect of rotavirus and/or Escherichia coli infection on the aggregated lymphoid follicles in the small intestine of neonatal gnotobiotic calves. Am J Vet Res 45: 652–660

Torres-Medina A, Underdahl NR (1980) Scanning electron microscopy of intestine of gnotobiotic piglets infected with porcine rotavirus. Can J Comp Med 44: 403–411

Totterdell BM, Patel S, Banatvala JE, Chrystie IL (1988) Development of a lymphocyte transformation assay for rotavirus in whole blood and breast milk. J Med Virol 25: 27–36

Tzipori S, Makin T, Smith M, Krautil F (1982) Enteritis in foals induced by rotavirus and enterotoxigenic Escherichia coli. Aust Vet J 58: 20–23

Uhnoo IS, Freihorst J, Riepenhoff-Talty M, Fisher JE, Ogra PL (1990a) Effect of rotavirus infection and malnutrition on uptake of dietary antigen in the intestine. Pediatr Res (in press)

Uhnoo I, Riepenhoff-Talty M, Dharakul T, Chegas P, Fisher J, Greenberg HB, Ogra P (1990b) Extramucosal spread and development of hepatitis with rhesus rotavirus in immunodeficient and normal mice. J Virol 64: 361–368

Vonderfecht SL, Huber AC, Eiden J, Mader LC, Yolken RH (1984) Infectious diarrhea of infant rats produced by a rotavirus-like agent. J Virol 52: 94–98

Vonderfecht SL, Eiden JJ, Torres A, Miskuff RL, Mebus CA, Yolken RH (1986) Identification of a bovine enteric syncytial virus as a non-group A rotavirus. Am J Vet Res 47: 1913–1918

Walsh JA, Warren KS (1979) Selective primary health care: an interim strategy for disease control in developing countries. N Engl J Med 301: 967–974

Ward RL, McNeal MM, Sheridan JF (1990) Development of an adult mouse model for studies on protection against rotavirus. J Virol 64: 5070–5075

Ward RL, McNeal MM, Scheridan JF (1992) Evidence that active protection following oral immunization of mice with live rotavirus is not dependent on neutralizing antibody. Virology 188: 57–66

Wenman WM, Hinde D, Feltham S, Gurwith M (1979) Rotavirus infection in adults: results of a prospective family study. N Engl J Med 301:303–306

Wilsnack RE, Blackwell JH, Parker JC (1969) Identification of an agent of epizootic diarrhea of infant mice by immunofluorescent and complement-fixation tests. Am J Vet Res 30: 1195–1204

Wolf JL, Cukor G, Blacklow NR, Dambrauskas R, Trier JS (1981) Susceptibility of mice to rotavirus infection: effects of age and administration of corticosteroids. Infect Immun 33: 565–574

Wright PF, Tajima T, Thompson J, Kokubun K, Kapikian A, Karzon DT (1987) Candidate rotavirus vaccine (rhesus rotavirus strain) in children: an evaluation. Pediatrics 80: 473–480

Wyatt RG, Kalica AR, Mebus CA, Kim HW, London WT, Chanock RM, Kapikian AZ (1978)
    Reovirus-like agents (rotaviruses) associated with diarrheal illness in animals and man.
    Persps Virol 10: 121–145
Yason CV, Summers BA, Schat KA (1987) Pathogenesis of rotavirus infection in various age
    groups of chickens and turkeys: pathology. Am J Vet Res 48: 927–938
Yolken RH, Wyatt RG, Mata L, Urrutia JJ, Garcia B, Chanock RM, Kapikian AZ (1978)
    Secretory antibody directed against rotavirus in human milk-measurement by means of
    enzyme-linked immunosorbent assay. J Pediatr 93: 916–921

Rotavirus Vaccines and Vaccination Potential

M.E. Conner[1], D.O. Matson[1,2,3], and M.K. Estes[1]

[1] Division of Molecular Virology, [2] and Department of Pediatrics, Baylor College of Medicine,
One Baylor Plaza, Houston, TX 77030, USA
[3] Current address: Center for Pediatric Research, Eastern Virginia Medical School, Norfolk, VA
23510, USA

1 Introduction

Rotaviruses were discovered to be human pathogens in 1973, and subsequent studies have shown that rotaviruses are the most common cause of severe gastroenteritis in young children worldwide (see Chap. 10). Rotavirus infections cause over 500 000 deaths each year among children less than 2 years of age in developing countries (INSTITUTE OF MEDICINE 1986). For children with rotavirus infections in developed countries, the case fatality rate is lower, but hospitalizations are frequent. In industrial and developing countries, most, if not all, children become infected with rotavirus during the first 2 years of life. This indicates that improvements in water supply, sanitation and hygiene do not eradicate or effectively control these infections (DEZOYSA and FEACHEM 1985). Indeed, in the United States, diarrhea remains the second most common illness in families (DINGLE et al. 1964). In addition, despite the effectiveness and availability of convenient oral methods to prevent dehydration, it is estimated that 11% of children with symptomatic rotavirus infections who seek medical care in the US become moderately dehydrated and require hospitalization (KOOPMAN et al. 1984; RODRIGUEZ et al. 1987). These data clearly indicate a need for an effective vaccine. This chapter provides a general overview of the history, achievements, and challenges of rotavirus vaccine development. Details of the status of and rationale for vaccine development have been reviewed recently (DEZOYSA and FEACHEM 1985; EDELMAN 1987; VESIKARI 1988; BISHOP 1988; CLARK 1988; BLOOM 1989; KAPIKIAN et al. 1989, 1990). This review does not reiterate details of specific vaccine trials found in earlier reviews unless they serve to illustrate a general feature about vaccine potential or problems.

2 Impact of Rotavirus Disease

While there is agreement on the need for rotavirus vaccines for children in both developing and developed countries, the priority which these vaccines should have among all needed vaccines has remained controversial. For example, a committee of the Institute of Medicine examined needed vaccines for the United States and did not list rotavirus vaccines among the top ten in priority (INSTITUTE OF MEDICINE 1985). One difficulty noted in this assessment was the paucity of information on the magnitude of the morbidity and economic impact caused by rotavirus infections in the United States. In 1980, hospitalization alone for rotavirus infections in the United States was estimated to cost $27 million per year (RODRIGUEZ et al. 1980). More recently, the Center for Disease Control (CDC) estimated that more than 210 000 children are hospitalized for gastroenteritis in the United States each year, at an annual inpatient cost of almost one billion dollars, and that more than half of these children have rotavirus-associated illness (HO et al. 1988a; LEBARON et al. 1990a). A surprising rate (about 500 children per year) of diarrheal deaths, most probably due to rotavirus infection, still occur in the United States (HO et al. 1988b). A recent analysis of the impact of rotavirus infections at a large pediatric hospital in Houston, Texas, estimated that the risk for hospitalization for rotavirus gastroenteritis during childhood in the US is 1 in 46; the extrapolated hospital bed costs alone were $352 million annually (MATSON and ESTES 1990). The economic burden of rotavirus infections due to outpatient care (90% of the cases), the cost of parent's absenteeism from work when they care for sick children, and the impact of disease in parents and the elderly remains unmeasured. Because rotavirus vaccines are being developed, precise mechanisms for measuring the incidence and impact of rotavirus infections in all age groups and different populations clearly are needed to allow an accurate assessment of vaccine efficacy. This is particularly important because a successful vaccine is likely to alter the epidemiology of disease.

The recent development of a case definition for hospitalization for rotavirus infection utilizing a combination of discharge diagnosis codes and laboratory-detected cases to measure rotavirus infections will facilitate monitoring rotavirus-associated illness (HO et al. 1988a; MATSON and ESTES 1990). The CDC has initiated a surveillance program of rotavirus infections in North America by monitoring the number of rotavirus detections in hospital laboratories (LEBARON et al. 1990b). A program based upon laboratory detections will grossly underestimate the impact of disease, because, currently, many clinicians do not perceive a benefit of a positive rotavirus test result, and they submit stool specimens from only a minority of children hospitalized for rotavirus diarrhea. A more accurate assessment of the number of hospitalizations comes from discharge diagnosis information. A monthly summary of diagnosis codes epidemiologically linked to rotavirus infection, combined with the laboratory summary of rotavirus detections, gives a better assessment of seasonal onset, duration, and total number of hospitalized cases than the current CDC surveillance method. This point is not trivial, for if a national vaccine program is initiated, physicians will be more likely to order a rotavirus test, in much the same way as they did when commercial assays for rotavirus detection became available (MATSON et al. 1990). Surveillance for rotavirus infections will be improved and simplified now that a specific category for rotavirus gastroenteritis is to be included in the 10th revision of the International Classification of Diseases (CDC 1991). The importance of understanding pre- and postvaccine epidemiology in different regions of the world for effective monitoring of vaccine efficacy is well-illustrated by poliovirus vaccine efforts (MELNICK 1990).

Rotavirus infections in animals also are significant. For cattle alone, the impact is great. Calf death rates due to neonatal calf diarrhea range from 5% to >20% in the US and economic costs in North America resulting from these deaths have been estimated to be at least $500 million annually (HOUSE

1978). The actual losses are much greater because of other economic factors, such as the cost of care and treatment of sick animals, poor growth and reduced beef production in some recovered calves, and possible enhanced susceptibility of rotavirus-infected calves to secondary infections. Trends in agriculture to intensify animal production and more closely confine calves increase the risk of enteric infections. The incidence and severity of diarrhea are greatest in calves within the first few weeks of life, the period when diarrhea due to rotavirus usually occurs (JANKE 1989). Rotaviruses also are important pathogens in other large domestic animals, including pigs (WOODS et al. 1976a; BOHL et al. 1978), lambs (McNULTY et al. 1976), and foals (CONNER and DARLINGTON 1980; EUGSTER and SCRUTCHFIELD 1980), and in smaller animals important for food production or research, such as rabbits (DIGIACOMO and THOULESS 1986; SCHOEB et al. 1986; BRYDEN et al. 1976), poultry (McNULTY et al. 1978), and mice (KRAFT 1957; ADAMS and KRAFT 1963). Such data indicate a need for effective veterinary rotavirus vaccines.

3 Vaccine Targets and Strategies for Making Rotavirus Vaccines

Several types of information need to be considered when developing a strategy to make a rotavirus vaccine (Table 1). Discussion of the knowledge gained from candidate rotavirus vaccine studies is useful to focus the history, achievements, and prospects of rotavirus vaccine development and to help identify areas for which additional information is needed. Vaccines for children and for animals have been developed according to different strategies. Therefore, this chapter considers the vaccine potential for children and for animals separately.

Table 1. Considerations for development of rotavirus vaccines

Natural history of the disease
   Peak age(s) of infection and clinical outcome of infection
   Frequency and severity of symptomatic and asymptomatic reinfections
   Evidence for development of natural immunity

Antigenic diversity of virus types

Host range of virus types

Existence of animal reservoirs

Knowledge of protective immune responses
   What type of immunity is protective and how can we monitor its development?
   Is immunity homotypic or heterotypic?
   Is immunity effective in preventing reinfection and/or disease?
   What is the duration of active immunity?

4 Rotavirus Vaccines For Children

4.1 Natural History of Rotavirus Disease

Severe rotavirus diarrhea occurs most frequently in children 6 months to 2 years old. The next highest frequency of severe disease is in infants less than 6 months of age. Newborns may become infected, but they are less likely to have clinical illness, which suggests that passively acquired maternal antibody is protective. Because the peak age of severe dehydrating illness occurs when transplacentally acquired maternal antibody is waning, it is thought that effective vaccination procedures will need to induce active immunity to mimic the protection conferred in early life by maternal antibody. In addition, the decrease in incidence of severe illness after 2 years of age indicates that acquired active immunity eventually results in protection against illness.

What remains unclear is whether protection against severe rotavirus disease results from an immune response to just one infection or to multiple symptomatic or asymptomatic rotavirus infections. Multiple infections with rotavirus are common in some settings (MATA et al. 1983; GRINSTEIN et al. 1989), but recognition that reinfections are more universal than originally appreciated has only been realized recently (FRIEDMAN et al. 1988; BEARDS and DESSELBERGER 1989; O'RYAN et al. 1990; COULSON et al. 1990a).

Reinfection may reflect sequential infections with viruses of distinct serotypes, e.g., VP7 or glycoprotein (G) types (see below), or viruses of the same serotype. The epidemiology of such reinfections and the influence of preexisting immunity on these reinfections are poorly understood, in part because studies of the influence of preexisting immunity on reinfections are difficult to perform. Longitudinal studies of children suggest that severe life-threatening illness may occur only when young children experience a primary infection (BISHOP et al. 1983; REVES et al. 1989; reviewed in BISHOP et al. 1990, 1991). Although children may be symptomatic when reinfected, second and subsequent infections rarely seem to be as severe as the initial infection. Complete natural immunity may result from the accumulated experience of mutiple infections. If this is the case, multiple doses of vaccine may be required to achieve adequate protective immunity.

Several studies in animals and children have suggested that protection against rotavirus illness is mediated primarily by local intestinal immunity. For example, the oral administration of anti-rotavirus antibody to low birth weight newborns or to children hospitalized for non-rotavirus illnesses can reduce the incidence and severity of symptoms of rotavirus infections (BARNES et al. 1982; EBINA et al. 1985; LOSONSKY et al. 1985; DAVIDSON et al. 1989; see below for animal studies). Increasing evidence also suggests that measurement of anti-rotavirus immunoglobulin A and/or neutralizing antibody coproconversions are the most sensitive markers of infection

(GRIMWOOD et al. 1988; LOSONSKY and REYMANN 1990; COULSON et al. 1990a), and the presence of jejunal neutralizing antibody or fecal IsA correlates with protection from illness (WARD et al. 1989; COULSON et al. 1992; MATSON et al. 1993).

These features of the natural history of rotavirus disease have directed most efforts of vaccine development toward the production of a live, attenuated, oral vaccine capable of stimulating intestinal immunity rather than a parenterally administered formulation. These observations also are the basis for the goal of immunizing young children with a rotavirus vaccine which induces active immunity against multiple virus serotypes prior to a first exposure at 2–6 months of age.

4.2 Importance of the Complexity of Rotavirus Antigenic Structure

Vaccine trials began before detailed knowledge of the nature of rotavirus antigens was acquired. However, a consideration of our current knowledge of rotavirus antigens and the development of immune responses to specific neutralizing epitopes on these proteins helps one understand: (1) why previous vaccine field trials have yielded variable results, (2) why mono-valent vaccines derived from animals fail to induce heterotypic immunity in children without previous rotavirus infections in spite of the development of cross-reactive antibodies, and (3) why antibody development to common antigens is not associated with neutralizing antibodies and protective immunity.

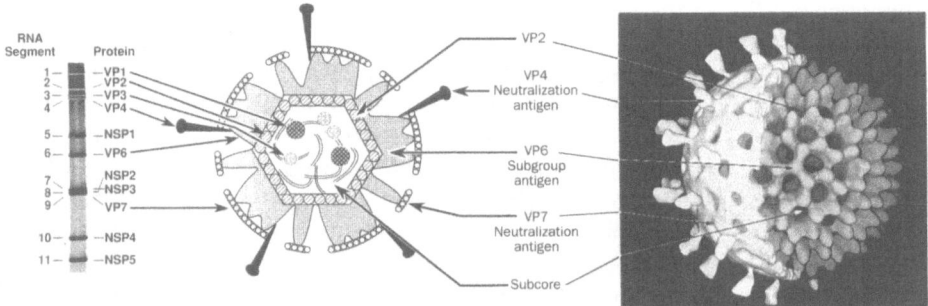

Fig. 1. Rotavirus proteins and antigens. Left, RNA segments and the gene coding assignments for the simian rotavirus SA11. The order of the genes for other virus strains can be different (reviewed in ESTES and COHEN 1989). Center, the viral structural proteins, VP1–VP7, have been localized to different shells of the virus particles. The major antigens of viral particles consist of two neutralization antigens (VP4 and VP7) and of subgroup antigens found on VP6. Right, the location of these proteins in the three-dimensional structure of virus particles is shown. This figure was composed in collaboration with V. Prasad (see Chap. 2 for more details on virus structure). The nonstructural proteins (NSP1–NSP5) are named following MATTION et al. (1993)

The rotaviral proteins have been characterized and gene-coding assignments are known (Fig. 1; reviewed in ESTES and COHEN 1989; MATTION et al. 1993). Analysis of differences between circulating virus strains and of epitopes on the viral proteins has demonstrated that rotavirus particles are antigenically complex. Virus particles contain six proteins and each of these proteins contains multiple epitopes capable of inducing specific antibody. However, only two proteins, VP4 and VP7, which make up the outer capsid of rotavirus particles, are known to induce neutralizing antibodies. Both outer capsid proteins contain at least three epitopes that induce antibody with neutralizing activity. Although neutralization epitopes on VP6 have been described (KILLEN and DIMMOCK 1982; SABARA et al. 1987), these reports have not been confirmed by others (reviewed in ESTES and COHEN 1989; MATSUI et al. 1989).

The existence of multiple human rotavirus serotypes and the unpredictable distribution of these serotypes have important consequences for the interpretation of vaccine trials and the strategy for vaccine development (EDITORIAL 1990). Rotavirus serotypes initially were differentiated using neutralization assays and hyperimmune antisera. Although neutralizing antibodies to both VP4 and VP7 are produced by hyperimmunization, the predominant antibody activity obtained after hyperimmunization seems to be directed to VP7. More recently, monoclonal antibodies specific for VP7 have been characterized that classify viruses in a manner identical to that obtained in plaque reduction neutralization assays with hyperimmune typing serum. Serotypes, as currently defined, represent different types of VP7. However, because both VP4 and VP7 can induce antibodies with neutralizing activity, it is probably important to classify viruses both by VP7 and VP4 types. For example, cross-reactive neutralization epitopes have been found on VP4 (SHAW et al. 1986; TANIGUCHI et al. 1988; COULSON 1993) and cross-neutralization observed in convalescent serum may be based on early immune responses to VP4. This hypothesis is consistant with early studies (ESTES and GRAHAM 1980; WYATT et al. 1982) which showed that convalescent sera do not permit differentiation of viruses by plaque reduction neutralization assays. Currently, six alleles and four antigenic types of VP4 (also called P types) have been described in human viruses, but rapid, convenient assays to detect the different antigenic types are not yet readily available (LARRALDE and FLORES 1990; GORZIGLIA et al. 1990a).

4.3 Influence of Virus Antigenic Diversity on Rotavirus Vaccine Development

More human rotavirus serotypes exist than originally was suspected. In 1980, only two serotypes were recognized; at present, a total of 14 rotavirus VP7 types (or G types) are known, and more may exist. Human viruses are found in nine of these. Currently, human serotypes 1–4 represent 95% of isolates from large collections (BEARDS et al. 1989; MATSON et al. 1990; BISHOP et al. 1991).

The existence of multiple human serotypes has raised several important questions about the antigenic diversity and clinical significance of different serotypes. For vaccine strategy, it has become important to know: (1) what is the worldwide and yearly distribution of these serotypes, (2) whether these serotypes are antigenically stable, (3) whether differences in age-specific attack rates or clinical outcome of infections with specific serotypes exist, (4) what is the host range of viruses in different serotypes, and (5) whether infection with a virus in one serotype can induce protective immunity to viruses only in the same serotype.

The development and incorporation of VP7 typing monoclonal anti-bodies (MAbs) into ELISAs has permitted direct characterization of viruses in clinical specimens and has facilitated studies of the epidemiology of specific serotypes. Most studies have screened relatively small numbers of samples collected from children hospitalized during 1 or 2 years, and most of these studies analyzed samples from outside the United States (HEATH et al. 1986; BEARDS et al. 1989; PONGSUWANNE et al. 1989; UNICOMB et al. 1989; URASAWA et al. 1989; AHMED et al. 1990; BELLINZONI et al. 1990; GERNA et al. 1990a; GOMEZ et al. 1990; GOUVEA et al. 1990; PADILLA-NORIEGA et al. 1990; WARD et al. 1991). However, when taken together with larger studies of serotype distribution conducted over 10 or more years in the United States and Australia (MATSON et al. 1990; BISHOP et al. 1991), the following information about rotavirus serotypes has emerged. First, viruses of six of the seven known human rotavirus serotypes appear to have a worldwide distribution (CLARK et al. 1987; BEARDS et al. 1989; GERNA et al. 1990a; NAKAGOMI et al. 1990). Information on the seventh, eighth and ninth serotypes is not available, because they have only been described recently (TANIGUCHI et al. 1990; URASAWA et al. 1990, 1993; GERNA et al. 1992). Second, in some regions, but not others, multiple serotypes cocirculate each year. In regions with multiple cocirculating serotypes, the predominant serotype, causing more than half the symptomatic infections, changes every 1–3 years (HEATH et al. 1986; BEARDS et al. 1989; URASAWA et al. 1989; O'RYAN et al. 1990). Serotype 1 viruses account for a significant portion of cases of rotavirus disease even when other serotypes predominate in a season. The predominant serotype varies among different cities within a country and in different countries in a single season. Finally, viruses in all of the serotypes can cause severe illness.

Concurrent studies of the viruses from hospitalized children and from nonhospitalized children monitored both for symptomatic and asymptomatic illness have found the predominant serotype of the season to be the same in both groups, indicating that changes of the predominant type are a community-wide event (PITSON et al. 1986; O'RYAN et al. 1990). The seasonal predominant type is more likely to cause symptomatic infection than other cocirculating types. Exposure to VP7 types in the first 2 years of life differs markedly between cohorts of children. Some groups of children are exposed to only one serotype in the first 2 years of life, while other groups are exposed to two or three serotypes (BEARDS and DESSELBERGER 1989; O'RYAN et al. 1990). These results demonstrate that individual children, infected with the same virus type at the same age, may differ with respect to the clinical outcome of infection. Therefore, because the distribution of serotypes is unpredictable, a succesful vaccine will need to induce protective immunity to all, or at least to the most prevalent, circulating serotypes. The occurence of multiple serotypes within different areas of a city indicates that local, rather than global, conditions are critical for the spread of individual virus strains, and vaccine efficacy trials must be carefully controlled.

Serotype 1 viruses appear to be more prevalent than other serotypes. Multiple strains within serotype 1 exist and these also change over time. It is important to determine what permits these type 1 viruses to persist. Persistence may result from antigenic drift, which has been recognized using MAbs to VP7 (COULSON 1987; GREEN et al. 1990a; RAJ et al. 1992); from changes in VP4; from changes within both VP7 and VP4; or from other currently unknown causes. Although still poorly understood, the appearance of monotypes may pose a previously unappreciated challenge to vaccine strategy. One puzzle in rotavirus vaccine trials has been variable heterotypic protection seen against natural serotype 1 virus challenges following some (but not other) trials with the serotype 6 bovine WC3 and with the serotype 3 rhesus RRV vaccines within the United States (CLARK et al. 1990a; BERSTEIN et al. 1990; KAPIKIAN et al. 1990). It is possible these different efficacies were due to variation within circulating serotype 1 viruses. This highlights the importance of carefully characterizing virus strains that cause illness in vaccinated children.

A number of similarities between rotaviruses and influenza viruses make influenza a useful model for describing and thinking about the antigenic diversity of rotaviruses and its impact on potential vaccination programs (RODGER and HOLMES 1979; GRAHAM and ESTES 1985; EDITORIAL 1990). For example, although rotaviruses and influenza viruses are very different in structure, replication, and biological characteristics, both have segmented genomes and both carry two surface proteins (HA, NA for flu; VP4, VP7 for rota), which are of significance for classification and for the host immune response. Both genera have a wide spectrum of different types/groups (flu A, B, C; rota A-E) and subtypes/types (flu A: H1–H14 with H1–H3 in humans, and N1–N9 with N1 and N2 in humans; rotavirus: VP7 types 1–14 with types 1–4, 8, 9, and 12 in humans and at least four VP4 types in humans). Both virus genera show antigenic drift and cocirculation of different types/subtypes at any given time. Both virus genera have a large circulation in animals which possibly is important for humans as a reservoir for new strains (proven for flu, likely for rotaviruses). Because VP4 of rotaviruses elicits neutralizing antibodies, justification of its inclusion in the classification of rotaviruses and consideration of its importance is at least as strong as for NA subtypes in influenza virus classification. In addition, rotavirus illness has an annual season occurring each winter in North America and that seasonal pattern is similar to the pattern of influenza (GLEZEN et al. 1982). Rotavirus displays a repetitive sequence of geographic spread across North America each winter which suggests the rapid spread of flu antigenic types (LeBARON et al. 1990b). However, this geographic sequence reflects the spread of rotavirus illness, not the movement of specific rotavirus types across the continent as with influenza (MATSON and ESTES 1990; GOUVEA et al. 1990). Finally, it is interesting that the winter timing of the annual rotavirus epidemics more closely resembles infections spread by the respiratory route (e.g., measles) than that of common enteric pathogens spread by the fecal-oral route (COOK et al. 1990).

4.4 Influence of Rotavirus Host Range on Vaccination

Animal and human strains may possess the same VP7 type, and in fact, porcine strains that cross-react with human VP7 types 1–4 now have been found (HOSHINO et al. 1984; BELLINZONI et al. 1990; NAGESHA et al. 1990). VP7 type 3 rotaviruses have been isolated from many species. Based on the apparent wide host range of different serotypes of virus, it is tempting to speculate that cross-species transmission occurs. Evidence for this includes the remarkable similarity of the sequences of the human and porcine VP7 genes. In fact, the nucleotide and amino acid sequences are so similar that it would be difficult to separate animal and human virus strains based on this information alone (HUANG et al. 1989). In addition, several human and animal strains exhibit a broad host range in experimental infections of mice and piglets. Replication of virus in children given vaccine viruses from rhesus monkeys and calves, including virus serotypes not yet isolated from humans, demonstrates the potential for transmission from animals to humans. Serotype 3 viruses from children also show an unexpected high degree of genetic homology to feline and canine rotaviruses (KITOAKA et al. 1987; NAKOGOMI et al. 1989; NAKOGOMI and NAKOGOMI 1989; GERNA et al. 1990b). Finally, a small outbreak of human disease due to a subgroup I rotavirus with a long electropherotype has been described in Manipur, northeastern India which could have been due to a virus from an animal reservoir (GHOSH and NAIK 1989). Because electropherotype and subgroup determinations cannot be used to definitively prove virus origin (reviewed in ESTES et al. 1984), direct evidence of natural cross-species transmission remains lacking. The strongest evidence comes from recent isolation from children of viruses serologically similar to viruses previously only isolated from cows (URASAWA et al. 1992; GENTSCH et al. 1993; DAS et al. 1993). Cross-species transmission remains a strong possibility, and such transmission theoretically could impede the control of infections in children unless concurrent measures can reduce successfully both virus load and transmission between and from animals.

4.5 Knowledge of Protective Immune Responses

The recognition that multiple human rotavirus serotypes exist raises the critical question of whether protective immunity is homotypic or heterotypic. This was first assessed directly among children in an orphange in Japan. During two outbreaks of infections with a serotype 3 virus, only homotypic serum neutralizing antibody titers exceeding 1:128 protected against symptomatic infection (CHIBA et al. 1986). Studies of the immune responses in several animal models almost uniformly concluded that heterotypic immunity does not develop following primary infections (WOODE et al. 1989; SNODGRASS et al. 1984, 1991; CONNER et al. 1991). These results suggested that rises of heterotypic immunity after infection with a single virus type probably represent boosting of preexisting antibodies, with a concomitant broadening of the immune response.

In spite of the data from animal models, heterotypic antibody responses measured in phase I trials of the animal vaccines in children were argued to reflect protective responses, and field trials were initiated (reviewed in KAPIKIAN et al. 1989). Following several vaccine failures in which serotyping indicated that the challenge virus was of a serotype distinct from the vaccine (KAPIKIAN et al. 1990), it began to be more readily accepted that heterotypic immunity may not be induced by the animal vaccine viruses. Use of an epitope-blocking immunoassay and neutralization assays to evaluate the serotype-specific serum antibody responses of infants and adults who received a live, attenuated, monovalent vaccine and who were not protected following natural challenge (often with a heterotypic virus) supported these conclusions (SHAW et al. 1987; BEARDS and DESSELBERGER 1989; GREEN et al. 1990b). It was shown that adult vaccinees exhibited both a homotypic response against the immunizing virus strain and heterotypic responses against other serotypes. In contrast, infant vaccinees below 6 months of age exhibited homotypic responses, and significantly fewer heterotypic responses (GREEN et al. 1990b). These results suggest that young infants do not mount a heterotypic antibody response to a monovalent vaccine because they have not been previously exposed to rotavirus. This conclusion is supported by a study of the prevalence of serum neutralizing antibodies to several serotypes among infants of different ages. Younger infants tended to have antibody to only one serotype whereas older infants had antibodies to multiple serotypes (BRÜSSOW et al. 1990b). These results help explain the failure of monovalent vaccines to induce protection against heterotypic virus challenges in young infants. These results also indicate that if the first rotavirus infection is most likely to be severe, then it will be important to deliver a multivalent vaccine before the first exposure.

The need for a previous rotavirus exposure to yield a heterotypic antibody response may not be absolute. Exceptions to the generalization that primary infections result only in homotypic responses have been documented when this question has been carefully addressed in animals whose exposure to rotavirus was well-controlled. Primary infection in rotavirus-naive animals with some, but not all, virus strains can induce heterotypic serum antibody responses (WOODE et al. 1987; NAGESHA et al. 1990; SNODGRASS et al. 1991; CONNER et al. 1991). For example, a majority of rabbits inoculated orally with a serotype 3 virus showed heterotypic serum neutralizing responses to serotype 8 and 9 viruses. It remains unknown if such heterotypic responses are due to responses to cross-reacting epitopes on the outer capsid VP4 or VP7. One proposed mechanism for such responses is that some virus strains in different serotypes share one or more neutralizing epitopes on their VP7s. Therefore, these strains can induce more broadly reactive antibody. An example of shared epitopes within region C of VP7 has been described among porcine viruses (HUM et al. 1989). It also has been shown that specific combinations of VP4 and VP7 may alter the biological and antigenic properties of the rotavirus outer capsid proteins (NAGESHA et al. 1990; CHEN et al. 1989, 1991). Although such alterations complicate the meaning of serotypes, they offer another explanation for the induction of heterotypic responses. A vaccine strategy for multiple serotypes might be simplified if the induction of heterotypic responses in unprimed individuals could be understood. One could then predict immune responses and protection against infections with heterotypic viruses. However, without this knowledge, at present, multivalent rotavirus vaccines must be considered.

Seroepidemiologic surveys of the prevalence and serotype specificity of rotavirus antibodies in different age groups of children also have reported the development of heterotypic neutralizing antibody responses following suspected primary infections, particularly following infections with serotype 4 viruses (GERNA et al. 1990c; PADILLA-NORIEGA et al. 1990; BRÜSSOW et al. 1990b, 1991). However, it is difficult to know when children are in fact experiencing a primary infection. This is particularly questionable in one study in which many of the primary infections were in children older than 6 months (GERNA et al. 1990c). COULSON et al. (1990a) have monitored methods to detect rotavirus infections in children and found that using seroconversion or virus excretion in stools to detect infections can underestimate the number of rotavirus infections by up to 200%. Therefore, the serum antibody status at the onset of infection may not correctly reflect previous infections. Instead, monitoring rotavirus immunoglobulin A coproconversion may be the most sensitive indicator of rotavirus infection and reinfection (COULSON et al. 1990a). Because measurable coproantibody may be transitory, no current humoral antibody detection assay conclusively documents the degree of previous exposure to rotavirus.

Another difficulty (not initially recognized) in measuring antibody responses is the choice of virus to use in the test assays. This is particularly true when the history of prior exposure and the serotype of the virus causing previous infections are not known. Studies in animals experimentally infected with viruses of known serotype have now shown clearly that the choice of test virus can change the test results (CONNER et al. 1991). Homotypic immune responses are detected most frequently and at higher titer when the test virus is the same as the infecting virus. The rate of detection of heterotypic immune responses following natural infection with human rotaviruses (GERNA et al. 1990c) and following vaccination also is influenced by the choice of the test strain. When heterotypic serum antibody responses were assessed after WC3 vaccination, tests performed using the serotype 3 SA11 virus resulted in higher rates of response than tests performed using the human serotype 3 virus P (CLARK et al. 1988; BERNSTEIN et al. 1990). A similar observation was made when antibody responses to rhesus-human reassortants were assessed (FLORES et al. 1989).

4.6 The First Rotavirus Vaccines: Testing of Rotavirus Strains from Animals

The development and testing of rotavirus vaccines from animal sources preceded much of the information reviewed above. Initially, live, attenuated, monovalent vaccines from animal rotaviruses were developed for oral administration to children (Table 2). Three candidate monovalent vaccines from animal rotaviruses (RIT 4237, a bovine serotype 6 virus; RRV MMU 18006, a rhesus monkey serotype 3 virus; and WC3, a bovine serotype 6 virus) were the first vaccines to progress to field trials.

Ten years after the description of rotaviruses as agents causing severe diarrhea in children, phase I safety and immunogenicity trials of the first live, attenuated, monovalent rotavirus vaccine (the bovine RIT 4237 vaccine) were reported (VESIKARI et al. 1983). Since that report, efforts to develop a live attenuated vaccine have continued steadily, and several efficacy trials

Table 2. Live attenuated rotavirus candidate vaccines

Formulations tested	Virus strains (G type)	Investigator or producer	Status of trials
Animal viruses	Bovine RIT 4237 (G6)	Smith Kline RIT	Field
	Rhesus RRV MMU 18006 (G3)	NIH and Wyeth	Field
	Bovine WC3 (G6)	Wistar Institute and Merieux Institute	Field
Animal/human reassortants	Rhesus RRV × HRV VP7 (G1, G2, and G4)	NIH and Wyeth	Field
	D × RRV (G1)		
	DS1 × RRV (G2)		
	ST3 × RRV (G4)		
	Bovine UK × HRV VP7 (G1–G4)	NIH	
	Bovine WC3 × HRV VP7 W179-9 (G1)	Wistar	Phase I
Human viruses	M37 (G1)	NIH	Field
	RV3 (G3)	Bishop	ND
	Cold adapted virus (G1)	Matsuno and NIH	Phase I

Table 3. Efficacy trials of rotavirus vaccine candidates in humans

Vaccine source/ VP7 type	Trial site	Age at enrollment (months)	Doses (n)	Period of observation (months)	Prevailing serotype(s)	Selected measures of efficacy against rotavirus						Efficacy reported	References
						Vaccine recipients			Placebo recipients				
						Total	Rotavirus diarrhea (number of cases)	Severe[b] rotavirus diarrhea (number of cases)	Total	Rotavirus diarrhea (number of cases)	Severe rotavirus diarrhea (number of cases)		
RIT 4237 bovine/ G6	Tampere, Finland	8-11	1	5	1	86	9	2[a]	92	18	18	Yes	Vesikari et al. (1984)
	Tampere, Finland	6-12	2	6	—[c]	168	5[a]	—	160	26	—	Yes	Vesikari et al. (1985)
	Butare, Rwanda	3-8	1	5	—	122	6	—	123	6	—	No	De Mol et al. (1986)
	Bakau, The Gambia	2.5	3	4	2	170	47[a,d]	43	83	34	26	No	Hanlon et al. (1987)
	Leoben, Austria	2-3	1	6	—	10	0	—	11	4	—	No	Mutz et al. (1989)
	Lima, Peru	2-18	1-3	18	1,2	291	82	17	100	35	12	Yes	Lanata et al. (1989)
	Tampere, Finland	0	1	25-32	1,4	~200[e]	40	19[a]	~200[e]	40	38	Yes	Ruuska et al. (1990)
	Navajo Reservation USA	2-5	1	0-18	1,2	106	11	6	107	9	8	No	Santosham et al. (1991)
	Tampere, Finland	0	2	28	1,4	~100[e]	15	1[a]	~100[e]	24	11	Yes	Vesikari et al. (1991a)
RRV rhesus/ G3	Baltimore, USA	4-24	1	7-8	1	14	0	—	10	3	—	No	Rennels et al. (1986)
	Caracas, Venezuela	1-10	1	12	2,3	151	8[a]	4	151	22	15	Yes Yes	Flores et al. (1987) Perez-Schael et al. (1990b)
	Rochester, USA	2-4	1	7-10	1	88	19	11	88	17	13	No	Christy et al. (1988)
	Umea, Sweden	4-12	1	24	1	53	9	2	51	17	10	Yes	Grothefors et al. (1989)
	Baltimore, USA	1-11	1	24	1	63	11	—	49	12	—	No	Rennels et al. (1990)
	Tampere, Finland	2-5	1	17-19	1,4	100	10	5	100	16	13	Yes	Vesikari et al. (1990)
	Navajo Reservation USA	2-5	1	0-18	1,2	108	11	8	107[f]	9	8	No	Santosham et al. (1991)

Vaccine	Location	Age range										Significant	Reference
WC3 bovine/ G6	Pennsylvania, USA Cincinnati, USA	3–12 2–12	1 1	6–10 7–10	1 1	49 103	3[a] 21	0[a] 9	55 103	14 25	11 15	Yes No	CLARK et al. (1988) BERNSTEIN et al. (1990)
WI79-9 ressor-tant/G1	Pennsylvania, USA	2–11	2	11	1	38	0[a]	—	39	8	—	Yes	CLARK et al. (1990b)

[a] Statistically significant protective effect of vaccine ($p \leq .05$).
[b] Definition of severe diarrhea differs between studies.
[c] Not studied or not reported.
[d] Four ill subjects were not examined by physicians.
[e] Number in group not reported.
[f] Same placebo group as for the RIT 4237 study.

have been performed (Table 3). Only one of the monovalent vaccine candidates (the rhesus MMU 18006 vaccine) represented a VP7 type found in human rotavirus strains. Initially, monovalent vaccines containing a heterologous VP7 were developed assuming that heterotypic protective immunity would be induced against disease caused by infection with viruses with VP7 types (serotypes) other than the vaccine serotype. This strategy was based on observations that animal and human rotavirus strains share some antigens and that children naturally infected with human rotaviruses respond serologically to both human rotaviruses and animal rotaviruses, such as bovine, simian or murine strains (FLEWETT et al. 1974; WOODE et al. 1976b; KAPIKIAN et al. 1976). In phase I trials in adults and children, each of the animal rotavirus vaccines was promising, based on the induction of antibody responses measured by several different assays.

Efficacy trials first were performed for the RIT 4237 vaccine, a high tissue culture passage, cold-adapted, bovine NCDV strain, in children in Finland (Table 3). RIT 4237 provided over 80% protection against clinically significant diarrhea in children vaccinated at 8–11 and 6–12 months of age (VESIKARI et al. 1984, 1985). These promising results were followed by additional trials in newborn infants in Finland and trials in other countries (see Table 3).

A field trial in Lima, Peru using RIT 4237 was one of several trials conducted in developing countries (LANATA et al. 1989). This trial evaluated the effect of multiple (one, two, or three) doses of vaccine, and etiologic agents in additions to rotavirus were sought for each episode of diarrhea. Efficacy for diarrhea in which rotavirus was the sole pathogen was demonstrated only after three doses of vaccine. When the results were analyzed according to the infecting rotavirus serotype, better protection was found against disease caused by serotype 1 than that caused by serotype 2 viruses (LANATA et al. 1989). A protective effect of RIT 4237 also was suggested by a small study in Austria (MUTZ et al. 1989).

In other trials in developing countries or in settings where enteric infections are highly endemic, such as Rwanda, The Gambia, and a Navajo reservation in Arizona in the United States, the RIT vaccine reportedly failed to induce protection against rotavirus diarrhea (DE MOL et al. 1986; HANLON et al. 1987; SANTOSHAM et al. 1991). The following points lead us to reinterpret the results of these trials. In the Gambian trial, a statistically significant protective effect of vaccination was observed ($p = .047$), but the reduction of the attack rate from 41% in the placebo group to 28% in the vaccine group was less pronounced than that observed in Finland. These data indicate that the wide consideration of this trial as a failure may be incorrect. In Rwanda, only 12 episodes of rotavirus diarrhea were detected among 245 study children and these were divided equally between the vaccine and placebo groups. The results of this trial were described in a brief letter so it is difficult to know what factors may explain such a low rate of detected illness at this site. In the trials in The Gambia and in Rwanda, it was postulated that either high levels of preexisting antibody or other local factors may have contributed to decreased vaccine "takes" that resulted in vaccine failure.

The trial on the Navajo reservation was a comparison of RIT 4237 and MMU 18006 (SANTOSHAM et al. 1991; Table 3 and see below). This trial was stopped after 31 episodes of rotavirus diarrhea had occurred among 321 study participants. The authors of this study concluded that neither vaccine was efficacious in preventing rotavirus diarrhea. However, the timing of rotavirus diarrhea after immunization was affected by vaccination (SANTOSHAM et al. 1991). For example, the majority (seven of nine) of placebo recipients had rotavirus diarrhea in the first 3 months after vaccination, and the other two episodes of illness in this group occurred 4–6 months after placebo administration. In the RIT 4237 study group, four children were ill in the first 3 months after vaccine administration, two in months 4–6, and five in months 7–13 (chi-squared for the trend = 5.3, $p = 0.02$). While these trials in The Gambia, Rwanda, and Arizona may be cited as failures, in at least two instances (The Gambia and Arizona), the assessment of failure is only by comparison to the striking results achieved in Finland. Overall, the differences in the trials conducted in Finland and in these developing countries demonstrate that the development of a successful vaccine suitable for many settings will be a complex problem. In some ways, this is not different from the problems of immunization of different populations with poliovirus vaccines (MELNICK 1990).

Subsequent trials of the RIT 4237 vaccine in 5–7 day old infants in Finland have provided some new ideas about vaccine strategy. While administration of this vaccine to neonates had no protective effect against the occurrence of rotavirus infection, it did protect against severe illness (RUUSKA et al. 1990). The protective effect was greater in a group given the vaccine just before the rotavirus season than in a group that received the vaccine several months before the rotavirus epidemic season. Protection lasted for 2–3 years after vaccination. Vaccine-induced clinical protection against rotavirus diarrhea did not correlate well with serologic response after vaccination, but it was hypothesized that the vaccine effect may have been amplified by exposure to wild rotaviruses during the season. A larger study in neonates in Finland concluded that immunization with the RIT 4237 vaccine in the neonatal period and at 7 months of age, with the second dose being given before the rotavirus epidemic season, yielded better protection (89% for cases of severe diarrhea) than administration of a single dose of vaccine to neonates (VESIKARI et al. 1991a). These studies suggest that the timing of vaccine administration, rather than vaccinee age or number of vaccine doses, may be critical for successful rotavirus vaccination.

The rhesus rotavirus candidate vaccine MMU 18006 was originally recovered from a 3 1/2 month old monkey with diarrhea (STUKER et al. 1980). This vaccine was developed as an alternative to RIT 4237 because the virus is of serotype 3, a serotype found in humans, and because it was readily cultivatable in DBS FRhL-2 cells, a semicontinuous, diploid, fetal, rhesus monkey lung cell strain (WALLACE et al. 1973). Phase 1 studies with 10^5 plaque-forming units (pfu) and 10^6 pfu of vaccine administered to adults, and subsequently to children, infants, and neonates, revealed a variable, and sometimes unacceptably high, rate of fever and diarrhea (reviewed in KAPIKIAN et al. 1989). A lower vaccine dose of virus of 10^4 pfu given to young infants with transplacentally acquired maternal antibodies caused transient fevers in at most one third of vaccinees and induced serum antibody responses in over 50% of vaccinees. These responses were considered to be acceptably safe and immunogenic to proceed to field trials (reviewed in KAPIKIAN et al. 1990). In general, the MMU 18006 vaccine has been more immunogenic and reactogenic than RIT 4237 (VESIKARI et al. 1986).

The largest efficacy trial of the MMU 18006 vaccine was performed in Venezuela and results of this trial were published in two parts (FLORES et al. 1987; PEREZ-SCHAEL et al. 1990b). Overall vaccine efficacy was 64% against rotavirus diarrhea and 90% against severe rotavirus diarrhea when given to infants in the first year of life. In this study, infants were vaccinated from 1 to 10 months of age. In children who were 6–10 months of age at enrollment, only five rotavirus diarrhea episodes occurred among 114 study subjects, and four of the five rotavirus episodes were among the vaccine recipients. A vaccine efficacy of 82% was observed for infants 1–5 months of age at enrollment, the group in this study population with a significant attack rate for rotavirus diarrhea. In this population, administering vaccine to infants 6 months of age or older apparently would be of no benefit. Therefore, in this type of population, vaccine would need to be administered to very young infants. Serotype 3 rotavirus diarrhea was more common than infection caused by other serotypes in this study, and only the protection of MMU 18006 against serotype 3 was statistically significant. However, MMU 18006 exhibited a protective effect against all serotypes detected.

In two efficacy studies of MMU 18006 in Baltimore, the vaccine conferred marginal protection against rotavirus diarrhea (RENNELS et al. 1986, 1990). In Finland, while fewer episodes of rotavirus diarrhea occurred among the vaccine recipients overall, most of the infections in the recipients occurred in the second season of rotavirus activity after vaccination. Infections were evenly divided between seasons in the placebo group, and the infections in the placebo recipients were more severe than in the vaccine group (VESIKARI et al. 1990). However, none of these trends was statistically significant. Vaccine efficacy was marginal in Sweden where study subjects also were monitored over two rotavirus seasons (GROTHEFORS et al. 1989). Rotavirus diarrhea ($p = 0.09$) and severe rotavirus diarrhea ($p = 0.11$) were less frequent in the vaccine group, and this protective effect against diarrhea was equally strong in both seasons of monitoring. In Rochester, New York, no protective effect was observed for MMU 18006 (CHRISTY et al. 1988).

In the comparative study of MMU 18006 and RIT 4237 on the Navajo reservation in Arizona, MMU 18006 delayed the occurrence of rotavirus diarrhea among the vaccine recipients and the pattern of that delay was indistinguishable from that caused by RIT 4237 vaccination (see above), but this pattern of infection was not statistically different from the placebo group (SANTOSHAM et al. 1991). Serotyping of the infecting rotaviruses during these trials indicated that the predominant serotype in the successful trial in Venezuela was serotype 3 (PEREZ-SCHAEL et al. 1990b), while serotype 1 was isolated in Arizona and Rochester, where the vaccine had failed to induce significant protection. These results indicated that RRV could product against clinically significant diarrhea caused by serotype 3 viruses but not by viruses of heterologous types (reviewed in KAPIKIAN et al. 1990).

The WC3 rotavirus vaccine strain, originally isolated from a calf in Pennsylvania, was evaluated in phase I studies at the twelfth cell culture passage using a dosage of 3.0×10^7 pfu. No adverse clinical effects have been noted after administration of this vaccine to children (CLARK et al. 1986). Fecal excretion of virus occurred in fewer than 30% of vaccinees and at a low titer. After a single vaccine dose, immune response rates were high; 71%–97% of vaccinees responded with a fourfold rise in serum neutralizing antibody to WC3 and about half responded to the serotype 3 simian rotavirus SA11. Only a few vaccinees ($<10\%$) developed antibody to human rotaviruses, including serotype 3 human rotaviruses. Responses to human serotype 1 viruses have been minimal and appear to occur only in infants with a prior exposure to type 1 virus (CLARK et al. 1986; BERNSTEIN et al. 1990). In an efficacy trial in suburban Philadelphia during a predominantly serotype 1 rotavirus season, a single dose of WC3 provided protection at a level of 76% against all rotavirus disease and 100% against severe disease (CLARK et al. 1988). Vaccination with WC3 did not show similar efficacy in Cincinnati (BERNSTEIN et al. 1990) or in the Central African Republic (H.F. Clark, personal communication). Another study of the WC3 vaccine in China also reported a protective efficacy against rotavirus illness, while measurements of seroconversion were low (R. Glass, personal communication).

Although the WC3 vaccine was ineffective in Cincinnati, the investigators sought to identify correlates of protection. The strongest correlate of protection in that study was natural infection itself. Infants who developed a high titer of neutralizing antibody to WC3 following vaccination or who had serum antibody of maternal origin to the predominant serotype during the trial were protected against illness. These results indicate that the WC3 vaccine was not ineffective, but that the response to vaccination was inadequate, most likely because WC3 replicates poorly in humans (BERNSTEIN et al. 1990).

In a related study, the serologic responses of infants vaccinated with WC3 were compared based on whether they had a natural primary rotavirus infection before vaccination. In children previously unexposed to rotavirus, vaccination with WC3 resulted in a low antibody response and minimal induction of immunologic memory. Subsequent reinfections with human viruses resulted in an immune response only to the human virus (WARD et al. 1990a). In contrast, WC3 vaccination of infants who had previously experienced a natural infection with a serotype 1 human rotavirus resulted in boosting of neutralizing antibody titers to all four human serotypes. The final titers were significantly higher to the human virus serotypes than to WC3 (WARD et al. 1990a). Because the humoral immune responses to the human viruses were poor in the two efficacious WC3 vaccine trials, it has been suggested that the vaccine protects by inducing cell-mediated immunity. This is unproven. However, experimental infections and transfer experiments of cytotoxic T lymphocytes (CTLs) in normal and immunocompromised mice support a role for CTLs in both protection from diarrhea and in clearance of virus following rotavirus infection (DHARAKUL et al. 1990; OFFIT and DUDZIK 1989a, 1990).

These numerous efficacy trials have shown that none of the three monovalent vaccines protect against infection, and protection against clinically significant disease has been variable (Table 3). One reason for this is that different populations of children have been studied for each vaccine. Therefore, results from each safety, antigenicity, and efficacy study must be examined separately. It is clear from these efficacy trials that the results of rotavirus vaccine studies in industrialized countries cannot be extrapolated reliably to predict the outcome in developing countries. Efficacy must be confirmed by trials carried out in developing countries. An alternative possibility may be to test vaccines in heterogeneous populations in large cities such as Houston, Texas. This may be feasible because recent epidemiology studies indicate that rotavirus infections in this type of population may more closely reflect the diversity of Third World countries than studies in smaller cities in the northcentral part of the United States (MATSON et al. 1990).

In spite of these differences in study populations, some trends are apparent among the trials of the monovalent vaccines. Overall, the monovalent vaccines reduced the incidence of rotavirus diarrhea about one third (odds ratio = 1.1–1.6, placebo vs vaccine group) and severe diarrhea was reduced about two thirds (odds ratio = 2.1–5.0, placebo vs vaccine group) at the test sites studied. However, the range of efficacy has been large, with a reduction in diarrhea from 100% to, in one instance, an increase of 23%, in the rate of illness among vaccinees. Statistically significant protective effects were observed in only a few trials. Small numbers of participants and of rotavirus diarrhea episodes are a shared characteristic of almost all the trials and this may be one cause of variability of vaccine efficacies.

Problems in study design and lack of uniformity of definitions also have confused the field. Definitions of severe diarrhea differed between studies and the choice of definition can affect the study interpretation (RUUSKA et al. 1990). The concept of "child-years" was used to substitute for "number of study subjects" in regions where exposure is not uniform year round. In two studies, the number of child-years or subjects cannot be determined from the published report. In addition, there has been no rigorous comparison of one vaccine candidate with another. This may be necessary to obtain reliable information about efficacy. A few trials have reported that vaccination reduces the rate of diarrhea, whatever the etiology (CLARK et al.1988). The mechanism by which this occurs is uncertain and worthy of further study.

Following the field trials summarized above, further evaluation of both monovalent bovine vaccines was stopped (VESIKARI 1988; KAPIKIAN et al. 1990; H.F. Clark, personal communication). However, because of the demonstrated efficacy against clinically severe gastroenteritis caused by serotype 3 rotaviruses, the rhesus vaccine continued to be tested as a component of quadrivalent reassortant vaccine. Since the WC3 vaccine strain exhibited an excellent safety record in infants, it has been used to produce single-gene reassortant viruses in which the outer capsid glyco-protein, VP7, for each of the human rotavirus serotypes is included (CLARK et al. 1990b; see below).

4.7 Second Generation Rotavirus Vaccines: Live Animal-Human Reassortant Rotaviruses and Human Rotavirus Vaccine Strains

The recognition that serotype-specific immunity is necessary for protection against rotavirus diarrhea in infants undergoing primary infection and the variable success of the monovalent vaccines (probably due in some cases to their inability to provide heterotypic protection in the youngest children) resulted in a modification of vaccine strategy. The goal of this modified strategy is to develop a multivalent live vaccine containing viruses with the VP7 specificity of each of the four rotavirus serotypes currently recognized to be epidemiologically most important (Table 2). Reassortant rotaviruses have been prepared by coinfecting cell cultures with a parental virus strain (either the RRV MMU18006, the bovine WC3, or the bovine UK) and a serotypically distinct human rotavirus. Under selective pressure of antibody against one of the parental rotavirus strains (see Chap. 6), reassortant progeny have been derived that contain ten genes of the parental virus and a single human rotavirus gene that codes for the VP7 of the human rotavirus serotypes 1, 2, 3, or 4 (MIDTHUN et al. 1985; CLARK et al. 1990a).

Phase I trials of individual RRV reassortants in adults and in progressively younger children revealed that the reassortant vaccine candidates were safe, in that they were similar in their reactogenicity to a 10^4 pfu dose of the parental RRV virus. Their immunogenicity also was similar to that of the RRV parent virus (HALSEY et al. 1988; FLORES et al. 1989; TAJIMA et al. 1990). However, the reassortant virus (RRV × D, serotype 1) differed from its parent (RRV) in that host susceptibility of children with prior rotavirus infection differed between the two strains. Preexisting antibody did not prevent infection and seroconversions with the serotype 3 RRV, but such antibody did block replication and excretion of the serotype 1 RRV × D reassortant (TAJIMA et al. 1990). These results raised the question of whether reassortant human vaccines (or other human vaccine strains) would be sufficiently immunogenic when given to infants with preexisting antibody. However, the excretion of reassortant vaccine virus by seronegative children and the induction of an antibody response in these children permitted further testing of the bivalent and multivalent vaccines for safety and immunogenicity.

A bivalent formulation was found to be as safe and immunogenic as the monovalent RRV vaccine (FLORES et al. 1989). Studies then were extended to a quadrivalent rotavirus vaccine formulation of 0.25×10^4 pfu of RRV and of three reassortants: D × RRV (containing human VP7 type 1), DSI × RRV (containing human VP7 type 2), and ST3 X RRV (containing human VP7 type 4; PEREZ-SCHAEL et al. 1990a). Additional infants were given an increased dose of 0.5×10^4 pfu or 10^4 pfu. When given to 2–5 month old infants, the number of neutralizing seroresponses to each serotype was disappointing even for the highest dose of each component tested (PEREZ-SCHAEL et al. 1990a). The results indicated that interference by one or more of the vaccine components occurred. This led to testing varied amounts of the individual components of the quadrivalent vaccine to overcome the interference and induce a higher rate of seroresponses to each component. A fivefold increase of the serotype 2 and 4 components (5×10^4 pfu) administered with 1×10^4 pfu of each of the serotypes 1 and 3 components did not increase the response rates to serotypes 2 and 4 (FLORES et al. 1990). Further increases in dose or the administration of multiple doses are being evaluated to try and increase the response rates to the serotypes 2 and 4 components of the vaccine. Problems of interference and lower VP7-specific neutralizing responses following administration of two RRV-human reassortant vaccines were confirmed in similar studies in Finland and the USA (VESIKARI et al. 1991b; WRIGHT et al. 1991).

A WC3 bovine-human reassortant, strain WI79-9, also was derived because the WC3 vaccine was inefficient at inducing specific antibody to serotype 1 virus. In phase I safety and immunogenicity studies, this reassortant yielded no side effects when given at $10^{7.3}$ pfu to 48 infants ages 2–11 months (CLARK et al. 1990a). The serotype 1 reassortant induced a type 1 seroresponse and a seroresponse to WC3 (type 6) in 39% of children of all ages. This represented an improvement over the type 1 serum antibody response rate of 10% observed previously with the WC3 vaccine. However, in contrast to the observation with the bovine WC3 vaccine, the infant response rate to WI79-9 was inhibited by the presence of preexisting antibody among 5–11 months old infants. Therefore, the enhanced capacity of WI79-9 to induce a serotype 1 immune response only was seen in infants seronegative to type 1 prior to immunization. A second dose was found to be effective in boosting the overall neutralization response rates to 59% among

2–4 months old infants and to 88% in those children originally 5–11 months old (CLARK et al. 1990).

Based on these studies, the WI79-9 reassortant vaccine was evaluated in a safety and efficacy trial in 2–11 months old infants and two doses were found to give complete protection against clinically significant illness following natural challenge (CLARK et al. 1990b; Table 3). In this study, rotavirus illness in children given placebo was caused by either serotype 1 or serotype 3 viruses. The vaccine did not protect against infection and the rate of serum neutralization antibody conversion to serotype 1 was lower (22%) than seen in the initial safety and immunogenicity study. Given the efficacy of the vaccine and the low level of neutralizing response, the mechanism of protection remains unclear. Surprisingly, fecal IgA responses were not detected following the first vaccine dose.

4.8 Third Generation Rotavirus Vaccines:
Live Attenuated Human Vaccine Strains

The variable success of monovalent and reassortant vaccines based on the animal virus strains led to the consideration of other approaches. It is possible that vaccines need to mimic naturally occurring human strains more closely than is possible with animal rotaviruses. Polyvalent vaccines representing the four or more epidemiologically important human rotavirus serotypes may be required to induce protection. This approach was not tested initially because of difficulties in cultivating human virus strains and of concern that human virus strains might not be sufficiently attenuated.

Safety and immunogenicity testing of the first candidate human rotavirus vaccine (M37) has been reported (FLORES et al. 1990). This virus originally was recovered from an asymptomatic newborn in Venezuela and adapted to grow in monkey kidney cells. M37 shares VP7 specificity with serotype 1 human rotaviruses and VP4 specificity with other human rotavirus strains of VP7 types 2, 3, and 4 (FLORES et al. 1986; GORZIGLIA et al. 1986). Therefore, the hope was that this strain would induce an immune response more cross-reactive with human rotaviruses than that induced by animal virus vaccine candidates (FLORES et al. 1990). M37 was thought to be naturally attenuated, because the neonate from whom it was isolated was asymptomatic at the time of infection. The rationale to use such a strain also was based on the observation that neonates who develop subclinical infections (with similar strains) within the first 2 weeks of life are protected against clinically significant rotavirus diarrhea during a 3 year period of follow-up (BISHOP et al. 1983).

When tested in 10–12 weeks old children at a dose of 10^4 pfu, the M37 strain showed some reactogenicity (fever) in 20% of immunized children (FLORES et al. 1990), but this effect was considered mild and acceptable. A total of 64% of children showed a neutralizing antibody response to M37 while only 27% of the children showed a response to WA, another serotype 1 virus. This low response to WA may not be surprising as other studies have shown that WA and M37 represent different serotype 1 monotypes (GREEN et al. 1990a; RAJ et al. 1991), and they also possess different VP4s (GORZIGLIA et al. 1990a). The neutralization responses to heterotypic serotype 2 (DS1) and serotype 3 (P) viruses were only 9% and 5%, respectively. More worrisome was the observation that the rate of serum neutralizing antibody responses against a given human rotavirus serotype was less among infants in whom preexisting antibody was detected. Virus excretion was noted only in 40% of vaccinees (FLORES et al. 1990). Similar vaccines representing all rotavirus serotypes could be prepared, because neonatal (possibly naturally attenuated) strains like M37 of serotypes 1–4 have been identified (HOSHINO et al. 1985).

It was somewhat surprising that the reactogenicity of this "asymptomatic" strain was 20% in the vaccinees. This human virus also was excreted at a rate lower then that observed among children given the quadrivalent RRV reassortant vaccine. This lower virus excretion rate was thought to be due to the poor growth of M37. In future studies it will be important to determine if this poor growth reflects attenuation to a point where the virus is not sufficiently immunogenic. Because the M37 strain is a serotype 1 strain and contains a VP4 of a type distinct from that on RRV, its ability to induce protection will be of interest. One efficacy trial in Finland using a single dose of vaccine at 10^4 or 10^5 pfu showed no efficacy (T. Vesikari, personal communication). Another efficacy trial is being conducted in Venezuela.

4.9 Questions About Administration of Live Attenuated Rotavirus Vaccines

Several questions have been raised concerning the use of live attenuated viruses as part of a rotavirus vaccine program. Answers to these questions are not yet completely known, but they should be sought to increase confidence in this approach.

The first question is whether a virus that is attenuated will replicate sufficiently to be immunogenic, because no precise markers of attenuation are known. Studies in animals suggest that virulence and disease severity may reflect the numbers of cells in the intestine destroyed by infection (TZIPORI et al. 1989; BRIDGER et al. 1990). If this also correlates with a good "vaccine take," then excretion of vaccine virus may be a critical parameter to predict success. Alternatively, it might be necessary to select attenuated virus strains with little or no ability to replicate in the intestine, but with sufficient quantity or antigenic integrity to stimulate the gut-associated lymphoid tissue. In such cases, it probably will be important to give more than one dose (TZIPORI et al. 1989). This question highlights the experience with the WC3 vaccine or WI79-9 reassortant vaccines, which are shed poorly in children but which were able to induce protection even in the absence of good seroresponses, in at least some efficacy trials.

A 2 year study of natural rotavirus infections in a small number of children suggests that natural immunity following serotype 1 infections can approach 100% protection from subsequent homotypic challenges (BERNSTEIN et al. 1991). In addition, asymptomatic primary infection appeared as protective as symptomatic homotypic primary infection (BERNSTEIN et al. 1991). More studies of this type are needed to assess the rate of protection from homotypic and heterotypic challenge. This study supports the idea that vaccination with attenuated animal viruses may not induce the same immune response and protection as natural infections with human rotaviruses. However, it remains questionable whether attenuated human rotavirus strains administered orally will remain sufficiently capable of replicating in the intestine and be sufficiently immunogenic to induce protection, particularly if given to neonates who have preexisting maternal antibody.

Perhaps the biggest challenge for the polyvalent second and third generation rotavirus vaccines is whether these vaccines will be able to induce protective heterotypic immunity and whether the four serotypes currently represented in these vaccines will remain those of greatest epidemiologic importance. Although the four human serotypes (1–4) previously were thought to be the epidemiologically important ones, and thus the targets of a multivalent vaccine. Recent serotyping of viruses in Italy, Thailand and India found that some which could not be characterized as VP7 types 1–4, were serotypes (6, 8 and 10) previously only reported in cows (GERNA et al. 1990c; URASAWA et al. 1992; DAS et al. 1993; GENTSCH et al. 1993). Interestingly, the Italian serotype 8 viruses do not possess supershort RNA patterns originally described for serotype 8 strains in Indonesia (GERNA et al. 1990c). The discovery of a seventh human serotype in the Philippines (URASAWA et al. 1990) also emphasizes the need to continue characterizing circulating strains, to determine how rotavirus antigenic diversity will affect the development of multivalent vaccines.

The question of whether attenuated viruses will be safe and effective in children in developing countries also remains partially unanswered. To date, in controlled conditions of administration, there have been no safety problems, but the bovine viruses clearly are less reactogenic than RRV and its derived reassortants. The reactogenic viruses produce fevers 3–4 days after vaccine administration. Such reactions have been more obvious in studies in developed countries (Finland, Sweden and the United States) or among children with lower levels of preexisting antibody than in studies in developing countries, e.g. Venezuela (FLORES et al. 1990), where reactions apparently are masked by higher levels of preexisting antibody. It has been noted that mild febrile reactions might have passed largely unnoticed if routine temperature readings had not been included in the clinical follow-up (FLORES et al. 1990).

Another concern about the safety of live attenuated vaccines has come from the recent demonstration that rhesus rotavirus and at least one human rotavirus strain (but not other virus strains such as the bovine WC3 virus) can produce hepatitis and death in immunocompromised mice and hepatitis to a lesser extent in immunocompetent mice (UHNOO et al. 1990b; RIEPENHOFF-TALTY et al. 1993). In addition, several simian rotavirus strains (SA11 and RRV), but not bovine (WC3, NCDV) or some human viruses (D, DSI, Price, ST3), are capable of replicating in liver-derived HepG2 cells (RAMIG and GALLE 1990; SCHWARZ et al. 1990; KITAMOTO et al. 1991). Evidence of rotavirus replication in the liver and kidney of immunocompromised children, including one with AIDS, suggests that these concerns about the safety of live rotavirus vaccines may not be simply problems for the laboratory (GILGER et al. 1991). The safety of live virus vaccines in malnourished children, who also often suffer from high parasite burden and are functionally immuno-compromised (FEIGIN and GARG 1987) must be proven, especially as malnutrition increases the severity of rotavirus illness (DAGAN et al. 1990; UHNOO et al. 1990a). This also is a concern because of the increasing number of children worldwide who are infected with the human immunodeficiency virus.

Because of the potential for any RNA virus to mutate and, in particular, for rotaviruses to reassort, an attenuated vaccine strain might revert to virulence or reassort with a field strain and produce a highly virulent new circulating virus strain. Although rotaviruses reassort efficiently both in vitro and in vivo (see Chap. 6), neither reversion nor reassortment has been reported in any of the vaccine trials. However, these possibilities have not been monitored closely.

Finally, is it safe to administer attenuated live viruses with oral polio vaccines (OPV)? This is an important question as it would be optimal if these vaccines could be administered simultaneously. Each vaccine candidate must be evaluated separately. Different studies with the bovine RIT 4237 vaccine and OPV have yielded conflicting results. A study in Yugoslavia indicated that polio antibody responses were unaffected by the rotavirus vaccine, but the rotavirus antibody response in children given combined rotavirus-OPV was significantly lower than the response in children given rotavirus vaccine alone (VODOPIJA et al. 1986). In Italy, a dose-response study of rotavirus vaccine and OPV administered together either once or twice concluded that, after one dose, the poliovirus responses to combined vaccination were lower than responses after polio vaccine alone. However, otavirus interference with OPV was overcome with administration of a second dose (GIAMMANCO et al. 1988). In The Gambia, administration of RIT 4237 and OPV simultaneously seemed to result in more OPV failure and lower neutralizing antibody titers to poliovirus types 1 and 3, but these differences were not statistically significant (HANLON et al. 1987). A fourth study of the simultaneous administration of OPV and RRV in infants 2–3 months of age in the United States did not find evidence for major interference between these viruses (HO et al. 1989). In general, these results are encouraging as they suggest the potential for incorporation of a rotavirus vaccine into the routine childhood immunization schedule. Whether the same will be true for the tetravalent vaccine remains to be seen. However, it may be premature to be testing these issues until an effective vaccine is first developed.

A related question is how will the large differences in exposure to enteropathogens between study sites affect the success of vaccination? In some areas, the rate of diarrhea is less than one episode per child-year. In Peru, the rate of diarrhea exceeded ten episodes per child-year during the vaccine trial. Some trials have included monitoring for diarrheal pathogens other than rotavirus. In some instances monitoring for infection with rotavirus and other enteropathogens has been passive, rather than active. The interference of naturally occurring infections with enteropathogens may have an important effect on vaccine take. Another issue is whether live attenuated vaccines will be able to be administered effectively to breast-fed children. This is particularly important for vaccine administration to children in developing countries where breast-feeding is being encouraged as part of the WHO Diarrheal Diseases Control Programme. An analysis of the effect of breast-feeding on oral RRV vaccination from studies conducted in the United States found a significant adverse effect of breast-feeding for a dose of 10^4 pfu given to infants 2–5 months of age (PICHICHERO 1990). This is disappointing, as ideally one would like to be able to administer the vaccine without limitation on concomitant breast-feeding. One variable not yet adequately assessed is the variability of breast milk antibody titers between mothers and how such variability affects the vaccine take in individual children. A large variability of rotavirus serotype-specific IgA antibodies in nursing mothers has been documented (HERRERA et al. 1991).

Other questions about live attenuated rotavirus vaccines remain to be resolved. It may be that multiple doses of vaccine will be necessary to overcome several of the problems described above. An interesting possibility raised recently is that vaccine efficacy can be affected by the timing of administration of the vaccine relative to the onset of the rotavirus epidemic season (RUUSKA et al. 1990; VESIKARI et al. 1991b). It would be interesting to know if coproantibody conversions correlated with vaccine-induced clinical efficacy in these studies. RUUSKA and VESIKARI (1990) also used a new method to score severity of clinical illness of diarrheal episodes and found this scoring system allowed clearer distinction between placebo and vaccine groups than previous systems of scoring clinical episodes of diarrhea (HJELT et al. 1987; FLORES et al. 1987). These recent studies are refreshing as they indicate that continued testing of existing vaccines in new schedules might lead to improved efficacy.

A disappointing fact from all the vaccine trials is that direct correlates of heterotypic protection have not been clearly defined which would allow for prediction of efficacy from the serologic data of phase 1 studies. The recent use of an epitope blocking assay to measure serotype-specific antibody responses has been useful to show that some vaccine failures were associated with a lack of seroresponse to the rotavirus serotype which later caused infection (SHAW et al. 1987; BEARDS and DESSELBERGER 1989; GREEN et al.

1990b; TANIGUCHI et al. 1991). Similar in-depth analyses need to be completed for each field trial in which efficacy was not obtained. In addition, molecular analyses of the infecting viruses probably need to be performed to determine whether antigenic drift in the circulating viruses is occurring. Finally, serotype-specific conversions of coproantibodies should be evaluated to determine if they will be a useful predictor of protection either from infection or clinically significant disease. This may be particularly important because preexisting maternal antibody, or the presence of antibody in breast milk, may reduce vaccine take as measured by serum responses, and protective antibody responses may only be detected by measuring the development of intestinal antibodies. Mucosal responses in the absence of serologic responses have been documeted following known rotavirus infections in animals and children (COULSON et al. 1990a; LOSONSKY and REYMANN 1990; CONNER et al. 1991). Therefore, it may be important to monitor responses directly in the intestine as predictors of protection, particularly if measurements of heterotypic serum responses do not indicate or correlate with the development of protective immunity. These analyses should be able to be performed on samples collected from the previous trials and, if the immunoglobulin is not degraded, perhaps before new efficacy trials are pursued. In addition, because of the observations of interference between different components of the quadrivalent reassortant viruses, evaluation of the inherent ability of these viruses to interfere with each other in cell culture might be useful. Such cell culture testing of the various formulations already evaluated in phase 1 trials could at least reveal if such in vitro assays correctly predict results in children.

Because most of the questions outlined above have not been resolved by previous vaccine trials, interest has heightened for evaluating alternative types of vaccines. Investigators are now considering developing cold-adapted strains of human rotavirus for eventual use as live oral attenuated vaccines. A serotype 1 strain developed in Japan (MATSUNO et al. 1987) has not yet reached phase I human trials. Inactivated human rotavirus strains also have been suggested (KAPIKIAN et al. 1990). Parenteral immunization with an inactivated vaccine has been shown to induce active protective immunity (CONNER et al. 1993).

5 Rotavirus Vaccines for Animals

5.1 Rotavirus Vaccines to Induce Active Immunity in Animals

Initial attempts to produce rotavirus vaccines for animals were performed in cattle. The first vaccine (Scourvax, Norden Laboratories) was a live attenuated bovine strain (NCDV) administered orally to calves at birth (MEBUS et al.

1972, 1973; NEWMAN et al. 1973; TWIEHAUS and MEBUS 1973; THURBER et al. 1977). Results in experimental and field trials appeared promising, but the vaccine was not efficacious when double-blind field trials were conducted (ACRES and RADOSTITS 1976; DELEEUW et al. 1980). Similar attempts to administer rotavirus of bovine or porcine origin orally to piglets have not been successful (LEECE and KING 1979; HOBLET et al. 1986).

The failure of these vaccines has been attributed to: (1) neutralization of vaccine virus in the intestine by passively derived maternal antibody, (2) exposure of the newborn animals to a virulent field strain of virus before a protective immune response could be induced by the vaccine, or (3) failure of the vaccine strain to induce a protective heterotypic immune response (CROUCH 1985; SNODGRASS 1986b; SAIF and JACKWOOD 1990). The inability to induce effective active immunity in calves by vaccination stimulated interest in protocols for providing passive immunity to neonatal animals. This is feasible because, in animals of economic importance, rotavirus infections usually occur early in life and often in the first month of life. In addition, offspring receive antibody only through colostrum, so vaccine strategy changed and focused on inducing lactogenic immunity.

5.2 Rotavirus Vaccines to Induce Passive Immunity in Animals

5.2.1 Background and Rationale of Passive Immunization of Animals

Lactogenic or passive immunity is a natural immune process that can be used to advantage, by parenteral or local (intramammary) rotavirus immunization of the dam, to stimulate prolonged secretion of high titer antibody in the colostrum and milk ingested by the neonate. The success of this strategy depends on the fact that most adult domestic animals have preexisting naturally acquired antibodies to rotavirus and these antibodies can easily be boosted to high titer by vaccination with rotavirus. The goal of inducing such lactogenic immunity is to provide the young with protection from clinical disease or at least to delay disease beyond the critical neonatal period when disease tends to be most severe. Studies with *E. coli* and transmissible gastroenteritis virus (TGE) indicated that this method was a feasible approach to use in domestic animals (reviewed in SAIF and JACKWOOD 1990). Complete protection from infection is not a goal of this vaccination strategy. In fact, if subclinical primary infections do occur during the suckling period, they can be beneficial by stimulating active immunity (or priming), which would prevent subsequent severe infections.

The development of effective vaccine strategies to boost passive lacto-genic immunity is based on species differences in the natural transfer of maternal antibody to the fetus and neonate (BRAMBELL 1970; OGRA et al.

1977; TIZARD 1982; APPLEBY and CATTY 1983). In ungulates (cows, sheep, pigs, and horses), the anatomical structure of the placenta prevents transplacental transfer of antibodies. Passive transfer of antibody in these animals occurs solely through ingestion of colostrum and postcolostral milk. This contrasts with humans, primates, mice, and rabbits in whom antibody (primarily IgG) is transferred across the placenta to the fetus. Additional immunologic differences between species are found in the immunoglobulin composition of colostrum and milk and in the length of time these immunoglobulins can be absorbed across the intestine. IgG of subclass 1, derived primarily from the serum, is the predominant immunoglobulin in the colostrum and milk of ruminants (cows and sheep), while in monogastrics (pigs and horses) IgG predominates in colostrum and IgA is the predominant immunoglobulin in milk. Gut closure resulting in a cessation of absorption of immunoglobulins in both the ruminant and monogastric farm animals occurs within the first 24–48 h of life. In contrast, in humans, primates, rabbits, and mice, IgA is the predominant immunoglobulin in both colostrum and milk, and only a minor portion of IgA is absorbed from these fluids. Rodents are able to selectively absorb IgG from colostrum and milk up to 3 weeks of age. In all species there is a dramatic decline in the antibody concentration from colostrum to milk.

These differences must be considered when passive immunization protocols are being developed to increase specific immunoglobulin titers in the colostrum and milk. In ruminants, IgG stimulation can be achieved with parenteral inoculation of pregnant animals which results in subsequent transfer of antibody to colostrum and milk (MEBUS et al. 1973; SNODGRASS et al. 1980; CASTRUCCI et al. 1984; SAIF et al. 1983). In monogastrics, IgA is produced locally in the mammary gland by plasma cells whose precursors are derived originally from the intestine (reviewed by SAIF and JACKWOOD 1990). Therefore, the stimulation of colostral and milk IgA in monogastrics may require primary intestinal antigen exposure. However, in situations where endemic infections occur, the majority of adult animals will have preexisting serologic and mucosal antibody due to natural exposure to rotavirus. Parenteral inoculation of such seropositive swine with rotavirus or TGE – and of humans with poliovirus or cholera – has been shown to induce both increased IgA and IgG in colostrum and milk (OGRA et al. 1977; SVENNERHOLM et al. 1980; SAIF and JACKWOOD 1990).

5.2.2 Experimental Studies with Rotavirus Vaccines to Induce Passive Immunity

Passive protection studies in sheep showed that protection from rotavirus infection was mediated by antibody in the intestine and not by circulating antibody (SNODGRASS and WELLS 1976). The protective role of lactogenic antibody was supported by additional studies in lambs, calves, pigs, and mice

(SNODGRASS et al. 1977; SNODGRASS and WELLS 1978a, b; SAIF et al. 1983; SHERIDAN et al. 1984; OFFIT and CLARK 1985a, b; OFFIT et al. 1986). An additional advantage of vaccines to increase passive immunity is the ability to stimulate both homotypic and heterotypic antibodies in colostrum and milk by the vaccination of the seropositive dam. The induction of heterotypic antibody was found to be dependent on the prior exposure of the dam to the heterotypic serotype (SNODGRASS et al. 1984; BRÜSSOW et al. 1987; SAIF and JACKWOOD 1990).

In ruminants, the requirement for lactogenic antibody in the intestine may not be absolute. Limited experimental data from lambs indicated that high titer rotavirus antibody administered by intraperitoneal injection (in the absence of lactogenic antibody) provided protection from rotavirus (SNODGRASS and WELLS 1978a). Circulating rotavirus antibody which can mediate protection has also been observed in colostrum-deprived calves administered high titer antibody by subcutaneous injection (BESSER et al. 1988a). This protection was shown to be mediated by the transfer of circulating IgG_1 to the intestine (BESSER et al. 1988a, b). Therefore, in ruminants it may be possible to provide partial protection from disease if very high titers of circulating antibody in the newborn animal are achieved. This may be important in dairy herds in which calves suckle their dams for very short periods of time and, therefore, are not provided with a continuous supply of immunoglobulin in the milk. Whether this type of transfer of immunoglobulins from the serum to the intestine occurs in nonruminants remains to be determined.

A number of experimental and field trials of passive vaccines conducted in cattle are summarized in Table 4. Although the dose, form, adjuvant, and virus strains of the vaccines varied in each trial, the majority of the vaccines induced significant increases in neutralizing antibody to rotavirus in serum, colostrum, and milk. Increased titers were not observed following immunization without adjuvants or when low dose vaccine preparations were used (SAIF et al. 1983; WALTNER-TOWES et al. 1985).

The efficacy of these vaccines in protecting calves against rotavirus infection and disease is less defined. In experimental studies in which calves nursed on immunized dams or calves were fed colostrum supplements from immunized cows, the number of calves with diarrhea, severity of diarrhea, duration of diarrhea, and length of virus excretion was decreased and a delayed onset of diarrhea was noted following virus challenge (see Table 4). Statistically significant differences of all parameters were found in some studies, and the majority of studies used small numbers of animals.

Relatively large field trials have been performed with several of the bovine vaccines. In two field trials in which disease was monitored both within the herd (vaccinated and unvaccinated animals in the same herd) and compared to historical data, vaccination was associated with a decreased incidence and severity of diarrhea (BELLINZONI et al. 1989; CASTRUCCI et al. 1989), decreased mortality rate (BELLINZONI et al. 1989), and decreased rotavirus isolation rate

Table 4. Experimental and field trial results of rotavirus vaccines to increase passive immunity in cattle

Type of study[a]	Virus strain	Virus titer	Inacti-vated	Adjuvant[b]	Number of cows		Significant increase in titer vaccinated cows vs controls			Number of calves ingesting colostrum		Significant clinical protection of calves for indicated parameter[c]		References
					Vaccine	Control	Serum	Colostrum	Milk	Vaccine	Control	Yes	No	
E	NR	$10^{4.8}$	+	IFA	7	9	+	+	+	7	9		D, E, DE, DO	Snodgrass et al. (1980)
E	OARDC	10^8	–	IFA	5	5	+	+	+	8	8	D, E, DE, DO		Saif et al. (1983, 1984)
E	NCDV[d]	10^4	–	–	5		–	–	–	8			D, E, DE, DO	Saif et al. (1983, 1984)
E	UK[e]	NR[f]	+	oil	10	6	+	+	+	10	6	D, E	DE	Snodgrass (1986a)
E	T67	10^7	+	oil	13	8	+	+	NR	13	8		D, E	Bellinzoni et al. (1989)
E[g]	81/36F	10^7	+	IFA	10	16	NR	+	NR	10	16	D		Castrucci et al. (1989)
E	Quebec 17	NR	–	oil	4	5	+	+	+	6	6	D, E, DE, DO		Archambault et al. (1988)
E	Quebec 17	NR	–	O/M	4		+	+	+	ND[h]	ND			Archambault et al. (1988)
E	Quebec 17	NR	–[i]	–	4		–	–	–	ND	ND			Archambault et al. (1988)
F	NCDV[d,e]	NR	NR	NR	182	95	NR	–	NR	182	95		D, E, DO, M	Waltner-Towes et al. (1985)
F	UK[e]	NR	+	oil	1638	1637	+[j]	NR	NR	1595[k]	1593	–[l]		Snodgrass (1986a)
F										25	29	D[m]		Snodgrass (1986a)
F										18	22	E[n]		Snodgrass (1986a)
F	UK[e]	NR	+	oil	14	17	NR	+	NR	19	29		D, E	McNulty and Logan (1987)
F	T67	10^7	+	oil	2526	–[o]	NR	NR	NR	2526	–[o]	D, M		Bellinzoni et al. (1989)
F					1540	2700	NR	NR	NR	1540	2700	D, M		Bellinzoni et al. (1989)
F	81/36F	10^7	+	IFA	248	210	NR	NR	NR	248	210	D, E		Castrucci et al. (1989)

[a] E, experimental trial; F, field trial.

[b] IFA, incomplete Freunds; O, oil, M, muramyl dipeptide.

[c] D, diarrhea: E, virus excretion; DE, duration of virus excretion; DO, delay in onset of virus excretion or diarrhea; M, mortality; asterisk, nonstatistically significant decrease observed. If abbreviation is not used then that parameter was not reported.

[d] Norden Scourgard 3.

[e] Vaccine in combination with K99 E. coli.

[f] NR, not reported.

[g] Summary of several studies.

[h] ND, not done.

[i] Inoculated orally.

[j] Total of 47 vaccinated cows tested.

[k] Total of 3275 cows, approximately 1/2 vaccinated.

[l] Less than 10% morbidity or nonrotavirus diarrhea occurred in herd.

[m] Rotavirus excretion was not evaluated.

[n] Diarrhea was not evaluated due to concurrent infections with crytosporidium.

[o] Comparisons were made from historical data from previous 3 years at same locations.

(CASTRUCCI et al. 1989). In one study, the efficacy of the vaccine was difficult to evaluate due to a low diarrhea morbidity rate (<10%) in many of the vaccinated herds and due to the occurrence of diarrhea caused by coronavirus and cryptosporidium in a number of the herds (SNODGRASS 1986a). In two herds studied by SNODGRASS (1986a), which did not have concurrent infections and were evaluated for diarrhea, the rates of diarrhea and rotavirus isolation were significantly reduced when compared to controls monitored within the herds. In two herds with concurrent cryptosporidium infections, there was a significant reduction in rotavirus infection in vaccinated vs control groups. A major question with these vaccines is whether sufficiently high antibody titers initially present in colostrum can be maintained in milk to protect animals throughout the period of susceptibility. This remains controversial. Recently, the use of prolactin in combination with rotavirus vaccination has been suggested as a method to stimulate and maintain high levels of antibody in milk (IJAZ et al. 1990).

An additional benefit of the ability to produce colostrum and milk containing high titers of rotavirus antibodies may be the use of such preparations as passive treatment for children or animals. Lactogenic immunity provided by colostrum or milk from hyperimmunized cows or serum immunoglobulin has provided protection from infection, ameliorated disease, or reduced virus excretion in children (BARNES et al. 1982; EBINA et al. 1985; LOSONSKY et al. 1985; BRÜSSOW et al. 1987; HILPERT et al. 1987; DAVIDSON et al. 1989). The use of such supplements, especially in children at high risk of severe or fatal infection (such as children with malignancy or combined immunodeficiency hospitalized during the peak of the rotavirus season) or in children in day care centers, may prove useful as a prophylactic measure.

The immunization of pregnant or lactating women has not been considered feasible to protect breast-fed infants from clinically significant rotavirus disease because the peak incidence of disease in children occurs at an age (6 months to 2 years) when breast feeding often has been discontinued. However, if maternal antibody is a primary correlate of protection early in life, boosting maternal antibody prior to birth might have benefits not currently appreciated.

5.3 Future Veterinary Rotavirus Vaccine Needs

Most efforts have focused on development of vaccines for calves. Future efforts will no doubt include efforts for foals because of the economic importance of race horses. In addition, there is interest in vaccinating piglets, because experimental evidence has shown that parenteral immunization of sows can result in increased levels of antibody in colostrum and milk, but vaccines have not been field tested (SAIF and JACKWOOD 1990).

6 The Next Generation of Rotavirus Vaccines: Vaccines Based on Advances in Molecular Biology

6.1 The Potential of Subunit Vaccines

The availability of molecular biology techniques has stimulated the pursuit of an effective rotavirus subunit vaccine. The recombinant DNA approach has several theoretical advantages as compared with live attenuated virus vaccines. The most important advantage is safety, especially in the immuno-compromised host. Additional advantages of subunit vaccines include potentially reduced manufacturing costs, stability during transport, and elimination of interference by other vaccines or enteric organisms. The evaluation of subunit vaccines is warranted, although there are inherent challenges facing the use of nonreplicating antigens as vaccines for enteric disease (Table 5). Several different approaches to make a rotavirus subunit vaccine are possible (Table 6). The status of the active pursuit of each of these approaches will be considered separately. These approaches could be used for development of vaccines for both children and animals.

6.1.1 Proteins from Purified Virus as Subunit Vaccines

The development of a vaccine from purified virus may be feasible, but not practical. Disruption of the viral particle, production of soluble proteins, and subsequent purification of individual proteins may result in denatured proteins incapable of inducing a protective immune response. An alternative

Table 5. Challenges for vaccination for enteric disease with non-replicating antigens

Greater amounts of antigen may be required.
Antigen stability is unknown.
Induction of intestinal immunity may require oral immunization.
Novel delivery systems may be required.
Long-lasting immunity is not assured.

Table 6. Types of potential subunit vaccines

Proteins from purified virus
  Viral capsids lacking RNA

Proteins synthesized from cloned gene(s)
  Purified proteins from prokaryotic expression vectors
  Purified proteins from eukaryotic expression vectors
  Proteins produced in the gastrointestinal tract with live bacterial or viral vectors

Virus-like particles produced from co-expression of cloned gene(s)

Synthetic peptides

Nucleic acid vaccines

approach is to use virus particles lacking RNA or inactivated virus. Viral particles lacking RNA were used to vaccinate mice dams subcutaneously prior to parturition, and mice pups were protected from challenge (SHERIDAN et al. 1984). The stimulation of high levels of antirotavirus IgG in the milk of the vaccinated dams provided this passive protection. Oral vaccination with empty capsids was not examined. In another study, adult female mice parenterally inoculated with inactivated rotavirus developed neutralizing serum antibodies, and newborn mice born to these dams were passively protected against RRV-induced gastroenteritis (OFFIT and DUDZIK 1989b). In addition, parenterally and orally inoculated adult mice developed virus-specific CTLs (OFFIT and DUDZIK 1989a).

Parenteral immunization of cows with empty virus capsids boosted preexisting neutralizing antibody titers (BRÜSSOW et al. 1990a). Attempts to induce a substantial neutralizing antibody increase with VP7 purified under denaturing or nondenaturing conditions or with synthetic peptides corresponding to two regions of VP7 were unsuccessful (BRÜSSOW et al. 1990a). These data indicated that such particles represent a minimal subunit vaccine, and they encourage attempts to produce empty virus particles by expression of recombinant rotavirus cDNAs.

6.1.2 Proteins Synthesized from Expression of Cloned Gene(s) as Subunit Vaccines

The nucleotide sequences of all 11 rotavirus genes, from different virus strains, have now been determined (reviewed in ESTES and COHEN 1989). Efforts have focused on the determination of the sequences of the genes that encode VP4 (gene 4) and VP7 (gene 8 or 9 depending on the strain) from a number of isolates and serotypes. VP4 is the hemagglutinin and VP7 is a glycoprotein. The cloning and expression of these two genes has been initiated in several systems in an effort to develop subunit vaccines, as described below.

6.1.2.1 Proteins Synthesized in Prokaryotic Expression Systems

In bacteria, results have been disappointing with expressed VP7 from human (RV-5, amino acids 69–158), bovine (UK, amino acids 15–151; NCDV, amino acids 50–265), and simian (SA11, amino acids 63–300) origin (ARIAS et al. 1986; FRANCAVILLA et al. 1987; McCRAE and McCORQUODALE 1987; JOHNSON et al. 1989; REEVES et al. 1990; SALAS-VIDAL et al. 1990). VP7 has been expressed as a fusion protein with β-galactosidase (β-gal), because VP7 expressed alone was found to be toxic in *E. coli*. Although all the fusion proteins were antigenic and immunogenic, the levels of neutralizing antibodies induced by the fusion proteins in two studies were low (titers of 1:100–1:1600 or 1:300; ARIAS et al. 1986; McCRAE and McCORQUODALE 1987). The failure of the fusion proteins to induce reasonable levels of neutralizing antibody was attributed to denaturation of the protein during purification, the production of an insoluble product, improper folding of the VP7 domains due to the presence of the β-gal domains, or failure to include gene sequences that might be responsible for proper folding. The inability of prokaryotic expression systems to glycosylate and express proteins with disulfide bonds may severely limit the usefulness of these systems for VP7, in which neutralizing epitopes are known to be dependent on conformation (reviewed in MATSUI et al. 1989). Attempts also have been made to express VP7 on the surface of bacterial hosts by insertion of antigenic regions into the lamB gene (REEVES et al. 1990). Evidence for the expression and export of a VP7 fusion protein through the cytoplasmic membrane to the outer membrane of bacteria was obtained, but this product was not organized within the outer membrane in the same manner as the parent LamB protein. Therefore, further work on these types of systems remains to be done.

VP4 (SA11 amino acids 42–387) has been expressed in *E. coli* as a fusion protein with MS2 polymerase in a thermoinducible expression plasmid under the control of the phage lambda P_L promoter (ARIAS et al. 1987). The VP4 fusion protein was antigenic and immunogenic, and it induced antibodies in mice that neutralized the virus (titers, 1:800–1:1600) and inhibited hemagglutination (titers, 1:1000–1:2000). The hemagglutination-inhibition titers were equivalent to those obtained with whole virus although the neutralizing titers obtained were 20-fold lower. No protection studies with this fusion protein have been reported.

The use of salmonella as a live bacterial vector expressing rotavirus proteins has been proposed, but to date there is little information on the success of this approach (GREENBERG et al. 1985; SALAS-VIDAL et al. 1990). The advantage of this system would be the ability of the vector to replicate in the intestine with subsequent presentation of the expressed protein to the mucosal immune system. Testing of the attenuated aro A strain of *Salmonella typhimurium* SL3261 as a vector to deliver a VP7-β-gal recombinant fusion polypeptide to the immune system of mice indicated that antibodies could be induced by the β-gal portion of the hybrid protein, but antibodies to SA11 were not detected (SALAS-VIDAL et al. 1990). These results are of interest as they indicated that transport from the cytoplasmic location of the recombinant polypeptide within salmonella does not appear to be a problem, because antibodies to β-gal were raised. However, it remains critical to optimize the immunogenicity of the antigen of interest and the stability of the recombinant plasmid. Expression of other rotavirus proteins with fewer conformational requirements (perhaps VP4) for their antigenicity, and introduction of these genes into the chromosome of salmonella to overcome the poor in vivo stability of the recombinant plasmids remain to be evaluated (SALAS-VIDAL et al. 1990).

The use of mycobacteria as a live bacterial vector to express proteins has several potential advantages (BLOOM 1989). The most important advantage is that vaccination with mycobacteria may only require one dose. These vectors are beginning to be evaluated but success with rotavirus may require expression of linear protective epitopes because it is doubtful whether conformational epitopes will be expressed properly.

6.1.2.2 Proteins Synthesized in Eukaryotic Expression Systems

The baculovirus system has been used to express successfully each of the 11 rotavirus genes (ESTES et al. 1987; MACKOW et al. 1989; NISHIKAWA et al. 1989; AU et al. 1989; COHEN et al. 1989; WELCH et al. 1989; GORZIGLIA et al. 1990b; MATTION et al. 1991; LABBE et al. 1991; REDMOND et al. 1991; Estes, Crawford, Liu and Cohen, unpublished). Generally, the expressed proteins are made in high yields and possess native conformation when compared immunologically and biochemically to native viral proteins. Expressed VP4s of RRV and SA11 possess hemagglutinating activity that can be inhibited by polyclonal hyperimmune serum to whole virus (MACKOW et al. 1989; Estes, unpublished). Expressed VP4s of SA11 and OSU also induce neutralizing and hemagglutinin-inhibiting antibody in guinea pigs and mice (NISHIKAWA et al. 1989; MACKOW et al. 1989, 1990; Estes, unpublished). Finally, the RRV VP4 induces lactogenic immunity sufficient to protect mouse pups against challenge with either RRV or a murine virus (MACKOW et al. 1990). SA11 VP7 has been successfully expressed, is of the correct size, reacts with a panel of monoclonal and polyclonal sera in a number of immunologic tests, and induces neutralizing antibody in immunized animals (Estes and Crawford, unpublished).

The advantage of the baculovirus system is that the proteins are expressed in a soluble native form and expressed proteins can be glycosylated (N-linked) (LUCKNOW and SUMMERS 1988; MILLER 1989). In addition, viral-like particles have been observed with expressed VP6, with co-expressed VP2 and VP6, and with co-expressed VP2, VP4, VP6 and VP7 (ESTES et al. 1987, 1991, 1993; LABBE et al. 1991; REDMOND et al. 1991), indicating the ability of expressed proteins to assume native structures.

6.1.2.3 Proteins Synthesized Using Live Viral Vectors

Virus vectors have the advantage that a large quantity of foreign genetic information can be inserted into the genome without affecting the replication of the vector. This provides the potential to express multiple genes in the same vector resulting in a multivalent vaccine. In some cases, extensive information on safety following immunization with such potential viral carriers (e.g., polioviruses, yellow fever) also is available. Little information currently is available on the possible success of these approaches with rotaviruses.

Vaccinia virus is one of the most widely used virus vectors to date, and experimental results obtained using this vector have been extremely useful in helping to understand the potential power of immunization with live recombinant viruses for many types of viral infections. Vaccinia vectors are likely to remain experimental tools because of the risk of complications in immunized individuals. Lethal infections in immunocompromised individuals are of particular concern in light of the current AIDS epidemic (REDFIELD et al. 1987) as well as the large number of patients with other acquired immunodeficiencies. However, modifications are being examined to define viral genes involved in neurovirulence and reactogenicity and it should be possible to create new, less reactogenic strains of vaccinia virus (RAMSHAW et al. 1987; BULLER, et al. 1985). An additional concern is that immunity to the virus could result from the first inoculation and inhibit infection, and therefore expression of the foreign genes, in subsequent inoculations. However, in spite of this concern, sequential immunizations with different expressed proteins have been successful (SMITH et al. 1984; PERKUS et al. 1985).

Vaccinia virus has been used successfully to express full-length wild-type SA11 VP7 and a deletion mutant of VP7 (ANDREW et al. 1987). The mutant VP7, which lacked amino acids 47–61, the anchor singal for retention of wild-type VP7 in the endoplasmic reticulum, had previously been shown to be secreted from transfected COS-1 cells (PORUCHYNSKY et al. 1985). Both the expressed wild-type and mutant VP7s were antigenic and immunogenic. Antibody produced in rabbits had low neutralizing titers (1:40–1:1000), and reacted in a serotype-specific manner by ELISA. Modification of the expressed VP7 by linkage to the transport region of the influenza hemagglutinin, to permit secretion and insertion into the plasma membrane of cells, resulted in an enhancement of immunogenicity (ANDREW et al. 1990). Cell surface expressed VP7 has provided passive protection in the suckling mouse model (ANDREW et al. 1992; BORH et al. 1993). This result suggests new methods of antigen presentation may improve subunit vaccines.

Regardless of the possible limitations of vaccinia recombinants for use as a vaccine, analysis of the cell-mediated immune responses in animals infected with these recombinants has been extremely worthwhile. Analysis of the cell-mediated immune response to rotavirus proteins expressed from vaccinia recombinants is in its infancy. To date, it has been shown that heterotypic CTLs are made to epitopes on VP7 and not to epitopes on VP6 (OFFIT et al. 1991).

Other live virus vectors might be considered. Adenovirus vectors seem attractive because adenovirus vaccines are currently given orally in enteric coated capsules and these vaccines protect against respiratory infections (reviewed in BLOOM 1989). While the safety of current adenovirus vaccines for children is not established, attenuated viruses might be produced and used if adequate expression and immunogenicity of the relevant antigens is maintained. The study of adenovirus vectors to express rotavirus antigens has been initiated (GORZIGLIA and KAPIKIAN 1992; BOTH et al. 1993).

6.1.3 Synthetic Peptides as Subunit Vaccines

Synthetic peptides of VP7 epitopes have been investigated for their ability to induce neutralizing antibody (GUNN et al. 1985). Only one peptide, corresponding to amino acids 247–259 of SA11, was found to induce neutralizing antibody at a very low titer (1:32). Evaluation of three VP4 synthetic peptides, corresponding to sequences of the trypsin cleavage region of SA11 (amino acids 228–241) and SA11 4fM (amino acids 220–223 and 258–271), showed that the peptide 228–241, when conjugated to rabbit serum albumin, induced low levels (titer of 1:800) of neutralizing antibody after hyperimmunization of rabbits (STRECKERT et al. 1988). Neither of the SA11 4fM peptides (individually cross-linked by glutaraldehyde) induced neutralizing antibodies after hyperimmunization of mice, although peptide antibodies were induced. However, VP4 peptide 220–223 was shown to prime for a neutralizing antibody response upon inoculation with either SA11 or ST3 (ARIAS et al. 1989).

The immunogenicity problem encountered with the rotavirus peptides is a general disadvantage of peptides. New techniques such as the colinear synthesis of a B cell epitope (using rotavirus VP6) and a T cell epitope (using the influenza hemagglutinin) in one peptide have been shown to increase immunogenicity (BORRAS-CUESTA et al. 1987). Peptide vaccines also will require the use of multiple epitopes to overcome MHC-restricted non-responsiveness in outbred populations. A synthetic peptide vaccine derived from the bovine C486 strain and composed of two highly conserved peptides from VP4 and VP7 coupled to keyhole limpet hemocyanin or to recombinant VP6 assembled into virus-like particles has been reported to confer passive protection against some strains of experimental rotavirus infection in neonatal mice (SABARA et al. 1987; IJAZ et al. 1991). Peptides could be useful in priming an immune response against a subsequent viral particle or sub-unit vaccine. A possible advantage of this approach is the use of significantly less material in the second immunization, as was shown for poliovirus (EMINI et al. 1983). The ability of VP4 peptides to prime the immune response has been demonstrated for at least one rotavirus strain (LIZANO et al. 1991).

Although the approaches mentioned above have encountered difficulties, potential advantages of their use continue to encourage research in this area. The successful production of a subunit vaccine may require: (a) the inclusion of a number of epitopes to overcome nonresponsiveness due to either lack of required B and T cell epitopes or MHC-restricted responses in outbred populations, (b) sufficient antigenic load, especially with nonreplicating antigens, (c) conservation of native conformational epitopes on proteins in order to induce protective neutralizing antibody, (d) efficient ways to target the vaccine to the mucosal immune system, (e) use of mucosal adjuvants, and (f) possible combinations of immunizations by the oral and parenteral routes.

7 Summary

The development of a successful rotavirus vaccine is a complex problem. Our review of rotavirus vaccine development shows that many challenges remain, and priorities for future studies need to be established. For example, the evaluation of administration of a vaccine with OPV or breast milk might receive less emphasis until a vaccine is made that shows clear efficacy against all virus serotypes. Samples remaining from previous trials should be analyzed to determine epitope-specific serum and coproantibody responses to clarify why only some trials were successful. Detailed evaluation of the antigenic properties of the viruses circulating and causing illness in vaccinated children also should be performed for comparisons with the vaccine strains. In future trials, sample collection should include monitoring for asymptomatic infections and cellular immune responses should be analyzed. The diversity of rotavirus serotype distribution must be monitored before, during, and after a trial in the study population and placebo recipients must be matched carefully to vaccine recipients. Epidemiologic and molecular studies should be expanded to document, or disprove, the possibility of animal to human rotavirus transmission, because, if this occurs, vaccine protection may be more difficult in those areas of the world where cohabitation with animals occurs. We also need to have an accurate assessment of the rate of protection that follows natural infections. Is it realistic to try to achieve 90% protective efficacy with a vaccine if natural infections with these enteric pathogens only provide 60% or 70% protection? Subunit vaccines should be considered to be part of vaccine strategies, especially if maternal antibody interferes with the take of live vaccines.

The constraints on development of new vaccines are not likely to come from molecular biology. The challenge remains whether the biology and immunology of rotavirus infections can be understood and exploited to permit effective vaccination. Recent advances in developing small animal models for evaluation of vaccine efficacy should facilitate future vaccine development and understanding of the protective immune response(s) (WARD et al. 1990b; CONNER et al. 1993).

Acknowledgements. Research in the authors' laboratories has been supported in part by grants from the National Institute of Health and the Veterans Administration. We thank Ulrich Desselberger and Robert Atmar for stimulating discussion and criticism and Denise Suzette Groner for help in making Fig. 1.

References

Acres SD, Radostits OM (1976) The efficacy of a modified live reo-like virus vaccine and an *E. coli* bacterin for prevention of acute undifferentiated neonatal diarrhea of beef calves. Can Vet J 17: 197–212

Adams WR, Kraft LM (1963) Epizootic diarrhea of infant mice: identification of the etiologic agent. Science 141: 359–360

Ahmed MU, Taniguchi K, Kobayashi N, Urasawa T, Wakasugi F, Islam M, Shaikh H, Urasawa S (1990) Characterization by enzyme-linked immunosorbent assay using subgroup and serotype-specific monoclonal antibodies of human rotavirus obtained from diarrheic patients in Bangladesh. J Clin Microbiol 27: 1678–1681

Andrew ME, Boyle DB, Coupar BEH, Whitfeld PL, Both GW, Bellamy AR (1987) Vaccina virus recombinants expressing the SA11 rotavirus VP7 glycoprotein gene induce serotype-specific neutralizing antibodies. J Virol 61: 1054–1060

Andrew ME, Boyle DB, Whitfeld PL, Lockett LJ, Anthony ID, Bellamy AR, Both GW (1990) The immunogenicity of VP7, a rotavirus antigen resident in the endoplasmic reticulum, is enhanced by cell surface expression. J Virol 64: 4776–4783

Andrew ME, Boyle DB, Coupar BEH, Reddy D, Bellamy AR, Both GW (1992) Vaccinia-rotavirus VP7 recombinants protect mice against rotavirus-induced diarrhoea. Vaccine 10: 189–191

Appleby P, Catty D (1983) Transmission of immunoglobulin to foetal and neonatal mice. J Reprod Immunol 5: 203–213

Archambault D, Morin G, Elazhary Y, Roy RS, Joncas JH (1988) Immune response of pregnant heifers and cows to bovine rotavirus inoculation and passive protection to rotavirus infection in newborn calves fed colostral antibodies or colostral lymphocytes. Am J Vet Res 49: 1084–1091

Arias CF, Ballado T, Plebanski M (1986) Synthesis of the outer-capsid glycoprotein of the simian rotavirus SA11 in Escherichia coli. Gene 47: 211–219

Arias CF, Lizano M, Lopez S (1987) Synthesis in Escherichia coli and immunological characterization of a polypeptide containing the cleavage sites associated with trypsin enhancement of rotavirus SA11 infectivity. J Gen Virol 68: 633–642

Arias CF, Garcia G, Lopez S (1989) Priming for rotavirus neutralizing antibodies by a VP4 protein-derived synthetic peptide. J Virol 63: 5393–5398

Au KS, Chan WK, Estes MK (1989) The role of the rotaviral nonstructural glycoprotein NS28 in virus morphogenesis. J Virol 63: 4553–4562

Barnes GL, Doyle LW, Hewson PH, Knoches AML, McLeilan JA, Kitchen WH, Bishop RF (1982) A randomised trial of oral gammaglobulin in low-birth-weight infants infected with rotavirus. Lancet 1: 1371–1373

Beards GM, Desselberger U (1989) Determination of rotavirus serotype-specific antibodies in sera by competitive enhanced immunoassay. J Virol Methods 24: 103–110

Beards GM, Desselberger U, Flewett TH (1989) Temporal and geographical distributions of human rotavirus serotypes, 1983 to 1988. J Clin Microbiol 27: 2827–2833

Bellinzoni RB, Mattion NM, Matson DO, Blackhall J, La Torre JL, Scodeller EA, Urasawa S, Estes MK (1990) Porcine rotaviruses antigenically related to human rotavirus serotype 1 and serotype 2. J Clin Microbiol 28: 633–636

Bellinzoni RC, Blackhall J, Baro N, Auza N, Mattion N, Casaro A, La Torre JL, Scodeller EA (1989) Efficacy of an inactivated oil-adjuvated rotavirus vaccine in the control of calf diarrhoea in beef herds in Argentina. Vaccine 7: 263–268

Bernstein DI, Smith VE, Sander DS, Pax KA, Schiff GM, Ward RL (1990) Evaluation of WC3 rotavirus vaccine and correlates of protection in healthy infants. J Infect Dis 162: 1055–1062

Bernstein DI, Sander DS, Smith VE, Schiff GM, Ward RL (1991) Protection from rotavirus reinfection: two year prospective study. J Infect Dis 164: 277–283

Besser TE, Gay CC, McGuire TC, Evermann JF (1988a) Passive immunity to bovine rotavirus infection associated with transfer of serum antibody into the intestinal lumen. J Virol 62: 2238–2242

Besser TE, McGuire TC, Gay CC, Pritchett LC (1988b) Transfer of functional immunoglobulin G (IgG) antibody into the gastrointestinal tract accounts for IgG clearance in calves. J Virol 62: 2234–2237

Bishop RF (1988) The present status of rotavirus vaccine development. Southeast Asian J Trop Med Public Health 19: 429–435

Bishop RF, Barnes GL, Cipriani E, Lund JS (1983) Clinical immunity after neonatal rotavirus infection. N Engl J Med 309: 72–76

Bishop RF, Lund J, Cipriani E, Unicomb L, Barnes G (1990) Clinical serological and intestinal immune responses to rotavirus infection of humans. In: de la Maza LM, Peterson EM (eds), Medical Virology, 9th ed. Plenum, New York

Bishop RF, Unicomb LE, Barnes GL (1991) Epidemiology of rotavirus serotypes in Melbourne, Australia, from 1973 to 1989. J Clin Microbiol 29: 862–868

Bloom BR (1989) Vaccines for the Third World. Nature 342: 115–120

Bohl EH, Kohler EM, Saif LJ, Cross RF, Agnes AG, Theil KW (1978) Rotavirus as a cause of diarrhea in pigs. JAVMA 172. 458–463

Borras-Cuesta F, Petit-Camurdan A, Fedon Y (1987) Engineering of immunogenic peptides by co-linear synthesis of determinants recognized by B and T cells. Eur J Immunol 17: 1213–1215

Both GW, Lockett LJ, Janardhana V, Edwards SJ, Bellamy AR, Graham FL, Prevec L, Andrew ME (1993) Protective immunity to rotavirus-induced diarrhoea is possively transferred to newborn mice from naive dams vaccinated with a single dose of a recombinant adenovirus expressing rotavirus VP7SC. Virology 193: 940–950

Brambell FWR (1970) The transmission of passive immunity from mother to young. North-Holland, Amsterdam

Bridger JC, Parsons KR, Varshney KC, Hall GA (1990) The biological basis of rotavirus virulence. Abstracts of the 8th international congress of virology, Aug 26–31, Berlin, Germany, p 71

Brüssow H, Hilpert H, Walther I, Sidoti J, Mietens C, Bachmann P (1987) Bovine milk immunoglobulins for passive immunity to infantile rotavirus gastroenteritis. J Clin Microbiol 25: 982–986

Brüssow H, Bruttin A, Marc-Martin S (1990a) Polypeptide composition of rotavirus empty capsids and their possible use as a subunit vaccine. J Virol 64: 3635–3642

Brüssow H, Sidoti J, Barclay D, Sotek J, Dirren H, Freire WB (1990b) Prevalence and serotype specificity of rotavirus antibodies in different age groups of Ecuadorian infants. J Infect Dis 162: 615–620

Brüssow H, Clark HF, Sidoti J (1991) Prevalence of serum neutralizing antibody to serotype 9 rotavirus WI61 in children from South America and Central Europe. J Clin Microbiol 29: 208–211

Bryden AS, Thouless EM, Flewett TH (1976) Rotavirus in calves and rabbits. Vet Rec 99: 322–323

Buller RML, Smith GL, Cremer K, Notkins AL, Moss B (1985) Decreased virulence of recombinant vaccinia virus expression vectors is associated with a thymidine kinase-negative phenotype. Nature 317: 813–815

Castrucci G, Frigeri F, Ferrari M, Cilli V, Caleffi F, Aldrovandi V, Nigrelli A (1984) The efficacy of colostrum from cows vaccinated with rotavirus in protecting calves to experimentally induced rotavirus infection. Comp Immun Microbiol Infect Dis 7: 11–18

Castrucci G, Frigeri F, Ferrari M, Aldrovandi V, Angelillo V, Gatti R (1989) Immunization against bovine rotaviral infection. Eur J Epidemiol 5: 279–284

CDC (1991) Rotavirus surveillance - United States, 1989–1990. MMWR 40: 80–81, 87

Chen D, Burns JW, Estes MK, Ramig RF (1989) The phenotypes of rotavirus reassortants depend upon the recipient genetic background. Proc Natl Acad Sci USA 86: 3743–3747

Chen D, Estes MK, Ramig RF (1992) Specific interactions between rotavirus outer capsid proteins VP4 and VP7 determine expression of a cross-reactive, neutralizing, VP4-specific epitope. J Virol 66: 432–439

Chiba S, Yokoyama T, Nakata S, Morita Y, Urasawa T, Taniguchi K, Urasawa S, Nakao T (1986) Protective effect of naturally acquired homotypic and heterotypic rotavirus antibodies. Lancet 2: 417–421

Christy C, Madore HP, Pichichero ME, Gala C, Pincus P, Vosefski D, Hoshino Y, Kapikian A, Dolin R (1988) Field trial of rhesus rotavirus vaccine in infants. Pediatr Infect Dis J 7: 645–650

Clark HF (1988) Rotavirus Vaccines. Plotkin SA, Mortimer E (eds) Vaccines. Saunders, Philadelpha

Clark HF, Furukawa T, Bell LM, Offit PA, Perrella PA, Plotkin SA (1986) Immune response of infants and children to low-passage bovine rotavirus (strain WC3). AJDC 140: 350–356

Clark HF, Borian FE, Bell LM, Modesto K, Gouvea V, Plotkin SA (1988) Protective effect of WC3 vaccine against rotavirus diarrhea in infants during a predominantly serotype 1 rotavirus season. J Infect Dis 158: 570–587

Clark HF, Borian FE, Modesto K, Plotkin SA (1990a) Serotype 1 reassortant of bovine rotavirus WC3, strain WI79-9, induces a polytypic antibody response in infants. Vaccine 8: 327–332

Clark HF, Borian FE, Plotkin SA (1990b) Immune protection of infants against rotavirus gastroenteritis by a serotype 1 reassortant of bovine rotavirus WC3. J Infect Dis 161: 1099–1104

Clark HF, Hoshino Y, Bell LM, Groff J, Hess G, Bachman P, Offit PA (1987) Rotavirus isolate WI61 representing a presumptive new human serotype. J Clin Microbiol 25: 1757–1762

Cohen J, Charpilenne A, Chilmonczyk S, Estes MK (1989) Nucleotide sequence of bovine rotavirus gene 1 and expression of the gene product in baculovirus. Virology 171: 131–140

Conner ME, Darlington RW (1980) Rotavirus infection in foals. Am J Vet Res 41: 1699–1703

Conner ME, Gilger MA, Estes MK, Graham DY (1991) Serologic and muscosal immune response to rotavirus infection in the rabbit model. J Virol 65: 2562–2571

Conner ME, Crawford SE, Barone C, Estes MK (1993) Rotavirus vaccine administered parenterally induces protective immunity. J Virol 67: 6633–6641

Cook SM, Glass RI, LeBaron CW, Ho M-S (1990) Global seasonality of rotavirus infections. Bull World Health Organ 68: 171–177

Coulson BS (1987) Variation in neutralization epitopes of human rotaviruses in relation to genomic RNA polymorphism. Virology 159: 209–216

Coulson BS, Grimwood K, Masendycz PJ, Lund, JS, Mermelstein N, Bishop RF, Barnes GL (1990a) Comparison of rotavirus immunoglobulin A coproconversion with other indices of rotavirus infection in a longitudinal study in childhood. J Clin Microbiol 28: 1367–1374

Coulson BS, Grimwood K, Hudson IL, Barnes GL, Bishop RF (1992) Role of coproantibody in clinical protection of children during reinfection with rotavirus. J Clin Microbiol 30: 1678–1684

Coulson BS (1993) Typing of human rotavirus VP4 by an enzyme immunoassay using monoclonal antibodies. J Clin Microbiol 31: 1–8

Crouch CF (1985) Vaccination against enteric rota and coronaviruses in cattle and pigs: enhancement of lactogenic immunity. Vaccine 3: 284–291

Dagan R, Bar-David Y, Sarov B, Katz M, Kassis I, Greenberg D, Glass RI, Margolis CZ, Sarov I (1990) Rotavirus diarrhea in Jewish and Bedouin children in the Negev region of Israel: epidemiology, clinical aspects and possible role of malnutrition in severity of illness. Pediatr Infect Dis J 9: 314–321

Das M, Dunn SJ, Woode GN, Greenberg HB, Rao CD (1993) Both surface proteins (VP4 and VP7) of an asymptomatic neonatal rotavirus strain (I 321) have high levels of sequence identity with the homologous proteins of a serotype 10 bovine rotavirus. Virology 194: 397–399

Davidson GP, Daniels E, Nunan H, Moore AG, Whyte PBD, Franklin K, McCloud PI, Moore DJ (1989) Passive immunisation of children with bovine colostrum containing antibodies to human rotavirus. Lancet ii: 709–712

De Mol P, Zissis G, Butzler JP, Mutwewingabo A, Andre FE (1986) Failure of live, attenuated oral rotavirus vaccine (letter). Lancet 2: 108

DeLeeuw PW, Ellens DJ, Talmon FP, Zimmer GN (1980) Rotavirus infections in calves: efficacy of oral vaccination in endemically infected herds. Res Vet Sci 29: 142–147

DeZoysa I, Feachem RG (1985) Interventions for the control of diarrhoeal disease among young children: rotavirus and· cholera immunization. Bull World Health Org. 63: 569–583

Dharakul T, Rott L, Greenberg HB (1990) Recovery from chronic rotavirus infection in mice with severe combined immunodeficiency: virus clearance mediated by adoptive transfer of immune CD8[+] T lymphocytes. J Virol 64: 4375–4382

DiGiacomo RF, Thouless ME (1986) Epidemiology of naturally occurring rotavirus infection in rabbits. Lab Anim Sci 36: 153–156

Dingle JH, Badger GF, Jordan WS (1964) Illness in the home. A study of 25,000 illnesses in a group of Cleveland families. Case Western University Press, Cleveland, pp 19–32

Ebina T, Sato A, Umezu K, Ishida N, Ohyama S, Oizumi A, Aikawa K, Katagiri S, Katsushima N, Imai A, Kitaoka S, Suzuki H, Konno T (1985) Prevention of rotavirus infection by oral administration of cow colostrum containing anti-human rotavirus antibody. Med Microbiol Immunol 174: 177–185

Edelman R (1987) Perspective on the development and deployment of rotavirus vaccines. Pediatr Infect Dis J 6: 704–710

Editorial (1990) Puzzling diversity of rotaviruses. Lancet 335: 573–575

Emini EA, Jameson BA, Wimmer E (1983) Priming for and induction of anti-poliovirus neutralizing antibodies by synthetic peptides. Nature 304: 689–703

Estes MK, Cohen J (1989) Rotavirus gene structure and function. Microbiol Rev 53: 410–449

Estes MK, Graham DY (1980) Identification of rotaviruses of different origins by the plaque-reduction test. Am J Vet Res 41: 151–152

Estes MK, Graham DY, Dimitrov DH (1984) The molecular epidemiology of rotavirus gastro-enteritis. In: Melnick JI (ed.) Progress in medical virology, vol. 29, Karger, Basel, pp 1–22

Estes MK, Crawford SE, Penaranda EM, Petrie BL, Burns JW, Chan W-K, Ericson B, Smith GE, Summers MD (1987) Synthesis and immunogenicity of the rotavirus major capsid antigen using a baculovirus expression system. J Virol 61: 1488–1494

Estes MK, Crawford SE, Labbe M, Cohen J (1991) Biologic properties of self-assembled rotavirus particles. Abstract presented at US-Japan meeting, Charlottesville, Virginia 23–25 Sep 1991

Estes MK, Arntzen C, Conner ME (1993) Progress towards new vaccines for pediatric diarrhea. In: Proceedings of the 104th Ross Conference of Pediatric Research, "Strategies for Pediatric Vaccines: Conventional and Molecular Approaches." (In Press)

Eugster AK, Scrutchfield WL (1980) Diagnosis of rotavirus infection in foals. 2nd international symposium of veterinary laboratory diagnosticians, pp 396–399

Feigin RD, Garg R (1987) Interaction of Infection and Nutrition. In: Feigin RD, Cherry JD (eds) Textbook of pediatric infectious diseases, 2nd edn. Saunders, Philadelphia

Flewett TH, Bryden AS, Davies H, Woode GN, Bridger JC, Derrick JM (1974) Relation between viruses from actute gastroenteritis of children and newborn calves. Lancet 2: 61–63

Flores J, Midthun K, Hoshino Y, Green K, Gorziglia M, Kapikian AZ, Chanock RM (1986) Conservation of the fourth gene among rotaviruses recovered from asymptomatic newborn infants and its possible role in attenuation. J Virol 60: 972–979

Flores J, Perez-Schael I, Gonzalez M, Garcia D, Perez M, Daoud N, Cunto W, Chanock RM, Kapikian AZ (1987) Protection against severe rotavirus diarrhoea by rhesus rotavirus vaccine in Venezuelan infants. Lancet 1: 882–884

Flores J, Perez-Schael I, Blanco M, Vilar M, Garcia D, Perez M, Daoud N, Midthun K, Kapikian AZ (1989) Reactions to and antigenicity of two human-rhesus rotavirus reassortant vaccine candidates of serotypes 1 and 2 in Venezuelan infants. J Clin Microbiol 27: 512–518

Flores J, Perez-Schael, Blanco M, White L, Garcfia D, Vilar M, Cunto W, Gonzalez R, Urbina C, Boher J, Mendez M, Kapkikian AZ (1990) Comparison of reactogenicity and antigenicity of M37 rotavirus vaccine and rhesus-rotavirus-based quadrivalent vaccine. Lancet ii: 330–334

Francavilla M, Miranda P, di Matteo A, Sarasini A, Gerna G, Milanesi G (1987) Expression of bovine rotavirus neutralization antigen in Escherichia coli. J Gen Virol 68: 2975–2980

Friedman MG, Galil A, Sarov B, Margalith M, Katzir G, Midthun K, Taniguchi K, Urasawa S, Kapikian AZ, Edelman R, Sarov I (1988) Two sequential outbreaks of rotavirus gastroenteritis: evidence for symptomatic and asymptomatic reinfections. J Infect Dis 158: 814–822

Gentsch JR, Das BK, Jiang B, Bhan MK, Glaso RI (1993) Similarity of the VP4 protein of human rotavirus strain 116E to that of the bovine B223 strain. Virology 194: 424–430

Gerna G, Sarasini A, Arista S, Di Matteo A, Giovannelli L, Parea M, Halonen P (1990a) Prevalence of human rotavirus serotypes in some European countries 1981-1988. Scand J Infect Dis 22: 5–10

Gerna G, Sarasini A, Di Matteo A, Zentilin L, Miranda P, Parea M, Baldanti F, Arista S, Milanesi G, Battaglia M (1990b) Serotype 3 human rotavirus strains with subgroup I specificity. J Clin Microbiol 28: 1342–1347

Gerna G, Sarasini A, Torsellini M, Torre D, Parea M, Battaglia M (1990c) Group—and type specific serologic response in infants and children with primary rotavirus infections and gastroenteritis caused by a strain of known serotype. J Infect Dis 161: 1105–1111

Gerna G, Sarasini A, Parea M, Arista S, Miranda P, Brüssow H, Hoshino Y, Flores J (1992) Isolation and characterization of two distinct human rotavirus strains with G6 specificity. J Clin Microbiol 30: 9–16

Ghosh SK, Naik TN (1989) Detection of a large number of subgroup 1 human rotaviruses with a "long" RNA electropherotype. Arch Virol 105: 119–127

Giammanco G, DeGrandi V, Lupo L, Mistretta A, Pignato S, Teuween D, Bogaerts H, Andre FE (1988) Interference of oral poliovirus vaccine on RIT 4237 oral rotavirus vaccine. Eur J Epidemiol 4: 233–236

Gilger MA, Matson DO, Conner ME, Rosenblatt HM, Finegold MJ, Estes MK (1991) Extraintestinal rotavirus infections in children with immunodeficiency. Manuscript (submitted for publication)

Glezen WP, Decker M, Perotta DM (1982) Survey of underlying conditions of persons hospitalized with acute respiratory disease during influenza epidemics in Houston. Am Rev Respir Dis 136: 550–555

Gomez J, Estes MK, Matson DO, Bellinzoni R, Alvarez A, Grinstein S (1990) Serotyping of human rotaviruses in Argentina by ELISA with monoclonal antibodies. Arch Virol 112: 249–259

Gorziglia M, Hoshino Y, Midthun K, Buckler-White A, Blumentals I, Glass R, Flores J, Kapikian AZ, Chanock RM (1986) Conservation of amino acid sequence of VP8 and cleavage region of 840-kDa outer capsid protein among rotaviruses recovered from asymptomatic neonatal infection. Proc Natl Acad Sci USA 83: 7039–7043

Gorziglia M, Larralde G, Kapikian AZ, Chanock KM (1990a) Antigenic relationships among human rotaviruses is determined by outer capsid protein VP4. Proc Natl Acad Sci USA 87: 7155–7159

Gorziglia M, Kapikian AZ (1992) Expression of the OSU rotavirus outer capsid protein VP4 by an adenovirus recombinant. J Virol 66: 4407–4412

Gothefors L, Wadell G, Juto P, Taniguchi K, Kapikian AZ, Glass RI (1989) Prolonged efficacy of rhesus rotavirus vaccine in Swedish children. J Infect Dis 159: 753–756

Gouvea V, Ho M-S, Glass R, Woods P, Forrester B, Robinson C, Ashley R, Riepenhoff-Talty M, Clark HF, Taniguchi K, Meddix E, McKellar B, Pickering L (1990) Serotypes and electropherotypes of human rotavirus in the USA: 1987-1989. J Infect Dis 162: 362–367

Graham DY, Estes MK (1985) Proposed working serologic classification system for rotaviruses. Ann Inst Pasteur/Virologie 136: 5–12

Green KY, James HD, Kapikian AZ (1990a) Evaluation of three panels of monoclonal antibodies for the identification of human rotavirus VP7 serotype by ELISA. Bull World Health Organ. 68: 601–610

Green KY, Taniguchi K, Mackow ER, Kapikian AZ (1990b) Homotypic and heterotypic epitope-specific antibody response in adult and infant rotavirus vaccinees: implications for vaccine development. J Infect Dis 161: 667–679

Greenberg H, Offit P, Kapikian A, Robinson W, Shaw R (1985) Vaccine strategies for prevention of rotavirus diarrhea. Microecol Ther 15: 47–54

Grimwood K, Lund JCS, Coulson BS, Hudson IL, Bishop RF, Bares GL (1988) Comparison of serum and mucosal antibody responses following severe acute rotavirus gastroenteritis in young children. J Clin Microbiol 26: 732–738

Grinstein S, Gomez JA, Bercovich JA, Biscotti EL (1989) Epidemiology of rotavirus infection and gastroenteritis in prospectively monitored Argentine families with young children. Am J Epidemiol 130: 300–308

Gunn PR, Sato F, Powell KFH, Bellamy AR, Napier JR, Harding DRK, Hancock WS, Siegman LJ, Both GW (1985) Rotavirus neutralizing protein VP7: antigenic determinants investigated by sequence analysis and peptide synthesis. J Virol 54: 791–797

Halsey NA, Anderson EL, Sears SD, Steinhoff M, Wilson M, Belshe RB, Midthun K, Kapikian AZ, Chanock RM, Samorodin R, Burns B, Clements ML (1988) Human-rhesus reassortant rotavirus vaccines: safety and immunogenicity in adults, infants, and children. J Infect Dis 158: 1261–1267

Hanlon P, Hanlon L, Marsh V, Byass P, Shenton F, Hassan-King M, Jobe O, Sillah H, Hayes R, M'Boge BH, Whittle HC, Greenwood BM (1987) Trial of an attenuated bovine rotavirus vaccine (RIT 4237) in Gambian infants. Lancet 1: 1342–1345

Health RC, Birch C, Gust I (1986) Antigenic analysis of rotavirus isolates using monoclonal antibodies specific for human serotype 1, 2, 3 and 4, and SA11. J Gen Virol 67: 2455–2466

Herrera IB, O'Ryan ML, Matson DO, Estes MK, Pickering LK (1991) Local and systemic immunity of postpartum women to human rotavirus. Abstract, American Pediatric Society and the Society for Pediatric Research. New Orleans, 29 Apr–2 May

Hilpert H, Brussow H, Mietens C, Sidoti J, Lerner L, Werchau H (1987) Use of bovine milk concentrate containing antibody to rotavirus to treat rotavirus gastroenteritis in infants. J Infect Dis 156: 158–166

Hjelt K, Grrauballe PC, Paerregaard A, Nielsen OH, Krasilnikoff PA (1987) Protective effect of preexisting rotavirus-specific immunoglobulin A against naturally acquired rotavirus infection in children. J Med Virol 21: 39–47

Ho MS, Glass RI, Pinsky PF, Anderson LJ (1988a) Rotavirus as a cause of diarrheal morbidity and mortality in the United States. J Infect Dis 158: 1112–1116

Ho MS, Glass RI, Pinsky PF, Young-Okoh N, Sapenfield WM, Buehler JW, Gunter N, Anderson LJ (1988b) Diarrheal deaths in American children. J Am Med Assoc 260: 3281–3285

Ho MS, Floyd RL, Glass RI, Pallansch MA, Jones B, Hamby B, Woods P, Penaranda ME, Kapikian AZ, Bohan G, Wilcox WD, Blumberg R (1989) Simultaneous administration of rhesus rotavirus vaccine and oral poliovirus vaccine: immunogenicity and reactogenicity. Pediatr Infect Dis J 8: 692–696

Hoblet KH, Saif LJ, Kohler EM, Theil KW, Bech-Nielsen S, Stitzlein GA (1986) Efficacy of an orally administered modified-live porcine-origin rotavirus vaccine against postweaning diarrhea in pigs. Am J Vet Res 47: 1697–1703

Hoshino Y, Wyatt RG, Greenberg HB, Flores J, Kapikian AZ (1984) Serotypic similarity and diversity of rotaviruses of mammalian and avian origin as studied by plaque reduction neutralization. J Infect Dis 149: 694–702

Hoshino Y, Wyatt RG, Flores J, Midthun K, Kapikian AZ (1985) Serotypic characterization of rotaviruses derived from asymptomatic human neonatal infections. J Clin Microbiol 21: 425–430

House JA (1978) Economic impact of rotavirus and other neonatal disease agents of animals. JAVMA 173: 573–576

Huang J, Nagesha HS, Dyall-Smith ML, Holmes IH (1989) Comparative sequence analysis of VP7 genes from five Australian porcine rotaviruses. Arch Virol 109: 173–183

Hum CP, Dyall-Smith ML, Holmes IH (1989) The VP7 gene of a new G serotype of human rotavirus (B37) is similar to G3 proteins in the antigenic C region. Virology 170: 55–61

Ijaz MK, Dent D, Babiuk LA (1990) Neuroimmunomodulation of in vivo anti-rotavirus humoral immune response. J Neuroimmunol 26: 159–171

Ijaz MK, Attah-Poku SK, Redmond MJ, Parker MD, Sabara MI, Babiuk LA (1991) Heterotypic passive protection induced by synthetic peptides croresponding to VP7 and VP4 of bovine rotavirus. J Virol 65: 3106–3113

Institute of Medicine (1985) New vaccine development, establishing priorities, vol 1. Diseases of importance in the United States. Prospects for immunizing against rotavirus. National Academy Press, Washington DC

Institute of Medicine (1986) New vaccine development, establishing priorities, vol 2, Diseases of importance in developing countries. National Academy Press, Washington DC

Janke BH (1989) Symposium on neonatal calf diarrhea. Vet Med 803–810

Johnson MA, Misra RM, Lardelli M, Messina M, Ephraums C, Reeves PR, Bolcevic Z, Noel JS, Hum CP, Van Mai H, Dyall-Smith ML, Holmes IH (1989) Synthesis in Escherichia coli of the major glycoprotein of human rotavirus: analysis of the antigenic regions. Gene 84: 73–81

Kapikian AZ, Cline WL, Kim HW, Kalica AR, Wyatt RG, Van Kirk DH, Chanock RM, James HD, Vaughn AL (1976) Antigenic relationships among five reovirus-like (RVL) agents by complement fixation (CF) and development of a new CF antigen for the human RVL agent of infantile gastroenteritis. Proc Soc Biol Med 152: 535–539

Kapikian AZ, Flores J, Hoshino Y, Midthun K, Green KY, Gorziglia M, Chanock RM, Potash L, Perez-Schael I, Gonzalez M, Vesikari T, Gothefors L, Wadell G, Glass RI, Levine MM, Rennels MB, Losonsky GA, Christy C, Dolin R, Anderson EL, Belshe RB, Wright PF, Santosham M, Halsey NA, Clements ML, Sears SD, Steinhoff MC, Black RE (1989) Rationale for the development of a rotavirus vaccine for infants and young children. Springer, Berlin Heidelberg New York, pp 151–180 (Progress in vaccinology, vol 2)

Kapikian AZ, Flores J, Midthun K, Hoshino Y, Green KY, Gorziglia M, Nishikawa K, Chanock RM, Potash L, Perez-Schael I (1990) Strategies for the development of a rotavirus vaccine against infantile diarrhea with an update on clinical trials of rotavirus vaccines. Adv Exp Med Biol 257: 67–90

Killen HM, Dimmock NJ (1982) Identification of a neutralization-specific antigen of a calf rotavirus. J Gen Virol 62: 297–311

Kitamoto N, Ramig RF, Matson DO, Estes MK (1991) Comparative growth of different rotavirus strains in differentiated cells (MA104, HepG2 and CaCo-2). Virology 184: 729–737

Kitaoka S, Nakagomi T, Fukuhara N, Hoshino Y, Suzuki H, Nakagomi O, Kapikian AZ, Ebina T, Konno T, Ishida N (1987) Serologic characteristics of human rotavirus isolate, AU-1, which has a "long" RNA pattern and subgroup 1 specificity. J Med Virol 23: 351–357

Koopman JS, Turkish VJ, Monto AS, Gouvea V, Srivastava S, Isaacson RE (1984) Patterns and etiology of diarrhea in three clinical settings. Am J Epidemiol 119: 114–123

Kraft LM (1957) Studies on the etiology and transmission of epidemic diarrhea of infant mice. J Exp Med 106: 743–755

Labbe M, Charpilienne A, Crawford SE, Estes MK, Cohen J (1991) Expression of rotavirus VP2 produces empty corelike particles. J Virol 65: 2946–2952

Lanata CF, Black RE, del Aguila R, Gil A, Verastegui H, Gerna G, Flores J, Kapikian AZ, Andre FE (1989) Protection of Peruvian children against rotavirus diarrhea of specific serotypes by one, two, or three doses of the RIT 4237 attenuated bovine rotavirus vaccine. J Infect Dis 159: 452–459

Larralde G, Flores J (1990) Identification of gene 4 alleles among human rotaviruses by polymerase chain reaction-derived probes. Virology 179: 469–473

LeBaron CW, Furutan NP, Lew JF, Allen JR, Gouvea V, Moe C, Monroe SS (1990a) Viral agents of gastroenteritis. MMWR 39: 1–24

LeBaron CW, Lew J, Glass RI, Weber JM, Ruiz-Palacios GM (1990b) Annual rotavirus epidemic patterns in North America. JAMA 264: 983–988

Lecce JG, King MW (1979) The calf reo-like virus (rotavirus) vaccine: an ineffective immunization agent for rotaviral diarrhea of piglets. Can J Comp Med 43: 90–93

Lizano M, Lopez S, Arias CF (1991) The amino-terminal half of rotavirus SA114fM VP4 protein contains a hemagglutination domain and primes for neutralizing anitbodies to the virus. J Virol 65: 1383–1391

Losonsky GA, Reymann M (1990) The immune response in primary asymptomatic and symptomatic rotavirus infection in newborn infants. J Infect Dis 161: 330–332

Losonsky GA, Johnson JP, Winkelstein JA, Yolken RH (1985) Oral administration of human serum immunoglobulin in immunodeficient patients with viral gastroenteritis. J Clin Invest 76: 2362–2367

Luckow VA, Summers MD (1988) Trends in the development of baculovirus expression vectors. Biotechnology 6: 47–55

Mackow ER, Barnett JW, Chan H, Greenberg HB (1989) The rhesus rotavirus outer capsid protein VP4 functions as a hemagglutinin and is antigenically conserved when expressed by a baculovirus recombinant. J Virol 63: 1661–1668

Mackow ER, Vo PT, Broome R, Bass D, Greenberg HB (1990) Immunization with baculovirus-expressed VP4 protein passively protects against simian and murine rotavirus challenge. J Virol 64: 1698–1703

Mata L, Simhon A, Urrutia JJ, Kronmal RA (1983) Natural history of rotavirus infection in the children of Santa Maria Cauque. Prog Food Nutr Sci 7: 167–177

Matson DO, Estes MK (1990) Impact of rotavirus infection at a large pediatric hospital. J Infect Dis 162: 598–604

Matson DO, Estes MK, Burns JW, Greenberg HB, Taniguchi K, Urasawa S (1990) Serotype variation of human group A rotaviruses in two regions of the USA. J Infect Dis 162: 605–614

Matson DO, O'Ryan ML, Herrera I, Pickering LK, Estes MK (1993) Fecal antibody responses to symptomatic and asymptomatic rotavirus infections. J Infect Dis 167: 577–583

Matsui SM, Mackow ER, Greenberg HB (1989) Molecular determinant of rotavirus neutralization and protection. Adv Virus Res 36: 181–214

Matsuno S, Murakami S, Takagi M, Hayashi M, Inouye S, Hasegawa A, Fukai K (1987) Cold-adaptation of human rotavirus. Virus Res 7: 273–280

Mattion N, Mitchell DB, Both GW, Estes MK (1991) Expression of rotavirus proteins encoded by alternative open reading frames of genome segment 11. Virology 181: 295–304

McCrae MA, McCorquodale JG (1987) Expression of a major bovine rotavirus neutralisation antigen (VP7c) in Escherichia coli. Gene 55: 9–18

McNulty MS, Logan EF (1987) Effect of vaccination of the dam on rotavirus infection in young calves. Vet Rec 120: 250–252

McNulty MS, Allan GM, Pearson GR, McFerran JB, Curran WL, McCracken RM (1976) Reovirus-like agent (rotavirus) from lambs. Infect Immun 14: 1332–1338

McNulty MS, Allan GM, Stuart JC (1978) Rotavirus infection in avian species. Vet Rec 103: 319–320

Mebus CA, White RG, Stair EL, Rhodes MB, Twiehaus MJ (1972) Neonatal calf diarrhea: results of a field trial using a reo-like virus vaccine. Vet Med/Small Anim Clin 67: 173–174

Mebus CA, White RG, Bass EP, Twiehaus MJ (1973) Immunity to neonatal calf diarrhea virus. JAVMA 163: 880–883

Melnick JL (1990) Poliomyelitis. In: Wareen KS, Mahmoud AAF (eds) tropical and geographical medicine, 2nd ed. McGraw-Hill, New York, pp 559–576

Midthun K, Greenberg HB, Hoshino Y, Kapikian AZ, Wyatt RG, Chanock RM (1985) Reassortant rotaviruses as potential live rotavirus vaccine candidates. J Virol 53: 949–954

Miller LK (1989) Insect baculoviruses: powerful gene expression vectors. Bio Essays 11: 91–95

Mutz ID, Krainer F, Deutsch J, Kunz C, Teuwen DE (1989) A trial of RIT-4237 rotavirus vaccine in 1-month-old infants. Eur J Pediatr 148: 634–635

Nagesha HS, Huang J, Hum CP, Holmes IH (1990) A porcine rotavirus strain with dual VP7 serotype specificity. Virology 175: 319–322

Nakagomi T, Nakagomi O (1989) RNA-RNA hybridization identified a human rotavirus that is genetically related to feline rotavirus. J Virol 63: 1431–1434

Nakagomi T, Matsuda Y, Ohshima A, Mochizuki M, Nakagomi O (1989) Characterization of a canine rotavirus strain by neutralization and molecular hybridization assays. Arch Virol 106: 145–150

Nakagomi T, Ohshima A, Akatani K, Ikegami N, Katsushima N, Nakagomi O (1990) Isolation and molecular characterization of a serotype 9 human rotavirus strain. Microbiol Immunol (JPN) 34: 77–82

Newman FS, Myers LL, Firehammer BD, Catlin JE (1973) Licensing and use of the calf scours vaccine, part II. An analysis of scourvax-reo. 77th annual meeting of the United States Animal Health Association Oct 1973, St. Louis

Nishikawa K, Fukuhara N, Liprandi, Green K, Kapikian AZ, Chanock RM, Gorziglia M (1989) VP4 protein of porcine rotavirus strain OSU expressed by a baculovirus recombinant induces neutralizing antibodies. Virology 173: 631–637

Offit PA, Clark HF (1985a) Maternal antibody-mediated protection against gastroenteritis due to rotavirus in newborn mice is dependent on both serotype and titer of antibody. J Infect Dis 152: 1152–1158

Offit PA, Clark HF (1985b) protection against rotavirus-induced gastroenteritis in a murine model by passively acquired gastrointestinal but not circulating antibodies. J Virol 54: 58–64

Offit PA, Dudzik KI (1989a) Rotavirus-specific cytotoxic T lymphocytes appear at the intestinal mucosal surface after rotavirus infection. J Virol 63: 3507–3512

Offit PA, Dudzik KI (1989b) Noninfectious rotavirus (strain RRV) induces an immune response in mice which protects against rotavirus challenge. J Clin Microbiol 27: 885–888

Offit PA, Dudzik KI (1990) Rotavirus-specific cytotoxic T lymphocytes passively protect against each parental serotype. J Virol 60: 491–496

Offit PA, Clark HF, Blavat G, Greenberg HB (1986) Reassortant rotaviruses containing structural proteins VP3 and VP7 from different parents induce antibodies protective against each parental serotype. J Virol 60: 491–496

Offit PA, Boyle DB, Borh GW, Hill NL, Svoboda YM, Cunningham SL, Jenkins RJ, McCrae MA (1991) Outer capsid glycoprotein VP7 is recognized by cross-reactive, rotavirus-specific, cytotoxic T lymphocytes. Virology 184: 563–568

Ogra SS, Weintraub D, Ogra PL (1977) Immunologic aspects of human colostrum and milk. III. Fate and absorption of cellular and soluble components in the gastrointestinal tract of the newborn. J Immunol 119: 245–248

O'Ryan M, Matson DO, Estes MK, Barlett AV, Pickering LK (1990) Molecular epidemiology of rotavirus in children attending day care centers in Houston. J Infect Dis 162: 810–816

Padilla-Noriega L, Arias CF, Lopez S, Snodgrass DR, Taniguchi K, Greenberg HB (1990) Diversity of rotavirus serotypes in Mexican infants with gastroenteritis. J Clin Microbiol 28: 1114–1119

Perez-Schael I, Blanco M, Vilar M, Garcia D, White L, Gonzalez R, Kapikian AZ, Flores J (1990a) Clinical studies of a quadrivalent rotavirus vaccine in Venezuelan infants. J Clin Microbiol 28: 553–558

Perez-Schael I, Garcia D, Gonzalez M, Gonzalez R, Daoud N, Perez M, Cunto W, Kapikian AZ, Flores J (1990b) Prospective study of diarrheal diseases in Venezuelan children to evaluate the efficacy of rhesus rotavirus vaccine. J Med Virol 30: 219–229

Perkus ME, Piccini A, Lipinskas BR, Paoletti E (1985) Recombinant vaccinia virus: immunization against multiple pathogens. Science 229: 981–984

Pichichero ME (1990) Effect of breast-feeding on oral rhesus rotavirus vaccine seroconversion: a meta analysis. J Infect Dis 162: 753–755

Pitson GA, Grimwood K, Coulson BS, Oberklaid F, Hewstone AS, Jack I, Bishop RF, Baines GL (1986) Comparison between children treated at home and those requiring hospital admission for rotaviruses and other enteric pathogens associated with acute diarrhea in Melbourne, Australia. J Clin Microbiol 24: 395–399

Pongsuwanne Y, Taniguchi K, Choonthanom M, Chiwakul M, Susansook T, Saguanwongse S, Jayavasu C, Urasawa S (1989) Subgroup and serotype distributions of human, bovine, and procine rotavirus in Thailand. J Clin Microbiol 27: 1956–1960

Poruchynsky MS, Tyndall C, Both GW, Sato F, Bellamy AR, Atkinson PH (1985) Deletions into an NH$_2$-terminal hydrophobic domain result in secretion of rotavirus VP7, a resident endoplasmic reticulum membrane protein. J Cell Biol 101: 2199–2209

Raj P, Matson DO, Bishop RF, Coulson B, Taniguchi K, Urasawa S, Greenberg HB, Estes MK (1992) Comparisions of rotavirus VP7 typing monoclonal antibodies by competition binding assays. J Clin Microbiol 30: 704–711

Ramig RF, Galle KL (1990) Rotavirus genome segment 4 determines viral replication phenotype in cultured liver cells (HepG2). J Virol 64: 1044–1049

Ramshaw IA, Andrew ME, Phillips SM, Boyle DB, Coupar BEH (1987) Recovery of immuno-deficient mice from a vaccinia virus/IL-2 recombinant infection. Nature 329: 545–546

Redfield RR, Wright DC, James WD, Jones TS, Brown C, Burke DS (1987) Disseminated vaccinia in a military recruit with human immunodeficiency virus (HIV) disease. N Engl J Med 316: 673–676

Redmond MJ, Ohmann HB, Hughes HPA, Sabara M, Frenchick PJ, Attah Poku SK, Ijaz MK, Parker MD, Laarveld B, Babiuk LA (1991) Rotavirus particles function as immunological carriers for the delivery of peptides from infectious agents and endogenous proteins. Mol Immun 28: 269–278

Reeves PR, Johnson MA, Holmes IH, Dyall-Smith ML (1990) Expression of rotavirus VP7 antigens in fustions with baterial proteins. Res Microbiol 141: 1019–1025

Rennels MB, Losonsky GA, Levine MM, Kapikian AZ (1986) Preliminary evaluation of the efficacy of rhesus rotavirus vaccine strain MMU188006 in young children. Pediatr Infect Dis 5: 587–588

Rennels MB, Losonsky GA, Young AE, Shindledecker CL, Kapikian AZ, Levine MM (1990) An efficacy trial of the rhesus rotavirus vaccine in Maryland. AJDC 144: 601–604

Reves RR, Mohassin MM, Midthun K, Kapikian AZ, Naguib T, Zaki AM, Du Pont HL (1989) An observational study of naturally acquired immunity to rotaviral diarrhea in a cohort of 363 Egyptian children. Calculation of risk for second episodes using age-specific person-years of observation. AM J Epidemiol 130: 981–988

Riepenhoff-Talty M, Schaekel K, Clark HF, Mueller W, Uhnoo I, Rossi T, Fisher J, Ogra PL (1993) Group A rotaviruses produce extrahepatic biliary obstruction in orally inoculated newborn mice. Pediatric Research 33: 394–399

Rodger SM, Holmes IH (1979) Comparison of the genomes of simian, bovine, and human rotaviruses by gel electrophoresis and detection of genomic variation among bovine isolates. J Virol 30: 839–846

Rodriguez WJ, Kim HW, Brandt CD, Bise B, Kapikian AZ, Chanock RM, Curlin G, Parrott RH (1980) Rotavirus gastroenteritis in the Washington DC area. Am J Dis Child 134: 777–779

Rodriguez WJ, Kim HW, Brandt CD, Schwartz RH, Gardner MK, Jeffries B, Parrott RH, Kaslow RA, Smith JI, Kapikian AZ (1987) Longitudinal study of rotavirus infection and gastroenteritis in families served by a pediatric medical practice: clinical and epdemiologic observations. Pediatr Infect Dis J 6:170–176

Ruuska T, Vesikari T (1990) Rotavirus disease in Finnish children: use of numerical scores for clinical severity of diarrhoeal episodes. Scand J Infect Dis 22: 259–267

Ruuska T, Vesikari T, Delem A, Andre FE, Beards GM, Flewett TH (1990) Evaluation of RIT 4237 bovine rotavirus vaccine in newborn infants: correlation of vaccine efficancy to season of birth in relation to rotavirus epidemic period: Scand J Infect Dis 22:269–278

Sabara M, Read KFM, Frenchick PJ, Babiuk LA (1987) Biochemical evidence for the oligomeric arrangment of bovine rotavirus nucleocapsid protein and its possible significance in the immunogenicity of this protein. J Gen Virol 68: 123–133

Saif LJ, Jackwood JJ (1990) Enteric Virus vaccines: theoretical considerations, current status and future approaches. In viral diarrhea of man and animals. CRC press, Boca Raton

Saif LJ, Redman DR, Smith KL, Theil KW (1983) Passive immunity to bovine rotavirus in newborn calves fed colostrum supplements from immunized or nonimmunized cows. Infect Immun 421: 1118–1131

Saif LJ, Theil KW (eds) Saif LJ, Smith KL, Landmeir BJ, Bohl EH, Theil KW, Todhunter DA (1984) Immune response of pregnant cows to bovine rotavirus immunization. Am J Vet Res 45: 49–58

Salas-Vidal E, Plebanski M, Castro S, Perales G, Mata E, Lopez S, Arias CF (1990) Synthesis of the surface glycoprotein of rotavirus SA11 in the aro A strain of Salmonella Tymphimurium SL3261. Res Microbiol 141: 883–886

Santosham M, Letson GW, Wolff M, Reid R, Gahagan S, Adams R, Callahan C, Sack RB, Kapikian AZ (1991) A field study of the safety and efficacy of two candidate rotavirus vaccines in a Native American population. J Infect Dis 163: 483–487

Schoeb TR, Casebolt DB, Walker VE, Potgieter LN, Thouless ME, DiGiacomo RF (1986) Rotavirus-associated diarrhea in a commercial rabbitry. Lab Anim Sci 36: 149–152

Schwarz KB, Moore TJ, Willoughby RE, Wee S-B, Vonderfecht SL, Yolken RH (1990) Growth of group A rotaviruses in a human liver cell line. Hepatology 12: 638–643

Shaw RD, Coulsen B, Offit PA, Vo PT, Greenberg HB (1986) Antigenic mapping of the surface proteins of rhesus rotavirus. Virology 155: 434–451

Shaw RD, Fong KJ, Losonsky GA, Levine MM, Maldonado Y, Yolken R, Flores J, Kapikian AZ, Vo PT, Greenberg HB (1987) Epitope-specific immune responses to rotavirus vaccination. Gastroenterology 93: 941–950

Sheridan JF, Smith CC, Manak MM, Aurelian L (1984) Prevention of rotavirus-induced diarrhea in neonatal mice born to dams immunized with empty capsids of simian rotavirus SA-11. J Infect Dis 149: 434–438

Smith GL, Mackett M, Murphy BR, Moss B (1984) Vaccinia virus recombinants expressing genes from pathogenic agents have potential as live vaccines. In: Chanock RM, Lerner RA (eds) Modern approaches to vaccines. Cold Spring Harbor Laboratories, Cold Spring Harbor

Snodgrass DR (1986a) Evaluation of a combined rotavirus and enterotoxigenic Escherichia coli vaccine in cattle. Vet Rec 119: 39–43

Snodgrass DR (1986b) Prevention of calf diarrhoea by vaccination. Practice : 139–140

Snodgrass DR, Wells PW (1976) Rotavirus infection in lambs: studies on passive protection. Arch Virol 52: 201–205

Snodgrass DR, Wells PW (1978a) Passive immunity in rotaviral infections. JAVMA 173: 565–569

Snodgrass DR, Wells PW (1978b) The immunoprophylaxis of rotavirus infections in lambs. Vet Rec 102: 146–148

Snodgrass DR, Madeley CR, Wells PW, Angus KW (1977) Human rotavirus in lambs: infection and passive protection. Infect Immun 16: 268–270

Snodgrass DR, Fahey KJ, Wells PW, Campbell I, Whitelaw A (1980) Passive immunity in calf rotavirus infections: maternal vaccination increases and prolongs immunoglobulin GI antibody secretion in milk Infect Immun 28: 344–349

Snodgrass DR, Ojeh CK, Campbell I, Herring AJ (1984) Bovine rotavirus serotypes and their significance for immunization. J Clin Microbiol 20: 342–346

Snodgrass DR, Fitzgerald TA, Campbell I, Browning GF, Scott FM, Hoshino Y, Davies RC (1991) Homotypic and heterotypic serological responses to rotavirus neutralization epitopes in immunologically naive and experienced animals. J Clin Microbiol 29: 2668–2672

Streckert H-J, Brussow H, Werchau H (1988) A synthetic peptide corresponding to the cleavage region of VP3 from rotavirus SA11 induces neutralizing antibodies. J Virol 62: 4265–4269

Stuker G, Oshiro L, Schmidt NJ (1980) Antigenic composition of two new rotaviruses from rhesus monkeys. J Clin Microbiol 11: 202–203

Svennerholm A-M, Hanson LA, Holmgren J, Lindblad BS, Nilsson B, Quereshi F (1980) Different secretory immunoglobulin A antibody responses to cholera vaccination in Swedish and Pakistani women. Infect Immun 30: 427–430

Tajima T, Thompson J, Wright PF, Kondo Y, Tollefson SJ, King J, Kapikian AZ (1990) Evaluation of a reassortant rhesus rotavirus vaccine in young children. Vaccine 8: 71–74

Taniguchi K, Maloy WL, Nishikawa K, Green KY, Hoshino Y, Urasawa S, Kapikian AZ, Chanock RM, Gorziglia M (1988) Identification of cross-reacitve and serotype 2-specific neutralization epitopes on VP3 of human rotavirus. J Virol 62: 2421–2426

Taniguchi K, Urasawa T, Kobayashi N, Gorziglia M, Urasawa S (1990) Nucleotide sequence of VP4 and VP7 genes of human rotaviruses with subgroup I specificity and long RNA pattern: implication for new G serotype specificity. J Virol 64: 5640–5644

Taniguchi K, Urasawa T, Kobayashi N, Ahmed MU, Adachi N, Chiba S, Urasawa S (1991) Antibody response to serotype-specific and cross-reactive neutralization epitopes on VP4 and VP7 after rotavirus infection or vaccination. J Clin Microbiol 29: 483–487

Thurber ET, Bass EP, Beckenhauer WH (1977) Field trial evaluation of a reo-coronavirus calf diarrhea vaccine. Can J Comp Med 41: 131–136

Tizard I (1982) An introduction to veterinary immunology. Philadelphia, Saunders

Twiehaus MJ, Mebus CA (1973) Licensing and use of the calf scour vaccine. 77th annual meeting of the United States Animal Health Association, October 1973, St Louis

Tzipori S, Unicomb L, Bishop R, Montenaro J, Vaelioja LM (1989) Studies on attenuation of rotavirus. A comparison in piglets between virulent virus and its attenuated derivative. Arch Virol 109: 197–205

Uhnoo I, Riepenhoff-Talty M, Chegas P, Fisher JE, Greenberg HB, Ogra PL (1990a) Effect of malnutrition on extraintestinal spread of rotavirus and development of hepatitis in mice. Nutr Res 10: 1419–1429

Uhnoo I, Riepenhoff-Talty M, Dharakul T, Chegas P, Fisher JE, Greenberg HB, Ogra PL (1990b) Extramucosal spread and development of hepatitis in immunodeficient and normal mice infected with rhesus rotavirus. J Virol 64: 361–368

Unicomb LE, Coulson BS, Bishop RF (1989) Experience with an enzyme immunoassay for serotyping human rotaviruses. J Clin Microbiol 27: 586–588

Urasawa S, Urasawa T, Taniguchi K, Wakasugi F, Kobayashi N, Chiba S, Sakurada N, Morita M, Morita O, Tokieda M, Kawamoto H, Minekawa Y, Ohseto M (1989) Survey of human rotavirus serotypes in different locales in Japan by enzyme-linked immunosorbent assay with monoclonal antibodies. J Infect Dis 160: 44–51

Urasawa S, Urasawa T, Wakasugi F, Kobayashi N, Taniguchi K, Lintag IC, Saniel MC, Goto H (1990) Presumptive seventh serotype of human rotavirus. Arch Virol 113: 279–282

Urasawa S, Hasegawa A, Urasawa T, Taniguchi K, Wakasugi F, Suzuki H, Inouye S, Pongprot B, Supawadee J, Suprasert S, Rangsiyanond P, Tonusin S, Yamazi Y (1992) Antigenic and genetic analyses of human rotaviruses in Chiang Mai, Thailand: Evidence for a close relationship between human and animal rotaviruses. J Infect Dis 166: 227–234

Vesikari T (1988) Clinical and immunological studies of rotavirus vaccines. Southeast Asian J Trop Med Public Health 19: 437–447

Vesikari T, Isolauri E, Delem A, D'Hondt E, Andre FE, Zissis G (1983) Immunogenicity and safety of live oral attenuated bovine rotavirus vaccine strain RIT 4237 in adults and young children. Lancet 11: 807–81

Vesikari T, Isolauri E, D'Hondt E, Andre D, Andre FE, Zisses G (1984) Protection of infants against rotavirus diarrhoea by RIT 4237 attenuated bovine rotavirus vaccine. Lancet i: 977–981

Vesikari T, Isolauri E, Delem A, D'Hondt E, Andre FE, Beards GM, Flewett TH (1985) Clinical efficacy of the RIT 4237 live attenuated bovine rotavirus vaccine in infants vaccinated before a rotavirus epidemic. J Pediatr 107: 189–194

Vesikari T, Kapikian AZ, Delem A, Zissis G (1986) A comparative trial of rhesus monkey (RRV-1) and bovine (RIT 4237) oral rotavirus vaccines in young children. J Infect Dis 153: 832–839

Vesikari T, Rautanen T, Varis T, Beards GM, Kapikian AZ (1990) Rhesus rotavirus candidate vaccine. Am J Dis Child 144: 285–289

Vesikari T, Ruuska T, Delem A, Andre FE, Beards GM, Flewett TH (1991a) Efficacy of two doses of RIT 4237 bovine rotavirus vaccine for prevention of rotavirus diarrhoea. Acta Paediatr Scand 80: 173–180

Vesikari T, Varis T, Green K, Flores J, Kapikian AZ (1991b) Immunogenicity and safety of rhesus-human rotavirus reassortant vaccines with serotype 1 or 2 VP7 specificity. Vaccine 9: 334–339

Vodopija I, Baklaic Z, Vlatkovic R (1988) Combined vaccination with live oral poliovirus vaccine and the bovine rotavirus RIT 4237 strain. Vaccine 4: 233–236

Wallace RE, Vasington PJ, Petricciani JC, Hopps HE, Lorenz DE, Kadanka Z (1973) Development of a diploid cell line from fetal rhesus monkey lung for virus vaccines production. In Vitro 8: 323–332

Waltner-Towes D, Martin SW, Meek AH, McMillan I, Crouch CF (1985) A field trial to evaluate the efficacy of a combined rotavirus-coronavirus/Escherichia coli vaccine in dairy cattle. Can J Comp Med 49: 1–9

Ward RL, Bernstein DI, Shukla R, Young EC, Sherwood JR, McNeal MM, Walker MC, Schiff GM (1989) Effects of antibody to rotavirus on protection of adults challenged with a human rotavirus. J Infect Dis 159: 79–88

Ward RL, Sander DS, Schiff GM, Bernstein DI (1990a) Effect of vaccination on serotype-specific antibody responses in infants administered WC3 bovine rotavirus before or after a natural rotavirus infection. J Infect Dis 162: 1298–1303

Ward RL, McNeal MM, Sheridan JF (1990b) Development of an adult mouse model for studies on protection against rotavirus. J Virol 64: 5070–5075

Ward RL, McNeal MM, Clemens JD, Sack DA, Rao M, Huda N, Green KY, Kapikian AZ, Coulson BS, Bishop RF, Greenberg HB, Gerna G, Schiff GM (1991) Reactivities of serotyping monoclonal antibodies with culture-adapted human rotaviruses. J Clin Microbiol 29: 449–456

Welch SKW, Crawford SE, Estes MK (1989) The rotavirus SA11 genome segment 11 protein is a nonstructural phosphoprotein. J Virol 63: 3974–3982

Woode GN, Bridger J, Hall GA, Jones JM Jackson G (1976a) The isolation of reovirus-like agents (rotaviruses) from acute gastroenteritis of piglets. J Med Microbiol 9: 203–209

Woode GN, Bridger J, Jones JM, Flewett TH, Bryden AS, Davies HA, White GBB (1976b) Morphological and antigenic relationships between viruses (rotaviruses) from acute gastroenteritis of children, calves, piglets, mice and foals. Infect Immun 14: 804–810

Woode GN, Zheng SL, Rosen BI, Knight N, Gourley NE, Ramig RF (1987) Protection between different serotypes of bovine rotavirus in gnotobiotic calves: specificity of serum antibody and coproantibody responses. J Clin Microbiol 25: 1052–1058

Woode GN, Zheng S, Melendy DR, Ramig RF (1989) Studies on rotavirus homologous and heterologous active immunity in infant mice. Viral Immunol 2: 127–132

Wright PF, King J, Araki K, Kondo Y, Thompson J, Tollefson SJ, Kobayashi M, Kapikian AZ (1991) Simultaneous administration of two human-rhesus rotavirus reassortant strains of VP7 serotype 1 and 2 specificity to infants and young children. J Infect Dis 164: 271–276

Wyatt RG, Greenberg HB, James WD, Pittman AL, Kalica AR, Flores J, Chanock RM, Kapikian AZ (1982) Definition of human rotavirus serotype by plaque reduction assay. Infect Immun 37: 110–115

Nongroup A Rotaviruses of Humans and Animals

L.J. Saif and B. Jiang*

1 Introduction

Rotaviruses are members of the *Reoviridae* family and possess a double capsid layer of concentric icosahedral shells surrounding a genome containing 11 segments of double-stranded (ds) RNA. Initial studies using various serologic techniques established that rotaviruses from diverse species share a common group antigen located on the inner capsid layer. Recently however, antigenically distinct rotaviruses lacking the common group antigen

Food Animal Health Research Program, Department of Veterinary Preventive Medicine, Ohio Agricultural Research and Development Center, The Ohio State University, Wooster, Ohio 44691, USA
* Present address: Viral Gastroenteritis Unit, Centers for Disease Control, Atlanta, GA 30333, USA

were identified which were morphologically indistinguishable from conventional rotaviruses. These viruses were first identified in diarrheic swine (SAIF et al. 1980; BRIDGER 1980) and subsequently in poultry (MCNULTY et al. 1981, 1984b; SAIF et al. 1985), calves (SAIF et al. 1982; SNODGRASS et al. 1984); humans (RODGER et al. 1982; SAIF and THEIL 1985; BRIDGER 1987; HUNG et al. 1987), lambs (CHASEY and BANKS 1984; SNODGRASS et al. 1984), rats (VONDERFECHT et al. 1984), and ferrets (TORRES-MEDINA 1987). These newly discovered rotaviruses have been referred to as pararotaviruses (PaRV), rotavirus-like viruses (RVLV), atypical, antigenically distinct, and novel rotaviruses.

It is now proposed that rotaviruses be separated into serogroups with members of each group sharing their own distinctive common antigen (PEDLEY et al. 1983, 1986; SAIF and THEIL 1985; BRIDGER 1987; SAIF 1990b). Currently, seven serogroups are recognized with the prototype group members as follows: group A (A/conventional rotaviruses); group B (B/porcine NIRD-1); group C (C/porcine Cowden); group D (D/avian 132); group E (E/porcine DC-9); group F (F/avian A4); and group G (G/avian 555) rotaviruses (PEDLEY et al. 1983, 1986; MCNULTY et al. 1984b; SAIF and THEIL 1985; BRIDGER 1987; SAIF 1990b). The originally recognized rotaviruses belong to group A, while the other rotaviruses will be referred to collectively as non-A rotaviruses, or by their respective groups, or, if not yet serogrouped, as unclassified non-A rotaviruses.

Although non-A rotaviruses possess 11 segments of dsRNA, their genome electrophoretic migration patterns differ from one another and from group A rotaviruses. These distinctive electropherotype patterns allow the non-A rotaviruses to be identified and separated into electropherogroups, most of which appear to be distinctive for each of the current rotavirus serogroups (PEDLEY et al. 1983, 1986; SAIF and THEIL 1985; SAIF 1990b).

The group A rotaviruses are common causes of enteric disease in the young of most species, but the prevalence and significance of non-A rotavirus infections are unclear. In comparison to group A rotaviruses, their detection is still rare (except in poultry and lambs), but they represent a possible source of confusion in the diagnosis of rotavirus infections based only on electron microscopy. During this past decade, the finding that non-A rotaviruses have been associated with major epidemics of diarrhea in China (group B rotaviruses, HUNG et al. 1987; HUNG 1988) and more recently in Japan (group C rotaviruses, USHIJIMA et al. 1989; MATSUMOTO et al. 1989) suggests they may be newly emerging pathogens for humans. The inability to isolate and serially propagate most non-A rotaviruses in cell culture severely limits the accessibility of these viruses and remains a major impediment to their further characterization and study. However application of molecular techniques to their study may circumvent this problem in the future provided adequate amounts of non-A rotaviruses are available directly from stool specimens.

2 Morphology and Morphogenesis

Non-A rotaviruses are characterized by single or double capsid particles indistinguishable in morphology from group A rotaviruses. The diameter for single and double capsid particles ranges from 50 to 65 nm and 65 to 80 nm, respectively (BRIDGER et al. 1982; HUNG et al. 1983, 1984; MCNULTY et al. 1981; SAIF 1990b; USHIJIMA et al. 1989). Using negative-contrast electron microscopy, ESPEJO et al. (1984) studied the arrangement of capsomeres radiated from the core and indicated that a human group C rotavirus was composed of an icosahedral arrangement of 720 subunits which formed 240 capsomeres assembled in a pattern characteristic of $T = 12$. This observation differs from that for a group A rotavirus ($T = 13$ with 132 capsomeres) defined by a more precise freeze-etching method (ROSETO et al. 1979). Non-A rotaviruses may further deviate from conventional rotaviruses in other aspects of their morphology. For example, thin-walled core-like particles (48–52 nm in diameter) are consistently found in humans and animals infected with group B rotaviruses (BRIDGER et al. 1982; HUNG et al. 1983; SAIF and THEIL 1985).

Non-A rotaviruses show variable appearances using different negative staining methods. SUZUKI et al. (1987) observed intact Chinese group B rotavirus particles when stained with uranyl acetate, but, when particles were stained with phosphotungstic acid (PTA, pH 4–9), different degrees of degradation were seen. NAKATA et al. (1987b) reported similar findings, but degradations occurred only using PTA at higher pH values with increased staining time.

Our laboratory also compared the effect of different staining techniques on the appearance of a porcine group C rotavirus and noted similar results. However, under identical staining conditions (3% PTA, pH 7) we observed that porcine group A and B and bovine group B rotaviruses were usually single capsid, whereas bovine group A, porcine group C, and turkey group D and F rotaviruses were primarily double-capsid (BOHL et al. 1982; SAIF et al. 1977, 1980, 1985; THEIL et al. 1985). The morphology of rotavirus serogroups A–D is illustrated in the electron micrograph in Fig. 1.

Studies of the morphogenesis of non-A rotaviruses have been conducted in small intestinal villous enterocytes of rats infected with group B rotaviruses (VONDERFECHT et al. 1984; HUBER et al. 1989); pigs infected with groups B, C, and E rotaviruses (CHASEY et al. 1986; HALL 1987; SAIF 1990b), and chickens infected with group D rotaviruses (MCNULTY et al. 1981). The morphogenesis of non-A rotaviruses resembled that of group A rotaviruses in that the incomplete virions or viral precursors in viroplasms obtained outer shells to form complete virions or they became enveloped by budding through the membrane of the endoplasmic reticulum into the cisternae. In addition, an important, perhaps unique, change of host cells was observed

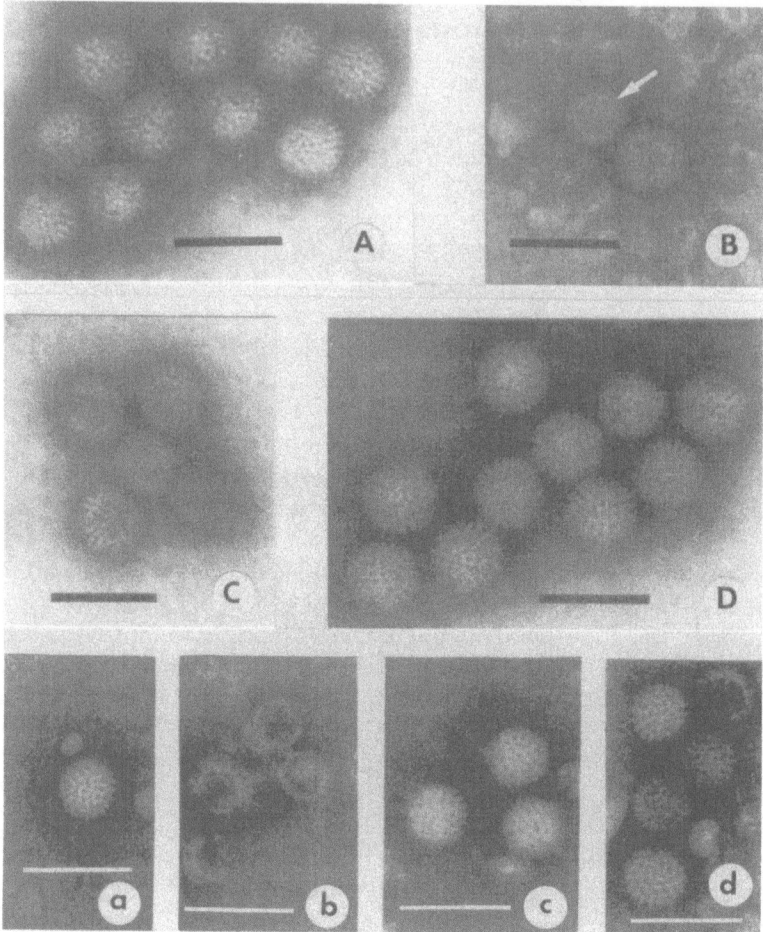

Fig. 1A–D. Immune electron micrographs of double-shelled rotavirus particles: **A** group A bovine rotavirus; **B** group B bovine rotavirus; **C** group C porcine rotavirus; and **D** group D turkey rotavirus. Single-shelled rotavirus particles are shown in **B** group B bovine rotavirus (*arrow*); and below in *a* group A bovine rotavirus; *c* group C porcine rotavirus; and *d* group D avian rotavirus. Core particles, characteristic of group B rotaviruses are shown in *b* for porcine group B rotavirus. *Bar* indicates 100 nm

during the process of group B viral morphogenesis, i.e., the formation of characteristic villous epithelial syncytia in the small intestine of infected rats (HUBER et al. 1989), infected pigs (HALL 1987; CHASEY et al. 1989), and infected calves (MEBUS et al. 1978). This distinguishing pathologic feature might provide pathognomonic evidence for the diagnosis of group B rotaviral infections.

3 Physicochemical Properties

Information about the buoyant desnsity of non-A rotaviruses is limited. Only group B and C rotaviruses have been purified and their densities characterized in CsCl gradients. FANG et al. (1989a) reported that double- and single-shelled group B human rotaviruses have a buoyant density of 1.373 g/ml and 1.435 g/ml, respectively. Other investigators reported a density of 1.377 g/ml in CsCl for another human group B rotavirus (DAI et al. 1987). ESPEJO et al. (1984) compared the physicochemical properties of a human group C rotavirus with those of a group A rotavirus and demonstrated a slightly higher buoyant density in CsCl for group C rotavirus. In our laboratory, double- and single-shelled particles of a porcine group C rotavirus purified from fecal specimens had densities in CsCl of 1.369 g/ml and 1.387 g/ml, respectively (JIANG and SAIF 1990, unpublished data). Accordingly, it appears that group B and C rotaviruses have similar, but not identical, densities to group A rotaviruses (1.36–1.38 g/ml).

The stability of porcine group C rotavirus (Cowden) to heat, acid, and ether treatments was compared with that of porcine group A rotaviruses. Similar to group A rotaviruses, group C rotavirus was labile to heat (56°C), only slightly sensitive to pH3, and was relatively stable to pH 5 and ether (TERRETT et al. 1987). These properties varied for group B rotaviruses, in which virus infectivity was lost after exposure to heat (56°C) or pH 3 (VONDERFECHT et al. 1984).

Research on the hemagglutinating activity of non-A rotaviruses is also limited. DAI et al. (1987) reported that a non-A rotavirus (antigenically similar to group B rotavirus) from an epidemic of diarrhea in China agglutinated erythrocytes of Rhesus monkey. However, a human group C rotavirus failed to agglutinate human erythrocytes (type O) (ESPEJO et al. 1984). Our laboratory conducted hemagglutination tests of a porcine group C rotavirus with erythrocytes from various species, including human (type O), guinea pig, sheep, and chicken. The virus from both the tissue culture supernatants and following purification on CsCl failed to agglutinate these erythrocytes (JIANG and SAIF 1990, unpublished data). These results suggest that this porcine group C rotavirus may lack hemagglutinating activity.

4 Viral Genome

As for group A rotaviruses, non-A rotaviruses possess a genome consisting of 11 segments of dsRNA with molecular weights ranging from 0.20 to 2.0×10^6 (McNULTY et al. 1981; PEREIRA et al. 1983). The genome segments can be separated by polyacrylamide gel electrophoresis (electropherotyping) and grouped into four size regions (I–IV) (SAIF and THEIL 1985; SAIF 1990b). However, the electropherotypes of non-A rotaviruses differ from group A rotaviruses in which the genomic profiles are often described as 4-2-3-2. During the past decade, at least six additional serogroups (B–G) of antigenically distinct rotaviruses have been discovered and each group has a distinct RNA electropherotype, often referred to as the electrophero-group (BRIDGER 1987; McCRAE 1987).

Although the genome segments (four or five) in region I migrate similar to those of group A rotaviruses, all non-A rotaviruses described to date lack the tight triplet of segments 7, 8 and 9 (region III) characteristic of the group A rotaviruses (Figs. 2, 3; PEDLEY et al. 1983, 1986; SAIF and THEIL

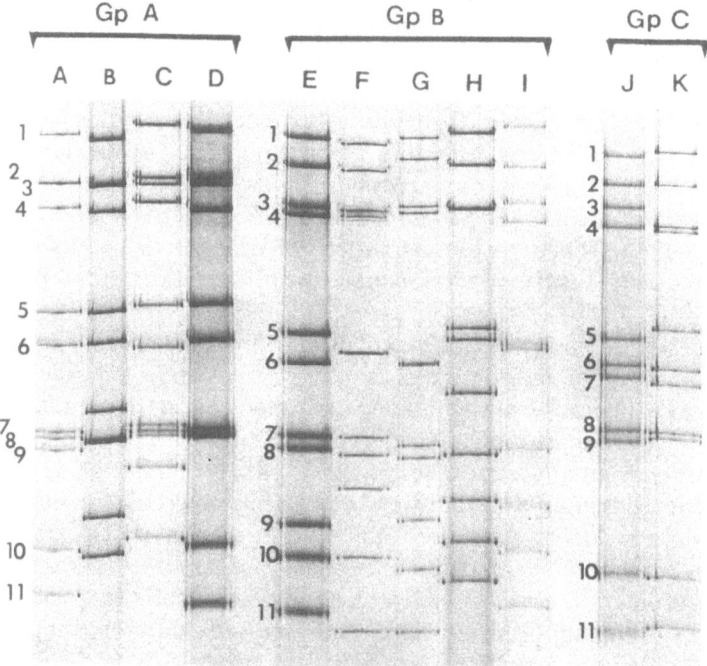

Fig. 2. Comparison of double-stranded (ds) RNA electropherotypes of mammalian group (*Gp*) A, B, and C rotaviruses in polyacrylamide gel electrophoresis. Migration is from *top* to *bottom*. Numbers to the *left* of each rotavirus group designate the segments of reference rotavirus genomes. *Lanes* (*left* to *right*): *A* and *B*, human gp A rotaviruses, WA and DS-1 strains, respectively; *C* and *D*, bovine gp A rotaviruses, ID and NCDV strains, respectively; *E* and *F*, porcine gp B rotaviruses, N338 and J strains, respectively; *G*, bovine gp B rotavirus, ATI strain; *H*, human gp B rotavirus, ADRV strain (dsRNA courtesy of Z. Y. Fang); *I*, rodent gp B rotavirus, IDIR strain (dsRNA courtesy of J. Eiden); *J* and *K*, porcine gp C rotaviruses, Cowden and HF strains, respectively. Note typical 7-8-9 triplet for gp A rotaviruses, faster migrating segment 9 for gp B rotaviruses, and slower migrating segment 7 for gp C rotaviruses. Like gp A rotaviruses (*lanes A–D*), minor differences in the migration patterns of genome segments for different strains of viruses within gps B (*lanes E–I*), or C (*lanes J* and *K*) were observed, although the overall patterns were similar within a group

1985; Todd and McNulty 1986; McCrae 1987; Saif 1990B). Group B rotaviruses isolated from humans, cattle, and pigs share similar electropherotypes, characterized often by two doublet segments, 3–4 and 5–6, but consistently by a faster migrating segment 9 (Fig. 2). In contrast to group B rotaviruses, group C rotaviruses have a slower migrating segment 7 and group D rotaviruses, slower migrating segments 5 and 7 (Figs. 2, 3). Consequently, groups B, C and D rotaviruses have dsRNA migration patterns of 4-2-2-3, 4-3-2-2, and 5-2-2-2, respectively (Figs. 2, 3). Recently, one new non-A rotavirus was isolated from pigs (Chasey et al. 1986; Pedley et al. 1986) and two from chickens (McNulty et al. 1984b; Todd and McNulty 1986), which were designated as serogroups E, F and G, respectively. Representative samples for electropherogroups A–D and F are shown in Figs. 2 and 3. These figures illustrate that major migrational differences for viruses within the same serogroup are conserved, but minor variations for each virus strain within a serogroup are evident.

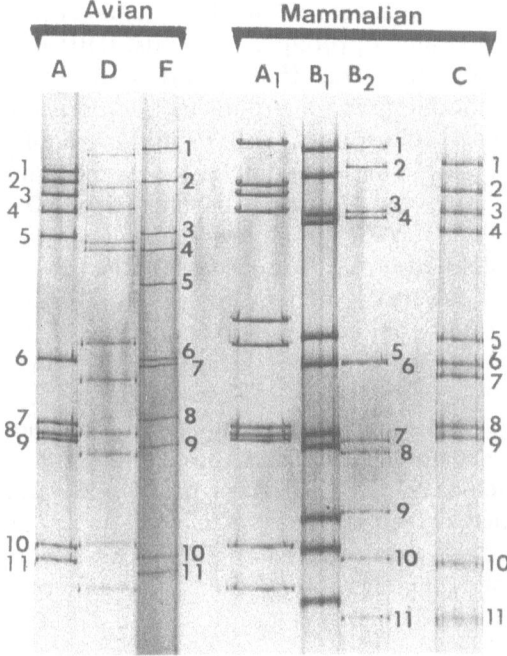

Fig. 3. Comparison of double-stranded (ds) RNA electropherotypes of avian and mammalian rotaviruses in polyacrylamide gel electrophoresis. Migration is from *top* to *bottom*. *Numbers* to the *left* indicate the segments of a reference avian group A rotavirus genome while the segments of reference non-A rotaviruses are indicated by *numbers* to the *right*. Lanes designated *Avian* (*left* to *right*): *A*, avian group A rotavirus, turkey Q strain; *D*, avian group D rotavirus, turkey RVLV strain; *F*, avian electropherogroup F rotavirus, turkey ATR strain, *Mammalian lanes*: *A₁*, porcine group A rotavirus, G strain; *B₁*, porcine group B rotavirus, N338 strain; *B₂*, bovine group B rotavirus, ATI strain; and *C*, porcine group C rotavirus, Cowden strain.

Although genome profile analysis is a relatively simple, practical technique that can be used to detect non-A rotaviruses, some non-A rotaviruses (serogroups B and E) have similar electropherotypes but are antigenically distinct. Therefore, electropherotyping should not be the only criteria for rotavirus grouping.

Single-dimension terminal fingerprint analysis and nucleic acid hybridization-based assays have been additional approaches to distinguish non-A rotaviruses from group A rotaviruses at the genomic level. In the group A rotaviruses, the absolutely conserved sequence at both terminals (eight to ten nucleotides) of the genomic RNA segments and a segment-specific conservation of G-nucleotide position at the internal region (35–45 nucleotides relative to the terminal) were observed (CLARKE and MCCRAE 1983; MCCRAE and MCCORQUODALE 1983). These conserved sequences are diagnostic for a particular RNA segment, irrespective of electrophoretic mobility shifts or species of origin of the virus. In non-A rotaviruses, a short region of absolute terminal sequence conservation was reported across all of the genomic segments of human and porcine group C rotaviruses (PEDLEY et al. 1986; MCCRAE 1987). However, a similar terminal sequence conservation was not observed for three strains of group B rotavirus, IDIR (rat), ADRV (human), and NIRD-1 (pig), and, therefore, group B rotaviruses could not be detected based on the terminal fingerprint technique (EIDEN et al. 1988). Nevertheless, PEDLEY and colleagues (1983, 1986) have applied terminal fingerprint techniques for grouping non-A rotaviruses and found that all of the genomic RNAs from four non-A rotavirus isolates (B/NIRD-1, C/Cowden or C/37030, D/132, and E/DC-9) differed in sequence from one another and also from well-defined group A rotaviruses. These results were consistent with serogroup analysis by indirect immunofluorescence and genome profile analysis by electropherogrouping; accordingly, five groups (A–E) of rotaviruses were proposed.

More detailed sequence homology within or between groups of rotaviruses can be analyzed by dot-blot or northern blot hybridization. Compared with the intensive labor and considerable technical expertise required for terminal fingerprint analysis, nucleic acid hybridization-based assays are more rapid and can simultaneously screen large numbers of virus isolates for the presence of particular viruses. Using this technique, cross-hybridizations based on high levels of sequence homology occur only within the same group of rotaviruses (MCCRAE 1987; EIDEN et al. 1986, 1989; NAGESHA et al. 1988).

Studies on nucleic acid sequences of group A rotaviruses have been extensive and the sequence of the 11 genes has been determined, while research on this area is still in its infancy for non-A rotaviruses. Recently, the entire third genomic segment of the IDIR strain of group B rotavirus was sequenced. This gene is 2309 bp in length and contains a single open reading frame of 2253 base pairs (bp) (SATO et al. 1989). As expected, the terminal sequences of this gene 3 were distinct from the well-conserved terminal sequences of group A rotavirus. Furthermore, the nucleic acid sequence of the IDIR gene 3 and the deduced polypeptide sequence were not complementary to its group A counterpart or the other group A rotavirus genes. CHEN et al. (1990) cloned and sequenced the eleventh gene of a human strain of group B rotavirus (ADRV). This gene contains one open reading frame and is 631 bp in length. It was predicted to encode a 19.9 kDa protein (170 amino acids) with a pI of 6.2. The protein of ADRV gene 11 showed distant amino acid sequence homology (25% identical amino acids) compared with the NS26 protein of a group A rotavirus.

Genomic reassortment has been documented to occur during in vivo and in vitro coinfection with distinct strains of group A rotaviruses (MIDTHUN et al. 1985, 1987; GOMBOLD and RAMIG 1986). It would be of great interest if the interchange of genetic materials between group A and non-A rotaviruses also occurs. With regard to this, YOLKEN et al. (1988a) used the infant rat model to investigate the possible generation of cultivable reassortments between the MMU (serotype 3) strain of group A rotavirus and the IDIR strain of group B rotavirus after coinfection of rats with these two viruses. Using ELISA, serum neutralization, PAGE, and DNA-RNA hybridization, the investigators were unable to demonstrate reassortment between the two rotavirus groups. Further information on the interchange of segments between group A rotaviruses and other non-A rotaviruses is lacking.

Limited research has been conducted on the in vitro transcription of non-A rotaviruses and the only information available in this regard pertains to a human group C rotavirus. Unlike the conventional rotaviruses, JASHES et al. (1986) found that the transcriptase of this non-A rotavirus was not activated by chelating agents such as EDTA or by thermal shock. In addition, enzymatic activation required the absolute presence of the ribonucleoside triphosphates and Mg^{2+} during the thermal shock.

5 Viral Proteins

The difficulty in growing non-A rotaviruses in vitro has hampered the study of viral proteins. To date only proteins of group B and C rotaviruses have been partially or fully identified (Table 1). Recently, FANG et al. (1989a) purified a human group B rotavirus (ADRV) from fecal specimens and identified three proteins (64 kDa 61 kDa and 41 kDa) in the outer shell, one protein (47 kDa) in the inner shell, and two proteins (136 kDa and 113 kDa) in the core (Table 1). The 41 kDa protein was glycosylated as demonstrated by endoglycosidase H treatment.

The structural proteins reported for human and porcine group C rotaviruses were similar. Six structural polypeptides, ranging in molecular weight from 35 000 to 93 000, were identified from purified human group C rotavirus

Table 1. Comparative composition of polypeptides of group A, B, and C rotaviruses[a]

	Polypeptides (kDa)		
	Group A (SA11)	Group B (ADRV)	Group C (Cowden)
Inner capsid and core	125 (VP1)	136	125
	94 (VP2)	113	93
	88 (VP3)		74
	41 (VP6)	47	41
Outer capsid	88 (VP4)	64	
	60 (VP5)[b]	61	
	28 (VP8)[b]		
	37 (VP7)	41	37
			33
			25[c]
NS	53	?	39
	35		35
	34		
	28		
	26 (VP9)		
	20 (VP10)[d]		22[c]
Modification	37[e]	41[f]	37[e]
	28[e]		
	26[g]		25[e]

[a] Cited from ERICSON et al. (1982); FANG et al. (1989a); JIANG et al. (1989, 1990); LIU et al. (1988); MASON et al. (1980, 1983); and WELCH et al. (1989).
[b] Cleaved products of VP4.
[c] Tentative assignments.
[d] Precursor of NS28.
[e] Glycosylated.
[f] Glycosylated determined only by endoglycosidase H assay.
[g] Phosphorylated.

(ESPEJO et al. 1984). Two (93 kDa and 44 kDa) of these were associated with the inner shell and the other four with the outer shell. BREMONT and colleagues (1988) reported that the structural polypeptide composition of a porcine group C rotavirus was similar to that of a group A rotavirus, with two structural proteins (52 kDa and 39 kDa) in the outer shell as judged by the ability of EDTA to remove them from the virion. One glycoprotein (39 kDa, analogous to VP7 of group A rotavirus based on molecular mass) was defined based on its sensitivity to endoglycosidase F treatment. The polypeptides of a porcine group C rotavirus were further characterized in our laboratory (JIANG et al. 1989, 1990). By analyzing purified virions, at least six structural proteins were identified (Table 1). Of these, the 37 kDa, 33 kDa and possibly 25 kDa were located in the outer shell and the other four (125 kDa, 93 kDa, 74 kDa and 41 kDa) in the inner shell as demonstrated by their loss after treatment with EDTA and $CaCl_2$, respectively. Glycosylation inhibition studies with tunicamycin in infected cells demonstrated that the 37 kDa and 25 kDa proteins were glycosylated and contained mannose-rich oligosaccharides, identified by radiolabeling with [^{3}H] mannose. Two proteins (39 kDa and 35 kDa) were nonstructural as determined by comparing the viral protein composition of radiolabeled, infected cell lysates with the proteins of purified virions. A kinetic study indicated that the maximal synthesis of viral polypeptides occurred at 6–9 h postinfection. The 41 kDa protein was the most abundant and had an oligomeric (possibly tetrameric) structure.

As summarized in Table 1, although the polypeptides of non-A rotaviruses are similar to those of group A rotaviruses, variations in the size, location, and post- or cotranslational modifications were observed. Additional studies are needed to identify the gene coding assignments for these proteins and determine their functions.

6 In Vitro Propagation

The non-A rotaviruses are fastidious in their in vitro growth requirements and until now only a few non-A rotaviruses have been adapted to serial passage in cell culture. They include porcine group C rotavirus (Cowden strain) and avian group D rotavirus (strain 132) (McNULTY et al. 1981; TERRETT and SAIF 1987; SAIF et al. 1988). A single cycle of replication of a porcine group B and an unclassified ovine non-A rotavirus (strain 761) in cell cultures (MA104) could only be detected by immunofluorescence and attempts to serially propagate these viruses were unsuccessful (THEIL and SAIF 1985; CHASEY and BANKS 1986). Multinucleate syncytia or focus formations occurred in cells infected in vitro by porcine group B or the unclassified ovine non-A rotavirus (CHASEY and BANKS 1986; HALL 1987).

By treating the inoculum with trypsin (5 µg/ml), McNULTY et al. (1981) were the first to propagate a non-A rotavirus (avian group D rotavirus) through 19 serial passages in primary chick embryo liver cells. However, the only mammalian non-A rotavirus successfully adapted to serial propagation in both primary cells and a continuous cell line (MA104) is the porcine group C rotavirus (TERRETT and SAIF 1987; SAIF et al. 1988). Key factors which were essential for the in vitro adaptation of non-A rotaviruses included: the initial use of primary cell cultures; use of roller tubes; and addition of proteolytic enzymes to the inoculum and the maintenance medium. Prior passage of the virus in primary porcine kidney cells seemed to be crucial for its subsequent adaptation to MA104 cells. Consideration should also be given to the concentrations and types of enzyme used. In our laboratory, trypsin and pancreatin were used to treat virus-inoculated cells and both enzymes facilitated the growth of group C rotaviruses

(TERRETT et al. 1987). However, pancreatin enhanced the infectivity of group C rotavirus greater than trypsin and was therefore chosen for routine use. Attempts to adapt group C rotavirus to serial passage in stationary cultures failed; however, low speed centrifugation prior to incubation appeared to facilitate virus growth. In addition, selection of a sensitive assay and the origin of cells used may affect the detection or growth of non-A rotaviruses. Recently, TSUNEMITSU et al. (1991) isolated a group C rotavirus (Shintoku strain) from cattle and adapted this virus to cultivation in MA104 cells by using a high concentration of trypsin in the inoculum.

7 Antigenic and Nucleic Acid Relationships

7.1 Antigenic Cross-Reactivity

7.1.1 In Vitro

The limited availability and small quantities of stool specimens from infected hosts, the present lack of an animal model for passage and amplification of human isolates, and the failure of most non-A rotaviruses to grow in cell culture has greatly hindered investigation of these viruses, including their antigenic relationships. Such problems have precluded use of many conventional serologic tests for examining the antigenic relatedness of non-A rotaviruses. However, the following assays have been adapted to detect and differentiate antigens common to each of the respective rotavirus serogroups (A–F): immunofluorescence (IF); immune electron microscopy (IEM); and ELISA (PEDLEY et al. 1983; SNODGRASS et al. 1984; MCNULTY et al. 1984b; SAIF and THEIL 1985; BRIDGER 1987; SAIF 1990b). Antigen sources for these tests included feces, intestinal contents, or small intestinal enterocytes from infected hosts. Antisera used in these tests were prepared in gnotobiotic or specific pathogen free (SPF) animals or were paired sera from recovered individuals.

Confirmation of the lack of reciprocal cross-reactivity between serogroup A–F rotaviruses has been reported in several reviews (PEDLEY et al. 1983; SAIF and THEIL 1985; BRIDGER 1987; SAIF 1990b). Non-A rotaviruses from different animal species can belong to the same serogroup, analogous to group A rotaviruses (SAIF 1990b). At present serogroup B rotaviruses have been confirmed in swine, cattle, humans, rats and sheep (PEDLEY et al. 1983; SNODGRASS et al. 1984; SAIF and THEIL 1985; BRIDGER et al. 1986; EIDEN et al. 1986; SAIF 1990b). Only one-way antigenic cross-reactions between the UK bovine (D522) and ovine (E1101) group B rotaviruses and UK porcine group B rotavirus (NIRD-1) were reported (SNODGRASS et al. 1984); however, two-way cross-reactions were detected between

Table 2. Summary of group B and C rotaviruses detected in humans

Country (region)	Year	Report[a]	Source (age)	Number of positive cases/number examined[b]
Group B rotaviruses				
China (Jinzhou and Lanzhou)	1982–83	Hung et al. (1984)	Children/adults	101/106 (5942/607,720)[b]
China (Guangxi)	1983–84	Wang et al. (1985)	Children/adults	15/34
China (Anhui)	1983	Su et al. (1986)	Infants/adults	13/32
China (Qinhuangdao)	1987	Fang et al. (1989b)	Children/adults	7/13 (75/487)[b]
Group C rotaviruses				
Australia	1981	Rodger et al. (1982)[a]	Child (20 months)	1/400
Netherlands	1981	Buitenwerf et al. (1983)[a]	Child (NR)	1/123
Brazil	1983	Pereira et al. (1983)	Infant	NR
France	1981	Nicolas et al. (1983)[a]	Child (2 years)	1/NR
	1984	Nicolas et al. (1983)[a]	Infants	3/1028
Bulgaria	1983	Dimitrov et al. (1983)[a]	Infant (4–5 months)	NR
	1986	Dimitrov et al. (1986)	Infants	5/691
Mexico	1984	Espejo et al. (1984)[a]	Child (2 years)	1/658
Italy	1985	Arista et al. (1985)[a]	Infant (4 months)	NR
Switzerland	1985	Breer et al. (1985)[a]	Adult	1/NR
Chile	1986	Jashes et al. (1986)[a]	Infant (6 months)	NR
Argentina	1986	Sorrentino et al. (1986)[a]	Child (1 year)	1/NR
Hungary	1987	Szucs et al. (1987)[a]	Children	2/105
China	1987	McCrae (1987)	Children/adults	60/NR
India	1988	Brown et al. (1988)	Children	4/NR
Finland	1988	Bonsdorff and Svensson (1988)	Children (4–7 years)	6/600
England	1989	Penaranda et al. (1989a)	Adult	1/NR
Nepal	1989	Penaranda et al. (1989a)	Child (8 years)	1/NR
Thailand	1989	Penaranda et al. (1989a)	Children/adult	4/NR
Japan	1985	Ushijima et al. (1989)	Child	
	1987	Ushijima et al. (1989)	Child (2 years)	4/572
	1988	Ushijima et al. (1989)	Infant (6 months)	
Japan	1988	Matsumoto et al. (1989)	Children (1, 9 years) Children/adults[b]	16/20 (675/3102)[b]

NR, not reported.
[a] Rotaviruses in these reports were classified only on the basis of electropherogrouping and their serogroup has not been confirmed.
[b] Numbers in parenthesis represent total number cases/total number at risk.

other group B rotaviruses (SAIF and THEIL 1985; CHEN et al. 1985; EIDEN et al. 1986; SAIF 1990b). To date only rotaviruses from pigs share the group C antigen with rotaviruses identified in humans in Brazil, Australia, and the UK (BRIDGER 1987; SAIF 1990b); Finland (BONSDORFF and SVENSSON 1988); Nepal and Thailand (PENARANDA et al. 1989a); and Japan (USHIJIMA et al. 1989; MATSUMOTO et al. 1989). However, the occurrence of electropherogroup C rotaviruses has been reported in humans in other countries (Table 2) and in ferrets (TORRES-MEDINA 1987). Preliminary evidence exists for a serogroup C rotavirus from cattle (TSEUNIMITSU et al. 1991). Serogroup E rotaviruses were identified only in pigs (CHASEY et al. 1986), while serogroup D, F, and G rotaviruses were reported only in poultry (MCNULTY et al. 1984b; SAIF et al. 1985; TODD and MCNULTY 1986; BRIDGER 1987).

7.1.2 In Vivo Cross-Infections and Protection

As for group A rotaviruses, non-A rotaviruses can infect more than one animal species under experimental conditions. This has been demonstrated with ovine and bovine group B rotaviruses which infected and caused diarrhea in gnotobiotic piglets and lambs (SNODGRASS et al. 1984; SAIF and THEIL 1985; SAIF et al. 1987). In the latter two studies, no cross-protection was evident between a bovine and porcine group B rotavirus in piglets, suggesting possible serotypic differences between non-A rotaviruses in the same serogroup. Lack of cross-protection between rotaviruses from distinct serogoups has also been confirmed in studies in swine (BOHL et al. 1982; BRIDGER et al. 1982). However at present, no one has succeeded in infecting piglets with human group B or C rotaviruses, suggesting that not all non-A rotaviruses are capable of cross-infections between species or the virus titers in the specimens used were too low to initiate infection (SNODGRASS et al. 1984; NAKATA et al. 1986; BRIDGER et al. 1986).

7.1.3 Nucleic Acid Homologies

Comparisons of nucleic acid homologies among the various rotavirus groups have been done by dot-blot or northern blot hybridization using cDNA probes of genomic segments or whole genomic RNA probes (MCCRAE 1987). To date, only limited cross-hybridization studies have been conducted, largely due to the limited availability of dsRNA or cDNA probes, particularly from human non-A rotavirus isolates. The following results have been reported using whole genomic RNA or cDNA probes. In the first study, a ^{32}P-labeled dsRNA probe prepared from an Australian human group C rotavirus hybridized with a Brazilian human group C rotavirus and two porcine group C rotaviruses (Cowden and S strains) by dot-blot hybridization (70% stringency), but only slight cross-reactions were seen with group A and B rotavirus RNA (BRIDGER et al. 1986). In a second study, a probe made against the Chinese (ADRV) group B rotavirus hybridized with all genome segments of a group B porcine rotavirus (NIRD-1) but not with group A or C rotaviruses (CHEN et al. 1985; MCCRAE 1987). Finally ^{32}P-labeled cDNA probes prepared against genomic RNAs of group A, group B (rat and bovine), and group C (porcine) rotaviruses were compared by northern blot hybridization for genome sequence homology with homologous and heterologous rotaviruses (EIDEN et al. 1986; MCCRAE 1987). In general, each probe hybridized only to viruses in the same serogroup and not to rotaviruses in distinct serogroups. An exception was the hybridization of group B and C rotavirus probes to genome segment 1 of group A rotaviruses. Although probes to group B rotaviruses cross-hybridized with all group B rotaviruses examined (rat, human, bovine, and porcine), the number of segments which hybridized between porcine and rat (eight) or bovine (ten) group B rotaviruses was greater than between rat and bovine group B rotaviruses (four). Such results suggest that these latter two rotaviruses may be more distantly related to each other, but this needs to be confirmed by sequence analysis.

A cloned cDNA probe from gene 3 of rat rotavirus (IDIR) hybridized to other group B rotaviruses (bovine and human) in a dot hybridization assay but not to group A rotaviruses (EIDEN et al. 1989). Similarly, a cDNA probe from gene 8 of a human group B rotavirus (ADRV) hybridized to the homologous virus and to porcine group B rotaviruses, but not to porcine group A or C rotaviruses (NAGESHA et al. 1988). Further generation of cloned cDNA probes for each of the genomic segments of the different rotavirus groups should permit more precise analysis of the sequence relationships between the corresponding genome segments of different rotavirus isolates.

8 Diagnosis

The failure of most non-A rotaviruses to grow in cell culture has hampered development of reagents needed for routine serologic diagnostic tests. At present, diagnosis relies on techniques similar to those initially used to detect fastidious group A rotaviruses. These include examination of fecal and intestinal samples by EM, IF staining of small intestinal sections or inoculated cell cultures, and ELISA of fecal or rectal swab fluids. Detection of viral nucleic acid in feces has been accomplished by polyacrylamide gel electrophoresis or use of hybridization probes.

Because non-A rotaviruses are morphologically identical to group A rotaviruses, they cannot be differentiated using routine negative staining EM techniques. However samples which are positive for rotavirus by EM, but negative for group A rotaviruses by other serologic tests (specific for group

A common antigens) may contain non-A rotaviruses. Further tests are necessary to confirm this possiblity and identify the rotavirus serogroups. One such test, IEM or solid phase IEM (SPIEM) using rotavirus serogroup-specific antisera, may be adapted to detect and distinguish rotavirus sero-groups (SAIF and THEIL 1985; CHEN et al. 1985; SAIF et al. 1985; NAKATA et al. 1986; MATSUMOTO et al. 1989; PENARANDA et al. 1989a; USHIJIMA et al. 1989; SAIF 1990b).

Both IF and ELISA have been used to detect non-A rotaviruses using hyperimmune antisera or monoclonal antibodies (MAbs) which recognize group-specific antigens common to each virus serogroup (MCNULTY et al. 1981, 1984b; SAIF and THEIL 1985; THEIL and SAIF 1985; VONDERFECHT et al. 1985; EIDEN et al. 1986; NAKATA et al. 1986; WANG et al. 1986; BRIDGER 1987; YOLKEN et al. 1988b; OJEH et al. 1991; BURNS et al. 1989; SAIF 1990b). The consensus of these studies and results of IEM suggest that homologous reactions (hyperimmune antiserum produced to homo-logous virus) were always stronger than heterologous reactions (hyper-immune antiserum produced against the same rotavirus serogroup from another species). Moreover, convalescent antisera reacted mainly with homologous virus strains and one-way antigenic reactions were observed, suggesting species or type-specific antigens also exist among non-A rota-viruses. Such findings demonstrate that the reactivity of various antisera must be carefully characterized prior to their routine use in diagnostic assays designed to detect particular rotavirus serogroups from various host species. Recently MAbs reactive with group-specific antigens of group B and C rotaviruses were produced (OJEH et al. 1988; YOLKEN et al. 1988b; BURNS et al. 1989). The MAbs generated to the IDIR group B rat rotavirus were broadly reactive with all strains of group B rotavirus tested including bovine, porcine, and human (YOLKEN et al. 1988b). Whereas the MAbs produced to a human strain of group B rotavirus (ADRV) were reactive with all human Chinese group B strains tested, their reactivity against animal group B rotaviruses was not characterized (BURNS et al. 1989). Similarly MAbs produced against the Cowden strain of porcine group C rotavirus were reactive with four other porcine strains of group C rotavirus, but their reactivities with human group C rotaviruses were not tested since no specimens were available (OJEH et al. 1988). Future availability of well-characterized immunoreagents such as MAbs for use in ELISA or IF tests should greatly facilitate diagnosis of non-A rotavirus infections in humans and animals.

The diagnosis of non-A rotavirus infections based on seroconversion has been accomplished using infected gut sections or cell cultures, or virus extracted from feces as antigen sources in IF, ELISA, or IEM. Because the prevalence of antibodies to non-A rotaviruses in humans is low, with the exception of epidemic areas of China, this method may be useful for diagnosis of human infections. However since a high percentage of certain species of animals possess antibodies to particular non-A rotaviruses (MCNULTY et al. 1984a; BRIDGER and BROWN 1985; THEIL and SAIF 1985; TERRETT et al. 1987; BROWN et al. 1987; NAGESHA et al. 1988), this techniques may not provide definitive evidence of recent infection. Furthermore, use of hetero- logous antigens in the antibody tests may decrease the sensitivity of such assays (NAKATA et al. 1986; YOLKEN et al. 1988b; MATSUMOTO et al. 1989; PANARANDA et al. 1989a).

Although it lacks serologic specificity, RNA electropherotyping is useful to detect and differentiate non-A rotaviruses. The migration pattern of rotavirus dsRNA segments, although slightly different for rotaviruses within each group, is distinctive for each electropherogroup (Figs. 2, 3; PEDLEY et al. 1983; SNODGRASS et al. 1984; SAIF and THEIL 1985; SAIF et al. 1985; TODD and McNULTY et al. 1986; McCRAE et al. 1987; HUNG et al. 1987). Thus electropherotyping may provide a tentative classification of non-A rotaviruses, pending final serologic definition of serogroups. Use of RNA electropherotyping may be especially helpful for detecting mixed infections with group A and non-A rotaviruses (not uncommon in swine or poultry) as evident by additional RNA bands in the gels.

As discussed in the section on nucleic acid relationships, hybridization probes consisting of genomic RNA or cDNA have been used to detect and differentiate non-A rotaviruses from humans, swine, cattle, and rats (CHEN et al. 1985; BRIDGER et al. 1986; EIDEN et al. 1986; McCRAE 1987; HUNG et al. 1987). However the specificity and sensitively of these probes for the routine detection of non-A rotaviruses from fecal specimens has not been evaluated. More recently a mAb assay for the detection of dsRNA viruses was devised, which when coupled with nucleic acid hybridization probes, permitted detection and differentiation of group A and B rotaviruses (KINNEY et al. 1989).

9 Epidemiology

9.1 Human Infections

Among non-A rotaviruses, only group B and C rotaviruses have been found in association with diarrheal disease in humans. As summarized in Table 2, group B rotaviruses have been detected from children and adults in China since 1984 (HUNG et al. 1983; CHEN et al. 1985). However, antibodies to group B rotaviruses have been reported in Hong Kong, Burma, Thailand, Australia, the USA, Canada, the UK, and Kenya, although their prevalence is much lower than in adults in epidemic areas of China (Table 3). In contrast, group C rotaviruses are more widely distributed and have been found in children or adults from the following geographical locations: Asia, Australia, Europe, and Central and South America (Table 2).

Table 3. Prevalence of antibody to serogroup A, B and C rotaviruses in human sera

Country	Year of collection	Report	Source (age)	Total no. sera tested	Rotavirus serogroup/host virus[a] (% positive)		
					A/simian	B/human	C/porcine
China Shanghai / Lanzhou	1977–1986	PENARANDA et al. (1989b)	adult glob. pool	10	100	100	ND[b]
	1983	PENARANDA et al. (1989b)	adult glob. pool	1	100	0	ND
Wuhan	1984–1987	PENARANDA et al. (1989b)	adult glob. pool	4	100	100	ND
	1984–1987	PENARANDA et al. (1989b)	adult glob. pool	4	100	100	ND
Chandu	1986	PENARANDA et al. (1989b)	adult glob. pool	1	100	0	ND
USA	1986	PENARANDA et al. (1989b)	adult glob. pool	3	100	0	ND
	1988	PENARANDA et al. (1989b)	adult	8	ND	0	ND
Burma	1980–1981	PENARANDA et al. (1989b)	adult	6	ND	17	ND
	1987	PENARANDA et al. (1989b)	adult	6	ND	17	ND
China	NR[b]	HUNG et al. (1987)	adults	249	ND	12–41	ND
Hong Kong	NR	HUNG et al. (1987)	adults	34	ND	18	ND
Australia	NR	HUNG et al. (1987)	adults	40	ND	15	ND
USA	NR	HUNG et al. (1987)	adults	202	ND	10	ND
Canada	NR	HUNG et al. (1987)	adults	0	ND	13	ND
Australia	1981–1983	NAKATA et al. (1987a)	adult	10	100	0	ND
Canada	1983–1985	NAKATA et al. (1987a)	adult	10	90	10	ND
Kenya	1985	NAKATA et al. (1987a)	adult	10	100	10	ND
Thailand	1984–1985	NAKATA et al. (1987a)	adult	20	100	5	ND
USA	1984–1986	NAKATA et al. (1987a)	adult	107	97	2	ND
USA	1984–1986	NAKATA et al. (1987a)	children	57	88	0	ND
U.K.	1983–1984	BROWN et al. (1987)	veterinarians	110	ND	4	ND
U.K.	1986	BROWN et al. (1987)	adults	100	ND	10	ND
					A/bovine	B/porcine	
U.K.	pre 1984	BRIDGER (1987)	adults	38	95	3	11
U.K.	1982	BRIDGER (1987)	children	15	93	0	0

Country	Year	Reference	Population	No.	A	B	C
Japan	1989–1992	Tsunemitsu et al. (1992)	adults	78	ND	ND	13
			children	53	ND	ND	3
USA	NR[b]	Eiden et al. (1985)	adults	33	ND	88 (B/rodent)	ND
USA	NR	Eiden et al. (1985)	children (1–5 y)	120	ND	63	ND
Mexico	NR	Espejo et al. (1984)	adults	12	100	ND	42 (C/human)
Mexico	NR	Espejo et al. (1984)	children	5	ND	ND	0
Japan	1988	Matsumoto et al. (1989)	children/adults (diarrhea)	13	ND	ND	85
Japan	1988	Matsumoto et al. (1989)	children/adults (healthy)	4	ND	ND	0

[a] Virus antigen used for immunofluorescence, immune electron microscopy, counterimmunoelectrophoresis or ELISA included A/simian (SA-11) or A/bovine (UK); B/porcine (NIRD-1) or B/human (ADRV); C/porcine (Cowden) or C/human.
[b] ND = not determined; NR = not reported.

Chinese ADRV is the prototype for human group B rotaviruses (CHEN et al. 1985). Because large-scale outbreaks of diarrheal disease with high morbidity occurred in certain parts of China from 1982 to 1983, extensive epidemiological studies of the disease were conducted. People in all age groups were susceptible to ADRV, but adults age 20–40 years were the highest risk group (HUNG et al. 1987; FANG et al. 1989b). Both males and females were susceptible. Although the pattern of disease outbreaks was influenced by ambient temperatures, a seasonal epidemic of the disease was not apparent.

The prevalence of antibody to ADRV in the healthy adult population and gamma globulin pools was screened by various methods, including counter-immunoelectrophoresis (CIE) and ELISA (HUNG et al. 1985; QIU et al. 1986). Antibodies to ADRV were prevalent principally in mainland China (Table 3), although seropositive samples were detected in other locations. Recent studies further demonstrated the low prevalence of antibodies to group B rotaviruses among human populations outside China, suggesting that human infections with group B rotavirus have not been widespread in areas other than China (Table 3). To determine whether ADRV was a new virus introduced during the outbreak of the 1982 epidemic or if it existed before that time, PENARANDA et al. (1989b) investigated seroepidemiology using gamma globulin pools prepared during 1977-1987 in four cities of China and found that antibodies to ADRV were present in most Chinese gamma globulin pools tested, indicating that the causative agent was present before the first epidemic in 1982 (Table 3).

Cross-species infections with group B rotaviruses between rats and humans have been proposed. EIDEN et al. (1985) reported that a group B rotavirus of rats (IDIR) caused diarrheal disease in humans. Seropositivity increased with increasing age, suggesting that the virus could also infect older children and adults. This finding agrees with the subsequent discovery that house rats in China were the most frequent animals infected with ADRV based on antibody seroprevalence (HUNG et al. 1987).

There has been limited research on the epidemiology of human group C rotaviruses and the prevalence of antibodies against group C rotaviruses is unclear. Many of the antibody surveys have been conducted using porcine group C rotavirus as antigen (Table 3, BRIDGER et al. 1986). Using immuno-fluorescence with this antigen, no antibodies to group C rotavirus were detected in immunoglobulin pools from the UK, Europe, North America, and Japan in a study reported in 1986, whereas antibodies to group A rotavirus were consistently identified (BRIDGER et al. 1986). A serologic survey of group C rotavirus antibodies in children and adults in the UK before 1984 indicated that 93% or more were positive for group A rotavirus in all age groups, whereas 0% and 11% were positive for group C rotavirus in children and adults, respectively (Table 3).

Recent outbreaks of gastroenteritis caused by group C rotavirus occurred in Japan (USHIJIMA et al. 1989) and in one study were characterized by a large-scale, sudden onset, and short duration (MATSUMOTO et al. 1989). In the latter outbreak, both schoolchildren and teachers were simultaneously infected, suggesting that group C rotaviruses may also be a major threat to adults. By EM, double-shelled virus particles were detected in samples of the patients. Antibodies to the rotavirus were found in convalescent serum samples but not in acute serum samples (Table 3).

As with other viral infections, environment also affects the spread or scale of the disease caused by non-A rotaviruses. The epidemics of acute diarrhea in China during 1982 and 1983, for example, were thought to be spread initially by fecally contaminated water (HUNG et al. 1984). In addition, close personal contact was also documented to play a role in the transmission of the disease (GRAHAM and ESTES 1985; HUNG 1988; FANG et al. 1989b). The epidemic of gastroenteritis in schoolchildren and teachers in Japan suggested that this group C rotavirus infection may have been introduced through the food-brone route (MATSUMOTO et al. 1989).

9.2 Animal Infections

Infections with non-A rotaviruses have been described in swine, poultry, cattle, sheep, rats, and ferrets in the USA, South America, Europe, Asia, and Australia (BRIDGER 1987; SAIF 1990b). Only limited information is available on the prevalence of non-A rotavirus infections in cattle, sheep, and rats.

Based on serologic evidence (summarized in Table 4) or the detection of virus, group B rotavirus infections occurred in rats and swine in the USA and China (VONDERFECHT et al. 1984; THEIL and SAIF 1985; HUNG et al. 1987) and swine, sheep and cattle in the USA and UK (SAIF et al. 1982; SNODGRASS et al. 1984; BRIDGER and BROWN 1985; BRIDGER 1987; BROWN et al. 1987; SAIF 1990b). In one study based on electropherotyping, non-A rotavirus infections (electropherogroup B) were more common in lambs in England than group A rotavirus infections (CHASEY and BANKS 1984) and a high percentage of sera from sheep and cattle in the UK had antibodies to group B rotavirus (BROWN et al. 1987; Table 4). Antibodies to group C rotaviruses have been detected in sera from adult cattle in the USA and UK (Table 4, 24%–40% positive), although no group C rotaviruses have been isolated from cattle in these two countries (BRIDGER 1987; JIANG and SAIF, 1989, unpublished data). However, a group C rotavirus was described in 1989 in cattle with diarrhea in Japan (TSUNEMITSU et al. 1991). At present, only group B rotaviruses have been identified in rats.

More extensive studies of antibody prevalence and shedding of non-A rotaviruses have been conducted in swine and poultry, the two species from which the most serogroups of non-A rotaviruses have been described. Serogroup B, C, and E rotaviruses have been identified in swine with detection of group E rotaviruses restricted to swine. There are few studies of antibody prevalence to group E rotavirus (CHASEY et al. 1986; NAGESHA

Table 4. Prevalence of antibody to serogroup A, B and C rotaviruses in sera from animals

Country	Year of sera collection	Report	Source (age)	Total no. sera tested	Rotavirus serogroup/host virus[a] (% positive)		
					A/porcine	B/porcine	C/porcine
USA	1984–1987	THEIL and SAIF (1985)	adult pigs	37	100	24	ND[b]
USA	1984–1987	THEIL and SAIF (1985)	young pigs (3–8 wks)	7	100	17	ND
USA	1984–1987	TERRETT et al. (1987)	adult pigs	68	100	ND	100
USA	1984–1987	TERRETT et al. (1987)	young pigs (0–8 wks)	69	100	ND	70
USA	1988–1992	TSUNEMITSU et al. (1992)	adult pigs	68	NR	NR	97
Japan	1988–1992	TSUNEMITSU et al. (1992)	adult pigs	80	NR	NR	93
USA	1988–1992	TSUNEMITSU et al. (1992)	adult cattle	32	NR	NR	47
Japan	1988–1992	TSUNEMITSU et al. (1992)	adult cattle	50	NR	NR	56
USA	1988–1989	JIANG and SAIF[b]	adult cattle	28	71	ND	40
UK	1981	BRIDGER and BROWN (1985)	adult pigs	39	A/bovine 97	92	79
UK	1981	BRIDGER and BROWN (1985)	young pigs (3–8 wks)	43	100	70	58
Australia	1988	NAGESHA et al. (1988)	adult pigs	12	100	58	100
UK	1973–1983	BRIDGER (1987)	adult cattle	118	91	20	24
UK	1981	BRIDGER (1987)	adult sheep	50	43	19	0
UK	1983	BROWN et al. (1987)	adult pigs	67	ND	B/human 97	ND
UK	1983	BROWN et al. (1987)	adult cattle	59	ND	71	ND
UK	1983	BROWN et al. (1987)	adult sheep	11	ND	91	ND
China	NR[b]	HUNG et al. (1987)	adult pigs	202	ND	36	ND
China	NR	HUNG et al. (1987)	house rats	60	ND	47	ND
China	NR	HUNG et al. (1987)	Wistaer rats	35	ND	17	ND
China	NR	HUNG et al. (1987)	cattle/sheep/horses	46/29/5	ND	0	ND

[a] Virus antigens used for immunofluorescence, immune electron microscopy, ELISA or counterimmunoelectrophoresis included: A/porcine (OSU) or A/bovine (U.K.) or A/simian (SA11); B/porcine (NIRD-1) or B/human (ADRV); and C/porcine (Cowden).
[b] ND = not determined; NR = not reported; Jiang and Saif 1990, unpublished data.

et al. 1988), but occurrence of antibodies to group B and C rotaviruses in swine has been the subject of investigations in the USA, UK, and Australia (Table 4). Whereas antibodies to group B rotavirus were less frequent in swine sera from one region of the USA (17%–24% positive) than in Australia (58% positive) or the UK (70%–97% positive), the incidence of antibodies to group A and C rotaviruses was similar in adult swine sera in these three countries (58%–100% positive). However antibodies to the non-A rotaviruses were less prevalent than antibodies to group A rotaviruses in younger than in older swine (Table 4). The widespread prevalence of antibodies to group C rotaviruses suggests these viruses may be enzootic in swine herds in these countries, analogous to the situation with group A rotaviruses. In contrast, the lower prevalence of antibodies to group B rotaviruses in the region of the USA surveyed may indicate this virus has a potential to cause epizootics in seronegative swine herds, similar to the epidemic caused by group B rotavirus in some segments of the Chinese population.

Group B, C, and E rotaviruses were isolated from cases of piglet diarrhea and produced diarrhea or abnormal stools in experimentally inoculated gnotobiotic pigs, suggesting these non-A rotaviruses may play a role in the etiology of diarrhea in pigs under field conditions (BOHL et al. 1982; BRIDGER et al. 1982; THEIL et al. 1985; CHASEY et al. 1986). However, at present the detection rate for non-A rotaviruses is low compared with that for group A rotaviruses. Only approximately 5% of rotaviruses identified from cases of pig diarrhea by a diagnostic lab in the UK or a study conducted on Brazilian swine were positive for non-A rotaviruses by dsRNA electropherotyping (CHASEY and DAVIS 1984; SIGOLO DE SAN JUAN et al. 1986). However, in the future, improved diagnostic techniques for non-A rotaviruses may alter this detection rate. Similar to human non-A rotaviruses, factors such as instability of the viruses, either upon storage or due to handling and processing procedures, or lower levels and shorter duration of virus shedding may greatly decrease non-A rotavirus detection rates.

The high seroprevalence rate observed for group C rotaviruses relative to their actual detection rate suggests that subclinical infections may be common in some herds, similar to group A rotaviruses. This was confirmed by a comparative study of the pattern of group A and C rotavirus excretion conducted in three closed swine breeding herds in Belgium using ELISA for antigen detection (DEBOUCK et al. 1984). Group A and C rotaviruses were enzootic in the herds examined, with peak group A and C rotavirus excretion at 3–5 weeks of age except in one herd in which group C rotavirus shedding occurred at 1–4 weeks of age. Another peak of group A rotavirus excretion was observed in nursing pigs, 3–4 weeks after the first, but pigs weaned at 6–10 weeks of age rarely excreted virus after weaning. In contrast, most pigs weaned at 4 weeks of age shed group C rotavirus at weaning although diarrhea was not noted. About 50% of fecal samples from nursing pigs with diarrhea contained group A rotavirus. However group C rotavirus excretion was associated with subclinical infections in most older pigs (3–10 weeks of age) in two herds and with diarrhea in 70% of virus-positive younger pigs (1–4 weeks of age) in the third herd.

Information on the epidemiology of non-A rotavirus infections in poultry is also availble. The prevalence of antibodies to group A and D rotaviruses in the sera of broiler-breeder chickens was similar (63% and 70% positive for group A and group D antibodies, respectively, MCNULTY et al. 1984b). Similarly in 10–21 day old diarrheic turkey poults, a RVLV was as prevalent (detected in 60% of flocks) as group A rotaviruses (detected in 58% of flocks) as assessed by IEM and electropherotyping (SAIF et al. 1984, 1985). The turkey RVLV is electropherotypically and antigenically similar to group D (A/132) chicken rotavirus (M. S. McNulty and L.J. Saif, 1988, unpublished data), but it is antigenically unrelated to rotaviruses in groups A, B, and C (SAIF et al. 1985; SAIF and THEIL 1985). Its role as a cause of diarrhea in turkeys is not well-defined, however, due to the presence of multiple enteric viruses in poults with enteritis and failure to separate these viruses for pathogenicity studies (SAIF et al. 1984).

In a subsequent study comparing the prevalence of electropherogroup A and D rotaviruses in diarrheic turkeys, 31% of rotaviruses detected were electropherogroup D and 69% were electropherogroup A (KANG et al. 1986). In another survey, electropherogroup D rotaviruses were detected most frequently in 3–4 week old turkey poults and occurred in 67% of diseased flocks but only 26% of normal flocks. By comparison, group A rotaviruses were detected in 22% of diseased flocks but 26% of normal flocks (REYNOLDS et al. 1987). A high percentage of turkey flocks examined (73%) had mixed infections with group D rotaviruses and other enteric viruses including group A rotaviruses (KANG et al. 1986; THEIL et al. 1986), reoviruses, astroviruses, adenoviruses, or enteroviruses (SAIF et al. 1985; REYNOLDS et al. 1987).

During a longitudinal survey of rotaviruses in broiler chickens, MCNULTY et al. (1984b) discovered two other serogroups of non-A rotaviruses besides groups A and D. They were designated A4 (group F) and 555 (group G). Recent studies suggest that these two avian rotaviruses are antigenically distinct from serogroup A–E rotaviruses and thus may be classified as two new serogroups F and G, respectively (BRIDGER 1987). Results of this survey in chickens further indicated that rotaviruses antigenically and RNA electropherotypically similar to group A and D rotaviruses were prevalent in birds at 22–29 days of age. Electropherogroup F (A4) rotaviruses were prevalent at 9–16 days of age, whereas between 43 and 50 days of age the electropherogroup G (555) rotavirus predominated. Thus, both group A rotavirus and non-A rotavirus infections occured in waves, with each wave of infection lasting about 1 week. However different patterns of rotavirus shedding were observed in another longitudinal survey of other broiler flocks, suggesting the patterns of rotavirus shedding may vary from flock to flock and at different sampling times (TODD and MCNULTY 1986). A second non-A rotavirus designated atypical rotavirus was recently described in diarrheic turkey poults and is electropherotypically similar to the chicken

A4, serogroup F rotavirus (Fig. 3, THEIL et al. 1986). As determined by electropherotyping it was detected in the lowest prevalence (4%) from 1-to 12-week-old turkey poults. An atypical rotavirus with an electropherotype similar to serogroup F avian rotavirus was also described in 7–10 day old chicks with diarrhea in Argentina (BELLINZONI et al. 1987). The antigenic relationship of these other avian electropherogroup F rotaviruses to the prototype avain serogroup F rotavirus (A4) has not been analyzed. The occurrence of group D, F, and G rotaviruses has not been reported in any species other than poultry.

10 Pathogenesis and Pathogenicity

Studies of the pathogenesis and pathology induced by non-A rotavirus have been reported in chickens, pigs, calves and rats. There are few reports describing the pathogenesis of human non-A rotavirus infections in the natural host or animal models. The in vivo replication of serogroup B, C, and D rotaviruses has been investigated by means of IF, immunoperoxidase labeling, transmission and scanning EM, PAGE, and dot hybridization assays. Non-A rotaviruses possess a tropism for the small intestine (HALL 1987; SAIF 1990a, b). Inoculation of SPF chickens or turkeys with group D rotaviruses (132 or RVLV, respectively; MCNULTY et al. 1981; SAIF et al. 1984) or gnotobiotic piglets with porcine group C rotavirus (BOHL et al. 1982) resulted in infection of villous enterocytes and diarrhea similar to that induced by group A rotaviruses (Table 5). However infection of gnotobiotic piglets, lambs and calves, or rats with their respective group B rotaviruses produced an acute transitory diarrhea; infection was often restricted to distinct foci of enterocytes near the villous tips or the apical third of the villi. The preferred site of replication for group B rotavirus in the pig was in cells throughout the small intestine (THEIL et al. 1985, Table 5); in the lamb, the virus replicated in cells in the mid-small intestine (CHASEY and BANKS 1984); and in the rat, in cells in the distal small intestine (VONDERFECHT et al. 1984; HUBER et al. 1989). In our experiments, infection of calves with group B rotaviruses or piglets with group B or C rotaviruses induced a more rapid onset of diarrhea but resulted in more transient shedding of fewer viral particles than group A rotavirus infections (Saif, unpublished data).

The comparative pathogenesis of porcine rotavirus infections (groups A, B, and C) in gnotobiotic piglets is summarized in Table 5. Whereas diarrhea was usually observed by 24–36 h postexposure to group A rotaviruses, with group B and often group C rotavirus infections, the onset of diarrhea usually occurred more rapidly, often by 12–18 h postexposure. Vomiting was occasionally observed after infection of piglets with group B rotaviruses.

Table 5. Comparative pathogenesis of porcine group A, B and C rotaviruses in gnotobiotic pigs and monkey kidney (MA104) cells

Rotavirus Serogroup/strain	Report	Infection of gut enterocytes					Principal site of villous atrophy	Diarrhea (onset)	Infect MA104 cells
		Vertical		Longitudinal		Extent			
		Villous	Region	D, J, I[a]	colon				
A/OSU, Gottfried	THEIL et a. (1978)	+	entire	D, J, I	±	patchy-continuous	J, I	moderate-severe (24–36 hr)	+
B/Ohio	THEIL et al. (1985)	+	tips	D, J, I	–	patchy	D, J	mild (12–24 hr)	+ (syncytia)[b]
C/Cowden	BOHL et al. (1982)	+	entire	J, I	–	patchy-continuous	J	mild-severe (18–36 hr)	+

[a] D = doudenum; J = jejunum; I = ileum.
[b] Group B rotaviruses infect and cause syncytia in MA104 cells but do not serially propagate.

Infection with group B rotavirus induced a transient mild diarrhea which persisted about 3 days but resulted in no mortality in piglets exposed at 5–6 days of age (THEIL et al. 1985). In comparison, group A and C rotaviruses caused more severe persisting diarrhea, dehydration, and frequently death in piglets under 6 days of age, but more transient diarrhea and little mortality in older piglets (THEIL et al. 1978; BOHL et al. 1982).

Recently, VONDERFECHT et al. (1988) utilized a murine model to study the kinetics of the in vivo replication of a group B rotavirus using various diagnostic methods. They found a biphasic pattern of viral antigen excretion, i.e., viral replication in the intestine peaked on days 1 and 4 postinfection. However, the actual intraepithelial replication of the virus and maximum shedding was observed at peak 1, indicating that early sample collection is critical for the diagnosis of viral infection and the maximal recovery of the virus.

A rodent (VONDERFECHT et al. 1988; HUBER et al. 1989), bovine (MEBUS et al. 1978; VONDERFECHT et al. 1986), and porcine group B rotavirus (HALL 1987; CHASEY et al. 1989) and unclassified non-A rotavirus from lambs (CHASEY and BANKS 1986) and one from pigs (ASKAA and BLOCH 1984) differed from other rotaviruses in their ability to induce syncytia in villous enterocytes, which may be a feature common to group B rotaviruses from all species. Both a porcine group B rotavirus and the unclassified rotavirus from lambs also induced syncytia in infected MA104 cell monolayers (Table 5, THEIL and SAIF 1985; CHASEY and BANKS 1986). Whether syncytia formation is unique to only group B rotaviruses and not the other rotavirus serogroups requires further investigation.

11 Immunity, Prevention, and Control

There is little information on immunity or methods to prevent or control infections with non-A rotaviruses. Further knowledge about their prevalence, severity, and epidemiology is needed to devise appropriate prevention and control strategies. This information as well as studies of immunity have been limited by failure of most non-A rotaviruses to grow in cell culture, hindering development of routine serologic assays to detect antibodies to non-A rotaviruses and possible future vaccine development.

The possibility of water-borne or food-borne spread of group B rotavirus in China (HUNG et al. 1984) and group C rotavirus in Japan (MATSUMOTO et al. 1989), respectively, suggests that proper sanitation and hygiene may play important roles in contolling or preventing epidemics. However, both person to person spread and asymptomatic infections with group B rotaviruses in China has also been reported (FANG et al. 1989b). Detection of antigenically related rotaviruses and antibodies in animals such as swine and rats raises unanswered questions about the zoonotic potential of non-A rotaviruses (groups B and C). However, no direct transmission from animals to humans has yet been documented.

The low prevalence of antibodies in human sera to non-A rotaviruses (B and C), except for areas in China or Japan where epidemics occurred, suggests that many adults and children may be at risk of acquiring these infections. This is in contrast to a high seroprevalence of antibodies to group A rotaviruses and the occurrence of clinical infections mainly in infants and young children. Group B rotaviruses caused more severe disease in adults than children, but both age groups were susceptible to infection with group B and C rotaviruses (FANG et al. 1989b; MASTSUMOTO et al. 1989). Serologic studies in Baltimore, Md (USA) revealed a high prevalence of antibodies to a rat group B rotavirus in humans (EIDEN et al. 1985), suggesting a possible occurrence of focal outbreaks of the disease. Additional studies are needed to access the prevalence and clinical significance of non-A rotaviruses in the human population and to provide further information on immunity to these viruses.

The results of limited serologic studies suggest non-A rotavirus infections are more prevalent in animals than humans. In swine and poultry, antibodies to group C and D rotaviruses were encountered at almost as high a frequency as antibodies to group A rotaviruses (Table 4; MCNULTY et al. 1984a; TERRETT et al. 1987; NAGESHA et al. 1988). The lower prevalence of antibodies to group B rotavirus in swine in one region of the USA (THEIL and SAIF 1985) and Australia (NAGESHA et al. 1988) compared with a higher prevalence in UK herds (BRIDGER and BROWN 1985) suggests regional variation may occur in the incidence of group B rotavirus infections in swine, and these infections may protentially occur as epizootics in some herds or areas. By comparison, antibodies to group C rotavirus were common in adult swine in the USA, Australia, and the UK, suggesting these infections, like group A rotaviruses, may be enzootic. Nursing pigs infected with group C rotavirus had a higher incidence of clinical diarrhea than pigs infected at an older age. Therefore factors which influence passive immunity may be important in controlling the severity of these infections. These include failure of pigs to nurse shortly after birth or the overwhelming of milk antibodies by high viral doses from a contaminated environment. Vaccination of the pregnant or lactating dam may enhance milk antibody titers thereby delaying or preventing diarrhea due to group C rotaviruses as was observed for group A rotavirus infections in pigs (SAIF 1985). Other factors useful in controlling the spread and severity of group A rotavirus infections among humans or animals may also be useful for preventing or controlling infections by non-A rotaviruses. Whether development of vaccines for non-A rotaviruses is warranted for use in humans or animals will depend upon additional information about the prevalence and severity of the diarrheas induced by these viruses.

Acknowledgements. The authors thank the following individuals: Mrs. Hannah Gehman, for her expert assistance and word processing skills in typing this manuscript and Mrs. Judy Mengel for technical assistance with the polyacrylamide gel electrophoresis.

References

Arista S, Giovannelli L, Titone L (1985) Detection of an antigenically distinct human rotavirus in Palermo, Italy. Ann Virol (Inst Pasteur) 136 E: 229–235

Askaa J, Bloch B (1984) Infection in piglets with a porcine rotavirus-like virus. Experimental inoculation and ultrastructural examination. Arch Virol 80: 291–303

Bellinzoni R, Mattion N, Vallejos L, LaTorre JL, Schodeller EA (1987) Atypical rotavirus in chickens in Argentina. Res Vet Sci 43: 130

Bohl EH, Saif LJ, Theil KW, Agnes AG, Cross RF (1982) Porcine pararotavirus: detection, differentiation from rotavirus, and pathogenesis in gnotobiotic pigs. J Clin Microbiol 15: 312–319

Bonsdorff CHV, Svensson L (1988) Human serogroup C rotavirus in Finland. Scand J Infect Dis 20: 475–478

Breer C, Wunderli W, Lee C, Weisser E, Schopfer K (1985) Rotavirus and pararotavirus infektionen bei erwachsenen. Analyse einer nosokomialen infektion sowie sporadisher folle. Schweiz Med Wochenschr 115: 1530–1535

Bremont M, Cohen J, McCrae MA (1988) Analysis of the structural polypeptides of a porcine group C rotavirus. J Virol 62: 2183–2185

Bridger JC (1980) Detection by electron microscopy of caliciviruses, astroviruses and rotavirus-like particles in the faeces of piglets with diarrhoea. Vet Rec 107: 532

Bridger JC (1987) Novel rotaviruses in animals and man. Ciba Found Symp 128: 5–23

Bridger JC, Brown JF (1985) Prevalence of antibody to typical and atypical rotaviruses in pigs. Vet Rec 116: 50

Bridger JC, Clarke IN, McCrae MA (1982) Characterization of an antigenically distinct porcine rotavirus. Infect Immun 35: 1058–1062

Bridger JC, Pedley S, McCrae MA (1986) Group C rotaviruses in humans. J Clin Microbiol 23: 760–763

Brown DWG, Beards GM, Chen GM, Flewett TH (1987) Prevalence of antibody to group B (atypical) rotavirus in humans and animals. J Clin Microbiol 25: 316–319

Brown DWG. Mathan MM, Mathew M, Martin R, Beards GM– Mathan VI (1988) Rotavirus epidemiology in Vellore, South India: group, subgroup, serotype, and electropherotype. J Clin Microbiol 26: 2410–2414

Buitenwerf J, Alphen MMV, Schaap GJP (1983) Characterization of rotavirus RNA isolated from children with gastroenteritis in two hospitals in Rotterdam. J Med Microbiol 12: 71–78

Burns JW, Welch SKW, Nakata S, Estes MK (1989) Characterization of monoclonal antibodies to human group B rotavirus and their use in an antigen detection enzyme-linked immunosorbent assay. J Clin Microbiol 27: 245–250

Chasey D, Banks J (1984) The commonest rotaviruses from neonatal lamb diarrhoea in England and Wales have atypical electropherotypes. Vet Rec 115: 326–327

Chasey D, Banks J (1986) Replication of atypical ovine rotavirus in small intestine and cell culture. J Gen Virol 67: 567–576

Chasey D, Davis D (1984) Atypical rotaviruses in pigs and cattle. Vet Rec. 114: 16–17

Chasey D, Bridger JC, McCrae MA (1986) A new type of atypical rotavirus and villous epithelial cell syncytia in piglets. J Comp Pathol 100: 217–222

Chen GM, Hung T, Bridger JC, McCrae MA (1985) Chinese adult rotavirus is a group B rotavirus. Lancet ii: 1123–1124

Chen GM, Hung T, Mackow ER (1990) cDNA cloning of each genomic segment of the group B rotavirus ADRV: molecular characterization of the eleventh RNA segment. Virology 175: 605–609

Clarke IN, McCrae MA (1983) The molecular biology of rotaviruses. VI. RNA species-specific terminal conservation in rotaviruses. J Gen Virol 64: 1877–1884

Dai GZ, Sun MS, Liu SQ, Ding XF, Chen YD, Wang LC, Du DP, Zhao G, Su Y, Li J, Xu WM, Li TH, Chen XX (1987) First report of an epidemic of diarrhea in human neonates involving the new rotavirus and biological characteristics of the epidemic virus strain (KMB/R85). J Med Virol 22: 365–373

De Bouck P, Callebaut P, Pensaert M (1984) The pattern of rotavirus and pararotavirus excretion in pigs in closed swine herds. In: Acres SD (ed) Proceedings of the 4th International Symposium on Neonated Diarrhea, VIDO, Saskatchewan, Canada, pp 77–89

Dimitrov DH, Estes MK, Rangelova SM, Shindarov LM, Melnick JL, Graham DY (1983) Detection of antigenically distinct rotaviruses from infants. Infect Immun 41: 523–526

Dimitrov DH, Shindarov LM, Rangeloa A (1986) Occurrence of antigenically distinct rotaviruses in infants in Bulgaria. Eur J Clin Microbiol 5: 471–473

Eiden JJ, Vonderfecht S, Yolken RH (1985) Evidence that a novel rotavirus-like agent of rats can cause gastroenteritis in man. Lancet ii: 8–11

Eiden JJ, Vonderfecht S, Theil KW, Torres-Medina A, Yolken RH (1986) Genetic and antigenic relatedness of human and animal strains of antigenically distinct rotaviruses. J Infect Dis 154: 972–982

Eiden JJ, Yolken RH, Vonderfecht SL, Hung T, McCrae MA (1988) Terminal fingerprint analysis of group B rotaviruses. J Infect Dis 158: 657–658

Eiden JJ, Firoozmand F, Sato S, Vonderfecht SL, Fang ZY, Yolken RH (1989) Detection of group B rotavirus in fecal specimens by dot hybridization with a cloned cDNA probe. J Clin Microbiol 27: 422–426

Ericson BL, Graham DY, Mason BB, Estes MK (1982) Identification, synthesis, and modifications of simian rotavirus SA11 polypeptides in infected cells. J Virol 42: 825–839

Espejo RT, Puerto F, Soler C, Gonzalez N (1984) Characterization of a human pararotavirus. Infect immun 44: 112–116

Fang ZY, Glass RI, Penaranda M, Dong H, Monroe SS, Wen L, Estes MK, Eiden J, Yolken RH, Saif L, Gouvea V, Hung T (1989a) Purification and characterization of adult diarrhea rotavirus: identification of viral structural proteins. J Virol 63: 2191–2197

Fang ZY, Ye Q, Ho MS, Dong H, Qing S, Penaranda ME, Hung T, Wen L, Glass RI (1989b) Investigation of an outbreak of adult diarrhea rotavirus (ADRV) in China. J Infect Dis 160: 948–953

Fijtman NL, Barrandeguy ME, Cornaglia EM, Schudel AA (1987) Variations and persistency of electropherotypes of bovine rotavirus field isolates. Arch Virol 96: 275–281

Gombold JL, Ramig RF (1986) Analysis of reassortment of genome segments in mice mixedly infected with rotaviruses SA11 and RRV. J virol 57: 110–6

Graham DY, Estes MK (1985) Rotavirus-like agent, rats, and man. Lancet ii: 886–887

Hall GA (1987) Comparative pathology of infection by novel diarrhea viruses. Ciba Found Symp 128: 192–217

Huber AC, Yolken RH, Mader LC, Strandberg, Vonderfecht SL (1989) Pathology of infectious diarrhea in infants rats (IDIR) induced by an antigenically distinct rotavirus. Vet Pathol 26: 376–385

Hung T (1988) Rotavirus and adult diarrhea. Adv Virus Research 35: 193–218

Hung T, Chen G, Wang C, Chou Z, Chao T, Ye W, Yao H, Meng K (1983) Rotavirus-like agent in adult non-bacterial diarrhea in China. Lancet ii: 1139–1142

Hung T, Chen G, Wang C, Yao H, Fang Z, Chao T, Chou Z, Ye W, Chang X, Den S, Liong X, Chang W (1984) Waterborne outbreak of rotavirus diarrhoea in adults in China caused by a novel rotavirus. Lancet ii: 1139–1142

Hung T, Fan R, Wang C, Chen G, Chou D, Chang J, McCrae MA, Wang W, Se W, Dan R, Ng MH (1985) Seroepidemiology of adult rotavirus. Lancet ii: 325–326

Hung T, Chen G, Wang C, Fan R, Yong R, Chang J, Dan R, Ng MH (1987) Seroepidemiology and molecular epidemiology of the Chinese rotavirus. Ciba Found Symp 128: 49–62

Jashes M, Sandino AM, Faundez G, Avendano LF, Spencer E (1986) In vitro transcription of human pararotavirus. J Virol 57: 183–190

Jiang BM, Sailf LJ, Kang SY (1989) Biochemical characterization of viral polypeptides of a porcine group C rotavirus. Conf Res Workers in Animal Dis, Chicago, abstract no 81

Jiang BM, Saif LJ, Kang SY, Kim J (1990) Biochemical characterization of the structural and nonstructural polypeptides of a porcine group C rotavirus. J Virol 64: 3171–3178

Kang SY, Nagaraja KU, Newman JA (1986) Electropherotypic analysis of rotavirus isolated from turkeys. Avian Dis 30: 797

Kinney JS, Viscidi RP, Vonderfecht SL, Eiden JJ, Yolken RH (1989) Monoclonal antibody assay for detection of double-stranded RNA and application for detection of group A and non-group A rotaviruses. J Clin Microbiol 27: 6–12

Liu M, Offit PA, Estes MK (1988) Identification of the simian rotavirus SA11 genome segment 3 product. Virology 163: 26–32

Mason BB, Graham DY, Estes MK (1980) In vitro transcription and translation of simian rotavirus SA11 gene products. J Virol 33: 1111–1121

Mason BB, Graham DY, Estes MK (1983) Biochemical mapping of the simian rotavirus SA11 genome. J Virol 46: 413–423

Matsumoto K, Hatano M, Kobayashi K, Hasegawa A, Shudo Y, Nakata S, Chiba S, Kimura Y (1989) An outbreak of gastroenteritis associated with acute rotaviral infection in school children. J Infect Dis 160: 611–615

McCrae MA (1987) Nucleic acid-based analysis of non-group A rotaviruses. Ciba Found Symp 128: 24–48

McCrae MA, McCorquodale JG (1983) Molecular biology of rotaviruses. V. Terminal structure of viral RNA species. Viology 126: 204–212

McNulty MS, Allan GM, Todd D, McFerran JB, McCracken RM (1981) Isolation from chickens of a rotavirus lacking the rotavirus group antigen. J Gen Virol 55: 405–413

McNulty MS, Allan GM, McFerran JB (1984a) Prevalence of antibody to conventional and atypical rotaviruses in chickens. Vet Rec 114: 219

McNulty MS, Todd D, Allan GM, McFerran JB, Greene JA (1984b) Epidemiology of rotavirus infection in Broiler chickens: recognition of four serogroups. Arch Virol 81: 113–121

Mebus CA, Rhodes MB, Underdahl MS (1978) Neonatal calf diarrhea caused by a virus that induces villous epithelial cell syncytia. Am J Vet Res 39: 1223

Midthun K, Greenberg HB, Hoshino Y, Kapikian AZ, Wyatt RG, Chanock RM (1985) Reassortant rotaviruses as potential live rotavirus vaccine candidates. J Virol 53: 949–954

Midthun K, Valdesuso J, Hoshino Y, Flores J, Kapikian AZ, Chanock RM (1987) Analysis by RNA-RNA hybridization assay of intertypic rotaviruses suggests that gene reassortment occurs in vivo. J Clin Microbiol 25: 295–300

Nagesha HS, Hum CP, Bridger JC, Holmes IH (1988) Atypical rotaviruses in Australian pigs. Arch Virol 102: 91–98

Nakata S, Estes MK, Graham DY, Loosle R, Hung T, Wang S, Saif LJ, Melnick JL (1986) Antigenic characterization and ELISA detection of adult diarrhea rotaviruses. J Infect Dis 154: 448–455

Nakata S, Estes MK, Graham DV, Wang S, Gary GW, Melnick JL (1987a) Detection of antibody to group B adult diarrhea rotaviruses in humans. J Clin Microbiol 25: 812–818

Nakata S, Petrie BL, Calomeni EP, Estes MK (1987b) Electron microscopy procedure influences detection of rotaviruses. J Clin Microbiol 25: 1902–1906

Nicolas JC, Cohen J, Fortier B, Lourenco MH, Bricout F (1983) Isolation of a human pararotavirus. Virology 124: 181–184

Ojeh CK, Jiang B, Tsunemitsu H, Kang SY, Weilnau P, Saif LJ (1991) Reactivity of monoclonal antibodies to the 41-kilodalton protein of porcine group C rotavirus with homologous and heterologous rotavirus serogroups in immunofluorescence tests. J Clin Microbiol 29: 2051–2055

Pedley S, Bridger JC, Brown JF, McCrae MA (1983) Molecular characterization of rotaviruses with distinct group antigens. J Gen Virol 64: 2093–2101

Pedley S, Bridger JC, Chasey D, McCrae MA (1986) Definition of two new groups of atypical rotaviruses. J Gen Virol 67: 131–137

Penaranda ME, Cubitt WD, Sinarachatanant P, Taylor DN, Likanonsakul S, Saif L, Glass RI (1989a) Group C rotavirus infections in patients with diarrhea in Thailand, Nepal, and England. J Infect Dis 160: 392–397

Penaranda ME, Ho MS, Fang ZY, Dong H, Sheng BX, Cheng DS, Ye WW, Estes MK, Echeverria P, Hung T, Glass RI (1989b) Seroepidemiology of adult diarrhea rotavirus in China, 1977–1987. J Clin Microbiol 27: 2180–2183

Pereira HG, Leite JPG, Azeredo RS, Farias VD, Sutmoller F (1983) An atypical rotavirus detected in a child with gastroenteritis in Rio de Janeiro, Brazil. Mem Inst Oswaldo Cruz (Rio de Janeiro) 78: 245–250

Qiu FX, Tian Y, Liu JC, Zhang XS, Hao YP (1986) Antibody against adult diarrhea rotavirus among healthy adult population in China. J Virol Methods 14: 127–132

Reynolds DL, Saif YM, Theil KW (1987) A survey of enteric viruses of turkey poults. Avian Dis 31: 89

Rodger SM, Bishop RF, Holmes IH (1982) Detection of a rotavirus-like agent associated with diarrhea in an infant. J Clin Microbiol 16: 724–726

Roseto A, EsCaig J, Delain E, Cohen J, Scherrer (1979) Structure of rotaviruses as studied by the freeze-drying technique. Virology 98: 471–475

Saif LJ (1985) Passive immunity to coronavirus and rotavirus infections in swine and cattle: Enhancement by maternal vaccination. In: Tzipori S (ed) Proc infectious diarrhea in the young, strategies for control in humans and animals. Elsevier, Amsterdam, pp 456–467

Saif LJ (1990a) Comparative aspects of enteric viral infections. In: Saif LJ, Theil KW (eds) Viral diarrheas of man and animals, CRC Press, Boca Raton, pp 9–31

Saif LJ (1990b) Nongroup A rotaviruses. In: Saif LJ, Theil KW (eds) Viral diarrheas of man and animals, CRC Press, Boca Raton, pp 73–95

Saif LJ, Theil KW (1985) Antigenically distinct rotaviruses of human and animal origin. In: Tzipori S (ed) Proc infectious diarrhea in the young: strategies for control in humans and animals, Elsevier, Amsterdam, pp 208–214

Saif LJ, Bohl EH, Kohler EM, Hughes JH (1977) Immune electron microscopy of TGE virus and rotavirus (reovirus-like agent) of swine. Am J Vet Res 38: 13

Saif LJ, Bohl EH, Theil KW, Cross RF, House JA (1980) Rotavirus-like, calicivirus-like, and 23-nm virus-like particles associated with diarrhea in young pigs. J Clin Microbiol 12: 105–111

Saif LJ, Theil KW, Redman DR (1982) Detection and pathogenicity of an enteric bovine rotavirus-like agent (RVLA). Proc Conf Res Workers Anim Dis. Chicago, IL, abstract no 98

Saif LJ, Saif YM, KW Theil (1984) Detection and pathogenicity of enteric viruses recovered from diarrheic turkeys: role of a rotavirus-like agent. Proceedings of 8th World's Poultry Congress, Helsinki, Finland, p 539

Saif LJ, Saif YM, Theil KW (1985) Enteric viruses in diarrheic turkey poults. Avian Dis 29: 798–811

Saif LJ, Redman DR, Mengel J, Estes MK (1987) Porcine and bovine group B rotaviruses: antigenic relationships and cross-infectivity in pigs and calves, Abstract US-Japan Co-operative Med Sci Prog, Viral Diseases Panel, Monterey, California

Saif LJ, Terrett LA, Miller KL, Cross RF (1988) Serial propagation of porcine group C rotavirus (pararotavirus) in a continuous cell line and characterization of the passaged virus. J Clin Microbiol 26: 1277–1282

Sato A, Yolken RH, Eiden JJ (1989) The complete nucleic acid sequence of gene segment 3 of the IDIR strain of group B rotavirus. Nucleic Acid Res 17: 10113

Sigola de San Juan C, Bellinzoni RC, Mattion N, La Torre J, Schodeller EA (1986) Incidence of group A and atypical rotaviruses in Brazilian pig herds. Res Vet Sci 41: 270–272

Snodgrass DR, Herring AJ, Campbell I, Inglis JM, Hargreaves FD (1984) Comparison of atypical rotaviruses from calves, piglets, lambs and man. J Gen Virol 65: 909–914

Sorrentino A, Scodeller EA, Bellinzoni R, Muchink GR, La Torre JL (1986) Detection of an atypical rotavirus associated with diarrhea in Chaco. Trans Soc Trop Med Hyg 80: 120–122

Su C, Wu Y, Shen H, Wang D, Chen Y, Wang D, He L, Yang Z (1986) An outbreak of epidemic diarrhea in adults caused by a new rotavirus in Anhui province of China in the summer of 1983. J Med Virol 19: 167–173

Suzuki H, Chen GM, Hung T, Beards GM, Brown DWG, Flewett TH (1987) Effects of two negative staining methods on the Chinese atypical rotavirus. Arch Virol 94: 305–308

Szucs G, Kende M, Uj M (1987) Atypical human rotaviruses in Hungary. Ann Inst Pasteur/Virol 138: 391–395

Terrett LA, Saif LJ (1987) Serial propagation of porcine group C rotavirus (pararotavirus) in primary porcine kidney cell culture. J Clin Microbiol 25: 1316–1319

Terrett LA, Saif LJ, Theil KW, Kohler EM (1987) Physicochemical characterization of porcine pararotavirus and detection of virus and viral antibodies using cell culture immunofluorescence. J Clin Microbiol 25: 268–272

Theil KW, Saif LJ (1985) In vitro detection of porcine rotavirus-like virus (group B rotavirus) and its antibody. J Clin Microbiol 21: 844–846

Theil KW, Bohl EH, Cross RF, Agnes AG (1978) Pathogenesis of porcine rotaviral infection in experimentally inoculated gnotobiotic pigs. Am J Vet Res 39: 213–220

Theil KW, Saif LJ, Morehead PD, Whitmoyer RE (1985) Porcine rotavirus-like virus (group B rotavirus): characterization and pathogenicity for gnotobiotic pigs. J Clin Microbiol 21: 340–345

Theil KW, Reynolds DL, Saif YM (1986) Comparison of immune electron microscopy and genome electropherotyping techniques for detection of turkey rotaviruses and RVLA in intestinal contents. J Clin Microbol 23: 695

Todd D, McNulty MS (1986) Electrophoretic variation of avian rotavirus RNA in polyacrylamide gel. Avian Pathol 15: 149

Torres-Medina A (1987) Isolation of an atypical rotavirus causing diarrhea in neonatal ferrets. Lab An im Sci 37: 167

Tsunemitsu H, Saif LJ, Jiang B, Shimizu M, Hiro M, Yamaguchi H, Ishiyama T, Hirai T (1991) Isolation, characterization, and serial propagation of a bovine group C rotavirus in a monkey kidney cell line (MA104). J Clin Microbiol 29: 2609–2613

Tsunemitsu H, Jiang B, Saif LJ (1992) Detection of group C rotavirus antigens and antibodies in animals and humans by ELISA. J Clin Microbiol 30: 2129–2134

Ushijima H, Honma H , Mukoyama A, Shinozaki T, Fujita Y, Kobayashi M, Ohseto M, Morikawa S, Kitamura T (1989) Detection of group C rotaviruses in Tokyo. J Med Virol 27: 299–303

Vonderfecht SL, Huber AC, Eiden J, Mader LC, Yolken RH (1984) Infectious diarrhea of infant rats produced by a rotavirus-like agent. J Virol 52: 94–98

Vonderfecht SL, Miskuff RL, Eiden JJ, Yolken RH (1985) Enzyme immunoassay inhibition assay for the detection of rat rotavirus-like agent in intestinal and faecal specimens obtained from diarrheic rats and humans. J Clin Microbiol 22: 726

Vonderfecht SL, Eiden JJ, Torres A, Miskuff RL, Mebus CA, Yolken RH (1986) Identification of a bovine enteric syncytial virus as a nongroup A rotavirus. Am J Vet Res 47: 1913

Vonderfecht SL, Eiden JJ, Miskuff RL, Yolken RH (1988) Kinetics of intestinal replication of group B rotavirus and relevance to diagnostic methods. J Clin Microbiol 26: 216–221

Wang CA, Yang RJ, Liang XG (1986) Development of ELISA for detection of adult diarrhea rotavirus antigen. Chin J Virol 2: 248

Wang S, Cai R, Chen J, Li R, Jiang R (1985) Etiologic studies of the 1983 and 1984 outbreaks of epidemic diarrhea in Guangxi. Intervirology 24: 140–146

Welch SKW, Crawford SE, Estes MK (1989) Rotavirus SA11 genome segment 11 protein is a nonstructural phosphoprotein. J Virol 63: 3974–3982

Yolken R, Arango-Jaramillo S, Eiden J, Vonderfecht S (1988a) Lack of genomic reassortment following infection of infant rats with group A and group B rotaviruses. J Infect Dis 158: 1120–1123

Yolken R, Wee SB, Eiden J, Kinney J, Vonderfecht S (1988b) Identification of a group-reactive epitope of group B rotavirus recognized by monoclonal antibody and application to the development of a sensitive immunoassay for viral characterization. J Clin Microbiol 26: 1853

Subject Index

Current Topics in Microbiology and Immunology

Volumes published since 1988 (and still available)

Vol. 161: **Racaniello, Vincent R. (Ed.):** Picornaviruses. 1990. 12 figs. X, 194 pp. ISBN 3-540-52429-0

Vol. 162: **Roy, Polly; Gorman, Barry M. (Eds.):** Bluetongue Viruses. 1990. 37 figs. X, 200 pp. ISBN 3-540-51922-X

Vol. 163: **Turner, Peter C.; Moyer, Richard W. (Eds.):** Poxviruses. 1990. 23 figs. X, 210 pp. ISBN 3-540-52430-4

Vol. 164: **Bækkeskov, Steinnun; Hansen, Bruno (Eds.):** Human Diabetes. 1990. 9 figs. X, 198 pp. ISBN 3-540-52652-8

Vol. 165: **Bothwell, Mark (Ed.):** Neuronal Growth Factors. 1991. 14 figs. IX, 173 pp. ISBN 3-540-52654-4

Vol. 166: **Potter, Michael; Melchers, Fritz (Eds.):** Mechanisms in B-Cell Neoplasia 1990. 143 figs. XIX, 380 pp. ISBN 3-540-52886-5

Vol. 167: **Kaufmann, Stefan H. E. (Ed.):** Heat Shock Proteins and Immune Response. 1991. 18 figs. IX, 214 pp. ISBN 3-540-52857-1

Vol. 168: **Mason, William S.; Seeger, Christoph (Eds.):** Hepadnaviruses. Molecular Biology and Pathogenesis. 1991. 21 figs. X, 206 pp. ISBN 3-540-53060-6

Vol. 169: **Kolakofsky, Daniel (Ed.):** Bunyaviridae. 1991. 34 figs. X, 256 pp. ISBN 3-540-53061-4

Vol. 170: **Compans, Richard W. (Ed.):** Protein Traffic in Eukaryotic Cells. Selected Reviews. 1991. 14 figs. X, 186 pp. ISBN 3-540-53631-0

Vol. 171: **Kung, Hsing-Jien; Vogt, Peter K. (Eds.):** Retroviral Insertion and Oncogene Activation. 1991. 18 figs. X, 179 pp. ISBN 3-540-53857-7

Vol. 172: **Chesebro, Bruce W. (Ed.):** Transmissible Spongiform Encephalopathies. 1991. 48 figs. X, 288 pp. ISBN 3-540-53883-6

Vol. 173: **Pfeffer, Klaus; Heeg, Klaus; Wagner, Hermann; Riethmüller, Gert (Eds.):** Function and Specificity of γ/δ T Cells. 1991. 41 figs. XII, 296 pp. ISBN 3-540-53781-3

Vol. 174: **Fleischer, Bernhard; Sjögren, Hans Olov (Eds.):** Superantigens. 1991. 13 figs. IX, 137 pp. ISBN 3-540-54205-1

Vol. 175: **Aktories, Klaus (Ed.):** ADP-Ribosylating Toxins. 1992. 23 figs. IX, 148 pp. ISBN 3-540-54598-0

Vol. 176: **Holland, John J. (Ed.):** Genetic Diversity of RNA Viruses. 1992. 34 figs. IX, 226 pp. ISBN 3-540-54652-9

Vol. 177: **Müller-Sieburg, Christa; Torok-Storb, Beverly; Visser, Jan; Storb, Rainer (Eds.):** Hematopoietic Stem Cells. 1992. 18 figs. XIII, 143 pp. ISBN 3-540-54531-X

Vol. 178: **Parker, Charles J. (Ed.):** Membrane Defenses Against Attack by Complement and Perforins. 1992. 26 figs. VIII, 188 pp. ISBN 3-540-54653-7

Vol. 179: **Rouse, Barry T. (Ed.):** Herpes Simplex Virus. 1992. 9 figs. X, 180 pp. ISBN 3-540-55066-6

Vol. 180: **Sansonetti, P. J. (Ed.):** Pathogenesis of Shigellosis. 1992. 15 figs. X, 143 pp. ISBN 3-540-55058-5

Vol. 181: **Russell, Stephen W.; Gordon, Siamon (Eds.):** Macrophage Biology and Activation. 1992. 42 figs. IX, 299 pp. ISBN 3-540-55293-6

Vol. 182: **Potter, Michael; Melchers, Fritz (Eds.):** Mechanisms in B-Cell Neoplasia. 1992. 188 figs. XX, 499 pp. ISBN 3-540-55658-3

Vol. 183: **Dimmock, Nigel J.:** Neutralization of Animal Viruses. 1993. 10 figs. VII, 149 pp. ISBN 3-540-56030-0

Vol. 184: **Dunon, Dominique; Mackay, Charles R.; Imhof, Beat A. (Eds.):** Adhesion in Leukocyte Homing and Differentiation. 1993. 37 figs. IX, 260 pp. ISBN 3-540-56756-9

Dear Reader,
Please help us to further improve this book series by answering the following questions:

This book was purchased
☐ for a general library
☐ for a university library
☐ for a university/research institute
☐ for a company
☐ for a government agency
☐ as a private copy
☐ for other use
(please specify): _____

Was this book purchased as part of a standing order for the series?
☐ yes ☐ no ☐ don't know

Which other volumes of the series have you personally worked with? Please give the volume numbers. (See volume listing in the back of this book)

Please give us your opinion on the following features:
(1 = very good, 2 = good, 3 = fair, 4 = poor)

	This book	The series
scientific quality	___	___
topicality	___	___
relevance to your own work	___	___
technical features (paper, printing etc.)	___	___
price	___	___

Your field(s) of interest:
☐ microbiology ☐ genetics
☐ general ☐ immunology
☐ virology ☐ hematology
☐ mycology ☐ biochemistry
☐ molecular biology ☐ other (please specify):

All answers arriving by June '94 will be included in a **prize draw**. The three winners may select a volume of their choice from the CTMI series.

Have you bought books from this series for your personal use?
Yes,
☐ more than 5 books
☐ 2 to 5 books
☐ one book
No,
☐ I can read them at the library/institute
☐ not important enough
☐ too expensive

Thank you!

All answers are strictly confidential

56761-5

Dear Reader,
Please help us to further improve this book series by answering the following questions:

This book was purchased
☐ for a general library
☐ for a university library
☐ for a university/research institute
☐ for a company
☐ for a government agency
☐ as a private copy
☐ for other use
(please specify): _____

Was this book purchased as part of a standing order for the series?
☐ yes ☐ no ☐ don't know

Which other volumes of the series have you personally worked with? Please give the volume numbers. (See volume listing in the back of this book)

Please give us your opinion on the following features:
(1 = very good, 2 = good, 3 = fair, 4 = poor)

	This book	The series
scientific quality	___	___
topicality	___	___
relevance to your own work	___	___
technical features (paper, printing etc.)	___	___
price	___	___

Your field(s) of interest:
☐ microbiology ☐ genetics
☐ general ☐ immunology
☐ virology ☐ hematology
☐ mycology ☐ biochemistry
☐ molecular biology ☐ other (please specify):

All answers arriving by June '94 will be included in a **prize draw**. The three winners may select a volume of their choice from the CTMI series.

Have you bought books from this series for your personal use?
Yes,
☐ more than 5 books
☐ 2 to 5 books
☐ one book
No,
☐ I can read them at the library/institute
☐ not important enough
☐ too expensive

Thank you!

All answers are strictly confidential

56761-5

Name_____

Adress_____

_____

_____

_____

Readership Survey

**Current Topics in
Microbiology and
Immunology**

Springer-Verlag
Corporate Development/
 Market Research
Tiergartenstr. 17

D-69112 Heidelberg
GERMANY

Name_____

Adress_____

_____

_____

_____

Please
affix
stamp

Readership Survey

**Current Topics in
Microbiology and
Immunology**

Springer-Verlag
Corporate Development/
 Market Research
Tiergartenstr. 17

D-69112 Heidelberg
GERMANY